195

95

INFRARED
SPECTROSCOPY

Experimental Methods and Techniques

INFRARED
SPECTROSCOPY

Experimental Methods and Techniques

JAMES E. STEWART

DURRUM INSTRUMENT CORPORATION
PALO ALTO, CALIFORNIA

1970

MARCEL DEKKER, INC., NEW YORK

To Gini

GENO GINELI

Preface

Twenty years ago I saw a small infrared spectrophotometer operating in the laboratory of Professor D. H. Rank. It was an early Perkin–Elmer Model 12 and I marveled at the complexity of the information produced by such a tiny instrument. I wondered, as other beginning spectroscopists must have wondered, how does this marvelous instrument work; what electrical and mechanical and optical devices are hidden under its cover?

Later, at the National Bureau of Standards, I used a succession of infrared spectrophotometers—always with a feeling that I did not fully understand the working of the instruments. In those days there were only a few books touching on experimental methods of infrared spectroscopy, and these did not help much in understanding the instrument itself.

In more recent years, in the applications and research departments of Beckman Instruments, I became convinced that the new spectroscopist needed a book devoted to the spectrophotometer so that he could understand his instrument and use it in an optimum manner. Moreover, I felt that the more experienced spectroscopist could also benefit from a better understanding of his instrument, its workings, and its limitations. Accordingly, I set out several years ago to write this book. As so often happens, the chief beneficiary of a book is its author. I found that in trying to explain certain features of infrared spectrophotometers my understanding was faulty and had to be corrected. I also found gaps in the literature of infrared spectro-photometers which required a little original investigation before a reasonably complete description could be given, particularly in the areas of slit function and noise filter effects and some aspects of photometric accuracy.

The book is divided into four parts. The first part consists of the first two chapters, which are brief introductory sections.

The second part covers the optics of the infrared spectrophotometer. Geometric and physical optics are developed to the extent that they are required in order to understand the rest of the book. The optical components of spectrophotometers and complete optical systems are discussed. The slit function and the spectral modulation transfer function are developed from the standpoint of linear filter theory. This approach certainly will be used more frequently in the future in view of the increased use of computers for data reduction. Interferometers are discussed in some detail because the interferometer is becoming a strong rival of the conventional monochromator, particularly in the far infrared. The last chapter of this part describes those mechanical parts of the spectrophotometer that are related to its optical character-istics.

The third part of the book treats the electronics of the spectro-photometer. An elementary introduction to electronics is followed by chapters on infrared detectors, electromechanical systems found in infrared spectrophotometry, the electromechanical transfer function, and noise filter properties.

The last part of the book is concerned with the experimental methods of infrared spectrophotometry. There are now many fine books covering experimental procedures for the laboratory infrared spectroscopist, although this was not the case several years ago. In keeping with the intent of this book the experimental methods are discussed from the standpoint of the instrument itself. Thus, the chapter on quantitative measurements has a discussion of the effects of stray radiation and extraneous radiation from various components. There are chapters on methods of absorption, reflection, and emission spectroscopy and a final chapter on advanced methods—all of these chapters emphasizing the spectrophotometer and associated apparatus.

The book is intended to be intelligible to advanced undergraduates and graduate students in the various disciplines where infrared spectro-photometers are used. It is also intended for practicing spectrosco-pists, both new and mature. Some portions of the treatment are neces-sarily fairly mathematical. No difficulty should be experienced by the reader with a working knowledge of calculus. The only mathematics likely to appear strange to some readers are the Fourier and Laplace transforms, and these are discussed briefly in the Appendix.

In retrospect, I find it difficult to determine the origin of many of the ideas I have discussed in this treatise. Over the years I have had the advantage of sharing discussions with many workers. I mention in particular my colleagues at the National Bureau of Standards: E. C. Creitz, F. A. Smith, and F. J. Linnig; and my former associates at Beckman Instruments: W. S. Gallaway, W. F. Ulrich, W. I. Kaye, and H. J. Sloane.

My thanks go especially to Dr. Gallaway, whose explanations and arguments did much to clear my own foggy ideas. Of course, the errors and mistakes remaining in this book are my own responsibility. I also would like to express my thanks to the many patient typists who labored over the manuscript. In particular my thanks go to Mrs. W. S. Gallaway of Beckman Instruments who typed the bulk of the first draft and to Mrs. K. Wingo who typed most of the final version.

Saratoga, California
June 1970

ACKNOWLEDGMENT

The following figures are reproduced from *Applied Optics* by permission of the American Institute of Physics: Figs. 5.33, 7.12, 7.19, 7.20, 7.21, and 14.2

The following figures are reproduced from *Infrared Physics* by permission of Pergamon Press: Figs. 13.1, 13.4, 13.11, 13.13, 13.15, 13.16, 13.18, 13.19, 13.21, 13.23, 13.25, 13.28, 13.29, and 13.30.

Contents

INFRARED
SPECTROSCOPY

Experimental Methods and Techniques

CHAPTER 1

Infrared Spectroscopy

1. INTRODUCTION

This book is devoted to the infrared spectrophotometer. We cannot very well discuss spectrophotometers without at least indicating the uses to which they are put. This first chapter is therefore a very abbreviated introduction to infrared spectroscopy.

II. PROPERTIES OF THE INFRARED REGION OF THE ELECTROMAGNETIC SPECTRUM

Radiant energy is emitted by matter and transmitted through space in the form of electromagnetic waves. *Radiation* is the process by which radiant energy is emitted. The term radiation is also used to describe that which is emitted. Radiation is characterized by the direction and velocity of propagation, the wavelength or frequency, the state of polarization, and the amount of energy carried.

The *velocity* of all electromagnetic radiation is the same in free space, namely

$$c = 2.99776 \times 10^{10} \text{ cm/sec}$$

The velocity in a medium other than free space is determined by the *refractive index*, n, of the medium

$$v = c/n$$

The refractive index varies throughout the electromagnetic spectrum.

The *wavelength* of electromagnetic radiation is the distance between peaks or troughs in the electromagnetic field. The *frequency* is the number of peaks passing a given point per unit time. The frequency of a wave is invariant as it propagates through space, but the wavelength depends on the refractive index. Frequency, f, and wavelength, λ, are related through the velocity of propagation,

$$v = \lambda f$$

Thus the wavelength in a medium is given by

$$\lambda = nc/f$$

The unit of wavelength is seldom taken to be centimeters in infrared spectroscopy because of the inconvenient size of the numbers this choice would lead to. The micron or micrometer, written μ or μm, is the most generally used unit for the infrared region. Infrared radiation is usually said to have wavelengths lying between 0.8 and 1000 μ. Longer wavelength radiation lies in the microwave region. Radiation of shorter wavelength, that is, occupying the region between 0.4 and 0.8 μ [400–800 mμ (or nm) or 4000–8000 Å], is in the visible region. The regions of these and other kinds of radiation are illustrated in Fig. 1.1.

Likewise the unit of frequency is seldom taken as cycles per second, again because of the inconvenient size of the numbers. Instead, the quantity

$$\nu = f/c$$

is taken as a measure of frequency. Indeed ν is often called "frequency." The units of ν are cm^{-1}, and the frequency expressed in this way is called *wavenumber*. It is in fact the number of waves per centimeter in free space. The frequency of infrared radiation lies between 12,500 and 10 cm^{-1}. Wavenumber regions of radiation are included in Fig. 1.1.

The infrared region is further classified somewhat arbitrarily into subregions. The very-short-wavelength region of about 0.8 to 1 μ is called the *photographic infrared* because there exist photosensitive materials which respond to this radiation. This brief extension of the

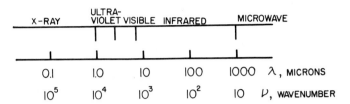

Fig. 1.1 The electromagnetic spectrum.

visible region was discovered by Sir William Herschel in 1800. He dispersed a beam of radiation from the sun with a prism and found that a thermometer placed in the dispersed beam responded to something falling beyond the red limit of visible light. The region from 1 to 5 μ is called the *near infrared*, the 5- to 50-μ region is the *mid infrared*, and the 50- and 1000-μ region is the *far infrared*. The near-infrared radiation can be detected by photoconducting detectors operating at room temperature, whereas the mid and far infrared require the use of other types of detectors. The requirements of the far infrared are so specialized, because of the properties of optical materials, that usually a separate instrument is necessary for this region. Thus the terms photographic, near, mid, and far infrared are associated with properties of instruments, not necessarily with the kinds of information obtainable from these spectral regions.

An *electromagnetic wave* consists of mutually perpendicular electric and magnetic fields transverse to the direction of propagation of the wave. An unpolarized wave has many such fields oriented in random directions. The electric fields of a polarized wave are aligned in one dominant direction and so are the magnetic fields. The interaction of an electromagnetic wave with matter depends strongly on its polarization.

The electromagnetic wave transmits energy from its place of origin to a point of absorption. It does this at a certain rate. Thus it is proper to speak of the power transmitted in a beam of radiation. The energy is stored during transmission in the fields associated with the wave. A beam of radiation can also be treated as a stream of individual particle-like photons. The energy transmitted by the beam is stored in the photons. Each photon carries an amount of energy

$$E = h\nu \text{ ergs}$$

where h is Planck's constant

$$h = 6.624 \times 10^{-27} \text{ erg-sec}$$

and ν is the frequency of the associated electromagnetic wave. Both views of radiation, electromagnetic waves and photon streams, are used by spectroscopists, depending on circumstances. They represent different incomplete approximations to nature. Neither view is correct, but within their appropriate boundaries neither is totally wrong.

Certain radiometric quantities are used by infrared spectroscopists. There are several sets of definitions and symbols in common use. These are discussed and compared by Wolfe and Nicodemus (52). In this book we shall use the term *radiance* to mean the radiant power emitted by a source per unit projected area of the source into a unit solid angle. This is sometimes called *steradiancy* or *radiant sterance*. The radiant power emitted by a point source into a unit solid angle is the radiant intensity of the source. The radiant power emitted in all directions per unit area of a surface is the *radiant emittance*, sometimes called *radiancy*. The radiant power per unit area incident on a surface is the *irradiance*, sometimes called *irradiancy* or *radiant incidence*. The adjective *spectral* preceding a radiometric quantity means that we are concerned with radiant power per unit interval of wavelength or wavenumber.

III. Infrared Studies of the Structure of Molecular Matter

Infrared radiation itself is not particularly interesting to the spectroscopist, but the interaction of radiation with matter can provide important clues to the structure of matter. The infrared spectroscopist measures the absorption, reflection, or emission of radiation by matter. Infrared radiation is especially suited to the study of molecular structure. Molecules are mainly thought of as rigid ball-like atoms connected by spring-like forces. The absorption of infrared radiation sets the molecule into vibrational or rotational motion, or a combination of the two. Similarly, emission of infrared radiation is accompanied by a diminution of the vibration or rotation. Only certain discrete frequencies of vibration or rotation are permitted, and these frequencies correspond to the frequencies of the radiation which are absorbed or emitted.

A molecule having N atoms can undergo $3N - 6$ fundamental modes of vibration—or $3N - 5$, if it is a linear molecule. Each of these fundamental modes has associated with it a ground-state energy

$$E = \tfrac{1}{2}h\nu$$

In addition vibrations of higher frequencies can be excited with energy

$$E = (n + \tfrac{1}{2})h\nu \qquad n = 1, 2, 3, \ldots$$

Thus vibrational energy states available to a molecule with three atoms might be plotted as in Fig. 1.2. The molecule is raised to the first excited level of the second normal mode of vibration by absorbing a photon having energy equal to the energy difference between levels 2 and 1. A photon of somewhat lower frequency is required to excite the first mode, and one of higher frequency excites the third mode. These three frequencies constitute the fundamental vibrational absorption spectrum of the molecule.

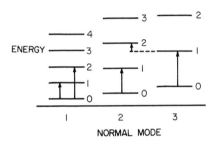

Fig. 1.2   Vibrational energy transitions.

Not all normal modes can be excited by the absorption of radiation. There are selection rules determined by the symmetry of the molecule which determine this. It also happens that a molecule can be excited by two or more changes at once, either by a change of two or more levels within a given normal mode or by simultaneous excitation of two or more modes. These give rise to *overtones* or *combination* absorption bands. Finally, a molecule already in an excited state can be further excited to a level of a different mode causing a difference band to be observed. These multilevel transitions are generally weak.

The spectroscopist observes the frequencies of absorbed radiation and assigns the corresponding molecular vibrational frequencies to particular modes of oscillation. He can then use the known geometry and atomic masses with the observed vibrational frequencies to compute the force constants of the interatomic bonds, that is, the stiffness of the springs.

The rotational energy levels of a molecule are also restricted to certain discrete values,

$$E = BJ(J+1)$$

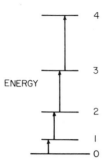

Fig. 1.3   Rotational energy transitions.

where B is a constant determined by the moment of inertia of the molecule about one of its axes. The spacing of these levels is seen to increase with increasing values of J, as shown in Fig. 1.3. There are selection rules prohibiting J from changing by more than one digit when radiation is absorbed or emitted. The spacing is generally quite small, so the pure-rotation spectrum is found in the far-infrared or microwave region of the spectrum. Further, because of the importance of perturbations by neighboring molecules, rotational spectra are observed only for gaseous molecules (with some possible rare exceptions).

In general, each vibrational state of a gaseous molecule is accompanied by a set of rotational levels, so a segment of the complete energy level diagram for one normal mode of vibration appears as in Fig. 1.4. A transition from one vibrational level to another caused by infrared absorption can occur without a change in rotational state or with a change of +1 or −1 in the J value, as illustrated. Each vibrational transition is therefore a complex band containing many possible rotational transitions. The selection rules for rotational transitions depend on the geometry of the molecule and the particular vibrational

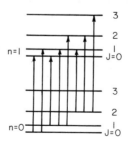

Fig. 1.4   Vibration–rotation transitions.

mode in question. Thus the appearance of a vibration–rotation band provides a clue to molecular structure.

The infrared absorption spectra of polyatomic molecules are quite complex, because of the multiplicity of fundamental vibrations and the additional overtones and sum and difference bands. The spectrum of a particular chemical compound is a unique characteristic of that compound alone, reflecting as it does the geometry, bond strength, and atomic masses of the substance. For this reason one of the important uses of infrared spectroscopy has been the identification of unknown chemicals. It is only necessary to record the spectrum of the compound and compare it with spectra of known materials until a match is found. It is imperative that a large library of spectra be available for success in this sort of work. Collections can be obtained from commercial suppliers such as the Sadtler Research Laboratories, and soon reference spectra will be made available by the National Bureau of Standards with the cooperation of the Coblentz Society. Several infrared laboratories make use of electronic computers in searching files of reference spectra.

The identification of chemicals is made somewhat more difficult if the unknown substance is a mixture of different components. The use of computers for analyzing the spectra of complex mixtures is just beginning to be exploited and should be of increasing usefulness in the future.

Certain structural or functional groups of atoms are found to vibrate with frequencies nearly unaffected by the rest of the molecule. For example, the methyl group generally causes infrared absorption bands at about 2800 and 3000 cm^{-1}, corresponding to the symmetric and antisymmetric C—H stretching vibrations, and also at about 1375 and 1460 cm^{-1} due to the symmetric and asymmetric bending or deformation vibrations. Such characteristic or group frequencies are extremely valuable to the spectroscopist, who, with their aid, is able to identify many of the structural groups present in a molecule, even though he cannot give the complete structure.

The intensity of an infrared absorption band is related to the geometric and electronic structures of the molecule. Of more practical usefulness is the relationship between the transmittance of a sample T at a specified frequency, the thickness b, the concentration c of a component of the sample, and the absorptivity a of the component at the specified frequency.

$$T(\nu) = e^{-a(\nu)bc}$$

With the aid of this equation, which is known as the Beer–Lambert law of absorption, it is possible to analyze a sample quantitatively by means of infrared absorption spectroscopy.

In crystalline solids molecules are aligned with respect to the crystal lattice. It is often possible to study the orientation of the molecules by observing the absorption bands with polarized radiation. This can be done frequently with quasicrystalline solids as well, such as stretched polymer films. Some absorptions observed in crystals are due to vibrations of the lattice itself. These are usually of very low frequency and are studied in the far infrared. Some crystalline transitions studied by semiconductor spectroscopists are not associated with vibrational motion at all, but are due to the electrons in the solid. Conduction electrons provide a continuum of absorption, increasing in intensity at low frequencies, and valence electrons give more or less discrete absorptions or absorption edges. Reflection spectroscopy is widely used in the study of solids, particularly in the determination of optical constants.

Infrared spectroscopy is by no means limited to absorption measurements. Reflection spectroscopy is widely used in the study of solids, particularly in determining optical constants. Emission spectroscopy provides information about reacting gases and data of use to heat transfer engineers. In the last chapters of this book we shall discuss some of the experimental methods used in the study of infrared absorption, reflection, and emission.

REFERENCES

In recent years a number of books of interest to infrared spectroscopists have appeared. We list many of them here. The classification is somewhat arbitrary in many cases and the list is not intended to be exhaustive.

GENERAL TREATMENTS

1. G. M. Barrow, *Introduction to Molecular Spectroscopy*, McGraw-Hill, New York, 1962.
2. R. B. Bauman, *Absorption Spectroscopy*, Wiley, New York, 1962.
3. G. H. Beaven, E. A. Johnson, H. A. Willis, and R. G. J. Miller, *Molecular Spectroscopy—Methods and Applications in Chemistry*, Macmillan, New York, 1961.
4. W. Brügel, *An Introduction to Infrared Spectroscopy*, Wiley, New York, 1962.
5. A. D. Cross, *Introduction to Practical Infrared Spectroscopy*, Butterworth, London, 1960.

6. M. Davies, ed., *Infrared Spectroscopy and Molecular Structure*, Elsevier, Amsterdam, 1963.
7. D. N. Kendall, ed., *Applied Infrared Spectroscopy*, Reinhold, New York, 1966.
8. J. Lecomte, "Spectroscopie dans l'infrarouge," *Handbuch der Physik XXVI*, Springer, Berlin, 1958.
9. C. E. Meloan, *Elementary Infrared Spectroscopy*, Macmillan, New York, 1963.
10. K. Nakanishi, *Infrared Absorption Spectroscopy—Practical*, Holden-Day, San Francisco, 1962.
11. W. J. Potts, Jr., *Chemical Infrared Spectroscopy*; Vol. I, *Techniques*, Wiley, New York, 1963.
12. C. N. R. Rao, *Chemical Applications of Infrared Spectroscopy*, Academic Press, New York, 1963.
13. H. A. Szymanski, *Theory and Practice of Infrared Spectroscopy*, Plenum Press, New York, 1964.
14. S. Walker and H. Straw, *Spectroscopy*. Vol. II. *Ultraviolet, Visible, Infrared, and Raman Spectroscopy*, Chapman and Hall, London, 1962.

THEORY OF VIBRATIONAL AND ROTATIONAL SPECTRA

15. G. Herzberg, *Infrared and Raman Spectra of Polyatomic Molecules*, Van Nostrand, Princeton, N.J., 1945.
16. E. B. Wilson, J. C. Decius, and P. C. Cross, *Molecular Vibrations*, McGraw-Hill, New York, 1955.
17. H. C. Allen, Jr., and P. C. Cross, *Molecular Vib-Rotors*, Wiley, New York, 1963.

FUNCTIONAL GROUP ANALYSIS

18. L. J. Bellamy, *The Infrared Spectra of Complex Molecules*, 2nd ed, Wiley, New York, 1958.
19. N. B. Colthup, L. H. Daly, and S. E. Wiberley, *Introduction to Infrared and Raman Spectroscopy*, Academic Press, New York, 1964.
20. M. St. C. Flett, *Characteristic Frequencies of Chemical Groups in the Infrared*, Elsevier, Amsterdam, 1963.
21. S. K. Freeman, ed., *Interpretive Spectroscopy*, Reinhold, New York, 1965.
22. D. W. Mathieson, ed., *Interpretation of Organic Spectra*, Academic Press, New York, 1965.
23. H. A. Szymanski, ed., *Infrared Band Handbook*, Plenum Press, New York, 1963.

INORGANIC COMPOUNDS

24. K. Lawson, *Infrared Absorption of Inorganic Substances*, Reinhold, New York, 1961.
25. K. Nakamoto, *Infrared Spectra of Inorganic Coordination Compounds*, Wiley, New York, 1963.

POLYMERS

26. C. J. Henniker, *Infrared Spectrometry of Industrial Polymers*, Academic Press, New York, 1967.

27. D. O. Hummel, *Infrared Spectra of Polymers in the Medium and Long Wavelength Regions*, Wiley, New York, 1966.
28. R. Zbinden, *Infrared Spectroscopy of High Polymers*, Academic Press, New York. 1964.

CARBOHYDRATES

29. R. G. Zhbankov, *Infrared Spectra of Cellulose and Its Derivatives*, Consultants Bureau, New York, 1966.
30. R. S. Tipson, *Infrared Spectroscopy of Carbohydrates*, Nat. Bur. Std. Monograph 110, 1968.
31. R. S. Tipson and F. S. Parker, "Infrared spectroscopy," Chap. 24, Part 5, *Carbohydrate Chemistry* (Pigman and Horton, eds.), Academic Press, New York, 1968.

STEROIDS

32. K. Dobriner, E. R. Katzenellenbogen, and R. N. Jones, *Infrared Absorption Spectra of Steroids — an Atlas*, Wiley-Interscience, New York, 1953.

SURFACE CHEMISTRY

33. M. L. Hair, *Infrared Spectroscopy in Surface Chemistry*, Dekker, New York, 1967.
34. L. H. Little, *Infrared Spectra of Adsorbed Species*, Academic Press, New York, 1966.

INTERNAL ROTATION

35. S. Mizushima, *Structure of Molecules and Internal Rotation*, Academic Press, New York, 1954.

EMISSION SPECTROSCOPY

36. H. H. Blau and H. Fischer, *Radiative Transfer from Solid Materials*, Macmillan, New York, 1962.
37. S. S. Penner, *Quantitative Molecular Spectroscopy and Gas Emissivities*, Addison-Wesley, Reading, Mass., 1959.
38. R. H. Tourin, *Spectroscopic Gas Temperature Measurement*, Elsevier, Amsterdam, 1966.

REFLECTANCE SPECTROSCOPY

39. W. W. Wendblandt and H. G. Hecht, *Reflectance Spectroscopy*, Wiley, New York, 1966.
40. N. J. Harrick, *Internal Reflection Spectroscopy*, Wiley, New York, 1967.

ASTRONOMY

41. H. L. Johnson, "Astronomical measurements in the infrared," *Ann. Rev. Astron. and Astrophys.* (1966).

41a. P. J. Brancazio and A. G. W. Cameron, *Infrared Astronomy*, Gordon & Breach, New York, 1968.

PHYSICS, INSTRUMENTS, AND TECHNIQUES

42. G. C. Cannon, ed., *Electronics for Spectroscopists*, Wiley-Interscience, New York, 1960.
43. G. K. T. Conn and D. G. Avery, *Infrared Methods — Principles and Applications*, Academic Press, New York, 1960.
44. H. L. Hackforth, *Infrared Radiation*, McGraw-Hill, New York, 1960.
45. J. T. Houghton and S. D. Smith, *Infrared Physics*, Oxford Univ. Press, London and New York, 1966.
46. J. A. Jamieson, R. H. McFee, G. N. Plass, R. H. Grube, and R. G. Richards, *Infrared Physics and Engineering*, McGraw-Hill, New York, 1963.
47. B. Kemp, *Modern Infrared Technology*, Sams, Indianapolis, Ind., 1962.
48. P. W. Kruse, L. D. McGlauchin, and R. B. McQuistan, *Elements of Infrared Technology: Generation, Transmission, and Optics*, Wiley, New York, 1962.
49. A. E. Martin, *Infrared Instrumentation and Techniques*, Elsevier, Amsterdam, 1966.
50. L. May, ed., *Spectroscopic Tricks*, Plenum Press, New York, 1967.
51. R. A. Smith, F. E. Jones, and R. P. Chasmer, *The Detection and Measurement of Infrared Radiation*, Oxford Univ. Press-Clarendon, London and New York 1957; 2nd ed., 1968.
52. W. L. Wolfe, ed., *Handbook of Military Infrared Technology*, U.S. Govt. Printing Office, Washington, D.C., 1965.
52a. A. Hadni, *Essentials of Modern Physics Applied to the Study of the Infrared*, Pergamon, New York, 1968.

WAVELENGTH STANDARDS

53. International Union of Pure and Applied Chemistry, *Tables of Wavenumbers for the Calibration of Infrared Spectrometers*, Butterworth, London, 1961.
54. K. N. Rao, C. J. Humphreys, and D. H. Rank, *Wavelength Standards in the Infrared*, Academic Press, New York, 1966.

MISCELLANEOUS

55. H. M. Hershenson, *Infrared Absorption Spectra Index for 1945–1957*, Academic Press, New York, 1959.
56. H. M. Hershenson, *Infrared Absorption Spectra Index for 1958–1962*, Academic Press, New York, 1964.

CHAPTER 2

The Infrared Spectrophotometer

I. THE SPECTROPHOTOMETER

In order to study spectral phenomena, the spectroscopist requires a device for separating electromagnetic radiation into its component frequencies. The form of the device naturally depends on the spectral region to be studied—x-ray, ultraviolet, visible, infrared, or microwave — as well as on the nature of the phenomenon under investigation — absorption, reflection, or emission.

A device which separates radiation into constituent frequencies and presents them for visual inspection is called a *spectroscope*; if it also permits a measurement of the frequency, it is called a *spectrometer*; if, furthermore, it provides a measure of the relative amount of energy associated with each wavelength, it is called a *spectrophotometer*. Evidently, spectroscopes are practically limited to the visible region, whereas spectrometers or spectrophotometers might be used in any part of the electromagnetic radiation spectrum for which they are designed. For the most part, the detectors, amplifiers, and recording systems used in constructing infrared spectrometers qualify them to be spectrophotometers.

A spectrophotometer must have a source of electromagnetic radiation, which may be the sample itself if an emission spectrum is being studied. It contains an arrangement for mounting and exposing the sample to the beam of radiation if reflection or absorption properties are to be measured. It contains a monochromator for isolating a narrow band of frequencies centered at some central frequency selected by the spectroscopist. Microwave spectroscopists enjoy the significant advantage of being able to use tunable sources of radiation that function as both source and monochromator. Finally, the spectrophotometer contains a detector and an indicator or recorder for measuring and displaying the power of the radiation beam passed by the monochromator.

II. The Infrared Spectrophotometer

We discuss the components of the infrared spectrophotometer and their function and performance in considerable detail in the course of this book. But the reader should have an overall understanding of the nature of the spectrophotometer before he begins a deeper study. To this end let us first describe a typical instrument with a few variations.

Figure 2.1 is a sketch of the optical parts of a simple hypothetical instrument which is used for measuring the transmittance of a gaseous sample in the spectral region lying between 2 and 15 μ. The source of radiation at S is a glowing rod of some refractory material, heated electrically, and emitting efficiently in the infrared. Radiation from the source is collected by a concave mirror M_1, which condenses the beam and forms an image of the source inside a tubular cell. The cell contains the sample. It is capped at each end by windows made of rock salt which transmits radiation of wavelength as long as 15 μ, but not much longer. A cell used for gases is typically 10 cm in length. Liquid cells

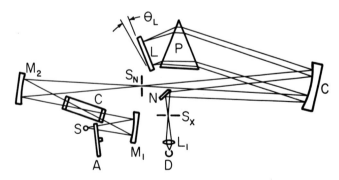

Fig. 2.1   A typical single-beam infrared spectrometer with prism monochromator.

are correspondingly thinner. The beam emerging from the cell is intercepted by a second mirror M_2 and refocused onto the entrance slit at S_N. The slit may be as narrow as perhaps 0.01 mm or as wide as several millimeters, and it is possibly 1- or 2-cm high. The radiation enters the monochromator through the entrance slit. It is collected by the collimating mirror C which sends it as parallel bundles of rays to the dispersing element P. As drawn in Fig. 2.1, P is a prism of rock salt which deviates the rays by different amounts depending on their wavelength. This enables the monochromator to discriminate different wavelengths. In some instruments a diffraction grating is used as the dispersing element. In our prism instrument, the rays are intercepted by a plane mirror L and returned through the prism a second time where they undergo deviation. This arrangement is called a Littrow arrangement and L is the Littrow mirror. There are other arrangements — the Ebert–Fastie, the Czerny–Turner, the Pfund, and so on. The radiation returns to mirror C which now acts as a condensing mirror. As a condensing mirror, C is often called a telescope or camera mirror. Sometimes the collimator and condenser are separate mirrors. The radiation is now brought to a focus at the exit slit S_X after undergoing a change in direction introduced by the plane mirror N. Rays of other wavelengths are focused in the plane of S_X but at different positions because they have been deviated through different angles by the prism. The nominal wavelength passed from the monochromator is selected by changing the angle θ_L of the Littrow mirror. The width of the range of wavelengths passing from the monochromator is determined by the width of the exit slit. The beam is finally refocused onto a detector by some element, such as a sodium chloride lens L_1.

Ordinary photocells and photographic plates do not respond to the feeble photons of infrared radiation. The detectors that are generally used in the infrared are temperature-sensing devices such as thermocouples and, less often, bolometers. The radiant power falling on the detector is usually only about 10^{-7} or 10^{-8} W, and the signal generated by the thermocouple is about 10^{-7} or 10^{-8} V. This signal must be amplified considerably before it is measured, and moreover it must be highly filtered to remove as much electrical noise as possible.

The detector is unable to discriminate between a temperature change due to absorption of radiation from the source and a change associated with a variation in the temperature of its surroundings. In the early days of infrared spectroscopy, it was necessary to control the temperature of the infrared laboratory very closely. The spectroscopist often had to take his place beside his instrument and remain there for some time while it adjusted its thermal condition to his presence. Only then could he record meaningful data, free of excessive temperature drift. Visitors to the spectroscopy laboratory were not welcome during measurements because their presence disturbed the temperature balance of the spectrometer.

The early infrared spectroscopist scanned a spectrum manually from one wavelength point to the next using a sensitive galvonometer or a galvanometer with a photoelectric cell amplifier. At each point, he observed the signal generated by the thermocouple with the sample cell removed. He then inserted the cell and made a new reading. Finally, he inserted a shutter and made a third reading. The difference between the first and third readings represents the signal I_0 due to the incident radiation. The difference between the second and third readings is the signal due to the incident radiation attenuated by the sample, I. The ratio of these is the transmittance of the sample at that wavelength, $T = I/I_0$. The whole procedure might have to be repeated several times to obtain a good average before the observer could move on to the next wavelength. As the process continued, the slit width was altered from time to time to allow for the change with wavelength of the spectral radiance of the source. Often, the collection of data was carried out at night. Not only was the temperature of the laboratory less variable after dark but the sensitive galvanometer was less likely to be perturbed by vibrations and stray electric fields. The pioneering giants of infrared spectroscopy — Coblentz, Langley, Paschen, Rubens, and many others — worked in this way.

The addition of dc amplifiers, recorders, and wavelength-scanning motors speeded up the collection of data in the 1930's and lead to the introduction of successful commercial spectrophotometers in the early 1940's. The breaker amplifier eliminated most of the notorious drift associated with dc amplifiers and permitted a high order of amplification, but the detector continued to drift under the influence of changes in ambient temperature. It was necessary to record the signal over a brief wavelength range, with and without the sample, then readjust zero with a shutter in the beam to compensate for temperature drift, change the slit width, and repeat over the next wavelength interval. When the desired spectral region was covered, the recorder chart was measured at close intervals, ratios were calculated, and finally the spectrum was plotted. The author employed this tedious procedure as recently as 1952. If all went well, two workers could obtain an infrared spectrum in a day, one worker recording and the other calculating and plotting.

The next major step in the evolution of the infrared spectrophotometer came with the use of radiation choppers in the mid 1940's. The thermal drift which plagued the early workers occurred because the detector responded to any temperature change, whether from exposure to the source or from changes in ambient conditions. A chopper, for example a rotating sector disk inserted between the source and the rest of the system (A in Fig. 2.1), converts the radiation beam from steady dc to pulsating ac. All the other sources of radiation that might affect the detector remain dc and, therefore, the slow drift of the signal from the detector due to the change of its mean temperature is essentially dc. The amplifier is a tuned ac amplifier responding only to ac signals of the same frequency as the chopped radiation. Only the radiation originating in the source is recorded, and thermal drift is entirely eliminated. The chopper in effect permits the spectroscopist to measure the differences in signal with the sample or the shutter (i.e., the chopper blade) in the beam and also the difference between source alone and shutter. With the problem of thermal drift solved, it was possible to scan for long periods of time without resetting the zero. The I_0 still changed continuously with wavelength. This difficulty was avoided by mechanically or electrically varying the slit width with wavelength according to a previously set program to maintain the I_0 at an approximately constant level.

At this stage in the development of the infrared spectrophotometer, the spectroscopist was able to record with good speed and accuracy the data required to determine the transmission spectrum of a sample. The raw recording of I served as a satisfactory qualitative spectrogram for the identification of samples or for functional group analysis. However, the curves still contained the effects of variations in I_0, including the very strong and sharp absorption lines of atmospheric water vapor and carbon dioxide. A transmission spectrum could be obtained from the I_0 and the I records only by resorting to the old-fashioned and time-consuming "read-calculate-and-plot" procedure.

A method for automatically recording transmission spectra was much to be desired and searches for such methods took various courses. Three general approaches were tried with some success. One of these is an extension of the slit width program. If a more precise and more flexible means of programming the slit width and amplifier gain could provide a straight I_0 record, then the I record would be a faithful recording of the sample transmission. Avery (1) had already built such an instrument in 1941, and the Beckman IR-3 operated on this principle with the program information stored on magnetic tape.

A few of these instruments are still in use, but the complexity and cost of the system prevented wide acceptance.

In contrast with the programmed single-beam instrument, the other two methods for automatic recording of transmittance use a double-beam approach. The radiation from the source is divided (Fig. 2.2), part of it traversing a sample beam in which the sample cell is placed and part of it traversing a reference beam in which an empty cell might be placed. The beams are recombined with an oscillating or rotating mirror M, and sent into the monochromator. Some early attempts $(2, 3)$ to use separate detectors for the two beams were made, but these were not highly successful because of the difficulty of matching infrared detectors.

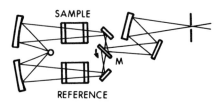

Fig. 2.2 Typical photometer optics of a double-beam spectrophotometer.

The most common form of double-beam infrared spectrophotometer operates on the optical null principle. A variable attenuator is installed in the reference beam, often in the form of a comb, as drawn in Fig. 2.3. The amount of attenuation depends on how far into the beam the comb is inserted. The rotating mirror permits the detector to sample radiation alternately from the sample and reference beams. If the power in the two beams is not equal, an ac signal is generated by the detector. After amplification this signal operates a servomechanism to drive the comb farther into or out of the beam until the ac signal vanishes or is nulled out. Thus as a spectrum is scanned and the sample beam becomes

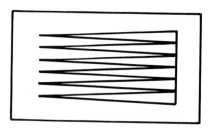

Fig. 2.3 Optical attenuator for a double-beam spectrophotometer.

brighter or darker, due to variations in sample absorption, the reference beam is made equally bright or dark. The position of the comb represents the transmission of the sample, and a recorder which follows the position of the comb also records the transmission spectrum. It is necessary to program the slit width to maintain more-or-less constant energy in the reference beam, but this is not nearly as critical as in the single-beam system. The first practical optical null infrared spectrophotometer was built by Wright and Herscher (4) in 1947 and Baird Associates (now Baird-Atomic, Inc.) soon began manufacturing a version of it (5). This was quickly followed by the Perkin–Elmer Model 21 in 1950 (6) and much later in 1957 by the Beckman IR-4. For many years the Perkin–Elmer and the Baird instruments were the standard infrared spectrophotometers in the laboratories of chemical spectroscopists. In recent years optical null instruments in much simplified and less expensive forms have made infrared spectroscopy available to chemical laboratories of the most modest requirements and budgets. We discuss the family trees of modern instruments in greater detail in the next section.

The second general class of double-beam instrument is the ratio recording spectrophotometer. Its function is to determine the ratio of energy in the sample and reference beams by electronic means. These instruments usually discriminate between sample and reference beams by use of different chopping rates or by phase discrimination. The Halford–Savitzky (7) method utilizes the top half of the slit for the sample beam and the bottom half for the reference beam. The Perkin–Elmer Model 13 is based on this approach. The modern Cary–White Model 90 uses $13\frac{1}{3}$ cps chopping rate in the sample beam and $26\frac{2}{3}$ cps in the reference beam. Some other ratio recording systems are described later in Chapter 12. The basic limitations of ratio recording systems lie in the slow response time of thermal detectors, which remain the standard detectors of infrared instruments. This restricts the chopping rates to somewhat less than 60 cps, usually no more than about 25 cps, and permits cross talk between sample and reference beam information.

While these advances in the electronic state of infrared spectrophotometers were taking place, equally significant improvements in the optical systems were being made. The quest for resolution and spectral purity led to the construction of the double prism monochromators of Leiss, Farrand, and others and the double-monochromator spectrophotometers of Beckman (the IR-3 and the IR-4). Walsh (8) devised a single prism monochromator with double prism resolution by double passing the radiation beam through the prism. Discrimination between single- and double-pass radiations is accomplished by chopping the beam between passes, so only double-passed radiation is modulated. This principle is used in the Perkin–Elmer Models 112 (single beam) and 321 (double-beam optical null).

Builders of high-resolution instruments had been using diffraction gratings as the primary dispersing element in their monochromators, frequently with a prism monochromator accompanying it to act as a tunable filter. The high cost of quality gratings and certain technical difficulties delayed the introduction of gratings into small instruments, although several workers had merely removed the prisms from their instruments and replaced them with gratings. Perfection of a reliable replication process for producing good copies of diffraction gratings made it inevitable that grating infrared spectrophotometers would be manufactured. The Beckman IR-4 was the first to appear, in 1958, with a prism–grating combination and a new name, the IR-7. This was followed by the Perkin–Elmer's modification of their Model 21, the Model 221, and by numerous European grating instruments such as the Unicam. Newer versions of these instruments appear from time to time. Even the simplified low-cost instruments have been adapted to grating operation. Presently, completely new instruments, designed from infancy as grating spectrophotometers, are appearing. Among these are the Perkin–Elmer Models 125 and 225 and the Cary–White 90.

One of the significant advances made by the new breed of instrument is a steady rollback of the long-wavelength limit. The traditional region of the pre-1940 chemical spectroscopist had been from 2 to 15 μ. Anything beyond 15 μ was the "far infrared," and was considered nearly inaccessible. New prism materials altered the feeling of inaccessibility about the longer wavelengths and the grating monochromator made the far infrared routinely available. Several modern instruments are capable of scanning automatically from 2 to 50 μ, and special far-infrared instruments can cover almost automatically the region beyond 50 μ out to 300 μ farther.

In the past decade strange infrared instruments have begun to appear. These use interferometers rather than monochromators to separate wavelengths, and the records they present are not spectra but *interferograms*, which are mathematically related to spectra and require a computer to unravel them. The workings of these instruments are described later in Chapter 8. Commercial versions are offered by Block, Grubb–Parson, and Beckman.

III. COMMERCIAL INFRARED INSTRUMENTS

The first infrared spectrometer to be commercially available was built over fifty years ago by the firm of Adam Hilger in England. Many commercial instruments have come and gone since then, mostly in the past two decades.

There is no point in reciting detailed specifications of commercially available instruments. Old instruments are modified and modernized and new instruments are introduced regularly, so specifications rapidly become obsolete and incomplete. In this section we shall describe briefly the products of some major manufacturers. The intent is to trace the development of the instruments, rather than completely characterize them.

The best of the grating instruments offer resolution approaching 0.2 cm^{-1} over most of their range, but only with extreme noise filtering and, consequently, very slow scanning speed. On the other hand, with more modest resolution even the simpler instruments can produce a useful record in about ten minutes.

A. Beckman Instruments, Fullerton, California

The first infrared instrument offered by Beckman was called the IR-1. It was patterned after an instrument built by R. R. Brattain, *et al.* (9), and used at the Shell Development Company. It was a simple, single-beam dc instrument. The IR-2 was a much improved spectrophotometer, still single beam but using a modulated radiation beam.

The IR-3 was a very sophisticated instrument. It made use of a magnetic tape memory to compensate for solvent and atmospheric absorptions. The IR-3 had a double monochromator with two prisms and it offered unusually good resolution and spectral purity in its day.

The IR-4 series of instruments (IR-4, 7, 9, 11, and 12) are the basis of Beckman's present line of high-performance instruments. They have monochromators of the Littrow type with spherical collimating mirrors of 50-cm focal length. Slits are 25 mm high. Nernst glower sources and thermocouple detectors are used except in the IR-11. All of these instruments are double beam with optical null photometer systems. The IR-4 has a double sodium chloride prism monochromator with provision for interchanging other prisms for extended wavelength range or greater resolution. With proper selection of prisms, it can be used from 10,000 to 280 cm^{-1}. The IR-7 was developed from the IR-4 by substitution of a grating monochromator for one section of the prism monochromator. The grating is used successively in four diffraction orders. The remaining prism serves as a filter to isolate selected spectral orders from the grating. With monochromator interchange units, it can be used to frequencies as low as 200 cm^{-1}. The IR-9 is an improved version of the IR-7. It has two gratings mounted on a turret and automatically selects one of these for the appropriate spectral region. Each grating is used in two orders. The IR-9 also incorporates numerous mechanical refinements, including an optically projected wavenumber scale for better accuracy and readability. It covers the 4000 to 400 cm^{-1} region in a single scan with no optical interchanges required. In the IR-12, the prism monochromator is replaced with a set of transmission filters mounted on a turret and automatically positioned. Four gratings are used on a turret. The spectrum is scanned from 4000 to 200 cm^{-1}. The IR-11 is a far-infrared version of the IR-7. It scans semiautomatically from 800 to 33 cm^{-1}. A mercury arc is used as a source and a Golay detector is used with a quartz or diamond window.

The IR-5 series of instruments (IR-5, 6, 8, 10) provide more modest performances and are intended for routine applications. The IR-5 (*10*) and the improved version IR-5a are double-beam instruments with sodium chloride prism monochromators of 35-cm focal length. There is also a "far-infrared" version of the IR-5 with a CsBr prism. These instruments use thermocouple detectors and coiled nichrome wire sources. The IR-6, no longer produced, was similar to the IR-5, but was a single-beam instrument. The IR-8 was derived from the IR-5 by conversion to a grating monochromator. It has two gratings on a turret and may be used from 4000 to 625 cm^{-1}. The IR-10 is also a grating instrument with its range extending to 300 cm^{-1}.

The IR-18 and IR-20 are the current Beckman instruments for the lower cost market. They are grating instruments covering the 4000 to 600-cm^{-1} and 4000 to 250-cm^{-1} regions, respectively. An unusual feature is the use of a circular tunable interference filter in place of discrete filters.

Beckman also produces a pair of instruments using tunable interference filters rather than conventional monochromators. One of these, the Microspec, is intended to record low-resolution survey spectra in the 2.5 to 14.5-μ region. The other, the 102, is a fast scanning instrument which can be coupled to the outlet from a gas chromatograph for identification of eluents as they appear.

Finally, a London subsidiary of Beckman, the former RIIC, produces a series of interference Fourier spectrometers for use in the very far infrared. Successive versions of these are labeled FS520, 620, 720, and 820.

B. Perkin–Elmer, Norwalk, Connecticut

The original Perkin–Elmer spectrophotometer was the Model 12-A which was the instrument developed at American Cyanamide by Barnes *et al.* (*11*). It was an unmodulated, single-beam instrument with a sodium chloride prism and an off-axis paraboloid collimator of about 27-cm focal length. It was used with a galvanometer. An improved model, the 12-B, provided for the recording of spectra with a potentiometric recorder and a breaker amplifier. The Model 12-C used a chopper to modulate the radiation. The Model 112 used the Walsh double-pass monochromator system (to be described later in Chapter 6) to increase dispersion and resolution. The Models 12G and 112G used gratings in place of the prisms. The monochromator portions of these instruments are made available as separate units, with various designations, such as Model 98 and Model 99. These appear in many special experimental systems and have contributed to a very great extent to the present advanced state of the infrared art.

The basic Model 12 monochromator has been used in two fundamentally different double-beam systems by Perkin–Elmer. One of these, the Model 13, uses the Savitzky–Halford ratio recording system. It has not been widely accepted as a general purpose spectrophotometer. An extensively modified form of this instrument, the Model 301, is used in the very far infrared out to about 400 μ. The other instrument, the Model 21 (*6*), incorporates an optical null photometer system and has been very widely accepted indeed. The addition or substitution of diffraction gratings in the monochromator has led to a succession of instruments: Models 221, 421, 521, and 621. The Model 621 uses a set of four interchangeable gratings on a turret, with filters for order separation. It covers the spectral region from 4000 to 200 cm^{-1}. A double-passed version of the Model 21, the Model 321, appeared only briefly.

The "Infracord" series of Perkin–Elmer infrared spectrophotometers was developed to provide the chemist with a low-cost instrument with adequate performance for most of his requirements. The original Infracord, the Model 137, offered a Littrow monochromator with a single sodium chloride prism and an optical null double-beam photometer system. The Model 137-B is a somewhat improved version and the Model 137-KBr has a KBr prism. The Model 137-G uses a grating instead of a prism. The Model 237 and the more recent Model 257 have turret-mounted dual gratings and offer additional refinements over the original basic versions. The latest version of the Infracord covers the region from 4000 to 250 cm^{-1} and features a flatbed recorder in place of the familiar Perkin–Elmer drum.

The Perkin–Elmer Models 125 and 225 are high-performance grating spectrophotometers produced by the German division of Perkin–Elmer. They are original designs, and are not based on traditional Perkin–Elmer monochromators.

A very recent development from Perkin–Elmer is the Model FIS-3 far-infrared spectrophotometer for the 400 to 30-cm^{-1} region. It is a double-beam optical null instrument with an evacuable housing.

C. Baird-Atomic, Cambridge, Mass. (Originally, Baird Associates)

The Baird 4-55 was the first commercial double-beam spectrophotometer. It used an unusually large prism in a Littrow monochromator. The infrared literature of the early 1950's contains many references to this instrument. A much improved version is the NK-1. The NK-3 is similar electronically and mechanically, but substitutes a two-grating turret in the monochromator. The Baird-Atomic KM-1 is a simplified, intermediately priced prism instrument.

D. Unicam Instruments, Cambridge, England

The monochromator of the Unicam SP-100 contains a turret on which up to four prisms can be placed and selected for the appropriate wavelength region. It has an optical-null photometer, but, unlike other such instruments, it utilizes a "star wheel" in place of a comb attenuator. The star wheel is a high-speed rotating sector with a starlike appearance. It is moved into and out of the beam to provide the desired amount of attenuation. It avoids some of the problems encountered when the discrete teeth of a comb attenuator are imaged on the slit and detector. The SP-100 can be evacuated to eliminate absorption by atmospheric water and carbon dioxide.

The Unicam SP-130 is similar to the SP-100, but is a double-monochromator system (12). The second monochromator has a turret containing two gratings.

The Unicam SP-200 is a simplified, low-cost instrument with a single prism monochromator and a conventional comb-type optical-null photometer. The SP-200G is a grating version.

E. Cary Instruments, Monrovia, California

The Cary Model 90 is a double-beam ratio recording spectrophotometer. It has a double monochromator with a prism section followed by a grating section.

F. Warner and Swasey, Flushing, New York

Warner and Swasey spectrometers are specialized instruments for use in observing spectra from fast reactions. They are confined to wavelengths shorter than 9 μ. The Models 301 and 401 observe emission or absorption at a selected wavelength. The Model 501 (13) sweeps the spectrum across the exit slit by means of rotating corner mirrors. Thus, it can record a wide region of spectrum in as short a period of time as 1 msec.

G. Block Associates, Cambridge, Mass.

The Block Associates Model 200 is an interferometer (14). It repeatedly scans the interferogram corresponding to an infrared spectrum, adds consecutive scans in order to improve the signal-to-noise ratio, and finally converts the interferogram into a spectrum. Various versions of the Model 200 are used between 1 and 40 μ.

There are several other commercial suppliers of infrared spectrophotometers, monochromators, and interferometers. We merely mention some of them: Jarrell–Ash, McPherson, White, and Farrand in the United States; Grubb–Parsons* in England; Cameca and Huet in France; Leiss and Zeiss in Germany; and Japan Spectroscopic Co., Shimadzu,† and Hitachi in Japan. The products of these organizations will not be described because they are not often seen in the United States and because the author is not familiar with their details.

REFERENCES

1. W. H. Avery, *J. Opt. Soc. Am.*, **31**, 633 (1941).
2. R. F. Wild, *Rev. Sci. Inst.*, **18**, 436 (1947).
3. G. B. B. M. Sutherland and H. W. Thompson, *Trans. Faraday Soc.*, **41**, 171 (1945).

*Grubb–Parsons spectrophotometers are now marketed by Wilks Scientific Corp.

†The Shimadzu Spectronic 270 IR is now being marketed by Bausch and Lomb. This instrument has a double-beam optical-null system and covers the 4000 to 400-cm^{-1} range with a filter–grating monochromator. Two gratings are used, but are not turret mounted. Grating selection is with a sliding mirror arrangement.

4. N. Wright and L. W. Herscher, *J. Opt. Soc. Am.*, **37**, 211 (1947).
5. W. S. Baird, H. M. O'Bryan, G. Ogden, and D. Lee, *J. Opt. Soc. Am.* 754 (1947).
6. J. U. White and M. D. Liston, *J. Opt. Soc. Am.* **40**, 29, 93 (1950); M. D. Liston and J. U. White, *ibid.*, 36 (1950).
7. A. Savitzky and R. S. Halford, *Rev. Sci. Inst.*, **21**, 203 (1950).
8. A. Walsh, *J. Opt. Soc. Am.*, **42**, 94 (1952).
9. R. R. Brattain, R. S. Rasmussen, and A. M. Cravath, *J. Appl. Phys.*, **14**, 418 (1943).
10. J. U. White, S. Weiner, N. L. Alpert, and W. M. Ward, *Anal. Chem.*, **30**, 1694 (1958).
11. R. B. Barnes, R. S. MacDonald, V. Z. Williams, and R. F. Kinnaird, *J. Appl. Phys.*, **16**, 77 (1945).
12. C. S. C. Tarpet and E. F. Daly, *J. Opt. Soc. Am.*, **49**, 603 (1959).
13. S. A. Dolin, H. A. Kruegle, and G. J. Penzias, *Appl. Opt.*, **6**, 267 (1967).
14. L. C. Block and A. S. Zachor, *Appl. Opt.* **3**, 209 (1964).

CHAPTER 3

Elements of Geometric Optics

I. Introduction

The natural laws that govern optical phenomena are grouped into three broad categories. Within each category are associated mathematical expressions that provide remarkably accurate descriptions of observable optical effects. Phenomena having to do with emission and absorption of light, from the far ultraviolet to the far infrared, are described by quantum optics which treats a light beam as a stream of particle-like photons. Spectroscopists are of course vitally concerned with this area, but it is outside the province of this book and will not be discussed in detail. Phenomena associated with the propagation of a light beam in a medium, including vacuum, are described by the laws of physical optics which consider light to be an electromagnetic field. Interference, diffraction, and polarization are treated in this area. These subjects and others are deferred to the next chapter. Finally, optical phenomena that are concerned only with the direction of propagation of reflected and refracted light beams are treated by geometric optics.

These three categories of optical phenomena are not segregated by nature, but are a fabrication created by man to conceal his inability to describe all optics within one unified theory. As more is learned of nature, the boundaries of such categories begin to crumble and it is found that some experiments can be described equally well by sets of equations from different areas, while other experiments remain unexplained by any known theories, a circumstance that we all feel is temporary. The boundary between geometric and physical optics disintegrated long ago, but it is nevertheless convenient to retain the terms and to file various optical effects under these labels.

In this chapter we discuss those optical phenomena that are describable by geometric optics. We develop equations that enable the user of a spectrophotometer to understand the optical design of his apparatus and to extend its usefulness by devising auxiliary optical systems for his experimental requirements.

A good introductory text on geometric optics is the classic of Hardy and Perrin (*1*). There are many other good sources. More advanced treatments may be found in the works of Born and Wolf (*2*), Luneburg (*3*), Herzberger (*4*), and Conrady (*5*), to cite a few examples.

II. THE LAW OF REFLECTION

Nature has not provided spectroscopists with materials for making good lenses for use in the entire infrared region. Consequently, infrared spectrometers are usually constructed with reflecting optics. For this reason, and others, we start our discussion of geometric optics with mirrors.

The basic principle of geometric optics is Fermat's principle. Indeed, in a very general form, this principel is basic to a great deal of physics. Fermat's principle can be shown to be a consequence of the theory of physical optics. Thus, the laws of geometric optics, which are derived from Fermat's principle, are an extension of the laws of physical optics, and the principle serves as a means of unifying the two. Let us start with the law of reflection.

Fermat's principle states that the length of the optical path taken by a light ray has a stationary value, that is, either a maximum or a minimum, but normally the latter, with respect to neighboring paths. It is not necessarily an absolute maximum or minimum. The term "stationary" implies only that the first derivative of the path length vanishes. Let us apply this to a plane mirror with a point source of light at P in front of it and consider possible paths from P to an observer at Q (Fig. 3.1). If we accept from geometry that a straight line is the shortest distance between two points, then the direct path PQ is clearly a minimum and a legal, if trivial, path for a light ray to take. A discussion

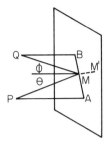

Fig. 3.1 Reflection from a surface: Fermat's principle.

of the path PMQ requires a little mathematics, although geometry again assures us that PM and MQ are straight line segments. For the moment, we insist that PM, MQ, and the normal to the surface, MN, are coplanar. We can write

$$PQ = PM + MQ = p/\cos\theta + q/\cos\theta \qquad (3.1)$$

We need another relation between θ and ϕ, such as

$$AB = AM + MB = p\tan\theta + q\tan\phi$$

Because AB is fixed by the placement of P and Q with respect to the mirror, we can write the derivative

$$\frac{d(AB)}{d\phi} = \frac{p}{\cos^2\theta}\frac{d\theta}{d\phi} + \frac{q}{\cos^2\phi} = 0$$

or $\qquad\qquad\qquad\qquad\qquad\qquad\qquad\qquad\qquad\qquad (3.2)$

$$\frac{d\theta}{d\phi} = -\frac{q\cos^2\theta}{p\cos^2\phi}$$

Now, if we differentiate Eq. (3.1) and introduce Eq. (3.2) we find

$$\frac{d(PQ)}{d\phi} = p\frac{\sin\theta}{\cos^2\theta}\frac{d\theta}{d\phi} + q\frac{\sin\phi}{\cos^2\phi} = \frac{-q}{\cos^2\phi}(\sin\theta - \sin\phi) \qquad (3.3)$$

If PQ is an extremal, its derivative is zero. Hence

$$\theta = \phi \qquad (3.4)$$

The second derivative, $d^2(PQ)/d\phi^2$, is easily shown to be positive, so $\theta = \phi$ defines a relative *minimum*. Now let us relax the temporary restriction to coplanarity of PM, MQ, and MN and move the point M in a direction perpendicular to AB to a new position M'. It is easily seen that PM' is the hypotenuse of a triangle and so $PM' > PM$. Similarly $QM' > QM$, so Eq. (3.4) is a general requirement. Equation

(3.4) is equally valid if the mirror is not a plane surface but a curved one. We need only construct a plane surface tangent to the mirror at *M*.

We have demonstrated a law of reflection that is a consequence of Fermat's principle, namely, that *the angle of reflection equals the angle of incidence and the incident and reflected rays are coplanar with the normal to the surface*. The incident ray and the normal to the surface lie in and define the *plane of incidence*.

Depending on our point of view, the law of reflection can be taken as an axiom, as a statement of an experimentally observed phenomenon, or as a consequence of the theory of physical optics. From it can be derived the equations of the optics of mirrors.

III. ELEMENTARY OPTICS OF MIRRORS

Now we are equipped with a law of reflection. Let us use it to derive equations describing the action of mirrors in altering the paths of light rays.

A. The Plane Mirror

First consider a family of rays emitted by an object at point *P* a distance *p* before a plane mirror, as in Fig. 3.2. Ray 1 is incident on the mirror at an angle θ_1, measured from the normal M_1N_1. It is reflected at the same angle on the opposite side of the normal. By extending the reflected ray backward to the right of the mirror, we construct a triangle OQM_1, similar to triangle OPM_1. The distance $OQ = q$ must equal the distance $OP = p$. Similarly, we see that all of the reflected rays when extended behind the mirror intersect at the point *Q*. Hence, an observer standing to the left of point *P* and viewing the reflected rays will see them as though they originated at *Q*. He sees the image of *P* at *Q* the same distance behind the mirror as *P* is in front of it. Certainly the rays

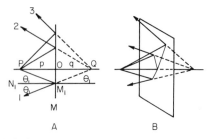

Fig. 3.2   Point image formation by reflection from a plane mirror.

need not be confined to a single plane, as in Fig. 3.2A, but can lie in any plane containing a normal to the mirror, as in Fig. 3.2B.

If the object is not a single point, but an extended object such as the arrow PP' of Fig. 3.3, the same argument can be used. In this case the observer sees an image of the extended object at QQ'. The image is once more at a distance behind the mirror equal to the distance of the object in front of the mirror. The image is the same size as the object and it has the same direction, that is, it is not inverted; it is imaged with unity magnification. What's more, the arrow PP'' is imaged at QQ'', also with unity magnification. The plane mirror is a rare device in optics; it forms perfect images.

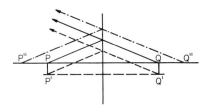

Fig. 3.3 Extended image formation by a plane mirror.

Images of the type discussed so far are called *virtual images*. They are formed by the intersection of constructed rays rather than real rays. Images formed by the intersection of real rays are called *real images*. A real image can be observed by placing a card at the place at which it is formed; a virtual image cannot be observed.

B. The Spherical Mirror

Now we shall consider a mirror surface which is not planar, but which has a concave spherical surface as shown in Fig. 3.4. The center of curvature is at C and the radius of curvature is R. The optic axis is the line CO through the center of curvature. It is normal to the surface at 0. In this section we shall consider only the simplified case where (1) the incident rays are *paraxial* (i.e., they are nearly parallel to the

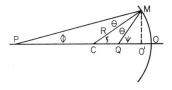

Fig. 3.4 Imaging by a concave mirror.

optic axis) and (2) the object is on the axis. Rays that intersect the optic axis, or would intersect it if extended, are *meridional* rays. Rays not intersecting the optic axis are *skew* rays. These simplifications permit us to make certain trigonometric approximations valid for small angles. They also permit us to make the assumption that $O'O$ in Fig. 3.4 is very small relative to PO' and QO'. The resulting equations constitute *Gaussian optics*.

We wish to find a relationship between the distance of the object from the mirror ($PO = p$), the distance of the image from the mirror ($QO = q$), and some property of the mirror such as its radius of curvature. For the moment we consider OP to be greater than OC. Our problem is one of trigonometry. The points P, M, O', Q, and C define six triangles, all having the constructed side MO' in common. By cleverly choosing the best triangles, we can shorten the solution of the problem. We must also confess that it is a considerable help to know the required relationship in advance. The angle ξ can be written, with a little trigonometry, as

$$\xi = \tfrac{1}{2}(\psi + \phi) \tag{3.5}$$

and we need deal only with the angles ψ and ϕ.

The distance $O'M$ can be written

$$O'M = q \tan \psi \tag{3.6}$$

or alternatively

$$O'M = p \tan \phi \tag{3.7}$$

where we have made use of the approximations $p = PO \approx PO'$ and $q = QO \approx QO'$. Equations (3.6) and (3.7) can be combined to form

$$\frac{1}{q} + \frac{1}{p} = \frac{\tan \phi}{O'M} + \frac{\tan \psi}{O'M} \tag{3.8}$$

But $O'M$ can be written using (3.5)

$$O'M = R \sin \xi = R \sin \tfrac{1}{2}(\psi + \phi)$$

This is introduced in (3.8) to give

$$\frac{1}{p} + \frac{1}{q} = \frac{1}{R \sin [(\psi + \phi)/2]} (\tan \phi + \tan \psi) \tag{3.9}$$

Now we make use of the small angle approximations $\tan \phi \approx \phi$, $\tan \psi \approx \psi$, and $\sin \tfrac{1}{2}(\psi + \phi) \approx \tfrac{1}{2}(\psi + \phi)$, which reduce (3.9) to

$$\frac{1}{p} + \frac{1}{q} = \frac{2}{R} \tag{3.10}$$

If the object is at a very large distance from the mirror, such that $1/p$ approaches zero, the image is formed at a distance $R/2$ from the mirror, or half the distance from the mirror to the center of curvature. For this reason the quantity $R/2$ is given the name *focal length*, for which we use the symbol f, and the point halfway from the mirror to the center of curvature is called the *focal point*. With this definition for f, Eq. (3.10) can be written in the equivalent forms

$$\frac{1}{p}+\frac{1}{q}=\frac{1}{f} \quad f=\frac{pq}{p+q} \quad p=\frac{fq}{q-f} \quad \text{and} \quad q=\frac{fp}{p-f} \qquad (3.11)$$

These equations provide a one-to-one relationship between a point of the object and a point of the image. Corresponding object and image points are called *conjugate points*. The manifold of points which might form an object defines the *object space*, and the points conjugate to the points in object space define the *image space*.

Let us return to Eq. (3.10) and Fig. 3.4 and consider some additional special cases. As the object is moved toward the mirror from infinity, θ decreases and the image moves away from the focal point toward the center of curvature. The object and the image reach the center of curvature simultaneously when $\theta = 0$. As the object approaches the focal point the image approaches infinity. The parallel bundle of rays headed for infinity forms a *collimated beam*, and a mirror with a light source at its focal point is acting as a *collimator*. If now we increase θ further by moving the object at the focal point slightly closer to the mirror, the reflected rays do not converge at all, or even form a collimated beam, but they diverge. In this case q in Eq. (3.11) is a large negative number. As the object is moved still closer to the mirror, the reflected rays diverge even more sharply and q becomes a smaller negative number, ultimately approaching zero when the object contacts the mirror. Just as in the case of the plane mirror, if the diverging reflected rays are extended behind the mirror, they will appear to an observer to originate at a common point behind the mirror. Thus, when the object is between the mirror and the focal point, a virtual image is formed. We see that a negative sign for q is naturally associated with a virtual image, while a positive sign implies a real image.

Equations (3.11) can be cast into another form which is often convenient. If we start with $q = fp/(p-f)$, for example, subtract f from both sides of the equation, and simplify, we obtain

$$q-f = f^2/(p-f) \qquad (3.12)$$

The quantities $q-f$ and $p-f$ are seen to be the object and image distances, respectively, measured from the focal point. If we make the definitions

$$x' = q - f \quad \text{and} \quad x = p - f$$

Eq. (3.12) takes the form

$$xx' = f^2 \tag{3.13}$$

This equation of mirror optics is said to be in the *Newtonian* form.

Now let us consider an object of finite but very small size, such as the arrow in Fig. 3.5, and determine the size of the image. Out of the multitude of rays which emanate from the arrowhead and converge to a point in space to form its image, we need to consider only three. Ray 1 is parallel to the optic axis and behaves as though it originated in a very distant object. Thus it passes through the focal point. Ray 2 passes through the center of curvature and impinges on the mirror normal to the surface. Thus ray 2 returns on itself, and passes once more through the center of curvature. Ray 3 intersects the mirror on the optic axis and is reflected symmetrically at the same angle. These three rays, and all others, intersect at the image. These rays are particularly useful in tracing imagery through a complicated optical system. For our present purpose, the magnification can be determined by considering only ray 3. Triangles $PP'O$ and $QQ'O$ are similar triangles. Therefore

$$m = \frac{QQ'}{PP'} = -\frac{q}{p} \tag{3.14}$$

is the equation for *lateral magnification* of the mirror. The minus sign implies that the image is inverted. When the image is a virtual one, q is negative and the magnification is a positive number, thus signifying that the image is erect. This can be verified by means of a ray sketch, as in Fig. 3.5B.

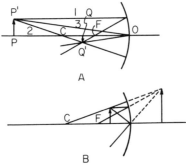

Fig. 3.5 Imaging of a finite object by a concave mirror.

The angular magnification of a mirror is defined as the ratio of the angles a ray makes with the axis as it arrives at the image and as it leaves the object. Referring to Fig. 3.4, the angular magnification μ is defined

$$\mu = \psi/\phi$$

and from Eqs. (3.6)–(3.8) and (3.14), we see that

$$\mu m = -1 \qquad (3.15)$$

This relationship is *Lagrange's law* for a reflecting surface.

A convex mirror can be treated in just the same way as a concave mirror. With reference to Fig. 3.6, we see that in place of Eq. (3.5) we have

$$\xi = \tfrac{1}{2}(\psi - \phi)$$

Equation (3.7) does not change, but because we have agreed to associate a negative q with a virtual image, Eq. (3.6) becomes

$$O'M = -q \tan \psi$$

With these changes we obtain

$$\frac{1}{p} + \frac{1}{q} = -\frac{2}{R} \qquad (3.16)$$

Equation (3.16) can be taken as a general equation for mirrors provided we adopt the convention that concave mirrors have a negative radius of curvature and convex mirrors have a positive radius of curvature. The Newtonian form of the mirror equation also holds for convex mirrors. An object at infinity has a virtual image at the focal point of the convex mirror, according to Eqs. (3-11). As the object is moved closer to the mirror, the virtual image approaches the mirror surface and reaches it as the object contacts the mirror.

The equation for lateral magnification of a convex mirror is derived in just the same way as for a concave mirror, referring to Fig. 3.7. The

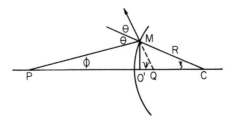

Fig. 3.6   The convex mirror.

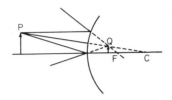

Fig. 3.7   Magnification by a convex mirror.

same equation (3.14) is found to hold for the convex mirror. As q is always negative, for an object to the left of the mirror surface the magnification is always positive, signifying an erect image.

C. Systems of Mirrors

We have distinguished real and virtual images according to whether rays actually or apparently convert to an image point. The same distinction is valid for objects. If the rays actually diverge from an object point, the object is virtual. Virtual objects can be included in the equations of the previous section by extending the sign convention already adopted for virtual images. A real object lies to the left of the mirror surface and its distance is given a positive sign. A virtual object lies to the right, and its distance is negative.

Virtual objects are important when we are dealing with an optical system having two or more elements. The image produced by the first element serves as the object for the second, and so on. It frequently happens that some of the intermediate objects generated in this fashion are virtual objects.

Let us illustrate this procedure with an example. The system of Fig. 3.8 is a Cassegrain condensing system. It might be used to image the exit slit of a spectrophotometer onto a detector. Light from object P_1 passes through a hole in mirror M_2 and encounters the convex surface of mirror M_1; M_1 produces a virtual image of P_1 at Q_1. The rays

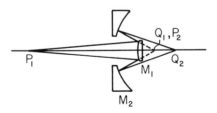

Fig. 3.8   Cassegrain mirror system.

diverging from Q_1 are refocused by mirror M_1, forming a real image at Q_2. Assume focal lengths of -1.05 and $+2.0$ cm for M_1 and M_2 respectively, place P_1 20 cm in front of the surface of M_1, and space M_2 and M_1 2 cm apart. Now the eqs (3.11) tell us that the image Q_1 occurs at $q_1 = -1$ cm, or 1 cm behind M_1. Its size is $-(-1)/20 = 0.05$ times smaller than the size of the object, from Eq. (3.14), and it is erect; Q_1 serves as a real object, P_2, for mirror M_2, and it is $p_2 = 3$ cm away. The image of P_2 is formed by M_2 at Q_2, a distance $q_2 = 6$ cm from M_2; Q_2 is $-6/3 = -2$ times bigger than P_2 and it is inverted. The overall magnification is $0.05 \times -2 = -0.1$.

Let us extend this illustration by supposing that the condenser of Fig. 3.8 is used to produce a reduced image of a spectrometer source, P_1, on a small sample placed at Q_2. A second identical Cassegrain system is added, as in Fig. 3.9, to reimage the sample on the entrance slit of the spectrometer at Q_4. Mirror M_3 now produces a real image of P_3 at Q_3, or at least it would if mirror M_4 were not in the way. However, Q_3 is 1.05 cm behind M_4, thus it is a virtual object with $p = -1.05$ cm.

A simple calculation procedure of this sort is often all that is required for the design of an experimental optical system. We have not yet discussed the question of apertures or how big the mirrors must be; nor have we discussed the nature of the defects in images which occur if we include rays that are not paraxial. These points are taken up in a later section, but first let us look into the equations for lenses.

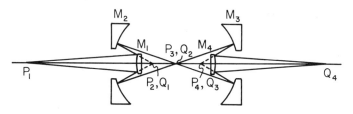

Fig. 3.9   Double Cassegrain system.

IV. THE LAW OF REFRACTION — SNELL'S LAW

In Section II, it was demonstrated that the law of reflection is a consequence of Fermat's principle. Now let us show that the law of refraction also follows from this basic principle. Recall that Fermat's principle states that a light ray selects that path between two points which has a stationary *optical* path length. We assumed, implicitly, that the optical path length was just the geometric path length; but this

assumption is correct only for a vacuum medium. Fermat's principle
can be stated alternatively: *the time taken by a light ray to pass
between two points has a stationary value.* Now the time required for
a signal to be propagated from an initial point in medium 1, through
several intermediate media, to a final point in medium N is

$$t = \sum_{k=1}^{N} \frac{x_k}{v_k} \qquad (3.17)$$

where x_k is the distance traveled across medium k, and v_k is the velocity
of light in medium k. We shall assume that each medium is homogeneous, so the velocity of propagation is constant throughout each
medium. The velocity can be written as a fraction of the velocity of
light in a vacuum, c,

$$v_k = \frac{c}{n_k} \qquad (3.18)$$

The parameter n_k is by definition the *refractive index* of the medium.
Equation (3.17) can be written

$$t = \frac{1}{c} \sum_{k=1}^{N} n_k x_k \qquad (3.19)$$

The condition that the time of propagation has a stationary value is,
according to Eq. (3.19), equivalent to the condition that the quantity
$\sum_{k=1}^{N} n_k x_k$ has a stationary value. The quantity $n_k x_k$ is the *optical path
length*. The refractive index of a vacuum is unity. Hence, in a vacuum
the optical path length equals the geometric path length. If the media are
not homogeneous and the refractive index varies from place to place,
the quantity $\int_A^B n(x) \, dx$ has a stationary value for the correct path from
A to B.

The derivation of the law of refraction from Fermat's principle can
proceed in a manner very much like that used in Section II, but for
variety let us use a somewhat different and slightly more sophisticated
approach.

Refer to Fig. 3.10. We are concerned with the ray PMQ which is
incident on a surface at M and makes an angle θ with the normal to the
surface. The ray penetrates the surface but is refracted so that it makes
an angle ϕ with the normal and proceeds to Q. We shall make several
assumptions to simplify the discussion, but these are not essential to a
more general treatment. Specifically, we assume: (1) the incident ray,
the refracted ray, and the normal to the surface are coplanar; (2) the
medium containing the incident ray and the medium containing the

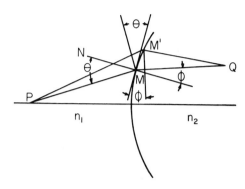

Fig. 3.10   Refraction at a curved surface.

refracted ray are each homogeneous, with refractive indices n_1 and n_2, respectively; (3) the surface has no discontinuities or steps.

Ray PMQ is the actual path taken by the light signal. Construct another path, $PM'Q$, which departs somewhat from PMQ. The optical path length of PMQ is

$$L = n_1(PM) + n_2(MQ)$$

The optical path length of $PM'Q$ differs by a small amount, and is

$$L + \delta L = n_1(PM) + n_1(MM')\sin\theta + n_2(MQ) - n_2(MM')\sin\phi \quad (3.20)$$

We have used the trigonometry of Fig. 3.10 and invoked the continuity of the surface in writing Eq. (3.20).

Fermat's principle tells us that in order for L to have a stationary path length, δL must vanish. Hence,

$$n_1(MM')\sin\theta - n_2(MM')\sin\phi = 0$$

or

$$n_1\sin\theta = n_2\sin\theta \quad (3.21)$$

This is the law of refraction, or *Snell's law*. A ray traversing the boundary between two optical media, from the less optically dense to the more optically dense ($n_2 > n_1$), is refracted closer to the normal. On the other hand, a ray traveling from a denser to a less dense medium is refracted further from the normal, and in fact when

$$\sin\theta = n_2/n_1 \quad (3.22)$$

we must have $\sin\theta = 1$ and $\theta = \tfrac{1}{2}\pi$. Equation (3.22) defines the *critical angle* of incidence for which the refracted radiation cannot escape from the initial medium but can only graze the surface separating the media.

When the angle of incidence exceeds the critical angle, all of the incident radiation is reflected back into the initial medium; it is *totally internally* reflected.

V. ELEMENTARY OPTICS OF LENSES

The law of refraction can now be used to derive equations that govern the optical behavior of lenses, analogous to the laws of mirrors.

A. The Plane-Parallel Window

Let us start with a window with planar parallel faces, as the refractive analog of the plane mirror. The window is not usually considered to have focusing properties. However, as shown in Fig. 3.11, the rays diverging from an object at P produce a virtual image at Q. The refractive index of the window is n and its thickness is t. Assume that it is

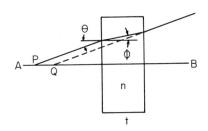

Fig. 3.11   Optics of the plane-parallel window.

suspended in a vacuum. The object is placed a distance p from the first surface, so a ray incident on the surface at an angle θ encounters the first surface a distance $p \tan \theta$ above the axis AB. This ray would leave the second surface a distance $(p+t) \tan \theta$ above the axis if it were not refracted. However, it actually leaves the second surface $p \tan \theta + t \tan \phi$ above the axis. This ray, when extrapolated back to the axis, crosses it at a distance $(p \tan \theta + t \tan \phi)/\tan \theta$ from the second surface or

$$q = \frac{p \tan \theta + t \tan \phi}{\tan \theta} - t = (p-t) + t \frac{\tan \phi}{\tan \theta} \qquad (3.23)$$

from the first surface.

For θ and ϕ very small, the tangents in Eq. (3.23) can be replaced by sines, and the ratio of the sines can be replaced by $1/n$ according to

Snell's law. The object P thus appears to shift by an amount

$$p - q = t\left(1 - \frac{1}{n}\right) \tag{3.24}$$

If a sample cell with a net path of 1 cm of salt having refractive index $n = 1\cdot5$ is placed in the beam of a spectrophotometer, the source appears to move 3·33 mm closer.

Moreover, if θ is not a small angle, so that the general equation (3.23) rather than the paraxial approximation equation (3.24) is valid, the intersection of a ray with the axis depends on the incident angle θ. That is, the rays do not all seem to originate in a common image point Q, and the image is blurred. The plane-parallel window, unlike the plane mirror, is not a perfect optical element.

B. Spherical Surfaces and the Thin Lens

Now let us derive an equation for the position of the image produced by a thin lens, as shown in Fig. 3.12. The lens is made of a material with refractive index n_2. The index is n_1 in the medium to the left of the lens and n_3 in the medium to the right. Just as in the case of the spherical mirror, the treatment is restricted to paraxial rays, and in addition we shall consider only lenses which are sufficiently thin so that the rays enter and leave them at the same height, h, above the axis. The angles α, β, γ, and ϵ can be written in terms of the distance h, the object and image distances, p and q, respectively, and the radii of curvature of the two spherical surfaces, R_1 and R_2

$$h = p \tan \alpha = q \tan \beta = R_1 \sin \gamma = -R_2 \sin \epsilon \tag{3.25}$$

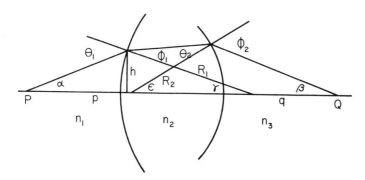

Fig. 3.12   Refracted rays in a lens.

In the paraxial approximation, Eq. (3.25) becomes

$$h = p\alpha = q\beta = R_1\gamma = -R_2\epsilon \tag{3.26}$$

R_2 is given a negative sign because we shall adopt the convention that the radius of curvature of a surface, whose center of curvature is to its left, is negative. This is consistent with the convention previously adopted for mirrors. ϕ_1 and θ_2 can be written in terms of θ_1 and ϕ_2 by Snell's law. Hence, for small angles,

$$\phi_1 + \theta_2 = \frac{n_1}{n_2}\theta_1 + \frac{n_3}{n_2}\phi_2 \tag{3.27}$$

θ_1 and ϕ_2 can be written from trigonometry as

$$\theta_1 = \alpha + \gamma \quad \text{and} \quad \phi_2 = \beta + \epsilon \tag{3.28}$$

Also $\phi_1 + \theta_2$ can be written

$$\phi_1 + \theta_2 = \epsilon + \gamma \tag{3.29}$$

If Eq. (3.29) and (3.28) are introduced into (3.27), we get

$$\frac{n_1}{n_2}\alpha + \frac{n_3}{n_2}\beta + \left(\frac{n_1}{n_2} - 1\right)\gamma + \left(\frac{n_3}{n_2} - 1\right)\epsilon = 0 \tag{3.30}$$

If we now introduce the relationships (3.26) into (3.30), we arrive finally at an equation from which h has been eliminated

$$\frac{n_1}{p} + \frac{n_3}{q} = \frac{n_2 - n_1}{R_1} - \frac{n_2 - n_3}{R_2} \tag{3.31}$$

In case the lens is mounted in a vacuum, or, to the degree of approximation we are considering, if it is mounted in air, we have $n_1 = n_3 = 1$, and Eq. (3.31) reduces to

$$\frac{1}{p} + \frac{1}{q} = \left(n_2 - 1\right)\left(\frac{1}{R_1} - \frac{1}{R_2}\right) = \frac{1}{f} \tag{3.32}$$

Equation (3.32) is of the same form as Eq. (3.11) for a spherical mirror. The focal length of a thin lens is given by

$$f = \frac{1}{n_2 - 1}\left(\frac{R_1 R_2}{R_2 - R_1}\right) \tag{3.33}$$

If the lens is convex-convex, $R_1 > 0$, $R_2 < 0$, and $f > 0$. This lens is said to be positive. If the lens is concave-concave, $R_1 < 0$, $R_2 > 0$, and $f < 0$. This lens is negative. If the first surface is convex and the second concave, the lens is positive for $|R_1| < |R_2|$ and negative for

$|R_1| > |R_2|$. If the first surface is concave and the second convex, the lens is positive for $|R_1| > |R_2|$ and negative for $|R_1| < |R_2|$.

If one of the surfaces is planar, its radius of curvature is infinite and it makes no contribution to these equations. It is easy to see that if we drop the second surface and take $n_2 = -1$, Eq. (3.32) is identical to the mirror equation (3.16) with the same sign convention for the radius of curvature.

The occurrence and properties of real and virtual images are just as in the mirror case. A positive image distance q is associated with a real image and a negative image distance q with a virtual image. A similar sign convention holds for object distances. The Newtonian form of the optical equation [Eq. (3.13)] is equally valid for lenses.

The lateral magnification of a thin lens can be calculated from Eq. (3.14) just as in the case of the mirror, as is readily verified by referring to Fig. 3.13. The ray passing through the center of the lens is undeviated, while a ray leaving the object in a direction parallel to the axis is refracted through the focal point. The angular magnification of a thin lens also is expressed by the same equation as for a mirror. This is readily shown by making use of Eq. (3.25) to calculate $\mu = \beta/\alpha$.

We derived earlier a form of Langrange's law for reflecting surfaces [Eq. (3.15)]. Now let us derive a similar equation for a refracting surface. If we consider a single spherical refracting surface, as in Fig. 3.13A, Eq. (3.31) reduces to

$$\frac{n_1}{p} + \frac{n_2}{q} = \frac{n_2 - n_1}{R}$$

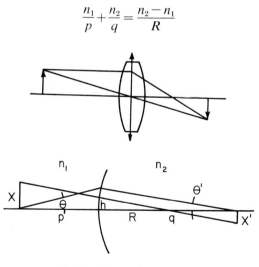

Fig. 3.13   Image formation by a lens.

which can be written

$$\frac{R-q}{R+p} = -\frac{n_1\,q}{n_2\,p} \qquad (3.34)$$

From the geometry of Fig. 3.13A, it is seen that

$$m = \frac{x'}{x} = \frac{R-q}{R+p} \qquad (3.35)$$

It follows from Eqs. (3.34) and (3.35) that the lateral magnification from refraction at a single surface is

$$m = -\frac{n_1\,q}{n_2\,p} \qquad (3.36)$$

We also have from Fig. 3.13A that

$$\theta = \frac{h}{p} \quad \text{and} \quad \theta' = \frac{h}{q}$$

or

$$\frac{\theta}{\theta'} = -\frac{q}{p} \qquad (3.37)$$

This equation may be combined with Eq. (3.36) to obtain *Lagrange's law* for lateral and angular magnification at a refracting surface

$$m = \frac{n_1}{n_2}\frac{\theta}{\theta'} \qquad (3.38)$$

In an optical system with several refracting (or reflecting) surfaces, Lagrange's law can be written successively for each surface. It will be seen that intermediate image sizes, angles, and refractive indices cancel, leaving an equation of the form of (3.38) for the initial object and the final image.

Consider the example illustrated in Fig. 3.14. This optical system is used to produce an image of the exit slit of a monochromator on the receiver of a thermocouple radiation detector. The slit has a height of 25 mm and a maximum width of 6 mm. The thermocouple has a target

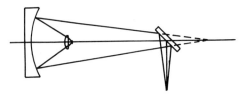

Fig. 3.14   Optical system for imaging on a detector.

which is only 3×0.6 mm, so we require a tenfold reduction in the size of the slit image. Let us say that we have elected to obtain a fivefold image reduction with a mirror and a further twofold reduction with a lens. This is a compromise. If a lens were to be used alone, image quality would suffer unless an expensive complex design were chosen. On the other hand, the use of a mirror alone would require either a greater object distance and consequently a larger mirror and a less compact design, or a shorter focal length mirror accompanied by more pronounced spherical aberration (to be discussed later). Besides, a window must be used to close the vacuum space in which the detector element is suspended, and it might as well serve a useful optical purpose by being shaped as a lens.

The planar diagonal mirror forms a virtual image of the slit $14.0 + 5.5 = 19.5$ cm in front of the concave mirror. This image is the same size as the slit, and it serves as a real object for the concave mirror. In order to be one-fifth the size of the object, the image formed by the concave mirror must be $\frac{1}{5} \times 19.5 = 3.9$ cm in front of the mirror. The focal length of the mirror must therefore be $19.5 \times 3.9/(19.5 + 3.9) = 3.25$ cm.

The lens can be used in two possible ways. The image produced by the mirror can serve as a real object for the lens, as in Fig. 3.15A. In order to obtain a twofold image reduction, we require $p = 2q$ and therefore $f = \frac{2}{3}q$. If the target is placed 3 mm behind the lens, this configuration requires a focal length $f = \frac{2}{3} \times 3 = 2$ mm. Alternatively, the mirror's image can act as a virtual object for the lens, as in Fig. 3.15B. In this case $p = -q$ and $f = 2q = 6$ mm. We choose the 6 mm lens because it is easier to produce. If the material from which the lens is to be made has a refractive index of 1.5 and the lens is to have a plano-convex shape, the radius of curvature of the convex surface, computed from Eq. (3.33) for a thin lens, is 3 mm. This lens is not thin at all; indeed, it is quite fat, and Eq. (3.33) is not strictly valid. We shall see later how to design a thick lens.

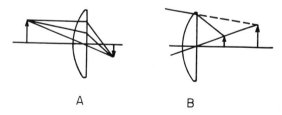

A   B

Fig. 3.15 Alternate ways of forming an image on a detector.

VI. Aperture Stops and Pupils

Now we are able to calculate the position and size of an image formed by an optical system, but we have not yet considered its irradiance. This is very important in infrared spectroscopy where instruments are energy limited and it is desired to put as much radiant power as possible into the spectrum. The irradiance of an image is determined by the amount of light from the object which is intercepted by the optical system and also by the efficiency of transmission of the optical system. Discussion of efficiency is deferred to a later chapter. The question of interception of light rays leads us into the topic of apertures.

Consider the simple system shown in Fig. 3.16. A self-luminous source is placed before an optical element (a mirror or lens). The source has a radiance N; that is, it radiates N watts of radiant power from each unit area of its surface into each unit steradian of solid angle surrounding it. The amount of power intercepted by the optical element is proportional to the solid angle subtended by the element as seen from the source. Hence

$$P = \frac{NA}{p^2} \tag{3.39}$$

where A is the area of the element and p is its distance from the source. Now the average irradiance of the image is simply the total power in the image divided by the area of the image. If a source of unit area is under study, then the irradiance of its image is

$$H = \frac{P}{m^2} \tag{3.40}$$

where m is the magnification. From Eq. (3.39) and the relationship for magnification, Eq. (3.40) can be written

$$H = \frac{NA}{q^2} \tag{3.41}$$

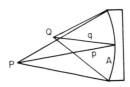

Fig. 3.16   Radiant power accepted by an optical element.

If the source is a great distance from the optical element, q can be replaced by the focal length f, and the irradiance is written

$$H = \frac{NA}{f^2} = NF^2 \qquad (3.42)$$

where F is the f/number of the optical system and is defined by

$$F = \sqrt{A}/f$$

If the optical element has a square shape of width d, the f/number is

$$F = \frac{d}{f}$$

This equation is also frequently used to define the f/number of an element with a round shape, where d is its diameter.

Equation (3.42) is well known to photographers who double the f/number of a camera in order to decrease the irradiance of the film by a factor of 4. It is essential to recognize that Eq. (3.42) is valid only for large object distances. When photographing close objects, corrections must be made to the "f/number" value shown on the lens of the camera. Spectroscopists should be particularly wary of the f/number as an indicator of the "speed" of a monochromator. We shall see in the next chapter why this is so.

In the simple single-element system just considered, the size of the element itself limits the amount of radiant energy accepted. We say that the element is the *aperture stop* of the system. But what happens if the optical system is more complex, as in Fig. 3.17A? With the object placed on the optic axis at P_1, a distance p_1 (or farther) from the first lens, all rays which are passed by the first lens are also passed by the second lens to form an image at Q_1. The first lens is the aperture stop of the system. If the object is moved closer to P_2 (Fig. 3.17B), the rays passed by the outermost zone of the first lens are just passed by the second lens. But for shorter object distances with the object at P_3 (Fig. 3.17C), the outermost rays from the first lens are not passed by the second lens and the second lens has become the aperture stop. Now the second lens is not seen directly by an observer at P_3, but rather its image is seen as formed by the first lens. The image of the aperture stop, formed by optical elements between it and the object, is the *entrance pupil.* Its size and position with respect to the object determines the limiting size of the cone of rays transmitted by the system. The smallest opening in the optical system which can be seen by an observer at the object point is the entrance pupil. Similarly, the *exit*

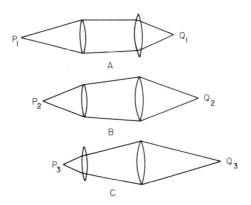

Fig. 3.17   Aperture stop for several object positions.

pupil is the image of the aperture stop formed by elements following the aperture stop. It is the smallest opening observable at the image point. Clearly, the exit pupil is the image of the entrance pupil, and vice versa. The pupils are illustrated in Fig. 3.18, where the center lens (*II*) is the aperture stop, its image produced by lens *I* at *II'* is the entrance pupil and its image formed by lens *III* at *II''* is the exit pupil. A ray that passes through the center of the aperture stop and, therefore, through the center of the entrance and exit pupils as well is a *chief ray* or *principal ray*.

We have shown that the aperture stop of a system may be different for different positions of the object on the optic axis. Similarly, if the object is placed at a point off the optic axis, the identity of the aperture stop is affected. In Fig. 3.19, for example, some of the rays from the off-axis point *P'* are limited by the second lens, while others are limited by the first lens, whereas all of the rays from the on-axis point *P* are limited by the second lens. Hence, for an extended object in the plane containing *P* and *P'* there is no single aperture stop. This is called *vignetting* and is generally undesirable. Often a diaphragm of suitable size is placed at an appropriate position in the optical system to serve

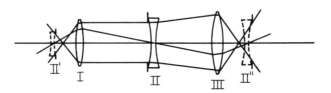

Fig. 3.18   Entrance and exit pupils.

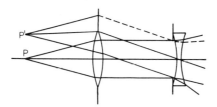

Fig. 3.19 Aperture for off-axis object points: vignetting.

as the aperture stop in order to ensure that there will be a definite stop for all off-axis points of interest and that the same stop will hold for a range of object distances.

The diaphragm aperture stop has another important function. If an optical element such as a lens serves as an aperture stop, rays which arrive at or beyond the outer limit of the lens are not always intercepted and stopped. They may be scattered in such a way so as to produce a background irradiance that degrades the contrast of the image, or in spectroscopy they may be observed as stray radiation of undesired wavelengths.

As an example, consider the elementary grating spectrometer shown schematically in Fig. 3.20A and shown in extended form with equivalent lenses in Fig. 3.20B. The source mirror has a focal length of 25 cm and it is placed two focal lengths from the source. It forms an

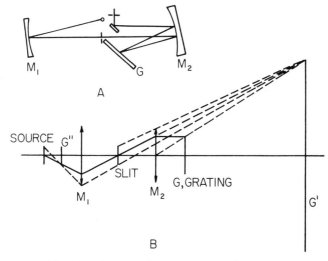

Fig. 3.20 Aperture of a grating spectrophotometer.

image of unit magnification of the source on the entrance slit. The slit is 25 mm high. The collimator has a focal length of 50 cm and is placed 50 cm from the slit in order to form a collimated beam of radiation which is incident on the grating. The grating is placed 40 cm from the collimator and it is 5 cm high. Because the grating (or prism) is likely to be the most expensive single element in the spectrometer, we generally desire it to be the element that limits the beam size so all of its surface will be used. That is, the grating should be the aperture stop of the spectrometer. Our problem is to determine how large the collimator must be in order for the grating to be the aperture stop with respect to an object at the entrance slit. The collimator forms a virtual image of the gratings at a distance $q = (50 \times 40)/(40-50) = -200$ cm, and this image is $200/40 = 5$ times larger than the grating itself. This image is to be the entrance pupil. If we were concerned with only a pinhole on the optic axis rather than a slit of finite height, the collimator would only need to be as large as the grating, as can be seen from Fig. 3.20 by noting that

$$\frac{\text{collimator height}}{\text{entrance-pupil height}} = \frac{\text{slit-to-collimator distance}}{\text{slit-to-entrance pupil distance}} = \frac{1}{5}$$

and

$$\frac{\text{entrance-pupil height}}{\text{grating height}} = \frac{5}{1}$$

However, with respect to a point at the end of the slit 12.5 mm from the optic axis, it is evident that the collimator must be larger than the grating if vignetting is to be avoided. The extra height is given by

$$\frac{\text{extra collimator height}}{\text{slit height}} = \frac{\text{collimator-to-entrance pupil distance}}{\text{slit-to-entrance pupil distance}} = \frac{200}{250}$$

Hence, the collimator must be 7 cm high in this case. Optical designers often state that the collimator must be the height of the grating plus the height of the slit. This is true if the grating is placed one focal length from the collimator because the entrance pupil will be at infinity and all rays from anywhere on the slit which arrive at the edge of the entrance pupil are parallel. In particular, the ray from the center of the slit and the one from the end of the slit are parallel and therefore must be intercepted by the collimator at points half of the slit height apart (see Fig. 3.20B). Therefore the collimator must be higher than the grating by a full slit height.

We wish to prevent any rays from striking the edge of the grating in order to minimize stray light in the monochromator. Furthermore, we prefer not to insert a diaphragm inside the monochromator because we fear that its edges might scatter. For this reason, we usually make the grating almost, but not quite, the aperture stop with respect to the source by inserting a diaphragm outside of the monochromator. This diaphragm is an aperture stop with respect to points at the spectrometer source. Refering again to Fig. 3.20B, we see that there is an image of the grating at G'' produced by the collimator M_2 and the source mirror M_1. The virtual image at G' acts as an object for M_1 at a distance of 300 cm. Mirror M_1 forms a real image G'' at a distance of 27.3 cm. The image is 2.28 cm high. If we place a diaphragm at G'' slightly smaller than this, 2.2 cm say, then this diaphragm will be the aperture stop of the spectrometer relative to points at the source, assuming mirror M_1 is sufficiently large. The entrance pupil coincides with the aperture stop. The stop is imaged on the grating and is 2.2/2.28 times as high as the grating. Thus, with respect to points on the slit, the locations of the aperture stop and entrance pupil have not changed, though they are slightly smaller.

It is very important in devising auxiliary optical apparatus for insertion in a spectrophotometer that the original pupils of the system remain the limiting apertures of the system. Otherwise, the performance of the spectrophotometer will be degraded. Let us suppose that the source in Fig. 3.20 is a special device whose emission is being studied and that the source mirror M_1 is an auxiliary condensing element mounted by the experimenter in front of the entrance slit of the monochromator. Proceeding just as we did above, it is easy to see that for a point on the source and on the optic axis

$$\frac{\text{source-mirror } (M_1) \text{ height}}{\text{diaphragm } (G'') \text{ height}} = \frac{\text{source-to-}M_1 \text{ distance}}{\text{source-to-}G'' \text{ distance}}$$

but relative to a point at the end of the source (Fig. 3.21), taking the source as the same height as the slit,

$$\frac{\text{extra } M_1 \text{ height}}{\text{source height}} = \frac{G''\text{-to-}M_1 \text{ distance}}{\text{source-to-}G'' \text{ distance}}$$

Using the distances found above, we calculate that the extra mirror height must be at least 27.3/22.7 = 1.2 times the source height, and so the total source-mirror height must be at least

$$2.2 \times \frac{50}{22.7} + 1.2 \times 2.5 = 7.85 \text{ cm}$$

A discussion parallel to this holds for the exit pupil. The experimenter

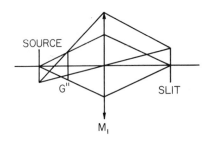

Fig. 3.21   Aperture imaged in the vicinity of the source.

must be equally cautious in devising apparatus to follow the exit slit of his monochromator. We go into this in a little greater detail in Chapter 6.

VII. FIELD STOPS AND WINDOWS

Principal rays intersect the aperture stop at the optic axis. They must also intersect the entrance and exit pupils at the optic axis because these are mutual images of the aperture stop. Certainly not all rays from any object point can reach the center of the pupils without being blocked by the edge of some optical element. Stated differently, there is a limit to the size of the cone of rays arriving at the entrance pupil with its apex at the intersection of the optic axis and the pupil. This limiting size (see Fig. 3.22) defines the *angular field of view*. The element which is responsible for determining the size of the field of view (lens *I* in the example shown) is the *field stop* and its image viewed from the entrance pupil is the *entrance window* (at *I'*). Similarly, the image of the field stop viewed from the exit pupil is the *exit window*.

Let us continue to discuss the spectrophotometer of Fig. 3.20. The angle between the optic axis and a principal ray from the edge of the

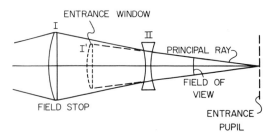

Fig. 3.22   Field stop and exit and entrance windows.

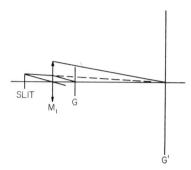

Fig. 3.23   Field stop of a grating monochromator.

collimator is easily calculated from Fig. 3.23 to be 0.0175 rad. The slit is imaged at infinity by the collimator, so all rays from the end of the slit are parallel after passing the collimator. The angle between the optic axis and a ray from the end of the slit through the center of the collimator is 0.025 rad. The ray from the end of the slit through the center of the grating makes the same angle with the optic axis. The entrance pupil is the image of the grating and is five times larger than the grating. Therefore, by using Lagrange's law [Eq. (3.15)], this ray makes an angle of $\frac{1}{5} \times 0.025 = 0.005$ rad when it intersects the entrance pupil. The source mirror is imaged 50 cm in front of the entrance pupil with a magnification of 2. Therefore, the principal ray from the edge of the source-mirror image is 0.157 rad from the optic axis. The source is imaged on the slit and is larger than the slit, so it cannot be the field stop. The grating and the aperture diaphragm are imaged on the entrance pupil, of course, so they cannot be field stops. Of the three possible elements in this example, slit, collimator, and source mirror, the slit image is found to subtend the smallest angle at the center of the entrance pupil. Thus the slit is the field stop and its image at infinity is the entrance window. Just as in the case of the entrance and exit pupils, the exit window is the image of the entrance window.

VIII. CARDINAL POINTS

The *cardinal points* of an optical system are certain pairs of points, one point of each pair associated with object space and the other with image space, from which distances can be measured advantageously. The focal points form one pair of cardinal points.

We have shown earlier how the image equations for a mirror or thin lens can be written in either Gaussian form, with distances measured from the lens or mirror surface, or in Newtonian form, with distances measured from the focal points. Let us now generalize this by relaxing the requirement that the focal lengths are measured from a single plane in the optical element; that is, we no longer will require the lens to be thin. In Fig. 3.24 are sketched the pertinent planes and points of a general optical system. There is an object point at A a distance X from the axis. Its image A' is X' from the axis. The distance from the object to the first focal point F is x and the distance from the image to

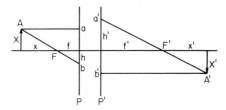

Fig. 3.24   Generalized optical system.

the second focal point F' is x'. For convenience we draw an inverted real image, but this is not a significant restriction. The planes P and P' are conjugate, that is, they are imaged on each other by the optical system, but otherwise they are chosen arbitrarily. The magnification of the image of P on P' is M. The distance from F to P is f and the distance from F' to P' is f'. Rays from A and A' that are parallel to the optic axis intersect P and P', as do rays from A and A' through the focal points F and F'. The points of intersection are a, a', b, and b' in the figure. From geometry we can write

$$\frac{x}{f} = \frac{-X}{h} \quad \text{and} \quad \frac{x'}{f'} = \frac{-X'}{h'} \tag{3.43}$$

We know that the point a on P is imaged at a' on P' because a ray from A parallel to the optic axis passes through F' and arrives at A'. Similarly, b is imaged on b'. Therefore we can write

$$h' = XM \quad \text{and} \quad h = X'/M \tag{3.44}$$

Equations (3.44) serve to simplify Eqs. (3.43) to give us

$$xx' = ff' \tag{3.45}$$

Equation (3.45) is the general Newtonian form of the paraxial optical equation.

The lateral magnification can be found by noting

$$X = \frac{-x}{f}h = -\frac{x}{f}\frac{X'}{M} \tag{3.46}$$

from which

$$m = \frac{X'}{X} = -M\frac{f}{x}$$

Alternatively

$$X' = \frac{-x'}{f'}h' = \frac{-x'}{f'}XM$$

which gives us

$$m = \frac{X'}{X} = -M\frac{x'}{f'}$$

For a thin lens or mirror, the planes P and P' coincide, $f = f'$, $M = 1$, and the equations reduce to

$$xx' = f^2 \quad \text{and} \quad m = -f/x = -x'/f'$$

Equation (3.46) can be converted to the general Gaussian form by introducing

$$s = x + f \quad \text{and} \quad s' = x' + f'$$

and simplifying to obtain the forms

$$\frac{f}{s} + \frac{f'}{s'} = 1 \qquad s = \frac{fs'}{s'-f'} \qquad s' = \frac{f's}{s-f} \tag{3.47}$$

This equation reduces to the thin-lens or mirror form when $f = f'$. The magnification is found by writing

$$m = -M\frac{f}{x} = -M\frac{f}{s-f}$$

and substituting the last form of the equations (3.47) to obtain

$$m = -M\frac{f}{f'}\cdot\frac{s'}{s}$$

This equation also reduces to the thin-lens or mirror form when $f = f'$ and $M = 1$.

The general equations are valid for any pair of conjugate reference planes, but are most useful for certain planes. The entrance and exit pupils are sometimes selected, in which case M is simply the ratio of the height of the exit pupil to the height of the entrance pupil.

A particularly useful pair of conjugate planes is the pair that is

imaged on each other with unit magnification, $M = 1$. These are the
principal planes of the system. The intersection of the optic axis with
the principal planes defines the *principal points* or *Gauss points*.

Let us find the relationship between f and f' measured from the
principal points. Figure 3.25 represents an optical system replaced by
its principal planes, for which $M = 1$. We see that the angular magnification for small angles can be found from

$$\theta = \frac{X'}{f} \quad \text{and} \quad \theta' = \frac{X}{f'}$$

or

$$\frac{\theta'}{\theta} = \frac{X}{X'} \frac{f}{f'}$$

But Lagrange's law [Eq. (3.38)] requires that $\theta'/\theta = X/X'$. Therefore,
$f = f'$ and the principal planes are the same distance from corresponding focal points. It follows that any optical system, no matter how
complicated, is described by an equation of the thin-lens or mirror
form when distances are measured from the principal planes.

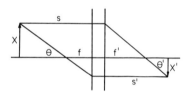

Fig. 3.25   Principal planes of an optical system.

The *nodal points* are another important pair of cardinal points.
These are conjugate points for which the angular magnification is
unity. If the initial and final media have the same refractive indices,
the nodal points coincide with the principal points. This follows from
Lagrange's law.

Now let us derive an optical relationship which is of particular
importance for monochromators. In Fig. 3.26, h is the height of the

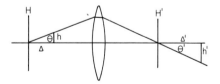

Fig. 3.26   Lagrange invariant.

image of the field stop (i.e., the slit) and H is the height of the aperture stop (i.e., the grating or prism). An optical element, such as the lens shown in the figure, produces images of these stops of height h' and H', respectively. The aperture and field stops are a distance Δ apart and their images are Δ' apart. The principal ray in object space from the end of the slit makes an angle θ with the optic axis. In image space, the angle of the principal ray is θ'. We can write

$$\theta' = h'/2\Delta' \quad \text{and} \quad \theta = h/2\Delta$$

From Lagrange's law we can also write

$$H'\theta' = H\theta$$

assuming $n = n'$. By combining these three equations in such a way as to eliminate the angles, we obtain

$$\frac{h'H'}{\Delta'} = \frac{hH}{\Delta} \tag{3.48}$$

Equation (3.48) states that the product of the image heights of the slit and the grating or prism divided by the distance between the images is invariant, regardless of the nature of the imaging element. A similar relationship holds for the widths of the slit and grating, and the relationships can be combined to give us one equation in the areas of the slit and the grating,

$$\frac{A'_s A'_G}{\Delta'^2} = \text{const.} \tag{3.49}$$

The constant in Eq. (3.49) is the *Lagrange invariant* or *Lagrange constant* of the monochromator.

IX. Ray Tracing and the Thick Lens

We have shown how optical image equations can be derived for simple systems, such as mirrors and thin lenses, to the approximation of paraxial rays. More complex systems can be studied by ray tracing. In its most straightforward form, this consists of calculating the point of intersection and direction of a ray as it progresses from surface to surface through the system. This procedure is by no means limited to paraxial rays, nor is it necessarily confined to meridional rays, although before the advent of digital computors skew rays were not often traced. We shall confine the discussion of this section to rays which are paraxial and meridional. The ray-tracing problem can be systematized by making use of matrix notation as a convenient short-hand. Consider two reference planes (Fig. 3.27) separated by a distance t. The planes are tangent to the boundaries of media of refractive index n_1, n_2, and n_3, respectively. A ray encounters the first plane a distance h from the optic axis, making an angle θ_1 with the optic axis.

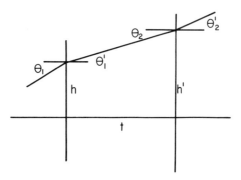

Fig. 3.27   Tracing a ray through successive media.

Its direction changes upon penetrating the plane to a new angle θ'_1 determined by the law of refraction (or reflection). The distance from the optic axis remains h after just entering the second medium. The ray traverses the second medium to the next reference plane with no further change in angle, but with a progressive change in distance from the optic axis. The ray then suffers an additional change in direction at the next surface, and continues in this way through the system. The system thus imposes successive changes in θ and h alternately. The change in the ray for a traversal between planes is easily written

$$h' = h + t\theta'_1 \qquad\qquad \theta_2 = \theta'_1 \qquad\qquad (3.50)$$

The parameters of the ray before traversing the space can be written as elements of a column matrix

$$r = \begin{bmatrix} h \\ \theta \end{bmatrix}$$

and after traversal, from Eq. (3.50),

$$r' = \begin{bmatrix} h' \\ \theta_2 \end{bmatrix} = \begin{bmatrix} h + t\theta' \\ \theta'_1 \end{bmatrix} = \begin{bmatrix} 1 & t \\ 0 & 1 \end{bmatrix} \begin{bmatrix} h \\ \theta'_1 \end{bmatrix} \qquad (3.51)$$

If the boundary between the first and second media is a spherical surface, a relationship between θ_1 and θ'_1 can be derived by referring to

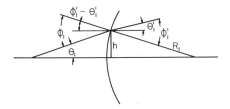

Fig. 3.28 Refraction at a spherical surface.

Fig. 3.28 and noting that

$$\phi_1 = \theta_1 + \phi_1' - \theta_1'$$

$$\phi_1' = \frac{h}{R_1} + \theta_1'$$

and

$$\phi_1 = \frac{n_2}{n_1} \phi_1'$$

from which

$$\theta_1' = \frac{n_1}{n_2} \theta_1 + \frac{h}{R_1}\left(\frac{n_1}{n_2} - 1\right) \qquad (3.52)$$

In matrix notation Eq. (3.52) can be written

$$\begin{bmatrix} h \\ \theta_1' \end{bmatrix} = \begin{bmatrix} 1 & 0 \\ \dfrac{1}{R_1}\left(\dfrac{n_1}{n_2} - 1\right) & \dfrac{n_1}{n_2} \end{bmatrix}\begin{bmatrix} h \\ \theta_1 \end{bmatrix} \qquad (3.53)$$

The traversal of a ray through a succession of surfaces can be described by the multiplication of a number of matrices of the types entered in Eq. (3.51) and (3.53).

Let us illustrate the procedure by an application to a thick lens. We could, of course, trace the appropriate rays for every desired object distance. However, we know from the previous section that if object and image distances are measured from principal planes, a very simple equation of the same form as the thin-lens equation can be used. Therefore, we shall find an equation for the focal length of the thick lens relative to measurements from principal planes. To do this, we consider a ray from infinity which encounters the first surface a distance h_1 from the optic axis (Fig. 3.29). The matrix representation of the initial ray is thus

$$r = \begin{bmatrix} h \\ 0 \end{bmatrix}$$

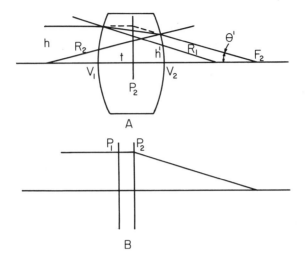

Fig. 3.29   A thick lens and its principal planes.

The refraction matrix for the first surface is

$$\begin{bmatrix} 1 & 0 \\ \dfrac{1}{R_1}\left(\dfrac{1-n}{n}\right) & \dfrac{1}{n} \end{bmatrix}$$

where R_1 is the radius of curvature of the first surface, n is the refractive index of the lens medium, and the lens is surrounded by a vacuum.

The traversal through the lens of thickness t is described by

$$\begin{bmatrix} 1 & t \\ 0 & 1 \end{bmatrix}$$

and the refraction at the second surface is given by

$$\begin{bmatrix} 1 & 0 \\ \dfrac{n-1}{R_2} & n \end{bmatrix}$$

The ray leaving the second surface is thus

$$\begin{bmatrix} h' \\ \theta' \end{bmatrix} = \begin{bmatrix} 1 & 0 \\ \dfrac{n-1}{R_2} & n \end{bmatrix}\begin{bmatrix} 1 & t \\ 0 & 1 \end{bmatrix}\begin{bmatrix} 1 & 0 \\ \dfrac{1-n}{nR_1} & \dfrac{1}{n} \end{bmatrix}\begin{bmatrix} h \\ 0 \end{bmatrix}$$

$$= \begin{bmatrix} h\left(1 - t\dfrac{n-1}{nR_1}\right) \\[2ex] h(n-1)\left[-\dfrac{1}{R_1} + \dfrac{1}{R_2} - \dfrac{(n-1)t}{nR_1R_2}\right] \end{bmatrix} \tag{3.54}$$

For the particular geometry shown in the figure $R_1 > 0$ and $R_2 < 0$. Equation (3.54) states that the ray leaves the second surface a distance $ht(n-1)/nR_1$ closer to the optic axis than the initial ray and that it makes an angle

$$-h(n-1)\left[\frac{1}{R_1} - \frac{1}{R_2} + \frac{(n-1)t}{nR_1R_2}\right]$$

with the optic axis, the minus sign showing that the ray is traveling downward as it moves toward the right. The point of intersection of the ray with the optic axis is the focal point of the system. It is a distance h'/θ' from the vertex of the second surface. The lens is equivalent to the pair of principal planes shown in Fig. 3.29B, in which the ray remains parallel to the optic axis until it passes the second principal plane. Thus the second principal plane is found from the intersection of the extended initial and final rays, as shown in Fig. 3.29A. The distance of the second principal plane from the second focal point is given by

$$\frac{P_2F_2}{h'/\theta'} = \frac{h}{h'}$$

By making use of these equations, we can write for the focal length measured from the principal plane,

$$\frac{1}{f} = (n-1)\left[\frac{1}{R_1} - \frac{1}{R_2} + \frac{(n-1)t}{nR_1R_2}\right] \tag{3.55}$$

and for the distance of the second principal plane from the vertex of the second surface,

$$P_2V_2 = \frac{(n-1)tf}{nR_1} = \frac{-tR_2}{n(R_1-R_2) - (n-1)t}$$

The sign of P_2V_2 is positive for P_2 to the left of V_2. The same procedure is used with the surfaces reversed, or with the ray incident from the right, to find the distance of the first principal plane from the vertex of the first surface,

$$P_1V_1 = \frac{-(n-1)tf}{nR_2} = \frac{-tR_1}{n(R_1-R_2) - (n-1)t}$$

The sign of P_1V_1 is positive when P_1 is to the right of V_1.

We can see from Eq. (3.55) how the focal length of a lens is affected by its thickness. Notice that the focal length of a plano-convex lens, for which $R_2 = \infty$, is independent of thickness and is just the focal length calculated from the thin-lens formula. Consequently, the plano-convex lens we designed earlier for use as a thermocouple condensing lens has the correct focal length. However, the second principal plane is a distance $t/1.5$ from the planar surface, inside the lens. If the lens is 4.5 mm thick, the thermocouple must be placed not 3 mm from the second surface but in contact with it. The first principal plane is tangent to the convex surface.

X. ABERRATIONS

We have purposely limited the discussion up to now to paraxial rays in order to take advantage of the small-angle approximation, $\sin \theta \approx \theta$. This approximation greatly simplifies the mathematics of ray tracing and permits the definition of unique relationships between object and image points. By admitting rays which make a larger angle with the optic axis we lose this advantage, and the ray-tracing procedure becomes more complicated. The angle of incidence and point of intersection on a surface of an optical element must be calculated by taking into consideration the curvature of the surface, and Snell's law must be used in its general form. In the past, optical designers have systematically traced a few selected rays through an optical system in order to judge its performance. Modern electronic computers have released the designer from this tedious practice and made it practical to trace many more rays. We cannot go into details here, but shall discuss in a qualitative way the nature of the imperfections in optical images.

Just as the approximation $\sin \theta \approx \theta$ is useful for very small angles, the next order of approximation, $\sin \theta \approx \theta - (\theta^3/3!)$, can be used for somewhat larger angles, and $\sin \theta \approx \theta - (\theta^3/3!) + (\theta^5/5!)$ for even larger angles. The use of these approximations lead to third-order geometric optics, fifth-order geometric optics, and so on. The first-order theory, of course, corresponds to the optics of paraxial rays. Deviations of rays and intercepts from the paraxial case are optical *aberrations*. Seidel derived third-order optical formulas, and classified the aberrations encountered. These are the Seidel or third-order aberrations and they are the basis of most descriptive discussions of optical systems.

As we mentioned earlier, and develop in greater detail in the next chapter, optical signals are propagated as electromagnetic waves. The rays of geometric optics represent the direction of propagation of the waves and are normals to the wave surfaces. Aberrations can be discussed equally well as distortions of the wave surfaces or as departures of the rays from their paraxial behavior. In Fig. 3.30 we see a bundle of rays and the corresponding wave surfaces converging on an image point Q. Because the rays converge to a common point and are normals to the surfaces, the surfaces must be sections of spheres. If the rays fail to converge on a common point, as in Fig. 3.31, the wave surfaces are not spheres, but depart from some reference spherical surfaces by an amount Δ. The reference sphere is often taken to lie in the exit pupil of the optical system and we shall adopt that convention here.

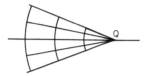

Fig. 3.30 Wave surfaces and rays converging to a point.

We shall need a relationship between the equation of the wave surface of nominal radius of curvature R and a ray that is normal to the wave surface at a point $A(x_p, y_p, z_p)$ and that passes through the image point $Q(x, y, z)$. The equation of the wave surface is

$$x_p^2 + y_p^2 + z_p^2 + (R + \Delta)^2 \approx R^2 + 2R\Delta$$

for small distortions. The direction cosines of a vector normal to the surface are

$$\cos \alpha = \frac{x_p - R(\partial\Delta/\partial x_p)}{1/2 \text{ grad } S}$$

$$\cos \beta = \frac{y_p - R(\partial\Delta/\partial y_p)}{1/2 \text{ grad } S} \qquad (3.56)$$

$$\cos \gamma = \frac{z_p - R(\partial\Delta/\partial z_p)}{1/2 \text{ grad } S}$$

where S is the equation of the wave surface. Notice that Δ is explicitly dependent on x_p and y_p, but not on z_p, so $\partial\Delta/\partial z_p = 0$. The direction cosines can also be written for the line connecting the point A on the reference surface to the image point Q,

$$\cos \alpha = \frac{x - x_p}{\overline{AQ}} \qquad \cos \beta = \frac{y - y_p}{\overline{AQ}} \qquad \cos \gamma = \frac{z - z_p}{\overline{AQ}}$$

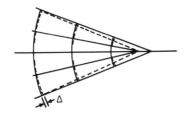

Fig. 3.31   Rays failing to converge to a point.

where $\bar{AQ}$ is the distance from A to Q. The direction cosines of Eqs. (3.56) and (3.57) are, of course, the same, so the equations can be combined to give the equation of the ray from A to Q

$$\frac{x-x_p}{x_p-R(\partial\Delta/\partial x_p)} = \frac{y-y_p}{y_p-R(\partial\Delta/\partial y_p)} = \frac{z-z_p}{z_p} \tag{3.57}$$

or

$$x = \frac{z}{z_p}\left(x_p - R\frac{\partial\Delta}{\partial x_p}\right) + R\frac{\partial\Delta}{\partial x_p} \quad \text{and} \quad y = \frac{z}{z_p}\left(y_p - R\frac{\partial\Delta}{\partial y_p}\right) + R\frac{\partial\Delta}{\partial y_p} \tag{3.58}$$

We are often interested in the x and y coordinates where the ray penetrates the Gaussian image plane, that is, where $z = 0$. These are given by

$$x_0 = R\frac{\partial\Delta}{\partial x_p} \quad \text{and} \quad y_0 = R\frac{\partial\Delta}{\partial y_p} \tag{3.59}$$

Δ depends on the coordinates of the point in question on the reference sphere, which we now denote by the polar coordinates ρ and θ, and the coordinates of the image point on the Gaussian image plane (see Fig. 3.32), r and ϕ. The transformation of Eqs. (3.58) into polar coordinates ρ, θ, z_p is elementary and gives us an alternate pair of equations

$$x = \frac{z}{z_p}\rho\cos\theta - R\left[1-\frac{z}{z_p}\right]\left(\cos\theta\frac{\partial\Delta}{\partial\rho} - \frac{\sin\theta}{\rho}\frac{\partial\Delta}{\partial\theta}\right)$$

$$y = \frac{z}{z_p}\rho\sin\theta - R\left[1-\frac{z}{z_p}\right]\left(\sin\theta\frac{\partial\Delta}{\partial\rho} + \frac{\cos\theta}{\rho}\frac{\partial\Delta}{\partial\theta}\right) \tag{3.60}$$

The intersection point of the rays with the Gaussian image plane, $z = 0$, is given by

$$x_0 = R\left(\cos\theta\frac{\partial\Delta}{\partial\rho} - \frac{\sin\theta}{\rho}\frac{\partial\Delta}{\partial\theta}\right) \tag{3.61}$$

and

$$y_0 = R\left(\sin\theta\frac{\partial\Delta}{\partial\rho} + \frac{\cos\theta}{\rho}\frac{\partial\Delta}{\partial\theta}\right)$$

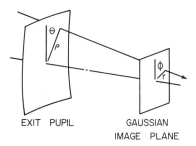

Fig. 3.32   Angles and distances in the exit pupil and image plane.

These are the coordinates of the point of intersection with the Gaussian image plane, measured from the Gaussian image point. They represent the departure from an ideal optical image. For an optical system that is symmetric about the optic axis, Δ depends on functional combinations of these coordinates of the form

$$x_1 = \tfrac{1}{2}\rho^2 \quad x_2 = r\rho \cos(\phi - \theta) \quad \text{and} \quad x_3 = \tfrac{1}{2}r^2 \tag{3.62}$$

Δ can be expanded as a power series in $x_1, x_2,$ and x_3

$$\Delta = a + a_1 x_1 + a_2 x_2 + a_3 x_3 + a_{11} x_1^2 + a_{12} x_1 x_2 + a_{13} x_1 x_3 + a_{22} x_2^2$$

$$+ a_{23} x_2 x_3 + a_{33} x_3^2 + \ldots \tag{3.63}$$

Termination of the series of Eq. (3.63) after the linear terms corresponds to the paraxial approximation of ray optics. Termination after the quadratic terms includes third-order aberrations.

The constant term a and the terms containing a_3 and a_{33} merely affect the initial choice of reference sphere, but not the quality of the optical image. The terms containing a_1 and a_2 represent focusing errors. They are not true aberrations when monochromatic radiation is used, because they can be eliminated with proper attention to focusing. With polychromatic radiation, however, the situation is different, because the optical properties of materials generally depend on wavelength and, therefore, the focal length of refracting optical systems can be wavelength dependent. In the visible spectral region this is called *chromatic aberration* and the term is carried over into the infrared region as well. The remaining five terms in Eq. (3.63) are the third-order aberrations which we shall now discuss using Seidel's nomenclature.

A. Spherical Aberration

The aberration associated with the term containing a_{11} is called spherical aberration. The contribution of this term is, from Eq. (3.62) and (3.63),

$$\Delta_s = \frac{a_{11}\rho^4}{4} \tag{3.64}$$

which states that the deformation of the wave front is proportional to the fourth power of the distance from the optical axis, but is not dependent on the location of image points. It is the only third-order aberration existing for object points on the optical axis.

In a system afflicted with spherical aberration, rays from different zones in the aperture are focused in different planes (see Fig. 3.33). The intersection with the Gaussian image plane of rays from a zone of radius ρ is found from Eqs. (3.64) and (3.61),

$$x_0 = Ra_{11} \cos \theta \rho^3 \quad \text{and} \quad y_0 = Ra_{11} \sin \theta \rho^3$$

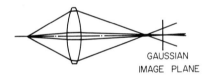

GAUSSIAN
IMAGE PLANE

Fig. 3.33   Spherical aberration in a lens.

Therefore, these rays form a ring of radius

$$r_s = (x_0{}^2 + y_0{}^2)^{1/2} = a_{11}R\rho^3$$

centered about the Gaussian image point. A point object imaged by an optical system whose exit pupil has a radius of $\rho_{\max}$ is thus viewed as a circular patch or circle of confusion of radius $a_{11}R\rho_{\max}^3$ in place of a perfect point in the paraxial image plane. This is often called *transverse* spherical aberration.

The rays from a zone of radius ρ converge to a common point a distance L_s from the paraxial image plane. It is readily seen from Fig. 3.34 that L_s is given by the approximation

$$L_s \approx a_{11}R^2\rho^2 \quad \text{for} \quad \rho \ll R$$

This is called *longitudinal* spherical aberration. If a system is defocused by placing the image plane midway between the paraxial image plane and the plane where the rays from the extreme edge of

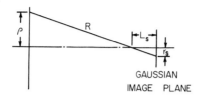

Fig. 3.34 Longitudinal spherical aberration.

the pupil (the *marginal rays*) are focused, a circle of confusion of radius $\frac{1}{2}a_{11}R\rho_{\mathrm{max}}^3$ will be found.

The other aberration coefficients are determined by the nature of the optical system. For a single spherical mirror with object at infinity, the value of a_{11} is $1/8f^3$. Thus the circle of confusion has a radius

$$r_s(\mathrm{mirror}) = \frac{1}{64}\frac{D^3}{f^2}$$

where D is the diameter of the mirror and f is its focal length. The marginal rays are focused closer to the surface of a concave mirror than are the paraxial rays. Equation (3.64), for the distortion of the wave front, defines a parabola. Conversely, if we use a mirror having a parabolic rather than a circular contour, the wave front corresponding to an object at infinity will not be distorted by spherical aberration. Such mirrors are more costly than spherical mirrors, but are sometimes required when it is necessary to construct optical systems with large apertures. We discuss this further in Chapter 5.

A simple lens has two surfaces, each of which contributes to the spherical aberration of the lens. The sign of a_{11} can be either positive

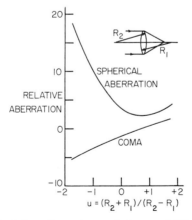

Fig. 3.35 Spherical aberration and coma for lenses of different curvatures.

or negative, so the contributions from the two surfaces can be made to oppose one another. In Fig. 3.35 is plotted the spherical aberration (circle of confusion) of a series of lenses having the same focal length but different curvatures of their surfaces. It is seen that the spherical aberration can be minimized, but not eliminated, in a simple lens. In more complex designs, with more than one element, the spherical aberration can be effectively eliminated. The experimenter seldom has the facility for designing and building an optimized lens, but must construct his optical system from readily available parts. Figure 3.35 shows that a plano-convex lens oriented with the plane surface closest to the image is a good choice of lens for imaging a distant object.

B. Coma

The aberration associated with a_{12} in Eq. (3.63) is called coma because of the characteristic appearance of the image of a point object produced by a comatic optical system. The wave front distortion due to coma is

$$\Delta_c = \tfrac{1}{2}[a_{12}\rho^3 r \cos(\phi - \theta)] \tag{3.65}$$

Coma depends not only on the radius of the zone in the exit pupil through which the rays pass, but also on the off-axis distance of the object or Gaussian image point and, in the case of skew rays, depends on the angular difference between a line from the axis to the point of intersection of a ray with the exit pupil and a line from the axis to the Gaussian image point. There is no coma for a point on the optic axis. The appearance of the distorted wave front is as sketched in Fig. 3.36.

The paths taken by the rays in a comatic optical system are illustrated in Fig. 3.37 for the outermost zone of the pupil. A point is

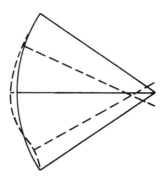

Fig. 3.36   Distorted wave front in the presence of coma.

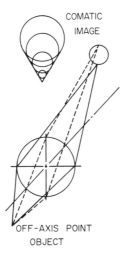

COMATIC
IMAGE

OFF-AXIS POINT
OBJECT

Fig. 3.37 Coma.

imaged as a ring in the Gaussian image plane by this zone. The rays through a slightly smaller zone of the pupil are imaged as a ring of slightly smaller diameter, somewhat closer to the optic axis, and so on for the remaining zones of the pupil. Consequently, a point is imaged by a comatic system as a comma-shaped patch, as sketched in Fig. 3.37. The paraxial image point is the tip of the patch.

We can use Eqs. (3.61) and (3.65) to find the distance of the point of intersection with the Gaussian image plane, measured from the Gaussian image point,

$$(x_0{}^2 + y_0{}^2)^{1/2} = \tfrac{1}{2}(a_{12}R\rho^2 r)(9\cos^2(\phi - \theta) + \sin^2(\phi - \theta))^{1/2}$$

The point farthest from the optic axis in the comatic pattern occurs for rays passing through the outermost zone of the exit pupil, with radius ρ_{max} at polar angle $\theta = 0$ or π and, of course, at $\phi = 0$ in the Gaussian image plane. Hence, for $\phi - \theta = 0$ or π

$$l_c = \tfrac{3}{2}(a_{12}R\rho_{max}^2 r)$$

is the length of the comatic pattern. The maximum width of the pattern (see Fig. 3.37) is given by

$$w_c = a_{12}R\rho_{max}^2 r$$

The quality of a comatic image is not improved by defocusing.

If a spherical mirror of focal length f and diameter D is used to image a distant point source and the principal ray makes an angle ω rad with the optic axis, the comatic pattern will have a length

$$l_c(\text{mirror}) = \frac{3}{16}\frac{D^2\omega}{f}$$

and width

$$w_c(\text{mirror}) = \frac{1}{8}\frac{D^2\omega}{f}$$

These equations must be modified somewhat if the mirror is not the aperture stop of the system. This is discussed further in Chapter 5.

The coma of a simple lens is plotted in Fig. 3.35, along with the spherical aberration. Fortunately, the configurations that are most favorable for minimum spherical aberration also tend to reduce coma to a minimum.

Because of the asymmetric shape of the comatic wave front, coma can be made to add or subtract as the wave front progresses from element to element in an optical system. The experimenter should keep this in mind as he lays out an optical train for whatever purpose he has. For example, if it is desired to produce a collimated beam of light from a small source and then refocus it to its original size, either of the two schemes shown in Figs. 3.38A and B might be used. However, system A is more desirable because the second mirror cancels the coma produced by the first, whereas in system B the second mirror doubles the coma from the first. If an intermediate image is produced

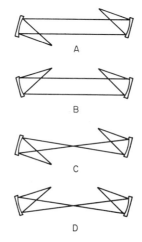

Fig. 3.38 Optical systems for canceling coma (**A** and **D**) and adding coma (**B** and **C**).

between the two mirrors, coma is canceled in the arrangement of Fig. 3.38D and doubled in Fig. 3.38C.

C. Distortion

An optical system that forms a curved line image of a straight line object suffers from distortion. This aberration is associated with the coefficient a_{23} in Eq. (3.63). The wave front deformation is

$$\Delta_D = \tfrac{1}{2}[a_{23}r^3\rho \cos (\phi - \theta)]$$

and the intercepts of the rays in the Gaussian image plane are

$$x_0 = \tfrac{1}{2}(Ra_{23}r^2 \cos \phi) \quad \text{and} \quad y_0 = \tfrac{1}{2}(Ra_{23}r^3 \sin \phi)$$

x_0 and y_0 are independent of ρ and θ. Thus, every object point is imaged sharply on the Gaussian image plane. But an off-axis object is imaged at a distance

$$(x_0{}^2 + y_0{}^2)^{1/2} = \tfrac{1}{2}(Ra_{23}r^3)$$

from the Gaussian image point. In other words, the magnification varies as the cube of the distance of the image point from the optic axis. If the distortion is positive, a square object is imaged as shown in Fig. 3.39A. This is termed "barrel" distortion. If the distortion is negative, "pincushion" distortion, as in Fig. 3.39B, is observed.

Neither the thin lens nor the spherical mirror produces distortion if the aperture stop coincides with the optical element.

OBJECT   A   B

Fig. 3.39   Barrel (A) and pincushion (B) distortions.

D. Field Curvature

If an optical system forms a sharp but nonplanar image of a plane object, it is afflicted with field curvature or nonflatness of field. The aberration which we shall call field curvature (sometimes called Petzval curvature) is associated with the coefficient a_{13} of Eq. (3.63). An additional field curvature is contributed by astigmatism, as we see in the next section. For clarity, we shall refer to it as astigmatic field curvature. The wave front distortion in an optical system with field curvature is

$$\Delta_F = \tfrac{1}{4}(a_{13}\rho^2 r^2)$$

We are now interested in the intersection of rays with planes other than the Gaussian image plane, so Eqs. (3.60) rather than Eqs. (3.61) are used. We find

$$x = \frac{z}{z_p}\rho \cos\theta + R\left(1 - \frac{z}{z_p}\right)\cos\theta \cdot \frac{a_{13}\rho r^2}{2}$$

and   (3.66)

$$y = \frac{z}{z_p}\rho \sin\theta + R\left(1 - \frac{z}{z_p}\right)\sin\theta \cdot \frac{a_{13}\rho r^2}{2}$$

Recall that z_p is the z coordinate of the point in the reference sphere with respect to the Gaussian image plane. Therefore

$$z_p = R \tag{3.67}$$

From Eqs. (3.66) we find the condition for $x = 0$ and also for $y = 0$,

$$z_F = \frac{R^2 a_{13} r^2/2}{R a_{13} r^2/2 - 1} \approx \frac{-a_{13}R^2 r^2}{2} \tag{3.68}$$

Note that this expression does not contain ρ or θ. It states that the rays from all parts of the reference sphere converge to a perfect point image, not on the Gaussian image plane but on a surface a distance z_F from it. The nonflatness of the field is proportional to the square of the distance of the image point from the optic axis.

In the Gaussian image plane the rays intersect at points

$$x_F = R \cos\theta\, a_{13}\rho r^2/2 \quad y_F = R \sin\theta\, a_{13}\rho r^2/2$$

which define a circle of radius

$$r_F = \tfrac{1}{2}(a_{13}R\rho r^2)$$

The thin lens and the spherical mirror have no field curvature if the aperture stop is the element itself.

E. Astigmatism

The remaining third-order aberration, associated with the coefficient a_{22}, is called astigmatism. The wave front deformation in the presence of this aberration is

$$\Delta_A = a_{22}r^2\rho^2 \cos^2(\phi - \theta)$$

and the intersection of rays in a stigmatic system with a plane displaced from the Gaussian image plane by a distance z is given by

$$x = \frac{z}{z_p}\rho \cos\theta - 2a_{22}r^2\rho R\left(1 - \frac{z}{z_p}\right)\cos\phi \cos(\phi - \theta)$$

and   (3.69)

$$y = \frac{z}{z_p}\rho \sin\theta - 2a_{22}r^2\rho R\left(1 - \frac{z}{z_p}\right)\sin\phi \cos(\phi - \theta)$$

The condition necessary for the image errors in the x direction to vanish is

$$z(x = 0) = \frac{2z_p a_{22}Rr^2 \cos\phi \cos(\phi - \theta)}{2a_{22}Rr^2 \cos\phi \cos(\phi - \theta) - \cos\theta}$$ (3.70)

and for vanishing image errors in the y direction

$$z(y = 0) = \frac{2z_p a_{22}Rr^2 \sin\phi \cos(\phi - \theta)}{2a_{22}Rr^2 \sin\phi \cos(\phi - \theta) - \cos\theta}$$ (3.71)

Unlike the case of curvature of field, these errors do not vanish simultaneously.

Let us consider off-axis objects with the off-axis direction defined by $\phi = 0$. From Eq. (3.71), image errors in the y direction vanish in the Gaussian image plane if the system has no field curvature while, from Eqs. (3.69), the maximum image error in the x direction in the Gaussian image plane amounts to

$$l_{A,S} = 2a_{22}r^2\rho R$$

That is, in the Gaussian image plane an off-axis object is imaged as a line of length $l_{A,S}$. The line lies in a direction normal to the optic axis and is called the *sagittal image* (see Fig. 3.40).

Equation (3.70) tells us that the image errors in the x direction vanish on the surface

$$z_T = \frac{2z_p a_{22}Rr^2}{2a_{22}Rr^2 - 1} \approx -2a_{22}R^2r^2$$ (3.72)

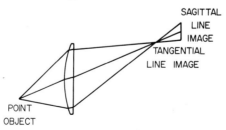

SAGITTAL
LINE
IMAGE
TANGENTIAL
LINE IMAGE

POINT
OBJECT

Fig. 3.40 Astigmatic line images.

where use has been made of Eq. (3.67). The maximum error in the y direction in the plane tangent to this surface is, from Eqs. (3.69),

$$l_{A,T} = 2a_{22}r^2\rho R$$

That is, in the surface given by Eq. (3.72), an off-axis point is imaged as a line of length $l_{A,T}$. This line lies in a direction tangential to circles on the image plane centered on the optic axis. It is called the *tangential image*. Note that $l_{A,T}$ has the same magnitude as $l_{A,S}$. In intermediate planes, Eqs. (3.69) define ellipses.

Thus we see that from an off-axis object in a stigmatic optical system (with no field curvature) the rays first form a tangential line image lying on a curved surface, then they form a sequence of ellipses of progressively varying eccentricity, and finally they form a sagittal line image in the Gaussian image plane. In an optical system having both field curvature and astigmatism, the sagittal image lies on a curved surface defined by z_F in Eq. (3.68), while the tangential image lies on a curved surface defined by

$$z_F + z_T = -R^2 r^2 (\tfrac{1}{2}a_{13} + 2a_{22})$$

The distances along the principal ray from the optical element to the sagittal and tangential images in a stigmatic system are given by (see Fig. 3.41)

$$q_S = q/\cos\omega \quad \text{and} \quad q_T = q\cos\omega$$

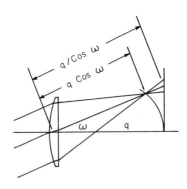

Fig. 3.41   Astigmatism.

respectively, where q is the Gaussian image distance and ω is the off-axis angle of the principal ray. An optical element with paraxial focal length f is said to have focal length

$$f_S = f/\cos\omega$$

for the sagittal image and focal length

$$f_T = f \cos \omega$$

for the tangential image.

A spherical mirror of focal length f and diameter D images an off-axis point object as a sagittal line of length

$$l_{A,S}(\text{mirror}) \approx D\omega^2$$

in the Gaussian image plane. The small off-axis angle ω is expressed in radians. The surface of the tangential line image is a sphere of radius

$$R_A = \tfrac{1}{2}f$$

and the distance between the tangential and sagittal images, called the astigmatic difference, is

$$d_{S,T}(\text{mirror}) \approx \omega^2 f$$

These equations are for a mirror configuration having the entrance pupil coincidental with the mirror and must be modified if the mirror is not the aperture stop.

REFERENCES

1. A. C. Hardy and F. H. Perrin. *Principles of Optics*, McGraw-Hill, New York, 1932.
2. M. Born and E. Wolf. *Principles of Optics*, Pergamon, New York, 1959.
3. R. K. Luneburg. *Mathematical Theory of Optics*, Univ. of California Press, Berkeley, Calif., 1964.
4. M. Herzberger. *Modern Geometrical Optics*, Wiley-Interscience, New York, 1958.
5. A. E. Conrady. *Applied Optics and Optical Design*, Dover, New York, 1957.

CHAPTER 4

Elements of Physical Optics

I. INTRODUCTION

In Chapter 3 we developed certain of the equations of geometric optics. These equations described the formation of images by considering only the direction in which optical information is transmitted. They offer a satisfactory description of the position, the size, and, to a large extent, the perfection of images, but do not provide any knowledge of the nature of the optical information or how it is transmitted. Nor do these equations provide us with an accurate description of the behavior of radiation beams in the presence of aperture screens or scattering objects.

In order to discuss these subjects, we need to study physical optics. In this chapter we review some of the elementary concepts of physical optics in sufficient detail to understand at least qualitatively the behavior of infrared spectrophotometers. More complete treatments of physical optics are to be found in standard textbooks, such as those of Jenkins and White (*1*) or Strong (*2*), and on a more advanced level in the treatises of Sommerfeld (*3*) or Born and Wolf (*4*) and in the treatment of modern optics by O'Neill (*5*).

II. Electromagnetic Waves

In the previous chapter we willingly accepted the premises of geometric optics, in spite of their acknowledged imperfections, because they were found to yield useful results in the design and analysis of image-forming instruments. Now we are asked to accept the proposition that radiation consists of electromagnetic energy propagated in the form of transverse waves. We do this because such a description provides a very satisfactory explanation of many optical phenomena not explainable by geometric optics. We shall not go further into the epistemology of this fascinating subject, but proceed immediately to write down the basic equations of physical optics. These equations are called *Maxwell's equations* and are given in terms of four fundamental field vectors which are defined according to experimental observations:

(1) The electric vector E is the force exerted on a unit electrical charge q_0 in free space according to Coulombs law: $F = q_0 E$.

(2) The electric displacement D is related to E through the dielectric constants of the medium, $D = \epsilon E$.

(3) The magnetic vector H is analogous to the electric vector E, but there is no unit magnetic charge in nature. Magnetic fields are generated by electric currents and H is defined in terms of a unit electric current according to the Biot–Savart law.

(4) The magnetic induction B is related to H through the magnetic permeability, $B = \mu H$.

Maxwell's equations, in the Gaussian system of units, are

$$\nabla \times H - \frac{1}{c}\frac{\partial D}{\partial t} = \frac{4\pi}{c}j = \frac{4\pi\sigma}{c}E$$

$$\nabla \times E + \frac{1}{c}\frac{\partial B}{\partial t} = 0 \tag{4.1}$$

$$\nabla \cdot D = 4\pi\rho \quad \text{and} \quad \nabla \cdot B = 0 \tag{4.2}$$

where ρ is the electrical charge density, j is the electrical current vector, σ is the electrical conductivity, and c is a universal constant, the velocity of light in a vacuum. The operators $\nabla \times$ and $\nabla \cdot$ are, respectively, the curl and divergence operators of vector analysis.

These four equations can be combined to give us two differential equations in E and H. Let us assume there are no free charges present, so $\rho = 0$. If we take the curl of Eqs. (4.1), the first term can be written

$$\nabla \times \nabla \times E = -\nabla^2 E + \nabla(\nabla \cdot E) = -\nabla^2 E$$

where ∇^2 is the Laplacian operator. The first step is an identity from vector analysis and the second is a consequence of Eqs. (4.2). The second term in Eqs. (4.1) becomes

$$\frac{1}{c}\nabla \times \frac{\partial B}{\partial t} = \frac{\mu}{c}\frac{\partial}{\partial t}(\nabla \times H) = \frac{\mu}{c}\left(\frac{1}{c}\frac{\partial^2 D}{\partial t^2} + \frac{4\pi\sigma}{c}\frac{\partial E}{\partial t}\right)$$

$$= \frac{\mu}{c}\left(\frac{\epsilon}{c}\frac{\partial^2 E}{\partial t^2} + \frac{4\pi\sigma}{c}\frac{\partial E}{\partial t}\right)$$

The final equation for E is

$$\nabla^2 E - \frac{4\pi\sigma\mu}{c^2}\frac{\partial E}{\partial t} - \frac{\mu\epsilon}{c^2}\frac{\partial^2 E}{\partial t^2} = 0 \qquad (4.3)$$

Similarly, an equation for H is

$$\nabla^2 H - \frac{\mu\epsilon}{c^2}\frac{\partial^2 H}{\partial t^2} = 0$$

In media having no electrical conductivity, $\sigma = 0$, the solutions of these equations have the form

$$E(t) = E_0 \exp\left[-j\omega\left(t - \frac{x\sqrt{\epsilon\mu}}{c}\right) + j\phi\right]$$

and $\qquad (4.4)$

$$H(t) = H_0 \exp\left[-j\omega\left(t - \frac{x\sqrt{\epsilon\mu}}{c}\right) + j\phi\right]$$

These equations describe E and H as waves of frequency ω rad/sec propagated through space in the x direction with velocity

$$v = c/\sqrt{\epsilon\mu} \qquad (4.5)$$

The quantity ϕ represents the phase of the waves and has an importance in physical optics which we discuss a little later. The terms E_0 and H_0 are, of course, the amplitudes.

Those who are nervous in the presence of complex variables, as in Eqs. (4.4), can make use of the identity

$$e^{-j\theta} = \cos\theta - j\sin\theta$$

along with the condition that $E(t)$ and $H(t)$ are real quantities. Equations (4.4) can thus be written in trigonometric form

$$E(t) = E_0 \cos\left(\omega t - \frac{x\sqrt{\epsilon\mu}}{c} - \phi\right)$$

and

$$H(t) = H_0 \cos\left(\omega t - \frac{x\sqrt{\epsilon\mu}}{c} - \phi\right)$$ (4.6)

However, the reader is advised to become familiar with the very useful exponential form. We use it without further apology in later sections.

By comparing Eq. (4.5) with Eq. (3.18), we see that the refractive index of a medium is given by

$$N = \sqrt{\epsilon\mu}$$

It can be shown that the vectors E and H are mutually perpendicular to each other and to the direction of propagation of the electromagnetic wave. Hence, the wave is described as a transverse wave. The power conveyed by the wave is given by the Poynting vector S

$$S = E_0 \times H_0 \frac{c}{4\pi}$$ (4.7)

E and H are proportional to one another in an electromagnetic field, $|H| = \sqrt{\epsilon/\mu}|E|$, permitting Eq. (4.7) to be written

$$|S| = \frac{c}{4\pi}\sqrt{\epsilon}E_0 E_0^* = \frac{c}{4\pi}\sqrt{\epsilon}|E_0^2|$$ (4.8)

E^* is the complex conjugate of E, obtained by replacing j with $-j$. It is conventional to describe electromagnetic radiation in terms of only the electric vector.

Naturally, this very brief discussion hardly serves as an adequate introduction to the theory of electromagnetic radiation. It is intended to outline in the sketchiest manner the nature and origin of basic equations which describe the propagation of radiation. We shall now make use of the properties of electromagnetic waves implied by these equations.

III. SUPERPOSITION, ATTENUATION, AND REFLECTION OF ELECTRO-
MAGNETIC WAVES

In this section we discuss addition, subtraction, and change in the direction of travel of radiation.

The superposition of two electromagnetic waves of the same frequency and amplitude but different phase is most conveniently treated by writing the amplitude equations in the complex form, as in Eqs. (4.4),

$$E = E_0 \exp\left[-j\omega\left(t - \frac{x\sqrt{\epsilon\mu}}{c}\right)\right](e^{j\phi_1} + e^{j\phi_2})$$

The factor in parenthesis can be written in terms of the phase difference $\phi_1 = \phi + \delta$ and $\phi_2 = \phi - \delta$,

$$e^{j\phi_1} + e^{j\phi_2} = e^{j\phi}(e^{j\delta} + e^{-j\delta}) = 2e^{j\phi} \cos \delta$$

So

$$E = 2E_0 \cos \delta \exp \left[-j\omega \left(t - \frac{x\sqrt{\epsilon\mu}}{c} \right) + j\phi \right] \qquad (4.9)$$

The combined wave has the same frequency as its components, a phase that is the average of the phases of the components, and an amplitude $2E_0 \cos \delta$ that varies from the sum of the amplitude to zero depending on the relative phases of the components.

Equation (4.9) for the superposition of two waves can, of course, be extended to any number of superposed waves. The rate of flow of energy associated with the combined waves is, according to Eq. (4.8),

$$|S|_c = \frac{c}{4\pi}\sqrt{\epsilon} \left(\sum_k E_k \right)^2 \qquad (4.10)$$

This description of the summation of superposed electromagnetic waves is valid for light waves of a particular kind, namely, those which are *coherent*. The power carried by a coherent wave is found from the square of the sum of the constituent waves. On the other hand, if the radiation is *incoherent*, the power is proportional to the sum of the squares of the constituent waves,

$$|S|_i = \frac{c}{4\pi}\sqrt{\epsilon} \sum_k E_k^2 \qquad (4.11)$$

The question of coherence of light plays an important role in physical optics, and we shall return to it later.

We have dealt with the phase term in Eqs. (4.4). Now let us turn to the distance-dependent part of the equation, written in the form

$$E = E_0 \exp \left[-j(\omega t - \phi) \right] \exp \left[j\omega xN/c \right] \qquad (4.12)$$

The refractive index is, in general, a complex quantity

$$N = n + jnk \qquad (4.13)$$

(some authors use the forms $N = n + j\kappa$ or $n - j\kappa$). The real part of N is n, the usual refractive index of nonabsorbing materials which we make use of in geometric optics. The imaginary part of N, $jn\kappa$, is the product of n and a quantity κ that we call the *attenuation index* (or extinction coefficient). By setting the conductivity equal to zero in Eq. (4.3), we limited ourselves to solutions containing the real part of the refractive index. These waves are not attenuated as they propagate through space. When σ does not vanish, the imaginary part of the refractive index occurs in the differential equations, and the waves, which are solutions to the equations, are attenuated as they travel. Instead of Eq. (4.12) we have

$$E = E_0 \exp\left[-j\left(\omega t - \phi - \frac{\omega x n}{c}\right)\right] \exp\left[-\omega n\kappa x/c\right] \qquad (4.14)$$

and this is recognized as an electromagnetic wave that is attenuated by the factor $\exp\left[-\omega n\kappa x/c\right]$ as it propagates in the x direction. We discuss later the relationship between κ and σ. The power in the wave is proportional to EE^*, and thus is attenuated at a rate given by

$$S = S_0 \exp\left[-2\omega n\kappa x/c\right] \qquad (4.15)$$

The power is reduced to $1/e$ of its original value after the wave has traversed a distance $x = c/2\omega n\kappa$. We shall give the quantity

$$K = \frac{2\omega n\kappa}{c} \qquad (4.16)$$

the name *absorption coefficient*.

Equation (4.16) can be modified by noting that ω is the frequency in radians per second. The frequency in cycles per second is $f = \omega/2\pi$, which is related to the wavelength λ (in vacuum) by $f\lambda = c$. Hence

$$K = \frac{4\pi\kappa}{\lambda}$$

and Eq. (4.15) becomes

$$S = S_0 e^{-Kx} \quad \text{or} \quad \ln\frac{S_0}{S} = Kx \qquad (4.17)$$

Equation (4.17) expresses *Bouguer's* or *Lambert's* law of absorption. It is very important for analytic chemists who perform quantitative analysis on the basis of the infrared absorption by a species of thickness b. They usually prefer, however, to use base 10 logarithms rather than natural logarithms. Equation (4.17) is then given the alternate form

$$A = \log\frac{S}{S_0} = 0.4343 \ln\frac{S}{S_0} = 0.4343 Kb = ab \qquad (4.18)$$

A is given the name *absorbance* and a is the *absorptivity*. In some older literature A is called optical density and a is referred to as the extinction coefficient. The term extinction coefficient is also sometimes used for the attenuation index κ, leading to some confusion. An even more serious source of confusion, however, is the relationship between absorptivity and absorption coefficient

$$0.4343 K = a$$

because chemists invariably use a, while physicists are usually inclined to use K, particularly in the literature of the solid state.

It is often the case that each component part of the solution absorbs independently of the other components in proportion to its concentration in molecules per unit volume or some equivalent measure. The Bouguer–Lambert law then can be written

$$\frac{S}{S_0} = \exp\left(-b \sum_{n=1}^{N} K_n c_n\right) \qquad (4.19)$$

or in the alternate form of Eq. (4.18)

$$A = b \sum_{n=1}^{N} a_n c_n \qquad (4.20)$$

with the sum extending over all N components of the solution, c_n the concentration of the nth component, and a_n the absorptivity of the nth component. Equation (4.19) or (4.20) is *Beer's law*. It is the basis of quantitative analysis of solutions by spectrophotometry.

Now let us consider the reflection of radiation. In order to keep the mathematics simple we shall treat the case depicted in Fig. 4.1. A beam of radiant energy is incident from the left in the x direction in a medium of real refractive index n_1 and attenuation index κ_1. It encounters a plane surface that is normal to the direction of propagation. A portion of the beam is reflected and a portion penetrates the surface into the second medium of real refractive index n_2 and attenuation index κ_2. The electric and magnetic vectors associated with the radiation are

incident: $\quad E_y{}^i = \left(\dfrac{\mu_1}{\epsilon_1}\right)^{1/2} H_z{}^i = a \exp\left\{-j\omega\left[t - \dfrac{x(\epsilon_1\mu_1)^{1/2}}{c}\right]\right\}$

reflected: $\quad -E_y{}^r = \left(\dfrac{\mu_1}{\epsilon_1}\right)^{1/2} H_z{}^r = a' \exp\left\{-j\omega\left[t + \dfrac{x(\epsilon_1\mu_1)^{1/2}}{c}\right]\right\}$

transmitted: $\quad E_y{}^t = \left(\dfrac{\mu_2}{\epsilon_2}\right)^{1/2} H_z{}^t = a'' \exp\left\{-j\omega\left[t - \dfrac{x(\epsilon_2\mu_2)^{1/2}}{c}\right]\right\}$

The subscripts y and z on the vector components E and H show them to be mutually perpendicular. The negative sign of $E_y{}^r$ and in the exponent of the second equation represent the reversal in direction of propagation, $S \propto E \times H$. The laws of electrodynamics tell us that there

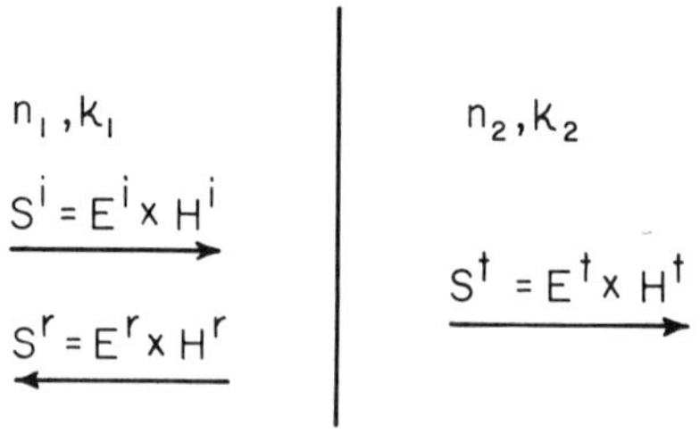

Fig. 4.1   Reflection at a plane boundary.

are no discontinuities in the tangential components of E or H at the surface provided that no current is flowing there. Hence,

$$a - a' = a'' \quad \text{and} \quad (\epsilon_1/\mu_1)^{1/2}(a + a') = (\epsilon_2/\mu_2)^{1/2}a''$$

We eliminate a'' between these two equations in order to find a', and make the substitutions

$$(\epsilon/\mu)^{1/2} = N/\mu = \frac{n}{\mu}(1 + j\kappa)$$

This provides us with the equation

$$a' = \frac{(n_2\mu_1 - n_1\mu_2) + j(n_2\kappa_2\mu_1 - n_1\kappa_1\mu_2)}{(n_2\mu_1 + n_1\mu_2) + j(n_2\kappa_2\mu_1 - n_1\kappa_1\mu_2)}a$$

The radiant power carried by the reflected beam is proportional to $E^r E^{r*} \propto a'a'^*$. Therefore, the reflectivity of the surface is

$$R = \frac{a'a'^*}{aa^*} = \frac{(n_2\mu_1 - n_1\mu_2)^2 + (n_2\kappa_2\mu_1 - n_1\kappa_1\mu_2)^2}{(n_2\mu_1 + n_1\mu_2)^2 + (n_2\kappa_2\mu_1 + n_1\kappa_1\mu_2)^2} \qquad (4.21)$$

Equation (4.21) is quite general for normally incident radiation, and it is a very useful relationship. Let us consider some special cases.

Except for ferromagnetic media, μ can be taken to be very nearly unity. If the initial medium is nonabsorbing, $\kappa_1 = 0$ and Eq. (4.21) becomes

$$R = \frac{(n_2 - n_1)^2 + (n_2\kappa_2)^2}{(n_2 + n_1)^2 + (n_2\kappa_2)^2} \qquad (4.22)$$

It is interesting that the reflectivity depends on both the refractive indices and on the attenuation index, and that a very high value of κ_2 leads to a reflectivity of unity.

If the second medium is also nonabsorbing, the reflectivity is

$$R = \left(\frac{n_2 - n_1}{n_2 + n_1}\right)^2 \qquad (4.23)$$

Thus, if the initial medium is air $(n_1 = 1)$ and the second is sodium chloride $(n_2 = 1.5)$, the reflectivity is $R = 0.04$. But if the second medium is silver chloride $(n_2 = 2.0)$, $R = 0.11$.

The reflectivity of metals is of particular interest to experimenters in infrared radiation, and leads us to a discussion of one of the classic experiments of infrared physics. Because metals characteristically are electrical conductors, the wave equation for E must be taken in the form of Eq. (4.5) and the solution for E is given by Eq. (4.14),

$$E = E_0 \exp\left[-j\omega\left(t - \frac{xN}{c}\right)\right]$$

where N is the complex refractive index of Eq. (4.13). When this solution is substituted back into Eq. (4.3), an equation containing both real and imaginary terms is obtained. The real and imaginary parts are separated, yielding two equations

$$\frac{n^2(1-\kappa^2)}{\mu} = \epsilon \tag{4.24}$$

from the real part and

$$\frac{2n^2\kappa}{\mu} = \frac{4\pi\sigma}{\omega} = \epsilon' \tag{4.25}$$

from the imaginary part. The quantity ϵ' is often called the imaginary part of the complex dielectric constant, while ϵ is called the real part.

Equations (4.24) and (4.25) are solved simultaneously for n^2 and $(n\kappa)^2$

$$n^2 = \frac{\mu}{2}\left[\left(\frac{\epsilon^2}{c^4} + 4\tau^2\sigma^2\right)^{1/2} + \epsilon\right] = \frac{\mu}{2}\left[\left(\frac{\epsilon^2}{c^4} + \epsilon'^2\right)^{1/2} + \epsilon\right] \tag{4.26}$$

and

$$(n\kappa)^2 = \frac{\mu}{2}\left[\left(\frac{\epsilon^2}{c^4} + 4\tau^2\sigma^2\right)^{1/2} - \epsilon\right] = \frac{\mu}{2}\left[\left(\frac{\epsilon^2}{c^4} + \epsilon'^2\right)^{1/2} - \epsilon\right] \tag{4.27}$$

where $\tau = 2\pi/\omega$ is the period of the wave, or alternatively $\tau = \lambda_0/c$, where λ_0 is the wavelength in vacuum. The physical constants of metals are such that $2\sigma\tau \gg \epsilon/c^2$; hence Eq. (4.26) and (4.27) take the approximate form

$$n \approx n\kappa \approx \sqrt{\sigma\mu\tau} \tag{4.28}$$

We substitute Eq. (4.28) into Eq. (4.22) to find an equation for the reflectivity of a metal (in vacuum) in terms of its electrical conductivity, its magnetic permeability, and the wavelength of the light,

$$R = \frac{2\sigma\mu\tau - 2\sqrt{\sigma\mu\tau} + 1}{2\sigma\mu\tau + 2\sqrt{\sigma\mu\tau} + 1} \approx 1 - \frac{2}{\sqrt{\sigma\mu\tau}}$$

or

$$R \approx 1 - 2 \left(\frac{c}{\sigma \mu \lambda_0} \right)^{1/2}$$

This is the equation of Hagen and Rubens (6). It was investigated experimentally by them very early in the present century and was found to hold remarkably well for sufficiently long wavelengths, namely in the infrared beyond $10\,\mu$ or so.

If a beam of radiation is incident on a rough surface, the rays are reflected in different directions, and the radiation is not specularly reflected, but is diffusely reflected or scattered. Similarly, radiation transmitted through a medium containing inhomogeneities or particles is scattered. In spectrophotometry, scattering by a sample represents attenuation even though no energy is absorbed. The theory of scattering is complicated, but if particle size is less than the wavelength, $d \ll \lambda$, the *Rayleigh scattering law* is valid.

$$\text{scattering} \propto \left(\frac{n_1 - n_2}{n_1 + n_2} \right)^2 \left(\frac{d}{\lambda} \right)^4$$

where n_1 and n_2 are the refractive indices of the particles and the surrounding medium. This expression is the leading term of a more general equation for scattering.

IV. POLARIZATION

The electric field associated with an electromagnetic wave is a vector quantity and must be specified by giving its direction as well as its magnitude. This is also true of the magnetic field, of course, but the following discussion will be given in terms of the electric vector. In the preceding section, we avoided the complications introduced by the vector nature of the field by considering the superposition of waves whose vectors are coplanar, the attenuation of waves by an isotropic medium, and the reflection of waves normally incident on a surface. Let us now relax these conditions and discuss the polarization properties of light.

Normally a beam of electromagnetic radiation is a collection of field vectors whose amplitude and direction vary from instant to instant in some random fashion. If, however, the direction of the electric vectors all lie in the same plane, the beam is said to be *plane polarized*. A device for producing plane-polarized radiation from natural radiation is called a *polarizer* and a device for analyzing the polarization properties of a beam of radiation is an *analyzer*. By properly orienting the axis of the analyzer relative to the direction of the electric vector of a plane-polarized beam, the radiation can be extinguished.

The terms *plane of polarization* and *direction of polarization* introduce some confusion into the nomenclature of optics. Traditionally, these terms have been applied to the magnetic vector with the direction of polarization referring to the direction of the magnetic vector and the plane of polarization containing the magnetic vector and the direction of propagation. Some authors, however, define these terms with reference to the electric vector. We shall avoid this confusion by defining the polarization of a beam by the *direction of the E vector* and the *plane of the E vector*.

Now let us consider the superposition of two waves whose electric vectors are not coplanar but lie in planes which are mutually perpendicular. This is quite general, because the electric vector of any beam can be represented as the sum of vectors along perpendicular axes. If the electric vectors of the two waves are in phase, that is, attain maximum and minimum values simultaneously in space and time, then it is easy to see that their sum is a plane-polarized wave whose electric vector lies in a plane which is the vector resultant of the planes of the component vectors. When the two vectors are not in phase, their superposition is slightly more complicated. Let us write the two waves in the form of Eq. (4.8).

$$E_y = E_{y_0} \cos\left(\omega t - \frac{x\sqrt{\epsilon\mu}}{c} - \phi_y\right) = E_{y_0} \cos(\Phi - \phi_y)$$

and

$$E_z = E_{z_0} \cos\left(\omega t - \frac{x\sqrt{\epsilon\mu}}{c} - \phi_z\right) = E_{z_0} \cos(\Phi - \phi_z)$$

where $\Phi = \omega t - (x\sqrt{\epsilon\mu}/c)$ has been introduced for simplification. These equations are written

$$E_y = E_{y_0}(\cos\Phi\cos\phi_y + \sin\Phi\sin\phi_y)$$

and

$$E_z = E_{z_0}(\cos\Phi\cos\phi_z + \sin\Phi\sin\phi_z)$$

We wish to eliminate Φ between these expressions in order to find an equation for the locus of the vector magnitude of the sum of the component vectors. This can be done by solving for $\cos\Phi$ and $\sin\Phi$

$$\cos\Phi = \frac{E_y/E_{y_0}\sin\phi_z - E_z/E_{z_0}\sin\phi_y}{\sin(\phi_z - \phi_y)}$$

and
$$\sin \Phi = \frac{E_z/E_{z0} \cos \phi_y - E_y/E_{y0} \cos \phi_z}{\sin(\phi_z - \phi_y)} \qquad (4.29)$$

Equations (4.29) are now squared and added

$$\sin^2(\phi_z - \phi_y) = \left(\frac{E_y}{E_{y0}}\right)^2 + \left(\frac{E_z}{E_{z0}}\right)^2 - \frac{2E_y E_z}{E_{y0} E_{z0}} \cos(\phi_y - \phi_z) \qquad (4.30)$$

Equation (4.30) is the equation of an ellipse, and it tells us that in general the superposition of two waves with perpendicular E-vector directions produces *elliptically polarized* radiation. This is illustrated in Fig. 4.2. As the beam is propagated through space, the amplitude of the

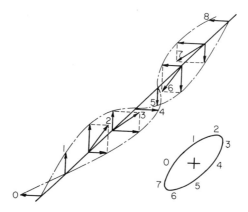

Fig. 4.2   Elliptically polarized radiation.

E vector traces out the ellipse given by Eq. (4.30). Notice that the axes of the ellipse do not necessarily lie in the y and z directions. In fact, the major axis of the ellipse makes an angle ψ with the direction of E_z with ψ determined from

$$\tan 2\psi = \cos(\phi_y - \phi_z) \tan\left(2 \tan^{-1}\frac{E_{z0}}{E_{y0}}\right)$$

and the ratio of the minor axis to the major axis is

$$\frac{\text{minor axis}}{\text{major axis}} = \tan \xi$$

where

$$\sin 2\xi = \sin(\phi_y - \phi_z) \sin\left(2 \tan^{-1}\frac{E_{z0}}{E_{y0}}\right)$$

Returning to Eq. (4.30), we see that if the phase difference between the E vectors is some multiple of π, that is,

$$\phi_z - \phi_y = m\pi \quad m = 0, \pm 1, \pm 2, \ldots$$

then

$$\frac{E_y}{E_{y_0}} = \frac{E_z}{E_{z_0}}$$

and the resultant is *plane-polarized* radiation. If the phase difference is some odd multiple of $\pi/2$ and $E_{y_0} = E_{z_0} = E_0$ then

$$E_y^2 + E_z^2 = E_0^2$$

and the resultant is *circularly polarized* radiation.

Now let us turn to the propagation of polarized radiation into and through some medium. We have seen that the propagation of radiation is controlled by the refractive index and the attenuation index of the medium. Some materials are nonisotropic and have a different attenuation index for E vectors in different directions. These materials are called *dichroic*. They partially polarize natural, unpolarized radiation by preferentially absorbing the component having its E vector perpendicular to the axis of polarization (see Fig. 4.3). In the visible

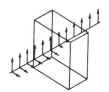

Fig. 4.3   Polarization by preferential absorption in a dichroic medium.

region of the spectrum such materials as Polaroid have this property over a wide spectral range and can be used as polarizers. In the infrared, the effect is usually confined to narrow portions of the spectrum, and the axis of polarization can lie in quite different directions for different wavelengths. Indeed, this property is often studied in the infrared, as we shall discuss later, in order to determine the direction of vibration of the atomic groups responsible for infrared absorption.

There are also anisotropic materials which refract radiation differently for E vectors lying in different planes. These materials are called *doubly refracting* or *birefringent* and can be uniaxial or biaxial. Uniaxial crystals are trigonal, tetragonal, or hexagonal systems. Biaxial crystals are orthorhombic, monoclinic, or triclinic. Cubic crystals and amorphous solids are optically isotropic. Uniaxial crystals have a unique direction of propagation, the direction of the optic axis, for which no birefringence is observed. All rays traveling along the optic axis have the same velocity of propagation and, therefore, the same refractive index. A ray along some other direction is split into two parts. The *ordinary ray* is refracted according to Snell's law and propagated as though the crystal has a refractive index n_0 that is the same regardless of the direction of propagation. This is the ordinary refractive index. The *extraordinary ray* is propagated with a velocity given by the extraordinary refractive index n_e that varies with direction in the crystal. Only in the direction of the optic axis does $n_0 = n_e$. The refraction of the extraordinary ray is not in accordance with Snell's law because the Huygens wavelets are ellipsoids rather than spheres. The value of n_e has a maximum or minimum value for radiation incident normal to the optic axis. This extreme value of n_e and the value of n_0 are the principal indices of refraction. In a positive uniaxial crystal the extreme value of n_e is greater than n_0; in a negative crystal it is less. The ordinary ray propagated normal to the optic axis is polarized with its E vector perpendicular to the optic axis. The extraordinary ray is polarized in the opposite sense.

Biaxial crystals have two optic axes. The optical phenomena observed in such crystals are necessarily more complicated than in uniaxial crystals.

The reflection of radiation at a surface is controlled by the attenuation index and the refractive index. We recall that the *plane of incidence* is defined as the plane that contains the incident ray and the

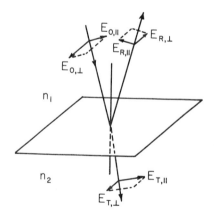

Fig. 4.4   Reflection and transmission at a plane surface.

normal to the reflecting surface. The incident light beam is considered to be resolved into two components having the E vector perpendicular and parallel to the plane of incidence (see Fig. 4.4). These components, $E_{o,\parallel}$ and $E_{o,\perp}$, are treated separately. The problem is similar to the boundary-value problem solved for the case of normal incidence, and once more we invoke the condition of continuity in the tangential component of the electric vector. We are considering nonabsorbing media for the present. Now, however, $E_{o,\parallel}$ is not parallel to the plane of the reflecting surface. The solution of this problem is expressed in the *Fresnel equations* for the amplitudes of the reflected and transmitted electric vectors, $E_{R,\parallel}, E_{R,\perp}$ and $E_{T,\parallel}, E_{T,\perp}$:

$$\frac{E_{R,\parallel}}{E_{o,\parallel}} = \frac{\tan(\theta_i - \theta_t)}{\tan(\theta_i + \theta_t)} \qquad \frac{E_{R,\perp}}{E_{o,\perp}} = -\frac{\sin(\theta_i - \theta_t)}{\sin(\theta_i + \theta_t)}$$

$$\frac{E_{T,\parallel}}{E_{o,\parallel}} = \frac{2\sin\theta_t\cos\theta_i}{\sin(\theta_i + \theta_t)\cos(\theta_i - \theta_t)}$$

$$\frac{E_{T,\perp}}{E_{o,\perp}} = \frac{2\sin\theta_t\cos\theta_i}{\sin(\theta_i + \theta_t)}$$

The angle of refraction is related to the angle of incidence by Snell's law. The reflectivities and transmissivities at the surface are found from these amplitudes:

$$R_{\parallel} = \frac{\tan^2(\theta_i - \theta_t)}{\tan^2(\theta_i + \theta_t)},$$

$$R_{\perp} = \frac{\sin^2(\theta_i - \theta_t)}{\sin^2(\theta_i + \theta_t)}$$

$$T_{\parallel} = \frac{\sin 2\theta_i \sin 2\theta_t}{\sin^2(\theta_i + \theta_t)\cos^2(\theta_i - \theta_t)}$$

$$T_{\perp} = \frac{\sin 2\theta_i \sin 2\theta_t}{\sin^2(\theta_i + \theta_t)}$$

The reflectivities are plotted for a typical choice of n in Fig. 4.5. We can make several pertinent observations about these equations:

(1) In the limit of vanishing θ_i and θ_t (normal incidence), these equations reduce to

$$R_{\parallel} = R_{\perp} = \left(\frac{n_2 - n_1}{n_2 + n_1}\right)^2 \qquad T_{\parallel} = T_{\perp} = \frac{4n_1 n_2}{(n_2 + n_1)^2}$$

in agreement with Eq. (4.23) and the requirement that $R + T = 1$.

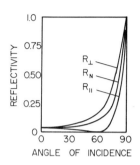

Fig. 4.5 Reflectivity of a dielectric for radiation polarized parallel and perpendicular to the plane of incidence.

(2) When $\theta_i + \theta_t = \frac{1}{2}\pi$, $R_{\parallel} = 0$ and only polarized light containing $E_{\perp}$ is reflected. The angle of incidence at which this occurs is the *Brewster's angle*. The condition on θ_i and θ_t can be stated in another form by noting $\cos\theta_i = \sin\theta_t = (n_1/n_2)\sin\theta_i$, and therefore

$$\tan\theta_i = n_2/n_1$$

which is the usual form of *Brewster's law*.

(3) At the *critical angle* when $\theta_t = \frac{1}{2}\pi$, both $T_{\parallel}$ and $T_{\perp}$ vanish and $R_{\parallel} = R_{\perp} = 1$, verifying our assertion in the previous chapter that at the critical angle all of the light appears in the internally reflected beam.

The Fresnel equations also apply to absorbing media, such as metals, provided the complex index of refraction is used.

V. INTERFERENCE

We have already introduced the subject of interference, in Section III, with the discussion of the superposition of two coherent light waves. As we have pointed out, the amplitude of the combined wave varies from the sum to the difference of the amplitudes of the super-posed waves, depending on their relative phases. This is what we mean by *constructive* and *destructive interference*. The requirement of coherency imposes certain restrictions on experimental procedures in interference studies. The two superposed waves must originate at a common source, and indeed from the same region of a common source, in order that a strong correlation will exist in their fluctuations, both spatial and temporal. Moreover, the source must have a very narrow wavelength distribution if interference is to be observable for sizable phase differences.

An instrument constructed for the purpose of observing interference phenomena is an *interferometer*. There are three general classes: (1) wave-front-dividing interferometers, in which radiation from a single source is sampled at two regions of space and recombined; (2) two-beam amplitude-dividing interferometers, in which a wave is split into two parts at a common region of space and recombined after a phase difference is introduced; and (3) multiple-beam amplitude-dividing interferometers in which the amplitude of the wave is split many times. We can best illustrate these with some examples.

The simplest example of a wave-front-dividing interference experiment is one which is also of great historical importance — the observation of Young's interference fringes. Two small holes (the hypothetical pinholes so often invoked by opticists) in an opaque screen and separated by a small distance s are illuminated by a plane wave from a common monochromatic source (see Fig. 4.6). Each hole acts as though it is a source sending out its own spherical waves, according to the principle of Huygens and as a consequence of diffraction, which is discussed in the next section. A screen is placed a distance S from the

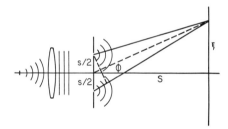

Fig. 4.6 Young's interference fringes.

pinholes. Because the waves issuing from the pinholes have a common origin, their fluctuations are highly correlated. That is to say, the pinholes act as coherent sources, and the electromagnetic disturbance at any point on the screen is the superposition of the electric and magnetic vectors from the two sources. Consider a point on the screen a distance ξ from the perpendicular bisector of the line connecting the pinholes and define an angle ϕ as shown in the figure. A ray from the lower hole has traveled approximately

$$\Delta = \sin \phi \approx s\xi /S$$

farther than a ray from the upper hole in reaching ξ. We have limited the discussion to small angles in order to take $\sin \phi \approx \tan \phi \approx \xi/S$.

This also permits us to ignore the difference in attenuation of the two beams by the inverse square of the distance to the screen. The phase difference 2δ between the two rays can be written in terms of the wavelength

$$2\delta = \frac{2\pi\Delta}{\lambda} = \frac{2\pi s\xi}{S\lambda}$$

From Eq. (4.9), we have the amplitude of the electric vector of the superposed disturbances

$$E = 2E_0 \cos\delta = 2E_0 \cos\frac{\pi s\xi}{S\lambda}$$

The power in the combined waves is, according to Eq. (4.10),

$$S_c = \frac{c}{4\pi}\sqrt{\epsilon} \cdot 4E_0^2 \cos^2\frac{\pi s\xi}{S\lambda}$$
$$= 2P \cos^2\frac{\pi s\xi}{S\lambda}$$

Thus, there are bright regions where $S_c = 2P$ on the screen wherever

$$\frac{\pi s\xi}{S\lambda} = \pm k\pi$$

or

$$\pm\xi = \frac{kS\lambda}{s} \qquad k = 0, 1, 2, \ldots$$

and there are dark regions wherever

$$\frac{\pi s\xi}{S\lambda} = \pm\tfrac{1}{2}(2k+1)\pi$$

or

$$\pm\xi = \frac{(2k+1)S\lambda}{s} \qquad k = 0, 1, 2, \ldots$$

At intermediate points the irradiance varies as the cosine squared. These alternating bright and dark regions are *interference fringes*. They were observed by Young in the very early 1800's.

If the two pinholes are really distinct sources, their fluctuations are uncorrelated, and the radiant power in the combined waves is found from Eq. (4.11) instead of (4.10)

$$S_i = \frac{c}{4\pi}\sqrt{\xi} \cdot 2E_0^2 = P$$

In this case, no fringes are observed on the screen. It is important to realize that the average irradiance over a range of ξ is P in both the coherent and incoherent cases. Interference neither destroys nor creates radiation, but merely rearranges its distribution.

Radiation is not necessarily either completely coherent or completely incoherent. If in Young's experiment we have only partial coherence in the light from the two pinholes, we find fringes occurring in the positions predicted by the equation for the coherent case but with reduced contrast between the bright and dark regions. Partial coherence is obtained, for example, by increasing the distance between the pinholes so that they are illuminated by different regions of the source. The contrast, or visibility, of interference fringes is related to the correlation of fluctuations in the superposed waves in a well-defined mathematical way, which we shall not pursue further.

The second general class of interferometer, the two-beam amplitude-dividing kind, is exemplified by the Michelson interferometer. One form is shown in Fig. 4.7. A collimated beam is divided into two more-or-less equal parts by a partially reflecting, partially transmitting mirror

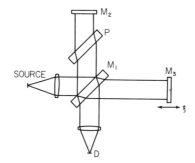

Fig. 4.7 Michelson interferometer.

M_1. One beam is reflected from a fixed mirror M_2, returned on itself through the partially transmitting mirror, and transmitted to a detector D. The other beam is transmitted by M_1, reflected by the movable mirror M_3, and reflected by M_1. With proper alignment, the two beams are brought to a common focus at D. The element P is a compensator plate of thickness equal to the thickness of M_1 and made of the same material. Thus we see that each beam has been reflected once from a partially reflecting mirror and once from a totally reflecting mirror and that each has passed through the same thickness of optical media. When the mirror M_3 is placed so that both beams traverse the same path length, the beams arrive at D in phase and they constructively interfere. If M_3 is displaced a distance ξ, an additional path length 2ξ is introduced into the path of the second beam and it arrives at D with a phase difference 2δ relative to the first beam, where

$$2\delta = 4\pi\xi/\lambda$$

The radiant flux observed at D is thus given by

$$S(\xi) = P \cos^2 \frac{2\pi\xi}{\lambda} \tag{4.31}$$

which varies between zero, when

$$\xi = \tfrac{1}{4}(2n+1)\lambda \qquad \pm n = 0, 1, 2, \ldots$$

and P, when

$$\xi = \tfrac{1}{2}n\lambda \qquad \pm n = 0, 1, 2, \ldots$$

A complete cycle of bright to dark to bright occurs with a movement of the mirror amounting to $\tfrac{1}{2}\lambda$.

Up to now, we have assumed that the source is monochromatic. But a truly monochromatic source is a mathematical fiction. Even a beam emitted by a laser covers a finite frequency range, although it may be extremely small. Indeed, we may be interested in studying a source which has a very broad spectral distribution. We must modify Eq. (4.31) to include the case of a nonmonochromatic source. This is easily done by considering P to be a spectral density function $P(\nu)$ and integrating over all spectral frequencies to find the total power at D. Notice that we have begun to speak of spectral frequency rather than wavelength. This is for mathematical convenience. The integrated power is

$$S(\xi) = \int_0^\infty P(\nu) \cos^2 2\pi\xi\nu \, d\nu$$

This equation can be reduced by means of a trigonometric identity to

$$S(\xi) = \tfrac{1}{2} \int_0^\infty P(\nu)(\cos 4\pi\xi\nu + 1) \, d\nu$$

$$= \tfrac{1}{2}P_0 + \tfrac{1}{2} \int_0^\infty P(\nu) \cos 4\pi\xi\nu \, d\nu \tag{4.32}$$

A plot of $S(\xi)$, which is the radiant power observed at D as a function of the difference in path length of the two beams, is called an *interferogram* and Eq. (4.32) shows how it is related to the spectral density function $P(\nu)$. The term P_0 is the total power in the incident beam, integrated over all wavenumbers. When the path lengths are equal, $\xi = 0$, the integral of Eq. (4.32) is also equal to the total power in the incident beam, and we have the not unexpected result $S(0) = P_0$.

The integral in Eq. (4.32) is one form of the *Fourier integral*.

Specifically, it is a Fourier cosine integral. We encounter this integral, or variations of it, repeatedly in optics and in electronics. For the present we shall make use of two of its properties, the sifting property and the Fourier transform reciprocal property. If we have a distribution that is exceedingly narrow, approaching zero in width, but is sufficiently large in magnitude to make up for its narrow width and give it unit area, and we write this distribution $\delta(\nu-\nu_0)$ to signify that it has the value zero everywhere except at ν_0, then the sifting property states that

$$\int_0^\infty \delta(\nu-\nu_0) \cos 4\pi\xi\nu \, d\nu = \cos 4\pi\xi\nu_0 \qquad (4.33)$$

The distribution $\delta(\nu-\nu_0)$ is usually called a Dirac delta function. The sifting property is not restricted to Fourier integrals, but is more properly described as a property of the delta function. In general,

$$\int_0^\infty \delta(\nu-\nu_0)f(\nu) \, d\nu = f(\nu_0)$$

where $f(\nu)$ is any suitably defined function.

Equation (4.33) states that the interferogram of a single, very narrow emission line is (apart from a constant term) a simple cosine whose periodicity is $\frac{1}{2}\nu_0$, in agreement with the discussion above which leads to a periodicity $\frac{1}{2}\lambda$. This relationship between periodicity of the interferogram and the wavenumber of a narrow emission line permits us to obtain the wavenumber directly from the interferogram. If the spectrum consists of a number of narrow lines

$$P(\nu) = \sum_{n=1}^{N} A_n \delta(\nu-\nu_n)$$

the interferogram consists of a sum of superposed cosine functions

$$S(\xi) = \tfrac{1}{2}P_0 - \tfrac{1}{2}\sum_{n=1}^{N} A_n \cos 4\pi\xi\nu_n$$

and the analysis is somewhat more complicated.

If the spectral distribution does not consist of narrow lines but has some general form, the spectrum is obtained from the interferogram by using the reciprocal property of Fourier transforms, namely:

If $$F(\xi) = \int_{-\infty}^{\infty} f(\nu) \cos 4\pi\xi\nu \, d\nu$$

is the Fourier transform of $f(\nu)$, then

$$f(\nu) = \tfrac{1}{2}\pi \int_{-\infty}^{\infty} F(\xi) \cos 4\pi\xi\nu \, d\xi$$

is the inverse Fourier transform of $F(\xi)$.

The spectrum can thus be calculated from the interferogram by means of the inverse Fourier transform. This method of spectroscopy is discussed further in a later chapter.

The third class of interferometer, the multiple-beam amplitude-dividing interferometer, is typified by the Fabry–Perot interferometer shown in Fig. 4.8A.

The Fabry–Perot interferometer consists of two plane-parallel plates mounted with their surfaces parallel and separated by a spacer of width b, which might be fixed or variable. We are interested only in the effect of the inner surfaces. The radiation is partly reflected and partly transmitted by each of these surfaces. In this way, each ray is decomposed into a number of components that have traversed the space between the plates different numbers of times. These components are superposed and they can constructively or destructively interfere. Each successive traversal of the space, assumed to contain vacuum, adds a path length of $2b \cos \theta$ to the ray, where θ is the angle of incidence. If the space contains a dielectric of refractive index n, the increment of optical path length is $2bn \cos \theta$. In terms of the wavelength of the radiation, this corresponds to a phase difference of

$$\phi = 4\pi bn \cos \theta / \lambda \text{ rad} \tag{4.34}$$

between successive traversals. We can discuss three cases:

(a) The incident wave front is plane parallel, the wavelength and angle of incidence are fixed, and the spacing is varied. A movement of

$$\Delta b = \lambda / 2n \cos \theta$$

of one plate is required to go through one cycle of interference or one fringe, that is, light to dark to light.

(b) The incident wave is plane parallel and the spacing and angle of incidence are fixed, while the wavelength is varied. A change in wavelength of

$$\Delta \lambda = 2bn \cos \theta$$

is required for one cycle of interference.

(c) The arrangement shown in Fig. 4.8b is used with fixed wavelength and spacing. Radiation from an extended source is collimated, passed through the interferometer, and focused on a screen.

The wave front from each point of the source is planar as it passes through the interferometer, but its angle of incidence depends on how far off axis it is. The pattern on the screen will therefore consist of concentric rings or fringes of alternating constructive and destructive interferences. From Eq. (4.34) is found the relationship between the angles subtended by successive rings on the screen

$$\cos\theta_k - \cos\theta_{k+1} = \lambda/2bn$$

The sharpness of Fabry–Perot fringes depends on the reflectivity of the surfaces of the plates. Uncoated plates of materials such as glass or sodium chloride produce fringes having a cosine-squared contour. As the reflectivity is increased, by coating with a very thin layer of metal for example, the fringes become sharper, but the transmittance of the interferometer decreases. Multilayer dielectric coatings do not impose a transmission loss. In the ideal but impossible limit of unity reflectance and also unity transmittance, the fringes become delta functions.

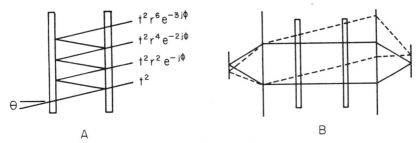

Fig. 4.8   Fabry–Perot interferometer.

This can be discussed on a more quantitative basis. With reference to Fig. 4.8A, consider the amplitude of the electric vector associated with successive transmitted rays of an incident monochromatic wave. The first ray has an amplitude t^2, where t is the amplitude transmission of each plate. We assume that the plate itself is transparent, but that the coating has a transmittance and reflectance. The second ray has amplitude $r^2 t^2 e^{-j\phi}$, where r is the reflectance of the coating and ϕ represents the phase relative to the first ray. The phase is partly due to the difference in optical path length of the two rays, and partly due to any phase change that occurs with reflection. If we continue in this way, we arrive at an equation for the total transmitted amplitude

$$E = E_0 t^2 (1 + r^2 e^{-j\phi} + r^4 e^{-2j\phi} + \cdots)$$

$$= \frac{E_0 t^2}{1 - r^2 e^{-j\phi}}$$

The transmitted *intensity* is

$$I = EE^* = \frac{I_0 t^4}{1 - r^2(e^{-j\phi} + e^{j\phi}) + r^4}$$

This equation can be written in terms of the intensity transmittance $T = t^2$ and reflectance $R = r^2$

$$I = \frac{I_0 T^2}{1 - 2R\sin(\phi/2) + R^2} = \frac{I_0 T^2}{(1-R)^2}\left[1 + \frac{4R}{(1-R^2)}\sin^2\frac{\phi}{2}\right]^{-1}$$

This is Airy's famous equation (see Fig. 4.9). The intensity has a maximum value of

$$I_{max} = \frac{I_0 T^2}{(1-R)^2} \qquad \text{when } \tfrac{1}{2}\phi = 0, \pi, 2\pi, \ldots$$

and a minimum value

$$I_{min} = \frac{I_0 T^2}{(1+R)^2} \qquad \text{when } \tfrac{1}{2}\phi = \tfrac{1}{2}\pi, \tfrac{3}{2}\pi, \ldots$$

The width of the fringes $\delta\phi$ at half-maximum intensity is found by setting $I = \tfrac{1}{2}I_{max}$ in Airy's equation and solving for ϕ. The result is

$$\sin\tfrac{1}{2}\phi = \pm\frac{1-R}{2\sqrt{R}}$$

and when the reflectivity is large,

$$\phi \approx \pm\frac{1-R}{\sqrt{R}} \quad \text{and } \delta\phi \approx 2 \cdot \frac{1-R}{\sqrt{R}}$$

The separation between fringes corresponds to $\Delta\phi = 2\pi$ and the ratio of $\Delta\phi$ to $\delta\phi$ is called the *finesse* $\mathscr{F}$ of the Fabry–Perot interferometer,

$$\mathscr{F} = \frac{\Delta\phi}{\delta\phi} = \frac{\pi\sqrt{R}}{1-R}$$

The resolving power of a Fabry–Perot interferometer is the product of

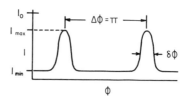

Fig. 4.9   Interference fringes produced by a Fabry–Perot interferometer.

the order of interference, $K = 2nb/\lambda$, and the finesse. We return to this point in the discussion of Fabry–Perot spectrometers in Chapter 8.

If the coating on the interferometer does not absorb, then $T = 1 - R$ and the maximum intensity is

$$\lim_{T \to 1-R} I_{max} = I_0$$

The minimum transmitted intensity is

$$\lim_{T \to 1-R} I_{min} = I_0 \left(\frac{1-R}{1+R}\right)^2$$

Thus, if the reflectivity of coatings is 90%, giving a finesse of about 30, the transmission midway between peaks is only about 0.25%.

VI. Diffraction

Geometric optics is based on the premise that light rays travel in straight lines in homogeneous media and change their direction only upon reflection or refraction at boundaries between different media. This, as we know, is merely a useful approximation to the facts of optical life. Radiation is propagated to some extent around corners, so a shadow cast by an obstruction is never an infinitely sharp boundary between light and dark. This phenomenon is termed *diffraction*. We are concerned here only with the effects of diffraction by the aperture on the resolution of a spectrometer.

Huygen's principle states that each point on the surface of a wave front of an electromagnetic disturbance acts as a source of spherical wavelets. The wavelets from each point add, with proper attention to

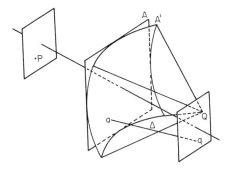

Fig. 4.10   Diffraction at an aperture.

phase, to produce the propagated wave front. The mathematical implementation of this principle is the Kirchhoff–Huygens theory.

Let us consider a wave front propagated through an aperture at A (see Fig. 4.10), which we suppose to be the exit pupil of an optical system. The wave front originates in a point source at P and if geometric optics were strictly valid, it would be focused to a geometric point image at Q. In Section 3.X we investigated how aberrations deformed the spherical wave front converging on Q and caused the light to be spread out somewhat in the Gaussian image plane. Let us now assume that the optical system is perfectly free of aberrations, at least with regard to objects on the plane G and images on the plane G'. We say that for these conjugate planes the system is *aplanatic*. We are interested in determining the diffraction pattern, that is, the distribution of radiation on the plane G' due to diffraction.

Refer back to Eqs. (4.4). The phase of the disturbance arriving at a point q on the G' plane from a point a in the wave front A' is $2\pi\Delta/\lambda$ where Δ is the length $\bar{a}\bar{q}$, λ is, of course, the wavelength, and we have for convenience set the phase of the wave front equal to zero. The amplitude of the disturbance at q due to a small area ds of wave front at a is then given by

$$dE = E_0(a) \exp(-j\omega t + 2\pi j\Delta/\lambda)\, ds$$

We are interested in the time average of the disturbance at q due to the entire wave front, and this is found by integrating over the wave front and dropping the time dependence,

$$\overline{E(q)} = \iint_A \overline{E_0(a)} \exp(2\pi j\Delta/\lambda)\, ds \qquad (4.35)$$

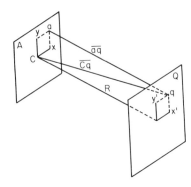

Fig. 4.11 Definition of distances and coordinates in the diffraction integral.

The path length Δ can be written by referring to Fig. 4.11, which serves to define coordinates x, y and x', y' of a and q, respectively, and the distance R between the surfaces A and G'. We simplify the analysis by assuming that the curvature of the wave front is slight, so we can substitute distances measured to the plane of the pupil A for distances to the wave front A'. We are free to define a reference phase on the plane G' and it is convenient to choose zero phase corresponding to path $\bar{C}\bar{q}$. Hence, we need only determine the path difference $\bar{a}\bar{q} - \bar{C}\bar{q}$ for every point a on the plane A and redefine Δ as $\Delta' = (\bar{C}\bar{q} - \bar{a}\bar{q})$. The distances x' and y' are small, so we can expand $\bar{C}\bar{q}$ and $\bar{a}\bar{q}$ in powers of x' and y' and retain only the lowest order terms. Moreover, we assume that the distance between the planes, R, is large relative to the distance of point a from the optic axis. Thus,

$$\bar{a}\bar{q} = \left[(x-x')^2 + (y-y')^2 + R^2 \right]^{1/2}$$

$$\approx R + \frac{xx' + yy'}{R}$$

and

$$\bar{C}\bar{q} = (x'^2 + y'^2 + R^2)^{1/2}$$

$$\approx R$$

So

$$\Delta' = \frac{xx' + yy'}{R}$$

Equation (4.35) can now be written

$$\overline{E(x', y')} = \int\int_{A'} \overline{E_0(x, y)} \exp\left[-2\pi j(xx' + yy')/R\lambda \right] dx \, dy$$

Finally, we can introduce a redefinition of the variables in the exit pupil with

$$\beta = 2\pi x/R\lambda \quad \text{and} \quad \gamma = 2\pi y/R\lambda$$

This gives us for the amplitude of the electric vector in the Gaussian image plane

$$\overline{E(x', y')} = \frac{R^2\lambda^2}{4\pi^2} \int\int_{A'} \overline{E_0(\beta, \gamma)} \exp\left[-j(\beta x' + \gamma y') \right] d\beta \, d\gamma \quad (4.36)$$

Relationships of the form of Eq. (4.36) occur often in theoretical physics and engineering. The integral is a Fourier transform, and the equation states that the amplitude distribution in the image plane is the Fourier transform of the amplitude distribution in the plane of the exit pupil. We make frequent use of the Fourier transform in later discussions. Its properties are summarized in the Appendix.

The irradiance is proportional to $(\bar{E})^2$, according to Eq. (4.8). As we have discussed above, if an extended object is being imaged, the manner in which radiation is combined in the Gaussian image plane depends on whether the object is emitting or reflecting coherently or incoherently. In most infrared spectroscopic work, a self-luminous incoherent source is used, so the radiant power from each point of the object is added where images overlap on the image plane. On the other hand, if the object is coherent, the electric vectors are added with proper consideration of their phases, and the sum of electric vectors is squared to obtain the radiant power.

Let us make use of Eq. (4.36) to calculate the diffraction pattern of a point object imaged by an optical system having an aperture of width A in the x' direction and of width B in the y' direction. The pupil is supposed to be uniformly illuminated, so

$$E_0(\beta, \gamma) = 1 \quad \text{for} \quad -\frac{\pi A}{R\lambda} < \beta < \frac{\pi A}{R\lambda} \quad \text{and} \quad -\frac{\pi B}{R\lambda} < \gamma < \frac{\pi B}{R\lambda}$$

$$= 0 \quad \text{for} \quad |\beta| > \left|\frac{\pi A}{R\lambda}\right| \quad \text{or} \quad |\gamma| > \left|\frac{\pi B}{R\lambda}\right|$$

We can write

$$\overline{E(x', y')} = \frac{R^2\lambda^2}{4\pi^2} \int_{-\pi A/R\lambda}^{\pi A/R\lambda} \exp(-j\beta x') \, d\beta \cdot \int_{-\pi B/R\lambda}^{\pi B/R\lambda} \exp(-j\gamma y') \, d\gamma$$

This expression can be evaluated by straightforward methods to obtain

$$\overline{E(x', y')} = AB\left[\frac{\exp(j\pi Ax'/R\lambda) - \exp(j\pi Ay'/R\lambda)}{2\pi jx' A/R\lambda}\right] \cdot$$

$$\cdot \left[\frac{\exp(j\pi By'/R\lambda) - \exp(j\pi By'/R\lambda)}{2\pi jy' B/R\lambda}\right]$$

$$= AB\left(\frac{\sin \pi Ax'/R\lambda}{\pi Ax'/R\lambda}\right)\left(\frac{\sin \pi By'/R\lambda}{\pi By'/R\lambda}\right) \tag{4.37}$$

The distribution of radiant energy in the image plane is obtained from the square of Eq. (4.37). We will drop the multiplicative constants and normalize the irradiance to unity at the Gaussian image point,

$$s(x', y') = \frac{\sin^2 \pi Ax'/R\lambda}{(\pi Ax'/R\lambda)^2} \cdot \frac{\sin^2 \pi By'/R\lambda}{(\pi By'/R\lambda)^2} \tag{4.38}$$

If the aperture is a circle instead of a rectangle, the integration yields the rotationally symmetric pattern

$$\overline{E(r')} = \frac{2J_1(2\pi ar'/\lambda R)}{2\pi ar'/\lambda R}$$

where J_1 is a Bessel function of first order, a the radius of the aperture, and r' the radius about the Gaussian image point. Also, the irradiance in the Gaussian image plane, normalized to unity at the maximum, is

$$s(r') = 4\left[\frac{J_1(2\pi ar'/\lambda R)}{2\pi ar'/\lambda R}\right]^2 \tag{4.39}$$

VII. SPREAD FUNCTIONS AND MODULATION TRANSFER FUNCTIONS

Equations of the type (4.38) and (4.39) describe the irradiance s in the image plane due to a point object. Electrical engineers study an electromechanical system by injecting an impulse and observing the variation of the output with time. The output is called the impulse response of the system. Similarly, opticists have begun to refer to s as the *impulse response* or the *point spread function*. We can prescribe a method for calculating the point spread function from the discussion of Section VI. An optical system is irradiated with radiation from a point source. The point spread function is found from the square of the two-dimensional Fourier transform of the electric field distribution in the exit pupil of the system. The operation of squaring, in the case of functions of a complex variable, implies the multiplication of a function by its complex conjugate.

We have been dealing only with the case of diffraction. The point spread function has a broader meaning in that it describes the image of a point source as affected by both diffraction and aberrations. In the foregoing discussion we introduced an approximation when we integrated not over the spherical wave front but over the plane of the exit pupil. This is by no means a necessary restriction and can be avoided at the expense of slightly more complicated mathematics by either integrating over the wave front or by introducing an additional path, the distance from the wave front to the pupil, and then integrating over the pupil. We have seen in the previous chapter that the aberrations of an optical system can be studied by considering the distortion of the ideal spherical wave front converging on an image point. Hence, at the

same time we are performing the integration to find the spread function
due to diffraction, we can generalize by including the aberrations and
find the general point spread function.

In defining the spread function, we have mentioned its electrical
engineering analog, the impulse response. This is more than just a
superficial analogy, and we shall see as we proceed a fundamental
similarity in the concepts and mathematical apparatus of electrical
engineering and optics.

If the object is not an ideal point source but rather an extended
incoherent object, the intensity distribution function in the image
plane can be calculated by superposing the point spread functions of
all of the points making up the object.

This is done by writing the superposition or convolution integral

$$i(x', y') = \iint o(x, y) s(x - x', y - y') \, dx \, dy \qquad (4.40)$$

The meaning of Eq. (4.40) is this: The ideal Gaussian image of an
object would be $o(x, y)$ if there were no diffraction and no aberrations.
The spread function is written about each point x, y in the image plane.
For example, the spread function for a square aperture [Eq. (4.38)] and
a Gaussian point image at x_0, y_0 is

$$s(x_0 - x', y_0 - y') = \frac{\sin^2 \pi A (x_0 - x')/R\lambda}{[\pi A (x_0 - x')/R\lambda]^2} \cdot \frac{\sin^2 \pi B (y_0 - y')/R\lambda}{[\pi B (y_0 - y')/R\lambda]^2} \qquad (4.41)$$

The quantities $(x - x')$ and $(y - y')$ are the distances of a point in the
image plane from the ideal point image at x, y. The irradiance at a point
x, y of the image plane is the summation of the diffraction- and aberra-
tion-smeared images of the points making up the object and is written
as the integral of Eq. (4.40). This same superposition integral is used by
electrical engineers to characterize the output of a *linear system* due
to any input signal. The superposition integral holds for any linear
system, be it of electrical, optical, acoustical, or any other discipline.

We can also think of the superposition equation (4.40) as an integral
over the intensity distribution of the object itself, rather than its ideal
image. Of course, the integral must be multiplied by the magnification
between the object and the image planes. Now the integral describes
the transformation of a radiance distribution $o(x, y)$ in the object
plane to an irradiance distribution $i(x', y')$ in the image plane.

If the object is an ideal point at coordinates x_0, y_0, we can express
its intensity as a delta function $\delta(x - x_0, y - y_0)$. Its image is, according
to Eq. (4.40),

$$i(x',y') = \int\int \delta(x-x_0, y-y_0)s(x-x', y-y')\,dx\,dy$$

$$= s(x_0-x', y_0-y')$$

which is the spread function itself, as we know it must be.

Spectroscopists are mostly concerned with slit-like objects rather than point objects. Accordingly, we define the *line spread function* as the function that determines the energy distribution along a direction normal to the image of a line object of infinite length (in order that we can ignore end effects). The intensity distribution for an infinite line object lying in the y direction and having an x coordinate x_0, is $\delta(x-x_0)$. The irradiance distribution in the image plane is

$$i(x', y') = \int\int \delta(x-x_0)s(x-x', y-y')\,dx\,dy = s(x_0-x', y-y')\,dy$$

If the point spread function is separable into x- and y-dependent factors, as in the case of the rectangular apertures [Eq. (4.41)], the factor containing x' is taken outside the integral, and the integration over the y coordinate is performed. We can lump this into a normalizing factor and write the line spread function for a rectangular aperture as

$$s_1(x_0-x') = \frac{\sin^2 \pi A(x_0-x')/R\lambda}{[\pi A(x_0-x')/R\lambda]^2} \tag{4.42}$$

The line spread function can also be found by squaring the one-dimensional Fourier transform of the electric field distribution in the exit pupil. Equation (4.42) is plotted in Fig. 4.12. If two line objects are

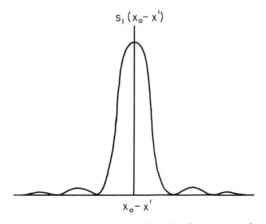

Fig. 4.12 Line spread function in the x direction for a rectangular aperture.

imaged side by side, their spread functions overlap, and if they are too close together, the two images cannot be resolved. Rayleigh's criterion for resolution states that two images are just barely resolved by an aberration-free system if the central maximum of the diffraction spread function of one falls on the first minimum of the spread function of the other. This is more than an arbitrary definition. (We shall see presently the more fundamental significance of the Rayleigh criterion.) Consequently, the smallest resolvable distance in the image plane is the distance from the central maximum to the first minimum of Eq. (4.42), or

$$\Delta x = R\lambda/A$$

If the object is at infinity, then R equals the focal length of the optical system and R/A is F, its f/number, so we have

$$\Delta x = F\lambda$$

Thus, an $f/10$ optical system viewing two parallel line objects with radiation of 10-μ wavelength can resolve their images if the objects are at least $10 \times 10 = 100\ \mu$ or 0·1 mm apart. Two such barely resolved overlapping images are plotted in Fig. 4.13A, and their sum is plotted in Fig. 4.13B. The equation for the curve in Fig. 4.13B is found by writing for the irradiance of the overlapping patterns

$$I = \frac{\sin^2 \pi A (x_0 - x)/R\lambda}{[\pi A (x_0 - x)/R\lambda]^2} + \frac{\sin^2 \pi A (-x_0 - x)/R\lambda}{[\pi A (-x_0 - x)/R\lambda]^2}$$

which becomes, with $x_0 = R\lambda/2A$,

$$I = \frac{2 \cos^2 \pi A x/R\lambda}{(\pi/2)^2 + (\pi A x/R\lambda)^2}$$

The value at the central minimum is found to be $8/\pi^2 \approx 81\%$ of the value at the maximum.

There are other resolution criteria. For example, Sparrow's criterion is that images are just resolved when the minimum at the center of the overlapping diffraction patterns just vanishes and only a central

A   B

Fig. 4.13   Resolved overlapping diffraction patterns.

maximum remains. In a practical sense, the question of resolution limit really depends on circumstances, such as how much noise is present and which aberrations degrade the optical system and how severely.

The superposition integral, Eq. (4.40), for a one-dimensional case is

$$i(x') = \int o(x)s(x-x')\, dx \tag{4.43}$$

Let us take the Fourier transform of Eq. (4.43), making use of the convolution theorem which states that the Fourier transform of a convolution is the product of the transforms of the convolved functions.

$$I(\omega) = O(\omega) \cdot S(\omega)$$

$I(\omega)$ and $O(\omega)$ are the Fourier transforms of $i(x')$ and $o(x)$. The term $S(\omega)$, the Fourier transform of the spread function, is called the *modulation transfer function*, the *optical transfer function*, or the *contrast transfer function*.

Now let us suppose that the optical system is called upon to image an object whose radiance varies sinusoidally in one direction, with a *spatial modulation frequency* ω (in radians per centimeter, for example). The irradiance distribution of the image, according to Eq. (4.43), is

$$i(x') = \int \sin \omega_0 x\, s(x-x')\, dx \tag{4.44}$$

Let us take the Fourier transform of Eq. (4.44),

$$I(\omega) = \mathscr{F}\{\sin \omega_0 x\} \cdot S(\omega) \tag{4.45}$$

The Fourier transform of $\sin \omega_0 x$ is a delta function

$$\mathscr{F}\{\sin \omega_0 x\} = \delta(\omega - \omega_0)$$

So Eq. (4.45) reduces to

$$I(\omega) = \delta(\omega - \omega_0)S(\omega) = S(\omega_0) \tag{4.46}$$

The inverse Fourier transform of Eq. (4.46) is

$$i(x') = S(\omega_0) \sin \omega_0 x'$$

From this relationship, the significance of the modulation transfer function becomes clear. A sinusoidal object of amplitude unity is reconstructed as a sinusoidal image, but with a loss in amplitude which depends on the spatial frequency of the object and the properties of the optical system. In fact, the property of the optical system which determines the amplitude of the image is the modulation transfer function. Now any object can be described by its Fourier transform in much the same way as a periodic function can be represented by a Fourier series. The Fourier transform describes the frequency content of the object. The transfer function then describes the efficiency with which the optical system transmits the various frequencies of the object. If many significant frequencies of the object are not transmitted efficiently, the image will be a poor representation of the object. The modulation transfer function is quite analogous to the system transfer function of electrical engineering, which describes the transfer of electrical signals of various frequencies.

The line spread function for an optical system having a rectangular aperture has been given by Eq. (4.42). The modulation transfer function is the Fourier transform of the spread function, which in this case is

$$S(\omega) = \begin{cases} \dfrac{(A/R\lambda) - |\omega|}{A/R\lambda} & 0 \le |\omega| \le A/R\lambda \\ 0 & |\omega| > A/R\lambda \end{cases} \qquad (4.47)$$

The function is an even function, but we are not concerned with negative frequencies. The transfer function is thus a triangle with its apex at $\omega = 0$ and a base width of $2A/R\lambda$ (Fig. 4.14). No frequencies greater than the limiting frequency

$$\omega_L = A/R\lambda \qquad (4.48)$$

are transmitted by the optical system. The period of a sinusoidal object of frequency ω_L is $R\lambda/A$, which is just the f/number of the system multiplied by the wavelength, and this is just the limiting resolution already given in conformity with Rayleigh's criterion. The modulation transfer function of an optical system with a rectangular aperture can have values no greater than those given by Eq. (4.48), but the values can be considerably lower if aberrations are present.

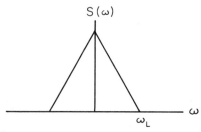

Fig. 4.14   Modulation transfer function of a rectangular aperture.

We have defined the modulation transfer function to be the Fourier transform of the spread function. The transfer function can be determined from the spread function by calculating the Fourier transform. There is an alternative calculation procedure, however, when we are dealing with an incoherently radiating line (or point) source and a symmetrically illuminated exit pupil. The spread function is the square of the distribution of the electric vector of the radiation field in the image plane. Moreover, the electric field distribution in the image plane is the Fourier transform of the electric field distribution in the exit pupil according to Eq. (4.36). The convolution theorem for Fourier transforms permits us to equate the product of the transforms of two functions with the transform of the convolution of the individual functions. The symmetry theorem tells us that the Fourier transform of a symmetric function has the same form as the inverse Fourier transform. Therefore, we can write

$$S(\omega) = \int E_0(\beta)E_0(\omega - \beta)\, d\beta \qquad (4.49)$$

which states that *the transfer function can be found by convolving the electric field distribution in the exit pupil with itself.* Equation (4.49) is written for the one-dimensional case, but the generalization is obvious. It is easy to see that the convolution of a uniformly illuminated rectangular aperture can be written by inspection to give us the triangle function of Eq. (4.47).

VIII. Emission

The radiation emitted by an object is called thermal radiation if it occurs as a consequence of thermal agitation of the molecules and atoms of the object. A blackbody is an ideal thermal emitter. No object can emit more profusely than a blackbody at the same temperature. The radiant power emitted by a thermal source, expressed as a fraction of the power emitted by a blackbody at the same temperature, is the emittance of the source. The spectral emittance is the emittance as a function of wavelength. The radiant power from a unit area of a blackbody in the wavelength interval between λ and $\lambda + d\lambda$ is given by the *Planck radiation law*

$$J(\lambda, T)\, d\lambda = \frac{c_1}{\lambda^5[\exp(c_2/\lambda T) - 1]}\, d\lambda$$

where $c_1 = 3.740 \times 10^{-12}$ W-cm^2, $c_2 = 1.438$ cm degrees, λ is the wavelength in centimeters, and T is the temperature in degrees Kelvin.

For sufficiently low values of λT the Planck equation is closely approximated by the *Wien equation,*

$$J(\lambda, T)\, d\lambda = \frac{c_1}{\lambda^5} \exp\left(-\frac{c_2}{\lambda T}\right) d\lambda$$

For large values of λT the *Rayleigh–Jeans equation* holds,

$$J(\lambda, T)\, d\lambda = \frac{c_1 T}{c_2 \lambda^4}\, d\lambda$$

The total power of all wavelengths radiated by a blackbody is given by the *Stefan–Boltzmann law*

$$J(T) = 5.669 \times 10^{-12}\, T^4$$

The maximum in the spectral distribution given by the Planck law is found at a wavelength λ_m expressed by the *Wien displacement law*

$$\lambda_m T = 0.2897$$

The absorptance of an object is the fraction of incident radiation that is absorbed. *Kirchhoff's law* states that the absorptance of an object is equal to its emittance at a given temperature.

REFERENCES

1. F. A. Jenkins and H. E. White, *Fundamentals of Physical Optics*, McGraw-Hill, New York, 1937.
2. J. Strong, *Concepts of Classical Optics*, Freeman, San Francisco, Calif., 1958.
3. A. Sommerfeld, *Optics*, Academic Press, New York, 1954.
4. M. Born and E. Wolf, *Principles of Optics*, Pergamon, New York, 1959.
5. E. L. O'Neill, *Introduction to Statistical Optics*, Addison-Wesley, Reading, Mass., 1963.
6. E. Hagen and H. Rubens, *Ann. Physik* **11**, 873 (1903); **14**, 936 (1904).

Optical Components
of Infrared Spectrophotometers

I. WINDOWS

Windows are used in spectrophotometers to contain samples in cells and to permit the isolation of compartments so they can be kept free of troublesome substances, notably water vapor, carbon dioxide, and dust.

A. Window Materials

A useful window material must be transparent over a desired spectral region. It must be available in a form consistent with the size required and at a reasonable cost. It must be capable of being shaped, ground, and polished to optical quality. Finally, it must be reasonably inert in the presence of the gases and liquids that will come in contact with it.

The most widely used window materials are crystalline alkali halides. These are produced by melting a batch of highly purified material in vacuum and permitting it to slowly crystallize from one end. The resulting crystals are large, free of flaws, and quite pure. The most widely used alkali halide window materials are sodium chloride and potassium bromide. Sodium chloride is transparent to wavelengths as long as 16 μ or so, in thicknesses of several millimeters. Potassium bromide is useful to 26 μ. When studies are to be carried out at longer wavelengths, cesium bromide can be used to 40 μ and cesium iodide to 50 μ. These four materials are hygroscopic and windows made from them must not be placed in contact with water. Even water vapor in the atmosphere will eventually fog their surfaces. Cells made of such materials should be stored in a dry atmosphere when not in use. A glass or plastic desiccator containing silica gel, preferably with a color indicator, is a convenient storage place. Some attempts have been made to waterproof alkali halide windows, for example, by coating them with films of selenium, but this is not an altogether satisfactory procedure.

There are alkali halides which are sufficiently impervious to attack by water to permit their use with water solutions. Of these, lithium fluoride transmits to 7 μ, calcium fluoride to 10 μ, and barium fluoride to 12 μ.

The mixed crystal thallium bromide-iodide, developed in Germany during World War II and called KRS-5, can be used at wavelengths as long as 40 μ and it is insoluble in water, but its high refractive index and attendant high reflectivity limit its use. Silver chloride is also water insoluble and can be used to 18 μ, but it too has a high refractive index. Moreover, it slowly darkens by photochemical action under room lights, particularly the fluorescent type, and it is highly corrosive to the nonnoble metals. The photochemical reaction can be avoided by coating the silver chloride with a layer of silver sulfide, stibnite, or plastic at the expense of transparency to visible and near-infrared radiations. Both silver chloride and thallium bromide-iodide are quite soft and easily scratched or deformed.

In the very far infrared, crystalline quartz windows are used. Quartz becomes opaque in the near infrared at about 3 μ, but begins to transmit again near 50 μ and remains reasonably transparent (about 80%) throughout the far-infrared region and into the microwave region at 1000 μ and beyond. Type II diamond transmits better than quartz in the far infrared, and begins to transmit at much shorter wavelengths. However, the high price and limited size usually restrict the use of diamond windows to detectors. Even so, applications of diamond windows to microcell construction have been made. In these cases samples are to be studied at high pressures and the ability of diamond to withstand such pressures makes its use desirable.

A series of window materials have been developed by Eastman Kodak. The Irtran materials are prepared by sintering powdered material under high pressure and temperature. Irtran 1 is made from magnesium fluoride, Irtran 2 from zinc sulfide, Irtran 3 from calcium fluoride, Irtran 4 from zinc selenide, Irtran 5 from magnesium oxide, and Irtran 6 from cadmium telluride. Irtran windows have very nearly the same optical properties in the infrared as single crystals of the same material. Their mechanical properties, particularly at elevated temperature, are superior. The Harshaw Chemical Company produces a material called T-12 that has properties similar to one of the Irtrans.

Somewhat less satisfactory windows can be prepared in the laboratory by pressing alkali halide powder in a die, in the same way as pellets are prepared. This procedure is discussed later in Chapter 15 on sample preparation. Such windows are inexpensive and can be discarded after use.

A variety of window materials are available for special purposes. These include sapphire, arsenic trisulfide glass, germanium, silicon, and many plastics. The properties of a number of window materials are given in Table 5.1. The transmittance spectra of some of them are shown in Fig. 5.1A and reflectance spectra are shown in Fig. 5.1B.

TABLE 5.1
Properties of Some Infrared-Transmitting Materials

Material	Practical wavelength limit[a] (μ)	Solubility in water (g/100 g)
Glass	2.5	Insoluble
Vycor	3.5	Insoluble
Fused silica	4.5	Insoluble
Crystal quartz	4.5 and 50	Insoluble
Calcite	5.5	0.0014
Rutile	6.2	Insoluble
Sapphire	6.5	Insoluble
MgF_2 or Irtran 1	7.5	Insoluble
MgO or Irtran 5	8.5	Insoluble
LiF	9	0.27
CaF_2 or Irtran 3	12	0.0017
Arsenic trisulfide	13	Insoluble
ZnS or Irtran 2	14.5	Insoluble
BaF_2	15	0.17
NaF	15	4.22
Silicon	15	Insoluble
PbF_2	16	0.064
Selenium	20	Insoluble
Germanium	23	Insoluble
ZnSe or Irtran 4	24	Insoluble

TABLE 5.1 (*cont'd.*)

Material	Practical wavelength limit[a] (μ)	Solubility in water (g/100 g)
NaCl	26	35.7
AgCl	28	Insoluble
KCl	30	34.7
CdTe or Irtran 6	30	Insoluble
KRS-6	35	0.32
KRS-5	40	0.05
KBr	40	53.5
KI	45	127.5
CsBr	55	124.3
CsI	80	44
Diamond	80	Insoluble

[a]About 10% T with 2-mm specimen.

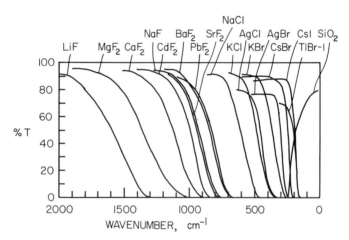

Fig. 5.1A   Transmittance of common optical materials in the infrared.

More extensive compilations of the optical and mechanical properties of materials may be found in several of the references cited at the end of Chapter 1. A useful series of papers has been published by McCarthy (*1*) and contains infrared transmission and reflection spectra and extensive bibliographies for numerous materials used in the infrared.

If a number of windows and window materials are to be used and stored in the laboratory, confusion can be reduced by painting a dot on the edges of crystals according to some prearranged color code.

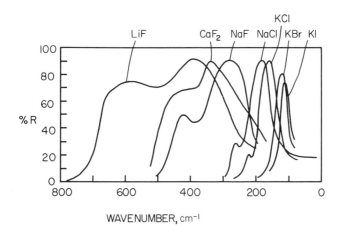

Fig. 5.1B Reflectance of common optical materials in the infrared.

B. Preparation of Windows

The cutting, grinding, and polishing of some window materials are best left to the experts because of the special skills and techniques required and because of the high cost of the material. However, the experimenter in infrared has frequent need of windows that need not be of high quality. Particularly in the infrared region of the spectrum, the surface quality of windows is not extremely critical because scattering of electromagnetic radiation is inversely proportional to the fourth power of the wavelength. Moreover, some alkali halide windows that are handled often or are in frequent contact with samples become fogged from attack by water and require occasional repolishing. Fortunately, the common window materials, sodium chloride and potassium bromide, are among those that are easily handled by relatively inexpert laboratory workers.

Window materials will usually be obtained in the form of small pieces or blanks that have been cut from large boules by the supplier. This initial cutting is done with an ordinary band saw, a diamond or abrasive saw, or a wet string, depending on the material. A very useful technique for shaping of smaller pieces is cleaving, and this should be mastered by the experimenter. It is especially applicable to the cubic crystals NaCl and KBr. The natural cleavage planes of the crystal are located by inspection and a sharp tool is placed on the material with its edge parallel to the cleavage planes. A single-edged razor blade is an ideal tool. When the back of the blade is struck sharply with a light hammer, the crystal separates along a cleavage plane, leaving clean flat surfaces, at least after some practice. This technique is worth learning because often the cleaved surfaces provide windows good enough to be used immediately with no further polishing. Moreover, cleaved surfaces are more resistant to fogging than polished ones, and polished surfaces that are natural cleavage planes are more resistant than other planes.

The surfaces of alkali halide crystals which are scratched, damaged by exposure to samples or moisture, or imperfectly cleaved can be ground to flatness with an abrasive. The abrasive can be suspended in alcohol, kerosene, or salt-saturated water and spread on a flat glass grinding plate, or a wet-or-dry sandpaper can be spread over the plate. It is best to move the crystal in a figure-eight pattern over the plate and to rotate the piece, or the plate, frequently. This minimizes nonflatness of the ground surface, although the edges are bound to become somewhat rounded. Indeed, the edges should be intentionally rounded or beveled to reduce the danger of chipping or flaking. Garnet or aluminum oxide are good choices of abrasive for the grinding of alkali halides. Generally, one starts with a coarse abrasive for rough grinding, and then follows with medium and fine grades to remove fine scratches.

The final polish is applied on a lap, using either a very fine grit abrasive for the harder window materials, or a material such as Barnesite or rouge for softer substances. The polishing can be done on a pitch lap, but most amateur workers use a cloth lap made by stretching a hard-surfaced cloth over a flat glass surface and clamped with a metal ring or sewed into place. The polishing strokes should also be made in a pattern to minimize nonflatness. Progressively dryer slurries of polishing agent are used until the last strokes are made on a nearly dry lap. Some suggested grinding and polishing agents are given in Table 5.2 for several window materials.

The beginner frequently and unnecessarily strives for perfection in polishing windows. It should be remembered that, even though a window appears badly fogged when viewed visually, it is often quite acceptable for infrared use, particularly at longer wavelengths. A sample cell that has been attacked by moist samples may appear translucent when empty, but when it is filled with a solvent having a refractive index nearly equal to that of the windows, it can become quite transparent. This is, of course, a consequence of the light-scattering properties of the irregularities in the surface with regard to wavelength, size, and relative refractive indices. Moreover, excessive

TABLE 5.2

Grinding and Polishing Agents[a]

Material	Rough grind	Fine grind	Polish
NaCl, KBr	400 grit	25-μ garnet	600 grit prepolish, finish with Barnesite in alcohol or kerosene
CaF$_2$, BaF$_2$	25-μ garnet	16-μ garnet	600 grit polish, on pitch use Linde A or B
AgCl	500 grit	600 grit	Barnesite in hypo mixture and methanol
KRS-5	400 grit	25-μ grit	600 grit prepolish, finish with Barnesite in methanol

[a]Optovac Optical Crystals, Bulletin No. 50.

polishing, especially on a cloth lap, produces an "orange-peel" texture on the surface of the window.

On occasion the polished surface of a window or prism is contaminated with sufficient embedded polishing agent to produce spurious absorption bands. McCarthy (2) has published reference spectra of common agents for use in identifying the origin of such bands.

It is often desired to drill small holes in windows when they are to be used in sample cells. Ordinary drills can be used, provided they are sharp. A cutting oil is helpful. Some workers prefer to use a pin vise and rotate a drill by hand, however a drill press can be used satisfactorily. There is danger of cracking the window because of heat and mechanical shock, so the drilling should be done slowly and cautiously.

The flatness of window materials can be measured interferometrically. The flatness is expressed in terms of the number of wavelengths or fringes, usually of visible light. The flatness requirements on window surfaces are not extremely severe. Sample cells for liquids have thicknesses of, typically, 20–100 μ. One fringe of sodium D light (of about 600-mμ wavelength) corresponds to a departure from ideal flatness of 1.2 μ, or 6% of the thickness of a 20-μ cell. In most applications, this is more than adequate flatness.

C. Optical Properties of Windows

We have already discussed an aberration introduced by a perfectly flat window (see Section 3.V.A). This aberration is a degradation of the image much as in spherical aberration. In addition, if a window of thickness t is placed at an angle θ to the beam, as in Fig. 5.2, a lateral

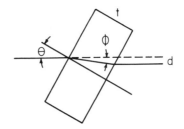

Fig. 5.2.   Lateral displacement by a window

displacement of

$$d = t \tan(\theta + \phi) \tag{5.1}$$

is introduced. With the use of Snell's law and a little trigonometry, Eq. (5.1) becomes

$$d = t \frac{\sin \theta (n^2 - \sin^2 \theta)^{1/2} + \cos \theta \sin \theta}{\cos \theta (n^2 - \sin^2 \theta)^{1/2} - \sin^2 \theta}$$

and if θ is not too large, we have

$$d \approx t \frac{n + \cos \theta}{n} \tan \theta \approx t \frac{n + 1}{n} \theta$$

Thus, a cell having two windows totaling 10 mm in thickness, made of a material with a refractive index of 1.5, displaces the beam laterally nearly 0.3 mm/deg. For this reason, care is required when inserting thick windows in a spectrometer, lest the beam be displaced excessively.

A plane window also introduces an off-axis aberration. We shall discuss it here and make use of the results when we turn to the aberrations introduced into a monochromator by a prism.

In Fig. 5.3, we define a plane surface separating a region of refractive index n on the right from a region with $n = 1$ on the left. A ray AO is incident on the surface at an angle θ measured from the normal ON. The refracted ray OA' is at an angle θ' from the normal, given by Snell's law. Define two mutually perpendicular planes containing OAB and OBN. These planes define the angles ϵ and β, as shown. Corresponding planes and angles are defined by the refracted ray. We wish to express the angle θ in terms of β and ϵ, and through Snell's law to find a relationship between β, β', ϵ, and ϵ'. The reason for this desire is as follows: Ray AO is an off-axis ray. We can define a whole family of off-axis rays passing through O and the line AB and having various values of ϵ, but all having the same value of β. We shall investigate the values of β' defined by the associated family of refracted rays, and thus determine whether the refracted rays all pass through the line $A'B'$, or whether they define some curved line instead.

From the figure we can write in terms of the auxiliary angle γ

$$\sin \theta = \frac{\bar{AN}}{\bar{OA}} = \frac{\bar{BA}}{\sin \gamma} \cdot \frac{\sin \epsilon}{\bar{BA}} = \frac{\sin \epsilon}{\sin \gamma} \tag{5.2}$$

Now

$$\tan \gamma = \frac{\bar{BA}}{\bar{NB}} = \frac{\bar{OB} \tan \epsilon}{\bar{OB} \sin \beta} = \frac{\tan \epsilon}{\sin \beta} \tag{5.3}$$

Also

$$\sin \gamma = \sin \tan^{-1}\left(\frac{\tan \epsilon}{\sin \beta}\right)$$

We shall restrict our argument to the case $\epsilon \ll \beta$, although it can be carried out formally without this restriction. This means that we consider only rays that are not too far from the plane OBN and not too close to normal to the surface. With this restriction, we can substitute the sine of a small angle for the tangent and write

$$\sin \gamma = \sin \tan^{-1}\left(\frac{\tan \epsilon}{\sin \beta}\right) = \frac{\tan \epsilon}{\sin \beta}$$

$$\approx \sin \sin^{-1}\left(\frac{\tan \epsilon}{\sin \beta}\right) = \frac{\tan \epsilon}{\sin \beta}$$

Then Eq. (5.2) becomes

$$\sin \theta = \frac{\sin \epsilon}{\tan \epsilon} \sin \beta = \cos \beta \sin \beta$$

This is one of our desired relations, and from Snell's law it can be written

$$n \cos \epsilon' \sin \beta' = \cos \epsilon \sin \beta \tag{5.4}$$

Equation (5.4) can be interpreted in this way: The projection of a ray on the plane OBN seems to be refracted according to Snell's law with a fictitious refractive index $\tilde{n}$ given by

$$\tilde{n} = n \frac{\cos \epsilon'}{\cos \epsilon} \tag{5.5}$$

We need a relationship between ϵ and ϵ'. We know that the refracted ray lies in the plane defined by the incident ray and the normal, so $\gamma = \gamma'$, and from Eqs. (5.2) and (5.3) we have

$$n \sin \epsilon' = \sin \epsilon$$

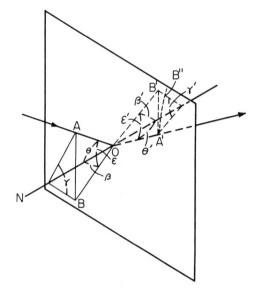

Fig. 5.3   Refraction of a skew ray.

Now the fictitious refractive index can be written

$$\tilde{n}^2 = n^2 \frac{\cos^2 \epsilon'}{\cos^2 \epsilon} = n^2 \left(\frac{1 - \sin^2 \epsilon'}{\cos^2 \epsilon} \right) = n^2 \left(\frac{1 - (\sin^2 \epsilon/n^2)}{\cos^2 \epsilon} \right)$$

$$= \frac{n^2 - 1}{\cos^2 \epsilon} + 1$$

from which

$$\tilde{n}^2 - 1 = (n^2 - 1)\sec^2 \epsilon = (n^2 - 1)(\tan^2 \epsilon + 1) = n^2 + (n^2 - 1)\tan^2 \epsilon - 1$$

and

$$\tilde{n} = [n^2 + (n^2 - 1)\tan^2 \epsilon]^{1/2} \tag{5.6}$$

The right side of Eq. (5.6) can be expanded in a binomial series, and if ϵ is sufficiently small only the first two terms need be retained

$$\tilde{n} \approx n + \frac{n^2 - 1}{2n}\tan^2 \epsilon \tag{5.7}$$

As ϵ increases, so does $\tilde{n}$. Equations (5.4), (5.5), and (5.7) tell us that β' does depend on ϵ, and therefore a plate tilted with respect to the optic axis not only displaces an image, but distorts it as well. Rays passing through line AB do not pass through $A'B'$, but rather they go through a curved line $A'B''$ as indicated in Fig. 5.3. The result is barrel-type distortion.

This result is not of any great significance in spectrophotometry because a thick plate would not normally be inserted inside the monochromator, while a certain amount of distortion can be tolerated in the optical path outside the monochromator. However, as we shall see, prisms produce a similar effect, which seriously affects the performance of a monochromator.

II. LENSES

Lenses are used extensively in some types of infrared instruments, such as infrared pyrometers and analyzers, various military trackers and cameras, and special-purpose spectrometers, including some designed for use in rockets. In general-purpose analytical spectrophotometers, the use of lenses is generally restricted to those areas where the angle requirements imposed by mirrors cannot be tolerated, such as focusing elements for detectors, refracting beam condensers for small samples, and some transfer optics systems where space is limited. Occasionally, such as in the original Perkin–Elmer Model 21 spectrophotometer, field lenses are used to form an aperture image at a convenient place without disturbing the slit image. In general, optical materials that are suitable for windows are also useful for the construction of lenses.

It is unfortunate that lenses are not better suited for use in infrared spectrophotometers. The fatal flaw in refracting optics is, of course, the chromatic aberration that occurs because of the dispersion in refractive index of optical materials. It is possible to design achromats by combining lenses made of materials having different dispersions. Examples may be found in the literature. However, the degree of achromatism that can be achieved over the long-wavelength ranges implicit in infrared spectroscopy does not make such lenses suitable for use in monochromators. Optical image quality is not so critical in applications outside the monochromator.

Lenses are ground and polished by using much the same materials and techniques as in the preparation of windows. However, the optician employs a lap with a curved surface to attain the desired spherical shape. The fabrication of lenses of good quality is a bit more difficult than the fabrication of windows, and probably should be left to the professionals. Lenses made of hygroscopic materials gradually become fogged from attack by atmospheric water vapor. A great deal of visible fog can be tolerated in the infrared and, until it becomes seriously objectionable, it should be ignored. The polish of the surface of a fogged lens can often be restored (shine might be a better term than polish) without greatly affecting the shape of the surface by dampening a paper tissue (e.g., Kleenex) with ethyl alcohol and rubbing the surface. Do not overdo it!

Lenses of some materials can be fabricated by pressing them in a mold. Silver chloride lenses have been made in this way using nickel molds; corrosion of the nickel by silver chloride is not excessive if dry conditions are maintained. More recently, lenses made of the various Irtran materials have been molded.

The third-order aberrations of lenses have been described in elementary terms in Chapter 3. Lenses are generally used more to transfer infrared energy than to transmit information in high-quality images, so the importance of aberrations is usually gauged in terms of energy loss. For example, a beam condenser is designed to produce a reduced image of the monochromator slit in order to obtain the absorption spectrum of a small sample. We are much more concerned with the proportion of energy transmitted through an opening of specified area than we are with the sharpness of the slit image.

III. Mirrors

Mirrors are the most widely used optical elements in spectrophotometers. They focus, collimate, or change the direction of radiation beams.

Mirrors are usually made of Pyrex because its relatively low thermal expansion coefficient limits distortion associated with temperature change. Pyrex is easy to work, at least for an experienced optician, it is inexpensive, and it has other desirable mechanical properties. There are newer materials (e.g., Cervit) having even lower thermal expansion. For stability a mirror should be at least about one-seventh as thick as its diameter. Mirrors are formed on a pitch lap in much the same way as salt windows and lenses, except that the harder materials require more vigorous grinding and polishing. The optical quality of a mirror surface, whether flat or curved, is measured interferometrically during the course of the mirror's production until the desired specifications are met. Most mirrors for spectrophotometers are required to have a surface that departs no more than one-fourth of the wavelength of visible light, usually the sodium D lines near 6000 Å. Thus, the quality of mirrors used in infrared instruments is extremely good when compared with the longer wavelength of infrared radiation.

Occasionally, mirrors that are used to collect radiation, but whose imaging properties are not highly critical, are fabricated from metal by forming aluminum sheet into a spherical surface. The imaging capability of such mirrors can be fairly good for infrared wavelengths, but poor imagery in the visible can make optical alignment difficult. Polished metal can be given a surface of optical quality equal to that of glass surfaces, and sometimes special applications demand such elements. However, they are expensive.

Mirrors of intricate shape can be constructed by cementing ground Pyrex blocks together with a suitable adhesive – epoxy resin is ideal – and applying a final polish to the proper surfaces of the assembly. The polishing action tends to roll the edges of optical surfaces. Ordinarily this is of no consequence because the entire surface of a mirror is rarely used. However, in the case of some fabricated mirrors this can be serious. If the reflecting facets have rounded edges, some of the reflected light will diverge. This can often be minimized by temporarily waxing glass strips between the permanent facets, thus forming a continuous glass surface that can be polished just as an ordinary mirror. After polishing, the assembly is gently heated to melt the wax, and the temporary strips are removed, leaving the permanent facets with sharp, unrolled edges.

The polished surface of a mirror is coated with a highly reflecting thin layer of metal. In earlier days metals were deposited chemically, but at present vacuum-evaporation techniques are used almost exclusively. The metal chosen for infrared mirrors is usually aluminum. Aluminum has a reflectivity that is satisfactory for most purposes, and, because of the hard, transparent oxide coating that forms on the aluminum surface, the quality of an aluminized mirror remains good over a long period of time. A fresh layer of silver has a higher reflectivity than aluminum at all wavelengths, but it tarnishes rapidly when exposed to some of the substances normally found in air. Gold has a higher reflectivity than aluminum in the infrared, only slightly inferior to fresh silver, and is often used when very high efficiency as well as chemical inertness is required, as, for example, in multipass gas cells.

The vacuum deposition of reflecting films is carried out in a bell jar under a pressure of no more than 10^{-5} mm Hg and preferably 10^{-6} mm Hg. The reflectivity of the surface is significantly greater if ultra-high vacuum conditions are attained, with pressures approaching 10^{-9}–

10^{-10} mm Hg. Cleanliness is very important. The mirror is carefully cleaned before it is placed in the bell jar and is further cleaned with a glow discharge under partial vacuum. The metal to be deposited is placed in a boat of metal having a much lower vapor pressure, such as tantalum or tungsten, and it is heated electrically or with an electron beam. The deposited layer must be thick enough to be nearly opaque, but if it is allowed to become too thick, its reflectivity suffers. A satisfactory coating transmits a barely visible amount of light from a strong source. The coating must be deposited very rapidly for best results.

Some mirror surfaces are required to have a nonspherical contour in order to minimize certain aberrations. If the asphericity is not too great, it can sometimes be attained by aluminizing a spherical mirror through a rotating sector disk mounted between the evaporation boat and the mirror. By properly controlling the size of the openings in the sector, it can be arranged, for example, to deposit a heavier layer at the center of the mirror than at the edge, thus transforming a sphere to a parabola.

Sometimes a thin film of a hard material, such as silicon monoxide, is vacuum deposited on top of the reflecting layer. The purpose of such an overcoating is to protect the soft metal surface against scratches and abrasions. Even with this protection, and more so without it, a front-surface optical mirror is a delicate thing and must be treated with care. It is usually best to ignore the dust particles which inevitably collect on mirror surfaces. If they become too objectionable, try to dislodge them with a stream of air from a squeeze bulb. Do not breathe on them, as the droplets of water from the breath will be more of a nuisance than the dust. Never use a brush. If a mirror is truly in need of cleaning, perhaps because of a fingerprint deposited by a clumsy technician, a rinse with alcohol will often restore its reflectance. Often it is best to remove the mirror from its mount, being careful not to disturb the kinematic adjusting screws so that the mirror can be returned to its exact former position. The surface can then be cleaned with the mirror in a horizontal position. In stubborn cases a coating of collodion dissolved in ether is poured onto the surface and allowed to harden. So-called flexible collodion is not suitable. When dried, the collodion is easily peeled from the mirror. Often it will part from the mirror of its own accord. Dirt particles will be embedded in the collodion, and fingerprints and other deposits will have been dissolved by the solvent and transferred to the collodion. Sometimes, two or more applications are required. If this treatment fails, there is nothing to be done but to remove the old metal layer from the mirror and replace it with a fresh surface. Incidentally, an aluminum layer cannot be removed with acid, as can most other metals, because of the protective aluminum oxide film which is always present. However, a sodium hydroxide bath will quickly strip it off.

There is another modern method for the production of mirrors which we have not yet discussed. That is the method of replication. A mold of the proper form is first constructed. It is coated with a "parting-agent," usually a silicone oil, and then filled with epoxy resin. The epoxy polymers cure with virtually no shrinkage, and thus are able to reproduce the desired shape faithfully. After the polymer is hardened the replica is easily separated from the master because the parting agent prevents the epoxy from bonding to the mold. The epoxy casting, after the film of oil is removed, can be coated with a reflecting metallic layer by vacuum evaporation in the usual way. Alternatively, the reflecting layer can be applied as part of the replication process by depositing the metal film on the film of parting agent and casting the epoxy on top of it. The reflecting film adheres to the epoxy when it is removed from the master.

High-quality mirrors, particularly aspheric, are costly because of the hand polishing and figuring required to produce them. Replication offers a means of mass producing many mirrors of excellent quality from a single master. From an individual concave mirror, numerous convex epoxy castings can be made. Each of these, in turn, can serve as a mold for a further generation of concave mirrors that duplicate the original mirror.

A *light pipe* is a special kind of mirror. If the inner surface of a tube is well polished and highly reflecting, a beam of radiation can be propagated through the tube with little loss, even though the beam is diverging. This is easily understood by considering Fig. 5.4. The beam leaving the exit end of the tube diverges at the same rate as the initial beam. The exit opening of the tube is uniformly illuminated. The original source cannot be reimaged. On the other hand, it is often an advantage in infrared spectrometers to scramble images before introducing them to the detector.

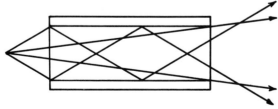

Fig. 5.4 A light pipe.

The light pipe can be a metal or glass tube, perhaps coated on the inside surface with aluminum or some other reflecting metal. Alternatively, the light pipe can be a solid polished rod which totally internally reflects the radiation at each encounter. The reflection losses of the aluminized tube are avoided, but both absorption and reflection losses at the entrance and exit ends may be serious. A bundle of fibers serves the same purpose. These fiber optic devices are used extensively in the visible region. Fiber optics have been used in the infrared, but are relatively rare in spectrophotometers.

Hollow-tube light pipes are sometimes used as absorption cells for gaseous samples, especially in instruments used in process control. Light pipes can be very useful for introducing or sampling radiation in difficult locations – for example, in transferring the beam of a spectrometer into or out of a low-temperature cryostat containing a sample or a detector.

A conical light pipe or *cone condenser* can be used to reduce the size of the cross section of a beam of radiation, particularly the beam leaving a cylindrical light pipe. It is important to notice that the rays approach the wall of the cone at a steeper angle with each encounter, as in Fig. 5.5. If the length and slope of the cone are not selected properly, some of the rays will not escape from the exit end of the cone, but will return out of the entrance. Notice also that the departing rays leave in a much wider cone than the incident rays. A useful procedure for designing light cones has been given by Williamson (*3*). The outline of the cone is repeated around a circle, as in Fig. 5.6. The fate of a given incident ray can be followed through successive reflections by merely extending it through the constructed cones, as shown. The equality of incident and reflected angles is automatically taken care of in this construction. Thus ray 1 cannot leave the cone, but ray 2 does after one reflection. Williamson also discusses the use of field lenses in the

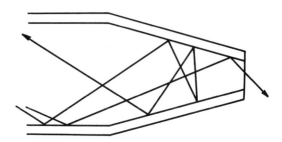

Fig. 5.5   A light cone condenser.

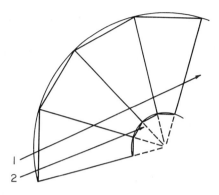

Fig. 5.6 Ray construction for a light cone.

entrance of the cone. Cone condensers placed on a detector can be used to reduce the beam size.

Now let us return to the discussion of Section 3.X and consider the magnitude of the third-order aberrations introduced with reflecting optics. Treatments of mirror aberrations have been given by Filler (4), Tarasov (5), and Rosendahl (6).

The spherical aberrations coefficient for a spherical mirror with object or image at infinity is

$$a_{11} = 1/(8f^3)$$

so the wave front deformation is

$$\Delta_s = \frac{\rho^4}{32f^3} \tag{5.8}$$

where, we recall, ρ is the radius of a zone in the exit pupil. The radius of the circle of confusion in the image plane is

$$r_s = \frac{1}{64} \frac{D^3}{f^2} \tag{5.9}$$

The coefficient for spherical aberration does not depend on the location of the aperture stop. However, the radius of the circle of confusion does indirectly, because the necessary size of the mirror (D) depends on the stop position. If the stop is moved away from the mirror surface, the mirror must be enlarged by an amount determined by the divergence of the beam.

The diameter of the circle of confusion due to spherical aberration of a collimating mirror of 500-mm focal length and 50-mm diameter ($f/10$) is 0.016 mm. A mirror of 200-mm focal length and 50-mm diameter

($f/4$) produces a circle of confusion 0.1 mm in diameter. The widths of the diffraction patterns for these two mirrors at a wavelength of $10\,\mu$ are 0.1 and 0.04 mm, respectively. Thus, spherical aberration is not significant in the first case, but dominates in the second.

In a monochromator, because two encounters with the mirror occur, we double the size of the circle of confusion calculated from Eq. (5.9). We have no right to add the magnitude of aberrations in this way, although we shall continue the practice as an approximation. The aberration patterns are really point spread functions and the cumulative effect from several optical elements is found by convolution. In fact, the variances or second moments of the aberrations are additive. This point has been discussed by Marchand (7). Hence, a better approximation to r_s for two successive reflections from identical mirrors is $(\sqrt{2}/64)D^3/f^2$.

In Chapter 3, we intimated, without proof, that Δ_s is a parabolic error that can be corrected by forming the surface of the mirror as a parabola rather than a sphere. Of course, we know from geometry that a parabola

$$x = \rho^2/4f$$

has the property of focusing all rays parallel to its axis at $y = f$, regardless of where the rays encounter the mirror. But it is instructive to see how this agrees with Eq. (5.8). We write the equation of the surface of a spherical mirror having a radius of curvature R_M, centered at $x = R_M$,

$$\rho^2 + (x - R_M)^2 = R_M^2$$

We ask how this surface can be modified to transform x into $x + \tfrac{1}{2}\Delta_s$ in order to cancel the wavefront distortion Δ_s. The modified mirror surface is thus described by

$$\rho^2 + \left(x + \frac{1}{2}\Delta_s - R_M\right)^2 = x^2 + x\left(\frac{\rho^4}{32f^3} - 2R_M\right) + \rho^2\left(1 - \frac{R\rho^2}{32f^3} + \frac{\rho^6}{(64)^2f^6}\right) = 0$$

$$(5.10)$$

For f/numbers not too small we can write $\rho \ll R_M = 2f$, and certainly we can write $x \ll R_M$, so Eq. (5.10) simplifies to

$$x = \frac{\rho^2}{2R_M} = \frac{\rho^2}{4f}$$

which is the equation of a parabola.

The parabolic correction required to eliminate spherical aberration,

$$\frac{1}{2}\Delta_s = \frac{\rho^4_{max}}{64f^3} \qquad (5.11)$$

is very small for modest-sized mirrors. For example, the $f/10$ collimator of 500-mm focal length considered above requires a $\frac{1}{2}\Delta_s$ of about 50 mμ. The 200-mm focal length mirror of the same diameter requires a $\frac{1}{2}\Delta_s$ of about 746 mμ.

Because a mirror folds a light beam back on itself, the process of introducing and removing the beam is somewhat more complicated than it is when using a straight-through lens system. The designer can use his mirror off-axis, as in the *Herschelian* system of Fig. 5.7, and suffer the off-axis aberrations (coma and astigmatism) thus introduced.

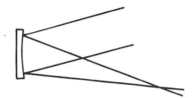

Fig. 5.7   Herschelian off-axis mirror.

He can use his mirror on-axis and intercept the beam with another element such as in the *Newtonian* mirror illustrated in Fig. 5.8. Or, he can introduce the beam through a hole in the main mirror as in the

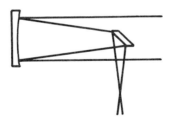

Fig. 5.8   Newtonian mirror.

Cassegrain system of Fig. 5.9. The latter two systems are free of off-axis aberrations for a point object on the optic axis, but the additional element obscures part of the beam. Moreover, the central obscuraton of the exit pupil affects the diffraction-limited resolution of the system, but not necessarily in a deleterious way.

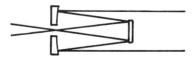

Fig. 5.9   Cassegrain mirror system.

A given spherical mirror can be used off-axis at any angle consistent with the tolerable aberrations. An off-axis parabolic mirror, on the other hand, should be limited to the angle for which it is intended. This is clear from Fig. 5.10, which shows a large parabolic mirror with sections used at various off-axis angles. Notice that the point object is located on the axis of the parabola in each case. Indeed, off-axis parabolic mirrors are often prepared by grinding a large parabola and cutting several smaller mirrors from it centered at the proper radius from the axis of the parabola. Notice that the distance ρ_{max} used in Eq. (5.11) to calculate the parabolic correction is measured from the axis of the parabola. It is not the radius of an off-axis parabolic mirror.

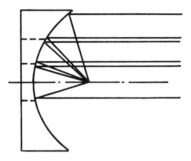

Fig. 5.10   On-axis and off-axis parabolic mirrors.

The parabolic mirror, as we have stressed, is free of spherical aberration for an object at its focus and an image at infinity (or vice versa), and so it is valuable as a collimator when large apertures (small $f/$ numbers) are required. Other aspheric mirrors are used when neither object nor image are at infinity. An ellipsoidal mirror is free of spherical aberration for an object at one focus of the ellipse and the image at the other focus, as in Fig. 5.11. An ellipsoidal mirror can also be used off-axis (Fig. 5.12).

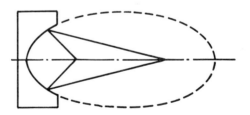

Fig. 5.11   On-axis ellipsoidal mirror.

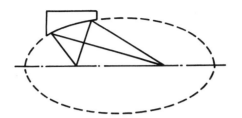

Fig. 5.12 Off-axis ellipsoidal mirror.

Now let us proceed to the second third-order aberration, coma. Unlike spherical aberration, the comatic aberration coefficient for a spherical mirror depends on the location of the aperture stop. It is given by

$$a_{12} = M/2f^3$$

where the distance of the aperture stop from the *center of curvature* of the mirror is $2fM$. Thus, if the stop is at the surface of the mirror or $2f$ from the center of curvature, $M = 1$. For the present, we restrict the aperture stop to be centered on the optic axis. The wave front deformation due to coma in a spherical mirror is

$$\Delta_c = \frac{M\rho^3\omega \cos(\phi - \theta)}{4f^2}$$

where, as before, ω is the field angle or off-axis angle and ϕ and θ are coordinate angles in the Gaussian image plane and the exit pupil, respectively. The length of the comatic pattern is

$$l_c = \frac{3MD^2\omega}{16f} \tag{5.12a}$$

and the width is

$$w_c = \frac{1}{8}\frac{MD^2\omega}{f} = \tfrac{2}{3}l_c \tag{5.12b}$$

The presence of M in these equations causes the coma to vanish if the aperture stop is placed at the center of curvature of the mirror, so that $M = 0$. For several reasons, including economy and mechanical convenience, the dispersing element of a monochromator is usually placed 0.8 to 1.0 focal lengths from the collimator, that is, $0.5 \leq M \leq 0.6$. We discuss the magnitude of comatic aberration in greater detail in Chapter 6 in connection with particular monochromator designs. For the present consider a monochromator having its grating one focal length from the collimator ($M = 0.5$) and an entrance slit h units tall. For simplicity we make the impractical assumption that the slit is centered on the optic axis. For a point at the end of the slit, $\omega = h/2f$. The comatic image of this point, after *two* reflections from the collimating mirror, has a width given approximately by

$$w_c = \frac{1}{16}\left(\frac{D}{f}\right)^2 h$$

which is twice the width produced by a single reflection. Let us take as an example the mirrors used in the above discussion of spherical aberration, that is $f/10$ and $f/4$, and take a slit height of 25 mm. With the $f/10$ system the comatic width is 0.016 mm, or just about half as large as the circle of confusion due to spherical aberration from two mirrors. The $f/4$ system with the same slit height has a comatic pattern 0.098 mm wide, again about half the spherical aberration. A taller slit will have a proportionally wider comatic aberration.

As we shall see later, the energy transmitted by a monochromator depends on the slit-height to focal-length ratio. Hence a fairer comparison of our hypothetical $f/10$ and $f/4$ systems demands that the slit height of the latter be scaled down to 10 mm. In this case, the comatic pattern has a width of 0.039 mm, or about the diffraction limit at 10 μ.

We have seen how aspheric mirrors can be used to eliminate spherical aberration. It is natural to ask what happens to the other aberrations when the spherical aberration is removed. If all of the spherical aberration is eliminated by means of an appropriately shaped mirror, the comatic aberration coefficient becomes

$$a'_{12} = 1/(2f^3)$$

which is independent of the position of the stop. Thus, if we start with a system having a spherical mirror and an aperture stop at the focal plane ($M = 0.5$) and substitute a parabolic mirror, we eliminate the spherical aberration, but double the comatic aberration. However, it is possible to eliminate the coma by using two mirrors in a symmetric arrangement. This was mentioned in Chapter 3 and is referred to again in our discussion of monochromator types in Chapter 6.

Astigmatism, like coma, also depends on the position of the aperture stop. The coefficient of astigmatism is

$$a_{22} = M^2/f^3$$

The wavefront deformation due to astigmatism is

$$\Delta_A = (M^2/f)\omega^2\rho^2 \cos^2(\phi - \theta)$$

It is seen that astigmatism also vanishes when the aperture stop is placed at the center of curvature of the mirror. The lengths of the astigmatic images are

$$l_{A,S} = l_{A,T} = M^2 D \sin^2 \omega \qquad (5.12c)$$

If the slit of a monochromator is placed on the optic axis (or very close to it), the importance of astigmatism is minimized because a point of the entrance slit is imaged as a line along the length of the exit slit. However, some monochromators have the slits far from the optic axis and astigmatism becomes important.

The elimination of spherical aberration by use of an aspheric mirror also affects astigmatism. The coefficient of astigmatism remaining after removal of spherical aberration is

$$a'_{22} = \frac{2M-1}{f^3}$$

This vanishes for $M = \frac{1}{2}$, that is, for the aperture stop one focal length from the mirror.

The tangential astigmatic line image falls on a surface of radius

$$R_{A,T} = \frac{R}{1-3M^2}$$

This surface is flat if $M = 1/\sqrt{3}$, that is, when the aperture stop is placed 0.84 focal length from the mirror.

A spherical mirror can distort an image. An approximate equation for the radius of curvature of a monochromator slit image distorted by two reflections from a spherical collimator mirror is given by Peisakhson (8),

$$r_{cs} = -\frac{f}{2M(1-M^2)\sin\omega}$$

The distortion is positive, or barrel type. It vanishes when the aperture stop is at the mirror surface or at the center of curvature. It is greatest for $M = 1/\sqrt{3}$, when the stop is 0.84 focal length from the mirror. For a slit of height h the sagittal distance between the curved image and a straight line is

$$x = \frac{(h/2)^2}{2r_{cs}} = \frac{-h^2 M(1-M^2)\sin\omega}{4f}$$

If an off-axis paraboloid is substituted for the spherical mirror, the sagittal distance is

$$x = \frac{h^2 \sin\alpha}{4f}$$

where α is the off-axis angle of the paraboloid and the stop is placed one focal length from the mirror. The contribution of the collimator to the slit image curvature in a monochromator is usually negligible in comparison to the effect of the prism or grating discussed in the next sections.

IV. PRISMS

A prism is used in a monochromator to spread a beam of radiation into a spectrum. It does this by virtue of the different refractive indices it presents to different wavelengths of light. Consequently, prism materials must satisfy all of the requirements met by window materials and, in addition, should have a high *dispersion*, by which we mean the rate of change of refractive index with wavelength. Because prisms are a great deal thicker than windows, the extent of wavelength region over which prism materials are sufficiently transparent is more limited. Moreover, the requirements on optical homogeneity and surface quality are much stricter than for windows.

A typical plot of refractive index and absorption versus wavelength is sketched in Fig. 5.13. Both the refractive index and the absorption are associated with oscillations of the atoms of the material. Recall

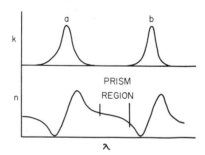

Fig. 5.13   Refractive index and absorption coefficient of typical optical material.

the meaning of the real and imaginary components of the refractive index of absorbing media as discussed in Chapter 4. It is typical of prism materials used in the infrared that there is a strong absorption in the ultraviolet, as at a in the figure, with a large negative rate of change of refractive index extending toward the visible region. Throughout the visible the material remains transparent, but its dispersion is too low to be of use as a good prism material. As the strong absorption at b in the infrared is approached, the refractive index curve becomes steeper and the material begins to be useful, but eventually as the dispersion becomes very good, the absorption increases and the usefulness of the material vanishes. Thus, the region of usefulness of a prism material is bounded by an area of too small dispersion at shorter wavelengths and too high absorption at longer wavelengths. Different prism materials have different regions of usefulness. In order to cover a wide spectral region with a prism monochromator it is not at all uncommon to use five or six different prisms. Limits of usefulness of prism materials are given in Table 5.3.

TABLE 5.3
Long-Wavelength Limit of Commonly
Used Prism Materials

Material	Wavelength limit (μ)
Silicon or quartz	3
LiF	6
CaF_2	9
NaCl	16
KCl	20
KBr	25
KRS-5	38
CsBr	40
CsI	50

The most widely used prism material is sodium chloride, which can be used to cover the so-called rock-salt region from 2 to 15 μ. Better spectra are obtained in the 2 to 4μ region with lithium fluoride prisms, and calcium fluoride improves the spectrum to wavelengths up to about 6.5 μ. Beyond 15 and out to 25 μ potassium bromide can be used. The region from 25 to 40 μ is covered by cesium bromide and from 40 to 50 μ by cesium iodide. Prism materials are not available for use in the far infrared beyond 50 μ. Other materials have been used in the 2 to 50-μ region in addition to the six alkali halides listed here. For a time the only prism material available for spectroscopy in the 25 to 50-μ region was thallium bromide-iodide. It was not ideal because of its tendency to cold flow even under its own weight on a prism table. Cesium iodide is also somewhat soft and should not be tightly clamped to its table; but it is stable in the absence of pressure.

Now let us turn to the effect of a prism on a beam of radiation passing through it. Assume that the radiation incident on the prism is collimated into an ideal plane wave front. Also, assume for the moment that the normal to the wave front is perpendicular to the apex of the prism formed by the intersecting faces, so we are dealing with principal rays. The nomenclature of the prism and rays is defined in Fig. 5.14. The incident and departing rays are at angles θ_1 and θ_2 with respect to the normals to the appropriate faces. The corresponding angles inside the prism, ϕ_1 and ϕ_2, are given by Snell's law

$$\sin \phi_1 = \frac{1}{n} \sin \theta_1 \tag{5.13}$$

$$\sin \phi_2 = \frac{1}{n} \sin \theta_2 \tag{5.14}$$

with n the refractive index of the prism material. The apex angle of the prism is α. The *angular deviation*, ψ, suffered by a ray in passing through the prism is found by writing the sum of the angles of the trapezoid *abcd*

$$(\pi - \alpha) + \theta_1 + (\pi - \psi) + \theta_2 = 2\pi \tag{5.15}$$

so

$$\psi = \theta_1 + \theta_2 - \alpha \tag{5.16}$$

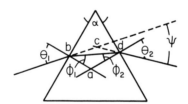

Fig. 5.14   Refraction by a prism.

Using Eqs. (5.13)–(5.16), the deviation can be calculated from the angle of incidence, the prism angle, and the refractive index. The prism angle is, of course, constant, and the angle of incidence in a monochromator is fixed. From the known value of the refractive index of the prism material at successive wavelengths, the optical designer can calculate the angle of departure or the deviation angle as a function of wavelength. In many prism monochromator designs, the radiation passed by the prism is intercepted by a mirror (the Littrow mirror) and returned through the prism a second time. Alternatively, the prism face itself can be aluminized to act as a return mirror. Because rays of different wavelength are deviated different amounts, only rays of one wavelength are returned exactly on themselves for a given setting of the Littrow mirror or prism. Hence, for each setting only one central wavelength and a narrow cluster of wavelengths around it are returned back to and through the entrance slit. One of the chores of the optical designer is to calculate the required angular setting for each wavelength. The practical requirement of separating the entrance slit from the exit slit complicates the calculation because in this case the paths of the returned rays no longer coincide with those of the incident rays. This is illustrated in Fig. 5.15.

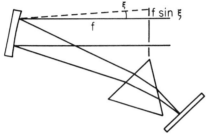

Fig. 5.15   Separation of entrance and exit slits in a prism monochromator.

Another function of importance to both the designer and the user of a monochromator is the *angular dispersion*, which is the rate of change of angular deviation with wavelength,

$$D_a = d\psi/d\lambda$$

Equations for calculating D_a can be found from Eqs. (5.13)–(5.16),

$$D_a = \frac{1}{n}\frac{dn}{d\lambda}\left[\tan\theta_2 + \frac{\cos\phi_2}{\cos\phi_1}\frac{\sin\theta_1}{\cos\theta_2}\right] = \frac{\sin\alpha}{\cos\theta_2\cos\phi_1}\frac{dn}{d\lambda} \qquad (5.17)$$

As $dn/d\lambda$ is negative (Fig. 5.13), long-wavelength radiation is deviated less than short-wavelength radiation. The angular dispersion D_a is readily calculated simultaneously with ψ and the Littrow-mirror angle or prism setting.

There is one particular setting of a prism for which the equations simplify considerably. This is the case of *minimum deviation*, which we define by setting

$$d\psi/d\theta_1 = 0$$

By again making use of Eqs. (5.13)–(5.16), we can write

$$\frac{d\psi}{d\theta_1} = 1 - \frac{\cos\theta_1}{\cos\theta_2}\cdot\frac{\cos\phi_2}{\cos\phi_1}$$

which is zero when the conditions

$$\theta_1 = \theta_2 \qquad \phi_1 = \phi_2 \tag{5.18}$$

are satisfied. It is necessary to demonstrate that these conditions do indeed define a *minimum* value for ψ. This is done by showing that $d^2\psi/d\theta_1^2$ is greater than zero when conditions (5.18) hold.

At minimum deviation Eqs. (5.13)–(5.16) reduce to

$$\sin \theta_1' = \sin\left(\frac{\psi' + \alpha}{2}\right) = n \sin \tfrac{1}{2}\alpha \tag{5.19}$$

and the angular dispersion is

$$D_a' = \frac{2}{n} \tan \theta_1' \frac{dn}{d\lambda} = \frac{2 \sin \alpha/2}{(1 - n^2 \sin^2 \alpha/2)^{1/2}} \tag{5.20}$$

The minimum-deviation condition is easily recognized when using a spectroscope and is useful in measuring the refractive index of a prism. In a spectrometer with a Littrow mirror the angle of incidence is fixed, so minimum deviation can occur for only one wavelength at most. This is also true of a spectrometer with an aluminized prism, for which the angle of incidence on the second face is fixed. However, Eq. (5.20) is often used as a useful approximate relationship for a monochromator at wavelengths for which the monochromator is not set at minimum deviation.

Of greater significance to the spectroscopist than angular dispersion is the *linear dispersion* at the exit slit

$$D_l = dx/d\lambda$$

D_l is a measure of how widely spread out the spectrum is as it leaves the monochromator. The width of the band of wavelengths, $\Delta\lambda$, falling on an exit slit of width w is

$$\Delta\lambda = w/D_l \tag{5.21}$$

The quantity $1/D_l$, the *reciprocal dispersion*, is usually plotted in the manuals accompanying commercial instruments.

The linear dispersion D_l can be calculated from D_a (see Fig. 5.15). The slits are placed at a distance from the collimator equal to one focal length. Two rays leaving the same point of the collimator in directions differing by an angle ξ will reach the plane of the exit slit a distance apart that is given by

$$\Delta x = f \sin \xi \approx f\xi$$

for small angles ξ. In a Littrow monochromator the beam passes through the prism twice, so the angular dispersion is very nearly twice the amount given by Eq. (5.17) for a single prism. It follows that the linear dispersion for a Littrow monochromator (or a similar type of monochromator) is

$$D_l = 2fD_a \qquad (5.22)$$

and therefore at the angle of minimum deviation,

$$D'_l = \frac{4f}{n} \tan \theta'_l \frac{dn}{d\lambda} = \frac{4f \sin \alpha/2}{(1 - n^2 \sin^2 \alpha/2)^{1/2}} \frac{dn}{d\lambda} \qquad (5.23)$$

The angle of incidence for minimum deviation θ'_l is found from Eq. (5.19).

Now let us investigate the spectral resolution that might be achieved by a prism monochromator without aberrations. That is, we shall find the diffraction-limited resolution. In Section 4.VII, in discussing spread functions and optical transfer functions, we pointed out that the width of the spread function associated with diffraction by a rectangular aperture, measured to the first zero of the diffraction pattern, is just the f/number of the optical system multiplied by the wavelength. This also defines the limiting period for a spatially modulated optical signal just passed by the optical transfer function. Hence, the slit width that corresponds to the least resolvable detail is

$$w_d = f\lambda/A \qquad (5.24)$$

where A is the width of the aperture. The slit width w_d is often called the *diffraction slit width*. It provides the greatest possible resolution obtainable from a diffraction-limited, aberration-free monochromator. Nothing is gained by narrowing the slit further. The diffraction-limited spectral resolution can now be found from the linear dispersion from Eq. (5.21)

$$(\Delta\lambda)_d = \frac{f\lambda}{AD_l} \qquad (5.25)$$

The *resolving power R* of a monochromator is defined by

$$R = \lambda/\Delta\lambda = \nu/\Delta\nu$$

which becomes, with Eq. (5.25) for the diffraction-limited case,

$$R = AD_l/f$$

The aperture A is found from the geometry of the prism, as in Fig. 5.16,

$$A = C \cos \theta_1 = \frac{B \cos \theta_1}{2 \sin \alpha/2}$$

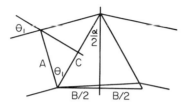

Fig. 5.16   Aperture of a prism.

The diffraction-limited resolution of a prism monochromator is thus

$$R = \frac{BD_l \cos \theta_1}{2f \sin \alpha/2} \tag{5.26}$$

If the prism is set at minimum deviation, Eq. (5.26) can be simplified by substituting from Eqs. (5.23) and (5.19)

$$R' = 2B \frac{dn}{d\lambda} \tag{5.27}$$

Equation (5.27) states that the diffraction-limited resolution of a prism monochromator set at minimum deviation is given by the product of the length of the *base* of the prism (with due allowance for prism masks, if any) and the dispersion of the prism material. The factor 2 appears in the equation because two passes are made through the prism. Ordinarily, optics textbooks give the equation for a single prism without the factor 2. If the instrument has a double monochromator, with two prisms each used twice, another factor of 2 is applied.

   Not all of the aberrations of a prism monochromator are associated with the collimating and condensing mirrors. The prism itself contributes a very important aberration, slit image distortion (sometimes called slit image curvature), in much the same way as a window introduces distortion as discussed in Section 5.I.C. This aberration can easily be the limiting factor in determining the practical resolution of a spectrophotometer. We have already shown, for an off-axis ray entering a plane surface, that the projection of the ray on the plane of the optic axis behaves as though it is refracted by a material having a fictitious refractive index $\tilde{n}$, given by Eq. (5.7), which also can be written

$$\Delta n = \tilde{n} - n = \frac{n^2 - 1}{2n} \tan^2 \epsilon \tag{5.28}$$

We wish to apply this to the off-axis rays in a prism monochromator. A ray entering the monochromator a distance y from the center of the

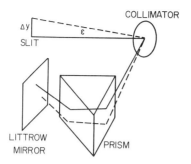

Fig. 5.17 Slit image distortion in a prism monochromator.

entrance slit (see Fig. 5.17) makes an angle ϵ with the optic axis which is given by

$$\tan \epsilon = y/f \qquad (5.29)$$

We now can write for a prism used twice, as in a Littrow system, from Eqs. (5.22) and (5.20) and using Δn from Eq. (5.28) and $\tan \epsilon$ from Eq. (5.29),

$$x = 2f\,\Delta\psi = 2f\frac{d\psi}{dn}\Delta n = \frac{2(n^2-1)}{fn^2}\tan\theta_1'\,y^2 \qquad (5.30)$$

Equation (5.30) tells us that *for a prism set at minimum deviation*, a line object, such as the entrance slit, is imaged as a parabola with the departure from a straight line image (i.e., the sagittal distance) proportional to the square of the distance from the optic axis. The parabola is often approximated by a circle. The radius of curvature at $y = 0$ is

$$r_{cp} = \frac{fn^2}{4(n^2-1)}\cot\theta_1' \qquad (5.30a)$$

Remember that this expression contains a factor $\frac{1}{2}$ that corresponds to the factor 2 in Eq. (5.22) because two passes are made through the prism. For a single pass, omit this factor. An equation for r_{cp} was given long ago by Kayser (9), but unfortunately with a $\tan\theta_1'$ instead of $\cot\theta_1'$. For more recent derivations see Sawyer (10), Gates (11), or Peisakhson (12).

The sense of the distortion of the slit image is readily noted by observing that an off-axis ray is refracted to a greater extent than the paraxial ray. Now a prism also refracts shorter wavelengths to a greater extent than long wavelengths, so the image of a straight entrance slit is a parabola having its ends bent in the direction of shorter wavelengths,

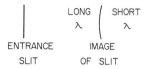

Fig. 5.18   Sense of slit image distortion in a prism monochromator.

as in Fig. 5.18. The distortion increases with increasing refractive index and, therefore, is greatest at shorter wavelengths.

The slit image distortion depends on the angle of incidence on the prism and the focal length of the collimator. Both of these factors are invariant for a given prism in a monochromator using a Littrow mirror. The curvature also depends on the refractive index and, therefore, the wavelength. In Fig. 5.19 we have plotted x against wavelength for rays from the end of an entrance slit 25 mm high ($y = 12.5$ mm), with

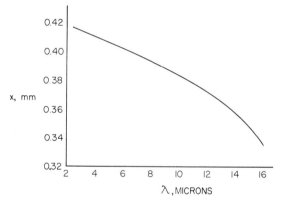

Fig. 5.19   Calculated slit image distortion from a sodium chloride prism (refractive index 1.5, slit height 25 mm total, prism angle 60°, angle of incidence 48°40′, collimator focal length 50 cm; x is the departure from the ideal line image at the slit ends).

an angle of incidence on a sodium chloride prism of 48°40′, and a collimator focal length of 50 cm. This is consistent with minimum deviation by a prism having an apex angle of 60° and a refractive index of 1.5.

The effects of slit image distortion can be eliminated at one wave length by intentionally curving the entrance slit in the opposite sense by just the amount calculated from Eq. (5.30) or, more commonly, by curving both the entrance and exit slits by half this amount. This compensates for the image distortion produced by the prism, so for this wavelength the entrance-slit image coincides exactly with the exit slit. Slit image distortion is normally corrected at a wavelength where the slit width and the bandwidth of the monochromator are usually small.

From Eq. (5.30) we can see what happens if we match slit image distortion at a certain wavelength (and refractive index) and then substitute a prism of a different material, keeping the other parameters the same. The slit image distortion will be matched at a new wavelength where the refractive index of the new prism is the same as the index of the old one at the wavelength of the original match. This is fortunate because in general the prism materials that are useful at longer wavelengths have higher refractive indices. Sodium chloride, for example, has a refractive index of about 1.52 at 4 μ, while potassium bromide has the same refractive index at 9 μ. Thus if a sodium chloride prism monochromator has its slit image distortion matched at 4 μ, we can substitute an identical potassium bromide prism to extend the useful range to longer wavelengths and find that with the same slits the slit images are matched at 9 μ.

A certain amount of the radiation incident on a prism face is reflected. This reduces the level of useful radiation somewhat, but not too seriously. Of considerably greater importance is the question of what happens to the reflected light. In a Littrow monochromator, as can be seen in Fig. 5.20, reflection occurs at two air-to-prism faces and two

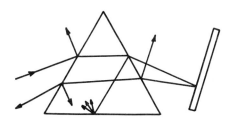

Fig. 5.20   Reflection and scatter at prism surfaces.

prism-to-air faces. The two internally reflected beams can be reflected, or scattered, at the prism base and eventually escape from the interior of the prism. Radiation is also scattered from optical inhomogeneities in the interior of the prism and from imperfections in the polished surfaces, particularly if the prism has been fogged by water vapor. All of this radiation bounces about inside the monochromator and eventually some of it finds its way out of the exit slit and onto the detector. It forms a background of polychromatic *stray radiation* underlying the desired monochromatic radiation delivered by the monochromator. Optical designers must design monochromator systems and components in such a way as to minimize this background. The use of carefully positioned baffles is often necessary.

We recall from Section 4.III that the reflectivity of a dielectric material is greatest for radiation polarized with its electric vector perpendicular to the plane of incidence (i.e., parallel to the reflecting surface). Consequently, the radiation reflected from the surface of a prism is partially polarized. The refracted radiation eventually reaching the exit slit is, therefore, also partially polarized, with the electric vector of the dominant component lying in the plane of incidence on the prism or perpendicular to the slit of the monochromator. The extent of the polarization is wavelength dependent.

We have tacitly assumed up to now that the prism is in a vacuum, and the refractive index is, therefore, the index relative to vacuum. This restriction is easily relaxed by redefining the refractive index, wherever it appears, to be

$$n = n_0/n_m \qquad (5.31)$$

where n_m is the refractive index of the prism's surroundings relative to vacuum. Now we would like to study the effect of a change of environment on the optical properties of the prism. Moreover, we would like to investigate the effect of a change in the refractive index of the prism itself due to a change in its temperature. The same approach is applicable to both problems. Let us consider first the problem of the changing environment. The refractive index of the prism material in vacuum is represented by the solid line in Fig. 5.21. The addition of a surrounding

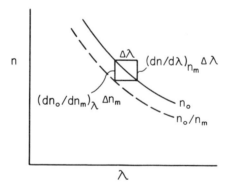

Fig. 5.21   Index of refraction and effect of surrounding medium.

medium of index n_m reduces the effective refractive index of the prism material, according to Eq. (5.31). This is represented by the dashed line in the figure. The change in effective refractive index is

$$\Delta n = \frac{dn}{dn_m}\Delta n_m = \frac{-n_0}{n_m^2}\Delta n_m$$

Suppose the monochromator in vacuum is adjusted to pass radiation of wavelength λ, for which the refractive index is n_0. If the adjustment is not altered with the admission of a new medium, the monochromator will still pass radiation for which the prism has the same refractive index, but this radiation has a different wavelength, $\lambda' = \lambda + \Delta\lambda$. At the new wavelength λ', the refractive index of the prism relative to vacuum is

$$n_0(\lambda') = n_0(\lambda) + \frac{dn_0}{d\lambda}\Delta\lambda$$

or

$$\Delta n_0 = n_0(\lambda') - n_0(\lambda) = \frac{dn_0}{d\lambda}\Delta\lambda$$

Clearly, we have

$$\frac{dn}{dn_m}\Delta n_m = \frac{dn_0}{d\lambda}\Delta\lambda$$

and so

$$\Delta\lambda = -\frac{n_0\,\Delta n_m}{n_m^2(dn_0/d\lambda)} \approx \frac{-n_0\,\Delta n_m}{dn_0/d\lambda} \tag{5.32}$$

is the equation for the wavelength shift corresponding to a change in the refractive index of the surroundings. The approximation equation is written on the basis of $n_m \approx 1$, which is certainly reasonable since the refractive index of air at $5\,\mu$ and at a temperature of 40°C is 1.00025. The refractive index of helium is even closer to unity, about 1.00003. Equation (5.32) tells us that the wavelength shift in a prism monochromator depends on the refractive index and on the dispersion of the prism material. It is independent of the geometry of the monochromator or prism. The shift is greatest where the dispersion is lowest. For example, consider a sodium chloride prism monochromator calibrated in air but flushed with helium. For NaCl, $n_0 \approx 1.52$ and $dn_0/d\lambda \approx -0.003$ at a wavelength of $4\,\mu$. Hence, a wavelength shift of about $+0.1\,\mu$ occurs with the introduction of helium. If the monochromator has a prism of potassium bromide, for which $n_0 \approx 1.535$ and $dn_0/d\lambda \approx -0.001$ at $4\,\mu$, the wavelength shift is about $+0.3\,\mu$. These are quite significant shifts indeed.

A sodium chloride prism monochromator might be calibrated in California or Connecticut where the atmospheric pressure is 760 mm

Hg, and then shipped to Colorado where the pressure is only 625 mm Hg. The wavelength shift accompanying this pressure change is about 0.01 μ at 4 μ.

In the same way, we can treat the wavelength shifts that occur with temperature change by writing

$$\left(\frac{dn}{d\lambda}\right)_T \Delta\lambda = \left(\frac{dn}{dT}\right)_\lambda \Delta T$$

from which

$$\Delta\lambda = \frac{(dn/dT)_\lambda}{(dn/d\lambda)_T} \Delta T$$

Once more, the greatest shift occurs at wavelengths where the dispersion is smallest. Sodium chloride has a temperature coefficient of refractive index of about $-2.5 \times 10^{-5}/°C$. Hence $\Delta\lambda/\Delta T$ is about 0.008. A 10°C temperature rise produces a wavelength error of nearly 0.1 μ at 4 μ.

V. Diffraction Gratings

Gratings, like prisms, are used in monochromators to disperse radiation into its constituent wavelengths. Gratings accomplish this by diffracting the radiation. The amount of diffraction depends on the wavelength. Diffraction gratings are of two basic types, transmission and reflection. Reflection gratings are either plane or concave, the advantage of the concave grating being that it serves as its own collimator and condenser. Modern grating monochromators used in the infrared generally have plane reflection gratings. Transmission gratings were once used in the far infrared, and are still occasionally used as filters.

An elementary reflection grating consists of a surface with alternating reflecting and nonreflecting strips called facets or lines. Similarly, an elementary transmission grating has alternating opaque and transparent strips. These are sketched in Figs. 5.22 and 5.23.

The choice of spacing between the lines, a, depends on the wavelength region in which the grating is to be used. The spacing is usually of the order of magnitude of the wavelength. A plane wave front incident on either type of grating may be thought of as being scattered into spherical Huygens wavelets whose envelope forms a new plane wavefront propagating in a new direction. The new direction is determined by the requirement that the wave front is the locus of constant phase of

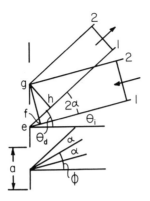

Fig. 5.22   Reflection diffraction grating.

the propagating electromagnetic radiation. In Fig. 5.22, we see an incident wave whose rays make an angle θ_i with the normal to the surface
of a reflection grating. The ray 1 encountering the grating at e has traveled a distance

$$\bar{fe} = a \sin \theta_i \qquad (5.33)$$

farther than the ray 2 arriving at the corresponding point g of the
neighboring strip. Consider a diffracted wave front whose rays depart
at an angle θ_d from the normal. Ray 1 leaving e travels a distance

$$\bar{eh} = a \sin \theta_d \qquad (5.34)$$

farther than ray 2 leaving g. Consequently, the difference in distances
traveled by rays 1 and 2 is

$$\Delta = \bar{fe} + \bar{eh} = a(\sin \theta_i + \sin \theta_d) \qquad (5.35)$$

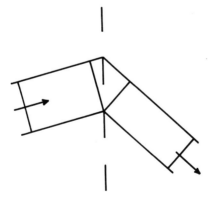

Fig. 5.23   Transmission diffraction grating.

This distance Δ is the same for any pair of corresponding points on adjacent facets.

Figure 5.22 is drawn with the diffracted wave front having its rays on the same side of the grating normal as the incident wave. Equation (5.35) holds equally well for diffracted waves on the other side of the normal, provided we define $\theta_d < 0$ in this case. Moreover, Eq. (5.35) holds for the transmission grating of Fig. 5.23 with the same convention, that is, $\theta_d > 0$ if the diffracted ray is on the same side of the normal as the incident ray, as drawn.

In order for the diffracted wave fronts to represent planes of constant phase, the distance Δ must contain an integral number of wave lengths. Therefore the fundamental equation for the diffraction grating is

$$n\lambda = a(\sin \theta_i + \sin \theta_d) \qquad n = 0, \pm 1, \pm 2, \ldots \qquad (5.36)$$

It is common in infrared monochromators for the angle 2α between the incident and the departing rays to be fixed, so in terms of the bisector of the two rays

$$\theta_i = \phi - \alpha \qquad \theta_d = \phi + \alpha \qquad (5.37)$$

With these relationships and a little trigonometry, Eq. (5.36) can be written in the equivalent form

$$n\lambda = 2a \sin \phi \cos \alpha \qquad n = 0, \pm 1, \pm 2, \ldots \qquad (5.38)$$

We use the symbol n here because it is traditional, as is the use of n for the refractive index. It is not likely that they will be confused.

With $n = 0$ in Eq. (5.36), $\theta_i = \theta_d$, which is just the condition for ordinary reflection. This case is referred to as zero-order diffraction. When $|n| = 1, 2, \ldots$, we have first-order, second-order, and so on, diffraction; nth-order diffraction thus implies an n-wavelengths path difference for adjacent rays. The sign of n is positive if the diffracted ray lies on the same side of the grating normal as the incident ray; otherwise n is negative. This nomenclature is diagrammed in Fig. 5.24.

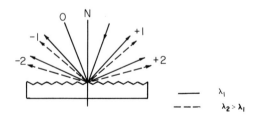

Fig. 5.24   Diffracted orders.

Observe that the longer the wavelength, the greater is the angle by which the light of a given wavelength is "bent" by a grating. This is just the opposite of dispersion by a prism. The maximum wavelength that can be observed with a given grating is found from Eq. (5.36) by setting both θ_i and θ_d equal to 90°. Both the incident and diffracted rays graze the surface of the grating. This gives us

$$\lambda_{max} = 2a/n$$

If the monochromator is such that there is a fixed angle between the incident and diffracted rays, Eq. (5.38) provides the relationship

$$\lambda_{max} = (2a/n) \cos \alpha$$

In practice, grazing incidence is not reached because the aperture stop of a monochromator is usually the grating, and at grazing incidence it presents vanishing width to the radiation. A practical rule of thumb is

$$\lambda_{max} \approx 1.4a/n$$

Thus, a 100-line/mm grating might be used to observe wavelengths as long as 14 μ, but in order to study the far infrared out to 300 μ, a coarse 4-line/mm grating is required.

From Eq. (5.36) or (5.38) we see that for a given choice of angles, the product $n\lambda$ is a constant. Therefore, when radiation of wavelength λ is observed in first-order diffraction, radiation of wavelength $\frac{1}{2}\lambda$ is diffracted in the same direction in second order, as is radiation of wavelength $\frac{1}{3}\lambda$ in third order, and so on. More generally, if radiation of wavelength λ is observed in a kth-order spectrum, simultaneous observation is made of wavelengths $k\lambda/n$ in their nth-order spectra. This overlapping order characteristic of diffraction gratings necessitates the use of some sort of filtering to eliminate all but the desired wavelength.

The gratings used in monochromators are really not of the elementary form sketched in Fig. 5.22. Rather than having flat surfaces, the facets are tilted at an angle B, as shown in Fig. 5.25. A grating with such a cross section is called an *echelette*. Originally, echelettes were

Fig. 5.25   Blaze angle.

made with rather coarse spacing and were intended for use in higher orders. The echelette is a *blazed* grating, with B the angle of blaze. A blazed grating used with the shallow side of the facets exposed to the radiation, as in Fig. 5.26, is called an *echelle*. It is used in very high orders.

Fig. 5.26   The echelle grating.

The reason for blazing a grating is to concentrate the diffracted radiation in a preferred direction. We return to a mathematical discussion of this point a little later. A blazed grating offers high efficiency at angles of incidence in the neighborhood of B, but the efficiency falls off when we depart too far from this angle. The wavelength of greatest efficiency is given by

$$\lambda_B = \frac{2a}{n} \sin B \cos \alpha \approx \frac{2a}{n} \sin B \qquad (5.39)$$

It is apparent that a grating blazed for a wavelength λ_B in the first order is also blazed for $\frac{1}{2}\lambda_B$ in the second order, $\frac{1}{3}\lambda_B$ in the third order, and so on. Hence it is common practice to use a grating in the first order to study spectra lying between λ_{max} and perhaps $0.6\lambda_B$, then reset the grating to study wavelengths between $0.6\lambda_B$ and $0.4\lambda_B$ in the second order, reset again to observe wavelengths in the neighborhood of $0.3\lambda_B$ in the third order, and so on.

Now that we understand the reasons for the form of reflection gratings, let us consider the methods by which gratings are produced. Modern reflection gratings are made by depositing a thick film of aluminum on an optical flat and then plowing evenly spaced parallel furrows into the aluminum surface with a diamond tool. The diamond is shaped and positioned to provide the required blaze angle. It is mounted on the carriage of a ruling engine which advances its position one line width after each stroke. This process must be carefully controlled with a precision lead screw. In modern engines, the process is controlled interferometrically through feedback mechanisms. Obviously, the ruling engine is a machine of high precision and there are only a few such devices in the world today.

Even the relatively coarse gratings required for infrared spectroscopy can contain many thousands of ruled lines. If original ruled gratings were required for the construction of monochromators, the cost would be very high. Fortunately, it is possible to make replica gratings from a ruled original. Replicas are made in a typical process by first depositing a film of parting agent, for example, silicone oil, on the original grating and evaporating a layer of aluminum on the oil. Then a layer of epoxy resin is poured on the aluminum and then topped by a plate of glass that bonds to the epoxy and provides the necessary rigidity and stability. After the epoxy has hardened, the replica is parted from the original, the parting agent is removed, and a final thin coating of aluminum is evaporated on the surface. The quality of a good replica is high, even higher than the original in some respects. Indeed, quality replicas can be made of replicas. Thanks to the replication process, good gratings are available at a price that makes them competitive with prisms, and gratings are largely supplanting prisms in modern commercial spectrophotometers.

Now let us return to the optical properties of diffraction gratings. The dispersion produced by a grating is calculated from the derivative of Eq. (5.36) in terms of the angle of diffraction,

$$D_a = \frac{d\theta_d}{d\lambda} = \frac{n}{a} \sec \theta_d = \frac{n}{a[1 - (n\lambda/a - \sin \theta_i)^2]^{1/2}}$$

For angles not too far from the normal, $\sec \theta_d$ is a slowly varying function. It follows that the dispersion is a slowly varying function of the wavelength. The linear dispersion at the exit slit is

$$D_l = \frac{dx}{d\lambda} = fD_a = \frac{fn}{a} \sec \theta_d \qquad (5.40)$$

There is no factor of 2 here, as there is in the Littrow prism monochromator, because we assume the grating is used only once. Actually, as we shall see, some monochromator designs do indeed make double or multiple use of the grating.

The diffraction-limited resolution of a grating monochromator, $R = \lambda/\Delta\lambda$, may be calculated from the diffraction slit width of Eq. (5.24) and the linear dispersion of Eq. (5.40),

$$R = \frac{\lambda}{D_l^{-1}(f\lambda/A)} = \frac{An}{a} \sec \theta_d$$

Now A, the width of the exit pupil, is determined by the width of the grating G and the angle its normal makes with the departing beam of radiation. Specifically,

$$G = A \sec \theta_d$$

So the resolution can be written

$$R = Gn/a = Nn$$

where N is the total number of lines ruled on the grating.

We have seen that the angle of diffraction is determined by the spacing of the grooves ruled in a grating. We have seen that the angular spread of a diffracted beam of monochromatic radiation is determined by the width of the grating. Finally, we have intimated that the distribution of diffracted radiation into various orders is determined by the shape of the facets of the grating. Let us consider these points in greater detail.

In Section 4.VI it was shown that the diffraction spread function of an optical system could be calculated by finding the Fourier transform of the amplitude of the electric field in the exit pupil and then multiplying the transform by its complex conjugate. Imagine a simple reflection grating made by cutting slots in a plane mirror and placing it at the aperture stop normal to the incident beam. The remaining reflecting facets have width b and are equally spaced with period a. The amplitude distribution in the aperture is given by

$$f(\beta) = \sum_{n=0}^{N-1} p_b\left(\beta - n \cdot \frac{2\pi a}{R\lambda}\right) \tag{5.41}$$

where $p_b(\beta)$ is a rectangular function of unit height and width b, and β is the coordinate in the exit pupil introduced in Eq. (4.36), namely, $\beta = 2\pi x/R\lambda$. The function $f(\beta)$ is shown in Fig. 5.27. The Fourier transform of $f(\beta)$ is

$$F(x') = P(x') \sum_{n=0}^{N-1} \exp\left(-j \cdot \frac{\pi n a x'}{R\lambda}\right) \tag{5.42}$$

where $P(x')$ is the transform of $p_b(\beta)$,

$$P(x') = \frac{\sin \pi x'b/R\lambda}{\pi x'b/R\lambda}$$

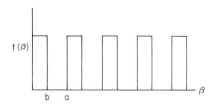

f(β)

b   a

β

Fig. 5.27   Amplitude distribution in the plane of an ideal grating.

Equation (5.42) can be written

$$F(x') = \frac{\sin \pi x' b/R\lambda}{\pi x' b/R\lambda}\left[\frac{1-\exp(-j\pi Nax'/R\lambda)}{1-\exp(-j\pi ax'/R\lambda)}\right]$$

The line spread function is found by calculating $F(x')F^*(x')$ and normalizing. We find

$$s(x') = \left(\frac{\sin \pi x' b/R\lambda}{\pi x' b/R\lambda}\right)^2\left(\frac{\sin \pi Nax'/R\lambda}{N \sin \pi ax'/R\lambda}\right)^2 \qquad (5.43)$$

Near the Gaussian image point, $x' = 0$, the second factor is very nearly

$$\left(\frac{\sin \pi Nax'/R\lambda}{\pi Nax'/R\lambda}\right)^2 = \left(\frac{\sin \pi Ax'/R\lambda}{\pi Ax'/R\lambda}\right)^2$$

where A is the width of the grating. This is the same function as the spread function associated with diffraction by a rectangular aperture of width A, according to Eq. (4.42). Unlike the case of the uniform aperture, however, the denominator of the second factor of Eq. (5.43) is really $N \sin \pi ax'/R\lambda$, and this causes the diffraction pattern to be repeated at intervals given by

$$\frac{\pi ax'}{R\lambda} = n\pi \qquad n = 0, \pm 1, \pm 2, \ldots \qquad (5.44)$$

Now $x'/R = \sin \theta_d$. Therefore, Eq. (5.44) reduces to the grating equation (5.36) for normal incidences, or

$$n\lambda = a \sin \theta_d$$

The first factor of Eq. (5.43) is recognized as the diffraction pattern of a rectangular aperture whose width is the width of the reflecting facets of the grating. Its effect is to modify the intensity distribution of the diffracted radiation. This is illustrated in Fig. 5.28. The intensity of order n is given by

$$I_k = \left(\frac{\sin n\pi b/a}{n\pi b/a}\right)^2$$

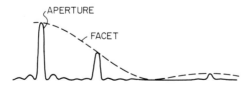

Fig. 5.28 Contributions to the intensity distribution of a grating by the aperture and the facets.

which is normalized to unity for order $n = 0$. Notice that diffracted rays of orders

$$\pm n = a/b, 2a/b, \ldots$$

have vanishing intensity. Furthermore, in the limit of vanishingly small grooves between the facets ($b \approx a$), only the zeroth order carries energy. This, of course, is the case of the plane mirror.

If the incident radiation approaches the grating in a nonnormal direction at angle θ_i, the phase of the wave front across the grating must be considered (see Fig. 5.29). The phase ϕ is given by

$$\phi = \frac{2\pi x}{\lambda} \sin \theta_i = R\beta \sin \theta_i$$

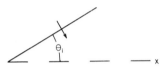

Fig. 5.29   Phase across the wave front for nonnormal incidence.

and in place of Eq. (5.41) we have

$$f(\beta) = \exp\left(-jR\beta \sin \theta_i\right) \sum_{n=0}^{N-1} P_b\left(\beta - n\frac{2\pi a}{R\lambda}\right)$$

The effect of the phase term is to introduce a coordinate shift in the Fourier transform. The line spread function becomes

$$s(x') = \left[\frac{\sin(\pi b/R\lambda)(x' + R\sin\theta_i)}{(\pi b/R\lambda)(x' + R\sin\theta_i)} \cdot \frac{\sin(N\pi a/R\lambda)(x' + R\sin\theta_i)}{N\sin(\pi a/R\lambda)(x' + R\sin\theta_i)}\right]^2 \tag{5.45}$$

Maxima occur when the denominator of the second term vanishes, that is, when

$$\frac{\pi a}{R\lambda}(x' + R\sin\theta_i) = n\pi \qquad n = 0, \pm 1, \pm 2, \ldots$$

or

$$a(\sin\theta_i + \sin\theta_d) = n\lambda$$

This is the general grating equation, Eq. (5.36). The first term of Eq. (5.45) tells us that maximum efficiency occurs for $\theta_i = \theta_d$. Notice also that in the case of the plane mirror ($b = a$), the zeroth order is the only nonvanishing diffracted order, and it is diffracted in the direction $\theta_d = -\theta_i$. This is, of course, the elementary law of reflection.

Fig. 5.30   Phase across the wave front in a blazed grating.

Now let us blaze our grating by tilting the facets out of the plane of the grating by an angle B. This introduces an additional phase (see Fig. 5.30)

$$\phi = -2\frac{2\pi(x-na)}{\lambda}\sin B$$

so the field distribution in the exit pupil is

$$f(\beta) = \exp(-jR\beta\sin\theta_i)\sum_{n=0}^{N-1}p_b\left(\beta - \frac{n\cdot2\pi a}{R\lambda}\right)\exp\left[2j\left(\beta R - \frac{2\pi na}{\lambda}\right)\sin B\right]$$

The line spread function is

$$s(x') = \left[\frac{\sin(\pi b/R\lambda)(x'+R\sin\theta_i-2R\sin B)}{(\pi b/R\lambda)(x'+R\sin\theta_i-2R\sin B)}\right]^2$$

$$\cdot\left[\frac{\sin(N\pi a/R\lambda)(x'+R\sin\theta_i)}{N\sin(\pi a/R\lambda)(x'+R\sin\theta_i)}\right]^2 \tag{5.46}$$

Once more, the second term defines the direction of the diffracted radiation. The first term tells us that maximum efficiency occurs for

$$\sin\theta_d+\sin\theta_i-2\sin B = 0$$

If the grating equation (5.36) is introduced into the first term of Eq. (5.46), we have for the diffraction efficiency of a blazed grating

$$s_e(\lambda) = \left[\frac{\sin(\pi b/a)(n-\lambda_B/\lambda)}{(\pi b/a)(n-\lambda_B/\lambda)}\right]^2 \tag{5.47}$$

where λ_B is the first-order blaze wavelength given by Eq. (5.39). Maximum efficiency occurs for observation at $\lambda = \lambda_B/n$ in the nth order.

Values of $s_e(\lambda)$ calculated from Eq. (5.47) are plotted in Fig. 5.31 for $b = a$. The intensity of the first-order diffracted radiation falls rapidly at wavelengths shorter than λ_B and becomes zero at $\frac{1}{2}\lambda_B$. At wavelengths longer than λ_B the rate of decrease is much slower. The distributions of higher order diffracted radiation become increasingly

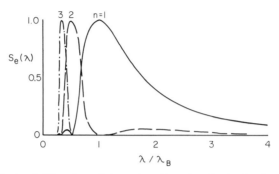

Fig: 5.31 Calculated diffraction efficiency for a blazed grating having facet widths equal to the spacing.

narrower and more closely spaced. By shifting from order to order, it is possible to cover a wavelength range from $\lambda = 0$ to $2\lambda_B$ with no less than 40% efficiency.

The theory discussed here and summarized in Fig. 5.31 serves as a useful guide and it approximates fairly well the properties of real gratings. It is called a scalar theory because it takes no account of the vector nature of radiation. It can be refined by the inclusion of the effects of obliquity and shadowing (13), but it remains an approximation. In reality the lobes of Fig. 5.31 shift somewhat for radiation polarized in different directions. More rigorous diffraction theory can, in principle at least, account more quantitatively for the behavior of gratings.

The actual performance of a grating, of course, falls somewhat short of the ideal if its construction is less than perfect. There are a number of defects in ruling that can afflict a grating. For the most part they are no longer observed with modern gratings, but we mention them if only for historical reasons. Errors in the spacing of the rulings of a grating can be aperiodic or periodic. Aperiodic errors might be random, they might vary slowly and smoothly across the grating width, or they might occur as sudden jumps in grating space once or several times across the grating. These errors can produce either extra satellite lines in a spectrum or multiple spectra. Aperiodic errors in spacing produce pairs of spurious lines symmetrically placed on either side of the real or parent lines at wavelength displacements given by

$$\Delta\lambda = \frac{k\lambda a'}{na} \qquad k = 1, 2, \dots$$

where a' is the spacing or periodicity of the periodic error, a is the average spacing of the rulings, and n, as before, is the order of the spectrum. These spurious lines are *Rowland ghosts*. The ghosts of higher orders are weakest. If there are two periodic errors in the grating, *Lyman ghosts* occur from the periodic interaction of the two errors, just as beat frequencies occur in acoustics. The Lyman ghosts are found far from the parent line and are quite weak.

There is another kind of so-called anomalous behavior not associated with grating defects at all, but occurring even with perfect gratings. The *Wood anomalies* appear as fairly narrow regions of abnormally high or low intensity in the radiation diffracted by reflection gratings. Their properties have been reviewed by Stewart and Gallaway (*14*). Wood (*15*) studied these anomalies as early as 1902. He found that the light missing from dark bands in the diffracted beam is found in the reflected zero-order beam, and conversely, bright anomalies in the diffracted beam have stolen their excess radiation from the zero order. He also established that the anomalies are highly sensitive to the state of polarization of the incident light. The anomalies usually observed are S anomalies that occur for radiation whose electric vector is perpendicular to the direction of the grating rulings. P anomalies have also been observed with the electric vector parallel to the rulings. The P anomalies are strongest in deeply grooved gratings and are not seen at wavelengths greater than the groove depth. The S anomalies, on the other hand, are generally stronger at longer wavelengths.

In 1907 Lord Rayleigh (*16*) associated the Wood anomalies with certain mathematical singularities in his "dynamical theory of the grating." These singularities occur at the wavelength for which radiation of the same wavelength is diffracted in some order so as to graze the surface of the grating. These are the Rayleigh wavelengths and they occur very close to the maximum disturbance of the Wood anomalies. It is as though a diffracted ray falls below the horizon of the grating and distributes its intensity among the other orders. This, of course, is a picturesque description only.

Equations for the Rayleigh wavelengths are easily derived with reference to Fig. 5.32. The diffracted ray of wavelength λ_R and order n under observation in the spectrophotometer is described by Eq. (5.38). Simultaneously, radiation of the same wavelength is diffracted so as to graze the surface of the grating. If the grazing order falls on the same side of the normal as the incident ray, its order must be greater than the order of the observed ray. Hence, we can write Eq. (5.36) for a ray of order $n + k$ and $\theta_d = \frac{1}{2}\pi$

$$(n+k)\lambda_R = a(\sin\theta_i + 1) \qquad k = 1, 2, \ldots \qquad (5.48)$$

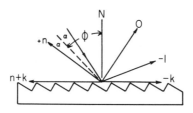

Fig. 5.32   Rayleigh wavelength for Wood's anomalies.

On the other hand, if the grazing ray falls on the opposite side of the normal, it must be a negative order, and we can write

$$-k'\lambda_R = a(\sin\theta_i - 1) \qquad k' = 1, 2, \ldots \tag{5.49}$$

The solution of Eq. (5.36) and (5.48) is

$$\lambda_R = \frac{2a}{n+2k}\left\{\frac{1 \pm [2(nk+k^2)^{1/2}/(n+2k)]\sin\alpha}{1 + [n\tan\alpha/(n+2k)]^2}\right\} \tag{5.50}$$

Exactly the same equation in k' is found by solving Eqs. (5.36) and (5.49).

If the angle between the incident and the observed rays is zero, as in an ideal Littrow monochromator, Eq. (5.50) simplifies to

$$\lambda_R = \frac{2a}{n+2k} \quad \text{or} \quad \frac{2a}{n+2k'} \tag{5.51}$$

Referring to Fig. 5.32, as the direction of the incident radiation is rotated counterclockwise (grating rotated clockwise), the diffracted rays move clockwise and longer wavelengths are observed. At the same time, rays of negative order approach grazing from the front of the grating while positive orders seem to appear from behind it. Equation (5.51) tells us that pairs of positive and negative rays approach grazing simultaneously in this case. For spectra observed in the first order, Rayleigh wavelengths occur at $\lambda_R = \frac{2}{3}a, \frac{2}{5}a, \ldots$, for spectra in the second order they are at $\lambda_R = \frac{2}{4}a, \frac{2}{6}a, \ldots$.

If α does not vanish, the double roots in Eq. (5.50) distinguished by the plus-or-minus sign indicate that pairs of rays do not approach grazing exactly simultaneously. If α is not too large, less than 15° say, we can set

$$1 + \left(\frac{n\tan\alpha}{n+2k}\right)^2 \approx 1$$

and the separation between pairs of Rayleigh wavelengths is

$$\frac{\Delta\lambda_R}{\lambda_R{}^0} \approx \frac{4(nk+k^2)^{1/2}}{n+2k}\sin\alpha$$

where $\lambda_R{}^0$ is calculated from Eq. (5.51). The wavelength separation thus amounts to about $1\frac{3}{4}\%$ of $\lambda_R{}^0$ per degree of angle 2α.

An example of a Wood anomaly in an infrared grating monochromator may be seen in Fig. 5.33. Notice first of all that this is an S anomaly because it is nearly unobservable with parallel polarized radiation. The monochromator has a 75-line/mm grating used in this case in the first order, so $n = 1$ and $a = 0.0133$ mm. Hence, from Eq. (5.51), a Wood

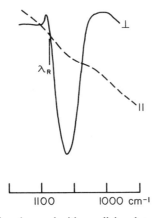

Fig. 5.33 Wood's anomalies observed with parallel and perpendicular polarized radiation in a Beckman IR-7 spectrophotometer with a 75-line/mm grating in first order.

anomaly is expected near 8.9 μ or 1120 cm⁻¹, as the second-order and the minus-first-order diffracted radiations of the same wavelength graze the grating surface. This monochromator actually has a small angle, $2\alpha = 4°$, between incident and observed rays. Equation (5.50) tells us that the anomaly associated with grazing second-order radiation shifts to about 1090 cm⁻¹, where the greatest rate of change of intensity is seen to occur. The other anomaly shifts to about 1150 cm⁻¹.

In the previous section on prisms we discussed the way in which a prism causes the image of the monochromator slit to be distorted. A grating also distorts the slit image. This has been discussed by Randall and Firestone (17), Minkowski (18), Rupert (19), and Peisakhson (12). In order to understand the origin of the distortion, return to Eqs. (5.33) and (5.34) and consider Fig. 5.34. If the incident and diffracted rays are off-axis by an angle ϵ, then the excess distances $\overline{fe}$ and $\overline{eh}$ become

$$\overline{f'e} = a \sin \theta_i \cdot \cos \epsilon \quad \text{and} \quad \overline{eh'} = a \sin \theta_d \cdot \cos \epsilon$$

With these modified relationships, the grating equation becomes, in place of Eqs. (5.36) and (5.38),

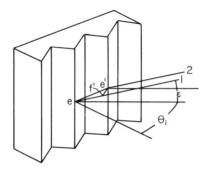

Fig. 5.34 Out-of-plane diffraction by a diffraction grating.

$$n\lambda = a \cos \epsilon (\sin \theta_i + \sin \theta_d) = 2a \cos \epsilon \cos \alpha \sin \phi \qquad (5.52)$$

From Eq. (5.52) it is seen that as ϵ increases, $|\theta_d|$ becomes larger. We know that for paraxial rays $|\theta_d|$ is larger for longer wavelengths. Therefore, a grating produces a distorted slit image whose ends are curved toward longer wavelengths. This is shown in Fig. 5.35. This is just the opposite of the action of a prism.

ENTRANCE SLIT   LONG λ SHORT λ IMAGE OF SLIT

Fig. 5.35 Sense of slit image distortion in a grating monochromator.

The magnitude of the slit image distortion can be calculated by differentiating Eq. (5.52)

$$\cos \theta_d \frac{d\theta_d}{d\epsilon} = \frac{n\lambda \sin \epsilon}{a \cos^2 \epsilon}$$

and making the following substitutions, with reference to Fig. 5.36:

$$\Delta \epsilon \approx \sin \Delta \epsilon \approx \Delta y / f \quad \Delta \theta_d = \Delta x / f \quad \text{and} \quad \cos^2 \epsilon \approx 1$$

Hence,

$$\frac{d\theta_d}{d\epsilon} = \frac{\Delta x}{\Delta y} = \frac{n\lambda y}{fa \cos \theta_d} = \frac{\lambda y D_l}{f^2} \qquad (5.53)$$

in terms of the linear dispersion of Eq. (5.40). Equation (5.53) is easily integrated to give the alternate forms

$$x = \frac{n\lambda y^2}{2fa \cos \theta_d} = \frac{\lambda D_l y^2}{2f^2} \qquad (5.54)$$

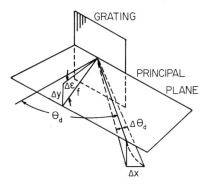

Fig. 5.36 Slit image distortion by a diffraction grating.

This equation, just as for a prism, defines a parabola. The radius of curvature in the vicinity of $\Delta y = 0$ is

$$r_{CG} = \frac{fa \cos \theta_d}{n\lambda} = \frac{f^2}{\lambda D_l} \qquad (5.54a)$$

The first form of Eq. (5.54a) tells us that if the slits are curved to match the slit image distortion for radiation of wavelength λ observed in order n at an angle of diffraction θ_d, the proper curvature will also be attained at the same angle for wavelength $\frac{1}{2}\lambda$ in order $2n$, and so on. Thus, a monochromator using the same grating in several orders can have properly curved slits at several wavelengths, unlike a prism monochromator whose slits are correctly curved for only one wavelength. The distortion introduced by a prism is generally somewhat greater than that introduced by a grating. For example, a 60° sodium chloride prism used with a 500-mm focal length collimator displaces the ends of the image of a 25-mm slit by about 0.4 mm at 3 μ. A 300-line/mm grating used in the first order with the same collimator displaces the image only 0.16 mm (in the opposite direction). Moreover, because the linear dispersion of the grating is roughly ten times larger than that of the prism at 3 μ the slit image curvature of the grating corresponds to a wavelength spread of only 4% of that of the prism.

A grating is not as sensitive to changes in environment as a prism. Changes of the grating with temperature are very slight and are due solely to thermal expansion which alters the value of a. Changes in the environment of the grating affect its optical properties by changing the wavelength of the radiation according to the relationship

$$\lambda' = \lambda_0/n_m$$

where λ_0 is the wavelength in vacuum and n is the refractive index of the medium surrounding the grating. Thus Eq. (5.38), for example, is written

$$n\lambda_0 = 2an_m \sin \phi \cos \alpha \qquad (5.55)$$

An expression for wavelength shift with changes in n_m is easily derived from Eq. (5.55)

$$\frac{\Delta\lambda_0}{\lambda_0} = \frac{\Delta\nu}{\nu} = \frac{\Delta n_m}{n_m}$$

The error is thus independent of the parameters of the grating or the monochromator. A change in environment from air ($n_m = 1.00025$) to vacuum introduces an error of 0.025%, or 0.25 cm^{-1} at 1000 cm^{-1}.

VI. OPTICAL FILTERS

Filters are used to attenuate radiation. Usually the attenuation is selective, with some wavelengths more strongly attenuated than others, in order to assist the monochromator in the purification of radiation. Thus a filter might be called upon to reduce stray radiation or unwanted grating orders. Because the properties of a filter as well as the function it performs are dependent on wavelength, it is usually necessary to change filters during the course of recording a spectrum. Often this must be done several times. The more sophisticated instruments perform the filter changes automatically.

Most filters might be classified loosely as either reflecting or transmitting optical elements, depending on whether they change the direction of propagation of the beam of radiation. Some filters, as we shall see, do not readily fit into either category. Filters can act by reflection, scatter, absorption, refraction, or interference. Following electrical engineering terminology, filters can also be classed as low pass or high pass, according to whether they permit the passage of low frequencies or high frequencies. Bandpass filters transmit radiation in a well-defined spectral region bounded at both long and short wavelengths by regions of high attenuation. Band-rejection filters attenuate strongly in a well-defined spectral region. In addition, there are neutral filters whose purpose is to attenuate radiation of all wavelengths by more or less the same factor. Useful neutral attenuators can be made of wire mesh, perforated screen, or metal louvers. The latter can be mounted in such a way that the transmission can be varied over a large range by varying the angle of the filter.

We observed in the discussion of reflection in Section 4.III, Eq.

(4.22), that very strong attenuation is accompanied by high reflectivity at a nearby wavelength. In Fig. 5.13 we sketched a typical plot of the behavior of the attenuation index and the refractive index in the vicinity of an absorption band. Notice that on the short-wavelength edge of the band the refractive index is very low, while on the long-wavelength edge it becomes large. Hence we expect a reflection maximum to occur at a somewhat longer wavelength than the absorption maximum. This is indeed just what is observed in the reflection spectra of solid materials in the infrared. The alkali halides, for example, are particularly strong reflectors near their vibration frequencies and poor reflectors elsewhere. They are, therefore, useful materials for the construction of reflecting bandpass filters. Such filters are called *residual ray* filters or, from the German, *Reststrahlen* filters. The reflection spectra of a number of these materials are included in Fig. 5.1B. Plates prepared from pressed powder can often be used in place of single crystals at considerable reduction in cost and with little sacrifice of performance. Reststrahlen filters have been of particular value in the far infrared in reducing stray radiation and overlapping grating orders. It is possible to select a sequence of crystals to isolate a succession of broad spectral bands. The reflecting properties, particularly the steepness of the boundary of the reflecting region, might be enhanced by application of a thin interfering film on the surface.

Often two or more crystals of the same material are used in series for more effective filtering. The reflectivity of k crystals in series is R^k. The very early far-infrared monochromators consisted of just these crystals, with no dispersing element. If the crystals are arranged in a crossed configuration, as in Fig. 5.37, the discrimination against the spectrally reflected short-wavelength radiation is more effective because of the polarization that accompanies reflection (20) (Section

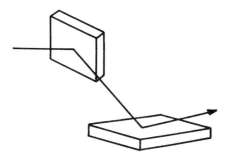

Fig. 5.37   Reflection from crossed Reststrahlen crystals.

4.III). The component of the incident radiation that is preferentially reflected by the first crystal is preferentially rejected by the second. The rejection is perfect if the radiation is incident on the crystal at Brewster's angle.

A certain amount of filtering of short wavelengths can be attained by selecting a mirror coating other than aluminum. Gold has a rather low reflectivity to visible radiation for a metal, but its reflectivity in the infrared is good. At 575 mμ the reflectivity of gold evaporated in ultrahigh vacuum is 0.871, and a system comprising ten reflections from gold-coated materials passes only $(0.871)^{10} = 25\%$ of the incident radiation. But at 5 μ the reflectivity of gold is 0.994, and $(0.994)^{10} = 94\%$ of the radiation is passed. A gold mirror is thus a low-pass filter.

A reflecting mirror can also be made to reject short wavelengths by rough grinding, omitting the polishing stage, before aluminizing it. Reststrahlen crystals sometimes are left rough ground in order that they can perform double duty as scatter filters, and filters are occasionally coated with soot that both scatters and absorbs. As we saw in Section 4.III, scattering by a particle, and therefore by the particle-like protrusions from a rough surface, is inversely proportional to the fourth power of the wavelength. Hence short-wavelength radiation is strongly scattered out of the beam of the optical system, leaving the long-wavelength radiation nearly unattenuated. A rough mirror may be mounted on a moveable member along with a polished mirror, thereby permitting either to be swung into the beam as desired.

Diffraction gratings can be used as low-pass or short-wavelength rejection filters (21). Consider a grating arranged as a mirror with an angle of incidence of 45°. The longest wavelength that can be diffracted by such a grating, $\theta_d = 90°$, is

$$\lambda_{max} = a(1 + \sin 45°) = 1.707a$$

Radiation having wavelengths longer than this can only be reflected into the zero-order beam. Shorter wavelengths, however, are partitioned into zero-order and diffracted-order beams and the latter are removed by means of baffles. Consequently, shorter wavelengths are attenuated. When the wavelength is very short relative to the grating spacing, the attenuation is quite efficient. We should be on the alert for Wood's anomalies in the radiation reflected from a grating filter. The Rayleigh wavelengths for zero-order radiation are found by setting $\phi_d = \pm\frac{1}{2}\pi$, and are given by

$$\lambda_R = \frac{a}{m_\pm}(\sin \phi_i \pm 1) \quad |m| = 1, 2, \ldots$$

where $+1$ is used when the sign of m is positive. Sometimes, in order to take advantage of the polarizing properties of the grating and the monochromator, it is advantageous to orient the grating so that its rulings are perpendicular to the monochromator slits. In this case it can be shown that the Rayleigh wavelengths occur at

$$\lambda_R = \pm a \cos \phi_i / m_\pm \qquad |m| = 1, 2, \ldots$$

Transmission gratings are also used as filters. They are discussed more fully in the next section on polarizers. Effective low-pass filters for far-infrared radiation have been reviewed by Ulrich (22).

It is fairly easy to devise band-rejection transmission filters centered at any wavelength by selecting one of the many thousands of chemical compounds whose absorption spectra are known. To be sure, there are some regions where strong absorption bands are scarce, such as the 4 to 5 μ region. The usefulness of such filters is usually restricted to establishing the level of stray radiation in the instrument at specified wavelengths. Unfortunately, strong infrared absorption bands are usually so plentiful that other spectral regions are rejected as well, resulting in a biased measurement.

High-pass transmission filters are readily available in the form of infrared transmitting crystals such as the alkali halides. A good selection of cutoff frequencies ranges from the near infrared (quartz) to 180 cm^{-1} in the far infrared (CsI). Low-pass transmission filters are not so plentiful. There are some semiconducting materials having absorption edges in the infrared. At frequencies higher than the absorption edge, attenuation is very nearly complete. At lower frequencies absorption is often quite low, usually being limited in pure materials to some weak lattice vibrations and absorption by free carriers which, although becoming stronger at lower frequencies, is generally not excessive. However, the high refractive indices of semiconductors results in quite high reflection losses. Germanium, for example, has its absorption edge at about 5500 cm^{-1}. Its refractive index is 4.07 at 4000 cm^{-1} so, although losses by absorption are essentially zero there, the reflectivity of a single surface is 37%. Reflection losses can be significantly lowered by properly coating the material, but this is restricted to a definite wavelength region.

The effect of low-pass filtering can be obtained with a high-pass filter in some spectrometer systems by means of a crystal chopper. Modern infrared instruments employ radiation choppers to convert steady radiation from the source to a pulsating beam. In a very real sense, the chopper is acting as part of a filter because it permits discrimination between the chopped radiation from the source and the steady background radiation from the surrounding world. If the chopper is constructed from a crystal that transmits at high optical frequencies but absorbs low frequencies, the low-frequency radiation from the source will be chopped and, therefore, detected, but the high frequencies will not. This technique is usually applied only to spectrophotometers operating as single-beam instruments, because in double-beam instruments the chopper usually also serves as a beam switcher.

Absorbing material in the form of powder can be suspended in a plastic medium to provide a convenient transmission filter. The alkali halides and other ionic crystals are good choices because of their strong attenuation in the Reststrahlen region. The use of plastic suspensions permits control of the amount of absorption and the bandwidth. Thin single crystals would generally have to be too thin and brittle to provide the desirable absorption, although they might be vacuum deposited on some rigid, transparent substrate. In addition, the choice of particle size permits a degree of filtration from scattering of short-wavelength radiation as well. Carbon black suspended in polyethylene is an excellent low-pass filter for the far infrared as it absorbs or scatters all visible and near-infrared radiation and most of the mid infrared. When used in combination with a crystal quartz window, no radiation of wavelength shorter than about 40 μ is transmitted. Powdered materials can be suspended in polyethylene by a technique in which the powder is milled into the polyethylene on a hot rotary mill. This method has been described by Yamada, et al. (23).

There is a phenomenon often observed in the absorption spectra of suspended particles which can be used to advantage in forming narrow bandpass filters. This is the *Christiansen filter effect*. Recall that the amount of radiation scattered by a particle of refractive index n_p in a medium of refractive index n_0 is proportional to $(n_p - n_0)^2/(n_p + n_0)^2$. In general both n_p and n_0 change with wavelength, but in the vicinity of a strong absorption band, the refractive index of the sample changes rapidly by an amount that can be large. If during this excursion n_p should take the same value as n_0 at some wavelength, the scatter will vanish at that wavelength. The spectrum will have the appearance sketched in Fig. 5.38. There is strong scatter at short wavelengths and strong absorption at long wavelengths, but at λ there is neither absorption nor scatter.

Filters can be constructed on the principle of separation of wavelengths by refractive dispersion. Indeed, a prism monochromator is a tunable filter of variable bandpass. When used as a foreprism mono-

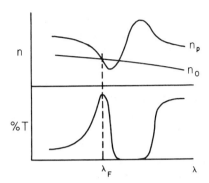

Fig. 5.38   Christiansen filter effect.

chromator preceding a grating monochromator, it provides ideal filtering of overlapping diffraction orders and stray radiation. Unfortunately, prism monochromators are not inexpensive and prism materials are not available for the very far infrared. To be sure, thin plates of crystal quartz and some other substances transmit far-infrared radiation, but in thicknesses necessary for effective prisms they are not sufficiently transparent. Rubens and Wood (24) have taken advantage of the dispersion of quartz to construct a far-infrared focal-isolation filter (see Fig. 5.39). Crystal quartz has a refractive index of about 1.54 to visible radiation and an index of about 2.1 in the far infrared. Thus a quartz lens has a shorter focal length for far-infrared radiation than for near-infrared and visible light. If it is used to focus an image of the source on an opening in a baffle, near-infrared and visible radiation from the source will converge at a much slower rate, or even diverge so that most of it will not pass through the opening. Mid-infrared radiation is absorbed by the quartz.

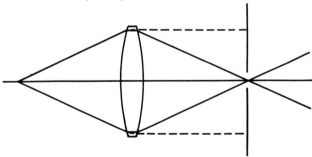

Fig. 5.39   Focal isolation.

The last type of filter which we shall discuss is the interference filter. This is a highly specialized subject and we can illustrate only a few examples. Interference filters might be either reflecting or transmitting. We have already mentioned the use of an elementary form of interference in connection with coatings applied to transmitting semiconductor filters and reflecting Reststrahlen filters. The purpose of such coatings is either to enhance or diminish the reflectivity of a surface. Consider a thin coating of thickness t and refractive index n_1 deposited on the surface of a substrate material having refractive index n_2 (Fig. 5.40). Some of the incident radiation will be reflected at the surface of the film with reflectivity

$$R_{01} = \left(\frac{n_0 - n_1}{n_0 + n_1}\right)^2 \tag{5.56}$$

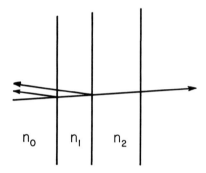

Fig. 5.40   Reflection from a coated surface.

More radiation will be reflected at the interface between the film and the substrate with reflectivity

$$R_{12} = \left(\frac{n_1 - n_2}{n_1 + n_2}\right)^2 \tag{5.57}$$

If the thickness of the film is just one-fourth of the wavelength of the radiation inside the medium, that is, $t = n_1\lambda/4$, and if the two reflected beams have the same intensity, they will totally destroy each other by interference. The condition for equal reflectance, from Eqs. (5.56) and (5.57), is

$$n_1 = (n_0 n_2)^{1/2}$$

We have ignored further internal reflections. Simple antireflection coatings are very useful if a restricted wavelength region is to be covered, such as in photographic optics. But if an extended range is

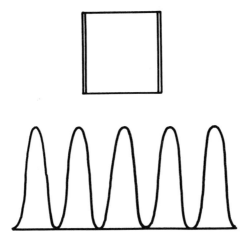

Fig. 5.41   Interference fringes from a thick etalon.

required, the film thickness obviously can be proper for only certain wavelengths.

The Fabry–Perot interferometer (Section 4.V) is an elementary interference transmission filter. If it is constructed in the form of a thin, transparent dielectric layer coated on both sides with a highly reflecting film, as in Fig. 5.41, its transmission spectrum will consist of a sequence of maxima as shown. Another coated layer of half the thickness of the first produces a similar sequence of maxima with twice the spacing (Fig. 5.42). By properly selecting the thickness,

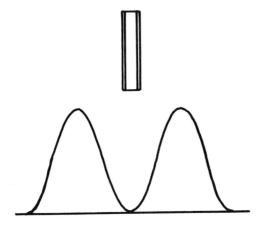

Fig. 5.42   Interference fringes from a thin etalon.

both interferometers may have a coincident maximum at frequency ν_0, but no other coincident maxima in the immediate neighborhood of ν_0. These interferometers can be mounted in tandem (Fig. 5.43) to produce a spectrum which is the product of the first two. Further elements can be added to suppress more of the remaining maxima and sharpen the desired one at ν_0 until only a single transmission maximum remains. Rather than place separate interferometers in tandem, the successive elements may be built upon the same substrate so that the reflecting films are shared as in Fig. 5.43 top. The analysis is now complicated because there are more possibilities of multiple internal

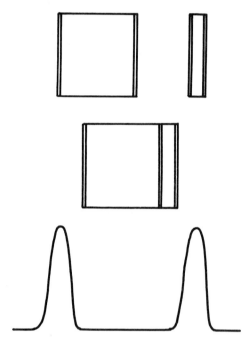

Fig. 5.43   Fringes from a thick and thin etalon in tandem (top) or combined (center).

reflections. Modern infrared interference filters are made of scores of layers of dielectric material deposited on a substrate by vacuum deposition. Alternate layers have high and low dielectric constants. Because only dielectric materials are used the filters do not absorb radiation as thin metal films necessarily do. Radiation is either reflected or transmitted. In some filter designs, all layers may have the same optical thickness, namely a quarter of the wavelength of the center of the reflection band. Other designs may have some layers with a half wavelength optical thickness. The design of complex filters is usually carried out with the aid of a digital computer.

Transmission interference filters can be of several types. These are illustrated in Fig. 5.44 which shows typical spectra of a spike filter, a band pass filter, a short wavelength pass filter, and a long wavelength pass filter.

Interference filters can be made to function anywhere in the infrared where substrate and dielectric materials are available. Typical dielectric materials are cryolite, germanium, and silicon monoxide.

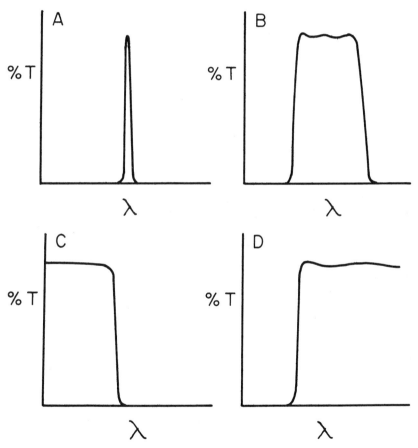

Fig. 5.44 Transmission spectra of several types of interference filters. A, spike filter; B, band pass filter; C, short wavelength pass filter; D, long wavelength pass filter.

A recent development is a narrow bandpass filter prepared on a circular substrate in such a way that the transmitted frequency varies uniformly around the disk. The desired frequency can therefore be tuned by properly orienting the disk in the beam. The bandwidth is narrow and the rejection at undesired frequencies is quite good, so the device can be used as a monochromator of modest resolution. The filters are produced by Optical Coatings Laboratory, Inc., and are used in the Beckman Microspec. and IR-20.

VII. POLARIZERS

Polarized infrared radiation is useful in the study of absorption spectra of oriented samples and in the investigation of reflectance spectra. Typical oriented samples are solids with a direction in which crystallites or fibres tend to be aligned, flowing liquids, and molecules in a magnetic field.

The function of a polarizer is to select the component of radiation whose electric vector lies in a preferred direction and to reject the component which lies in a perpendicular direction. Polarizers discriminate between components by preferential absorption, reflection, refraction, or diffraction. We shall discuss several examples of these.

A dichroic material absorbs radiation that is polarized in one direction and transmits radiation polarized in the perpendicular direction. Polaroid sheet is a well-known dichroic polarizer which is extremely useful from the near ultraviolet, through the visible region, and extending to about 2.7 μ in the near infrared (Polaroid HR) (25). Many materials exhibit dichroic behavior over fairly narrow regions in the infrared and could be used as polarizers of limited spectral range. A polarizer of crystal calcite (26), 1 mm thick and cut with its optic axis parallel to the surface, has been used at wavelengths near 3.45, 4.0, 9.35, and 10.5 μ and between 12.5 and 16 μ. This is not a particularly useful general-purpose polarizer, but can be quite valuable, for example, in studying laser radiation in the 3.4-μ region.

Dichroic polarizers have been made of thin sheets of pyrolytic graphite which is prepared by pyrolyzing hydrocarbons at a very high temperature and depositing the carbon thus produced on a plane surface (27). The carbon collects in a graphite form of strata of honeycomb-like structures. The strata tend to pile up parallel to the collecting surface. This is illustrated in Fig. 5.45. A polarizer is made from this block by cutting or grinding a thin foil from it (along the dashed

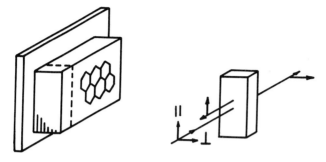

Fig. 5.45   Pyrolytic graphite polarizer.

lines in the figure). The thickness is about $10\,\mu$ or less. The foil is polished and mounted in a frame for structural support. Radiation incident normally on the polarizer in the direction shown is strongly reflected when the electric vector is parallel to the graphite layers, and is transmitted when the electric vector is perpendicular. In this way the foil acts as an efficient transmission polarizer, although it actually polarizes by preferential reflection. It has been used from wavelengths of 4 or 5 μ to the very far infrared beyond 0.5 mm.

Let us consider why a sheet of pyrolytic graphite acts as it does. The electrons associated with the bonds of the condensed graphite rings are relatively free to migrate along the plane of the rings, but cannot jump easily from plane to plane. Consequently, the material acts as a conductor of electric current in a direction parallel to the planes, and as an insulator in a direction perpendicular to the planes. The ratio of electrical conductivities in these two directions is more than 1000 to 1. Therefore, pyrolytic graphite acts as a metallic reflector to radiation with its electric vector in the plane of the graphite layers, and as a dielectric in the other direction.

We can imagine other structures that function in the same way. For example, a block made up of many thin evaporated layers, alternately metal and dielectric, should have electrical and optical properties greatly resembling pyrolytic graphite. Such a structure would be hard to fabricate, of course. A similar structure is a grid of closely and evenly spaced parallel wires, as in Fig. 5.46. The diameter and spacing of the wires must be small relative to the wavelength. Wire grids were used by Hertz in 1888 to polarize radio waves and by duBois and Rubens in 1911 in the far infrared. Bird and Parrish (*28*) devised a method of fabricating grids for use over the entire infrared. They start with an echelette grating replicated in a plastic which is as transparent

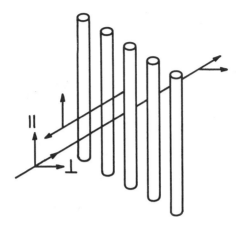

Fig. 5.46 Grid polarizer.

as possible over the infrared region of interest. Aluminum is deposited on the grating from a near grazing angle, as shown in Fig. 5.47. The result is a strip of aluminum only on the edge of each facet of the grating. This is equivalent to a grid of very fine wires. A blaze angle of about 20° gives a good groove shape. A very fine spacing is necessary for use at shorter wavelengths. Bird and Parrish used polyethylene or polychlorotrifluoroethylene replicas of a 2160-line/mm grating. Aside from the regions of absorption by the plastic, the polarizers were found to be very effective from $2\,\mu$ to the far infrared, with a high degree of polarization and low losses. There are other procedures for fabricating such polarizers. Gratings have been ruled in the surface of plates of Irtran 2 or 4, and aluminized to make grid polarizers (29). This avoids the discrete absorption bands of plastic substrates. Also, electroformed grids can be produced on various substrates (30).

We have seen that the reflectivity of a dielectric surface vanishes at the Brewster angle for radiation polarized with its electric vector lying in the plane of incidence. It is finite for the other plane of polarization. This effect can be used to construct polarizers of either the reflection or the transmission type. A transmission polarizer of the pile-of-plates type is shown in Fig. 5.48. In this example six plates are mounted at

Fig. 5.47 Evaporated grid polarizer.

Fig. 5.48   Pile-of-plates polarizer.

the Brewster angle. At each surface a portion of the perpendicular component is reflected out of the main beam, but none of the parallel component is reflected. Consequently, the transmitted beam is predominantly polarized parallel to the plane of incidence. A material of high refractive index is chosen to increase the reflectance for the perpendicular component. If internal reflections between the plates are ignored, the *proportion of polarization* is given in terms of the number of plates N and the refractive index n by

$$P = \frac{I_{\parallel} - I_{\perp}}{I_{\perp} + I_{\parallel}} = \frac{1 - [2n/(n^2 + 1)]^{4N}}{1 + [2n/(n^2 + 1)]^{4N}}$$

Polarizing plates of silver chloride, selenium, mylar, or polyethylene have been used. Polyethylene is especially useful for the far infrared. Sheets of the plastic are mounted on frames at elevated temperatures and as the plastic cools it becomes taut. Selenium can be used in the form of self-supporting (but fragile) films or deposited films on plastic substrates.

The form of polarizer sketched in Fig. 5.48 has the disadvantage of displacing the beam laterally by refraction in the plates. This can be compensated with a suitably tilted plate of some transparent material. A self-compensating configuration, devised by Makas and Schurcliff (*31*), is shown in Fig. 5.49. This arrangement is long and occupies a great deal of sample space. A variation due to Harrick (*32*) is shown in Fig. 5.50. It is limited to two plates, requiring a high refractive index for high polarization. Greenler *et al.* (*33*), constructed the

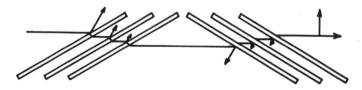

Fig. 5.49   Makas–Schurcliff pile-of-plates polarizer.

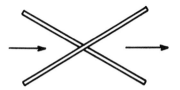

Fig. 5.50   Harrick polarizer.

form of polarizer shown in Fig. 5.51. Pleated films of plastic — very thin cellulose nitrate — are cast on forms and coated with selenium. Several of these sheets are mounted in a holder. This polarizer occupies much less space than the others.

Polarizers can also be made to work in reflection by collecting the radiation reflected from a single dielectric surface at the Brewster angle. A material of very high refractive index is necessary to provide good reflectivity. Germanium has been used by Harrick (*34*) to make such a polarizer. Harrick also constructed a polarizer utilizing the reflectance at a germanium–mercury interface (*35*).

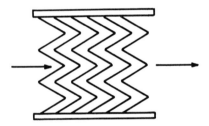

Fig. 5.51   Greenler–Adolf–Emmons pleated polarizer.

In the visible and ultraviolet regions of the spectrum, very efficient polarizers are made by using the double-refractive properties of birefringent materials. Birefringent substances that are transparent in regions of the infrared are known, and include crystal quartz, rutile, sapphire, cadmium sulfide, and lanthanum fluoride. A rutile birefringent crystal has been used to construct an infrared polarizer (*36*), and calcite has been used in the near infrared, but such designs are rare.

The radiation diffracted by an echelette grating is partially polarized, the percentage and predominant direction varying with the wavelength, blaze angle, and grating space. Thus the beam emerging from a grating monochromator is partially polarized. This can be used in some special cases, and the need for a polarizer relaxed. The radiation in a prism monochromator also tends to be polarized, in this case with the electric vector perpendicular to the slit because of preferential reflection at the prism faces.

In addition to polarizers, some infrared polarization experiments require the use of retardation devices to introduce a phase shift between components of plane-polarized light. A shift of a quarter wavelength produces circularly polarized radiation. A half-wavelength shift effectively rotates the plane of polarization by 90°. Such devices can be based on reflection or transmission phenomena.

Birefringent crystals are commonly used in the visible and ultraviolet regions and can be used in the infrared also when suitable materials are available. A crystal of thickness b is cut with the optic axis parallel to the surface, as in Fig. 5.52. Incident radiation is plane

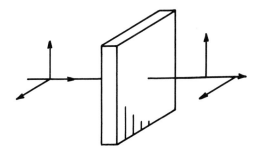

Fig. 5.52   Phase plate.

polarized at an angle of 45° with the optic axis. The component parallel to the optic axis passes through the crystal at a velocity determined by the ordinary refractive index, and the velocity of the other component is determined by the extraordinary index. The phase difference of the emerging rays is

$$\phi = 2\pi b \frac{(n_o - n_e)}{\lambda}$$

If the thickness is chosen to make $\phi = \frac{1}{2}\pi$, the crystal is a quarter wave plate and the transmitted radiation is circularly polarized. Of course, the retardation changes continuously and rapidly with wavelength, which can be a serious limitation of the device. A Soleil compensator can be constructed, as in Fig. 5.53, of two prisms that can be translated relative to each other. This permits the thickness to be varied, so any desired retardation can be selected at a given wavelength. Palik and Henvis (*37*) have constructed a Soleil compensator of CdS for use in the spectral region from 0.57 to 14 μ. Crystal quartz can be used in the far infrared beyond 60 μ (*38*).

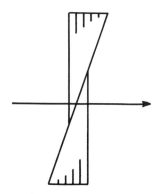

Fig. 5.53   Soleil compensator.

The Fresnel rhomb is a well-known device for the production of circularly polarized radiation. Radiation that is totally internally reflected suffers a phase shift

$$\phi_s = 2 \tan^{-1}\left[\frac{(\sin^2 \theta - n^2)^{1/2}}{\cos \theta}\right]$$

for the component perpendicular to the plane of incidence and

$$\phi_p = 2 \tan^{-1}\left[\frac{(\sin^2 \theta - n^2)^{1/2}}{n^2 \cos \theta}\right]$$

for the parallel component, where θ is the angle of incidence. The Fresnel rhomb, Fig. 5.54, internally reflects the radiation twice and, by properly selecting the angle of incidence, a net phase change of $\frac{1}{2}\pi$ rad. can be produced to make circularly polarized radiation. Palik has used Fresnel rhombs made of NaCl and CsI (*39*). Other materials can be used, of course. The radiation is perfectly circularly polarized for only one wavelength, and then only for ideally collimated radiation. But the radiation remains reasonably well circularly polarized over wide wavelength regions and a fairly large cone of incident radiation can be tolerated.

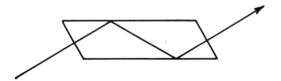

Fig. 5.54   Fresnel rhomb quarter wave plate.

A phase difference is introduced between the two components of plane-polarized radiation by reflection from a metal or semiconductor surface. From the known optical constants of a semiconductor, an angle of incidence might be selected for the production of circularly polarized radiation. Reflecting mirrors normally should not be inserted between polarizer, sample, and analyzer in polarization measurements, unless the plane of polarization is parallel or perpendicular to the plane of incidence on the mirror. Otherwise, the difference in phase introduced by reflection will elliptically polarize the radiation.

VIII. SOURCES

Probably the two most commonly used sources of radiation in spectrometers intended for use in the 2 to 50-μ region are the Nernst glower and the globar. Nernst glowers are rods of rare-earth oxides, largely cerium oxide, several millimeters in diameter and perhaps 2 or 3 cm long. Wires are attached to each end so that they may be heated by electrical conduction. Normally 0.3 to 0.6 A are passed through the glower, raising its temperature to about 1500°K and causing it to emit brilliantly. The glower is a semiconductor. Its resistance at room temperature is very high and its temperature must be raised with an auxiliary heater before it begins to conduct appreciably. Once conduction has started, however, the temperature rises, the resistance drops, and unless the current is limited, the glower runs away until it burns out. An incandescent light bulb in series with the glower is sometimes used as a limiter and regulator. A phototube in a feedback circuit can be used to monitor the glower temperature by sensing the emission of visible light.

A globar is a rod of silicon carbide, also heated electrically. The globar is usually not used at such high temperatures as the Nernst glower, and the electrodes must be cooled with circulating water. This makes the globar somewhat less convenient than the Nernst glower, and it is presently not used as much as in the past. Under typical operating conditions, a globar emits more radiation than a Nernst glower at longer wavelengths, and less at shorter wavelengths.

Bare metal filaments are not good emitters of infrared radiation, but are sometimes used in low-cost instruments because of their low price and long life. If the filament is coated with a ceramic, the emittance of the source can be enhanced considerably.

The most frequently used source of far-infrared radiation is a medium-pressure mercury arc. A very useful source is a General Electric AH-4 projection lamp with the outer glass envelope removed. The mercury arc is contained in a fused quartz inner envelope. At the shorter wavelengths of the far infrared, quartz is opaque. It becomes very hot and emits good quantities of thermal radiation. At the longer wavelengths, radiation from the mercury plasma is transmitted by the quartz and replaces the thermal radiation. Low-pressure gaseous discharges are not useful as general-purpose sources because they emit discrete line spectra. This makes them valuable for wavelength calibration, and standard wavelength values have been published (40) for argon, neon, krypton, and zenon.

The Welsbach mantle was used often in the past as a source of far-infrared radiation. It is prepared by soaking a loosely woven fabric in a solution of thorium nitrate. After it is dried, the fabric is ignited. A delicate lattice of thorium oxide remains, which, when heated in a flame, emits profusely in the infrared. The thorium oxide can also be prepared on a metal support that can be heated electrically.

Carbon arcs have also been used as far-infrared sources. The crater of a carbon arc is exceedingly hot, but the arc tends to be unstable. Zirconium arcs have also been proposed as infrared sources, but are not widely used.

Perhaps someday lasers may serve as useful sources in spectrophotometers. At present, because they cannot be tuned over large wavelength regions, their usefulness is limited to a few special studies. The subject of infrared sources has been reviewed by Cann (41).

REFERENCES

1. D. E. McCarthy, *Appl. Opt.*, **2**, 591, 596 (1963); **4**, 317, 507, 878 (1965); **6**, 1896 (1967).
2. D. E. McCarthy, *Appl. Spectry.*, **22**, 66 (1968).
3. D. E. Williamson, *J. Opt. Soc. Am.*, **42**, 712 (1952).
4. A. S. Filler, *Am. J. Phys.*, **29**, 687 (1961).
5. K. I. Tarasov, *Opt. Spectry.*, **8**, 380 (1960).
6. G. R. Rosendahl, *J. Opt. Soc. Am.*, **51**, 1 (1961); **52**, 408, 412 (1962).
7. E. W. Marchand, *J. Opt. Soc. Am.* **54**, 263 (1964).
8. I. V. Peisakhson, *Opt. Spectry.*, **11**, 65 (1961).
9. H. Kayser, *Handbuch der Spectroscopie*, S. Hirzel, Leipzig, 1900.
10. R. A. Sawyer, *Experimental Spectroscopy*, Prentice-Hall, Englewood Cliffs, N. J., 1951.
11. D. M. Gates, *Am. J. Phys.*, **20**, 275 (1952).
12. I. V. Peisakhson, *Opt. Spectry.*, **4**, 481 (1958).

13. R. P. Madden and J. Strong, in J. Strong, *Concepts of Classical Optics*, Freeman, San Francisco, Calif., 1958.

14. J. E. Stewart and W. S. Gallaway, *Appl. Opt.*, **1**, 421 (1962).

15. R. W. Wood, *Phil. Mag.*, **4**, 396 (1902); **23**, 310 (1912); *Phys. Rev.*, **48**, 928 (1935).

16. Lord Rayleigh, *Proc. Roy. Soc. (London)*, **A79**, 399 (1907); *Phil. Mag.*, **14**, 60 (1907).

17. H. M. Randall and F. A. Firestone, *Rev. Sci. Inst.*, **9**, 404 (1938).

18. R. Minkowski, *Astrophys. J.*, **96**, 306 (1942).

19. C. S. Rupert, *J. Opt. Soc. Am.*, **42**, 779 (1952).

20. R. C. Lord and T. K. McCubbin, Jr., *J. Opt. Soc. Am.*, **47**, 689 (1957).

21. J. U. White, *J. Opt. Soc. Am.*, **37**, 713 (1947).

22. R. Ulrich, *Infrared Phys.*, **7**, 65 (1967).

23. Y. Yamada, A. Mitsuishi, and H. Yoshinaga, *J. Opt. Soc. Am.*, **52**, 17 (1962).

24. H. Rubens and R. W. Wood, *Phil. Mag.*, **21**, 249 (1911).

25. R. P. Blake, A. S. Makas, and C. D. West, *J. Opt. Soc. Am.*, **39**, 1054 (1949).

26. T. J. Bridges and J. W. Klüver, *Appl. Opt.*, **4**, 1121 (1965).

27. G. Rupprecht, D. M. Ginsberg, and J. D. Leslie, *J. Opt. Soc. Am.*, **52**, 665 (1962).

28. G. R. Bird and M. Parrish, Jr., *J. Opt. Soc. Am.*, **50**, 886 (1960).

29. J. B. Young, H. A. Graham, and E. W. Peterson, *Appl. Opt.*, **4**, 1023 (1965).

30. M. Hass and M. O'Hara, *Appl. Opt.*, **4**, 1027 (1965).

31. A. S. Makas and W. A. Schurcliff, *J. Opt. Soc. Am.*, **45**, 998 (1955).

32. N. J. Harrick, *J. Opt. Soc. Am.*, **54**, 1281 (1964).

33. R. G. Greenler, K. W. Adolf, and G. H. Emmons, *Appl. Opt.*, **5**, 1468 (1966).

34. N. J. Harrick, *J. Opt. Soc. Am.*, **49**, 860, 1122 (1959).

35. N. J. Harrick, *J. Opt. Soc. Am.*, **49**, 376, 379 (1959).

36. R. Duverney and A. M. Vergnoux, *J. Phys. Radium*, **18**, 527 (1957).

37. E. D. Palik and B. W. Henvis, *Appl. Opt.*, **6**, 2198 (1967).

38. E. D. Palik, *Appl. Opt.*, **4**, 1017 (1965).

39. E. D. Palik, *Appl. Opt.*, **2**, 527 (1963).

40. K. N. Rao, C. J. Humphreys, and D. H. Rank, *Wavelength Standards in the Infrared*, Academic Press, New York, 1966.

41. M. W. P. Cann, *Appl. Opt.*, **8**, 1645 (1969).

CHAPTER 6

Optical Systems of Infrared Spectrophotometers

I. INTRODUCTION

We have discussed the important concepts of geometric and physical optics and described the properties of the optical components of infrared spectrophotometers. Now we are ready to consider the performance of complete optical systems. We shall treat separately the monochromator, source, photometer, detector optics, and optical systems involved in sample handling and special experiments.

II. MONOCHROMATOR OPTICS

The monochromator is the vital center of a spectrophotometer. Its essential function is to isolate selected bands of wavelength, and the success of an instrument depends on the resolving power and efficiency of its monochromator. In previous chapters we showed how the ultimate diffraction-limited spectral resolving power of a monochromator is determined by the size of the aperture stop (the prism or grating in most instruments) and by the dispersion of the dispersing element. We discussed the aberrations imposed on the radiation beam by the focusing elements, and an aberration, namely slit image distortion, introduced by the prism or grating. The effect of these factors on the performance of an instrument depends on the particular monochromator design.

There are several classes of monochromators, depending on the placement of slits, mirrors, and dispersing element. Let us describe them and discuss their relative merits and faults. Welford (*1*) has recently given a similar but less detailed discussion.

A. In-Plane Littrow Monochromator

A detailed ray-trace analysis of the in-plane Littrow monochromator with parabolic collimator and either a grating or a prism has been given by Yoshinaga, *et al.* (*2*). Kudo (*3*) has given appropriate calculations for the aberrations of a grating Littrow monochromator with a parabolic mirror. A typical Littrow monochromator with a prism is sketched in Fig. 6.1A. A grating Littrow monochromator is shown in Fig. 6.1B. Littrow systems are characterized by their asymmetric arrangement, with both the entrance and the exit slits on the same side of the dispersing element. In the in-plane Littrow arrangement the centers of all optical elements, including slits, lie on a common plane. The radiation enters through the entrance slit S_n. It is collimated by mirror M_1 which may be a spherical or paraboloidal mirror. The off-axis angle is α, and incident and reflected rays are separated by 2α. The disperser P or G is usually placed from 0.8 to 1.0 focal length from the collimator, partly in the interest of compactness. The dispersed beam in the prism instrument is returned by the Littrow mirror M_2 through the prism a second

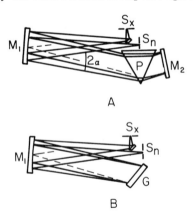

Fig. 6.1 A and **B**   In-plane Littrow monochromators with short-wavelength radiation dispersed farthest from the dispersing element: (**A**) prism; (**B**) grating.

time. The prism is usually fixed in place. The wavelength is selected by rotating the Littrow mirror. In some visible-ultraviolet monochromators, a thin (e.g., 30°) rotatable prism is aluminized on its back surface so that it acts as both prism and Littrow mirror. This arrangement is seldom used in the infrared. In the grating instrument, wavelength is selected by rotating the grating. The dispersed radiation is now returned to M_1, which acts as a condensing mirror, thus bringing the radiation to a focus at S_x and forming an imaging of the entrance slit on the exit slit. Often a small plane mirror M_3, the so-called *Newtonian* is inserted before the exit or entrance slit, or both, to permit physical separation of the two slits and the associated mechanism.

In the monochromators of Figs. 6.1A and B, the dispersed radiation extends along the plane of the entrance slit (or its image when the Newtonian is used) with the short-wavelength radiation farther from the prism (dotted rays) and long-wavelength radiation closer to it. Monochromators constructed as in Figs. 6.1C and D, on the other

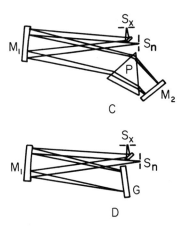

Fig. 6.1 C and D In-plane Littrow monochromators with long-wavelength radiation dispersed farthest from the dispersing element: **(C)** prism; **(D)** grating.

hand, disperse the radiation in such a way that the short wavelengths fall close to the disperser. There is a long-wavelength limit to the radiation passed by the system which is imposed by the optical transmission limit of the prism material or other elements. There is usually no short-wavelength limit, and furthermore there is many times more energy in the short-wavelength spectrum of the source than in the long wavelengths. Short-wavelength radiation falling on the nose of the prism or on the grating is reflected, scattered, or further dispersed, and some of it finds its way through the exit slit, where it contaminates the desired radiation observed there. This effect is often called double dispersion, and it is an important contribution to stray radiation in a monochromator. Therefore, the arrangements of Figs. 6.1A and B are much to be preferred over those of Figs. 6.1C and D for an infrared monochromator However, just the reverse is true for a monochromator intended for use in the visible-ultraviolet region. For this reason it is seldom satisfactory to convert an instrument from the visible ultraviolet to the infrared, or vice versa, by merely trading dispersing elements. Double dispersion can be avoided if a large off-axis angle of the collimator can be tolerated. This is illustrated in Fig. 6.2. A line OM is drawn from the center of curvature of the collimator mirror to the extreme edge of the mirror. The mirror lies entirely on one side of this line and the grating is mounted on the other side. Any ray from the grating that is reflected by the mirror must lie on the side of OM opposite the grating. Therefore, no ray can be returned to the grating and there is no double dispersion. The off-axis angle must be at least as large as $\sin^{-1} D/S$ where D is the width of the disperser and S is its distance from the collimator.

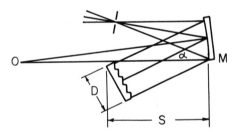

Fig. 6.2   Avoiding double dispersion.

The Littrow monochromator is vulnerable to stray radiation originating in another way. The area of the collimator mirror used to intercept radiation from the entrance slit is nearly coincident with the area used to send radiation to the exit slit. This is in fact one of the advantages of the Littrow system, as it permits use of the smallest possible mirror. Unfortunately, the mirror inevitably has imperfections, dust, fingerprints, and the like, which scatter a certain amount of the incident radiation. Radiation from the entrance beam which is scattered in the overlap region of the mirror common to both entrance and exit beams can proceed directly to the exit slit. Consequently, stray radiation in a Littrow monochromator is likely to be high.

The magnitude of the aberrations in any given optical system can be determined only by a thorough ray trace. We can discuss in a general way what to expect by way of image degradation by using the formulas given in Section 5.III.

We showed that spherical aberration from the collimating mirror is small in infrared instruments having relatively small mirrors and large f/numbers, that is, f/10 or larger. For monochromators of smaller f/number, spherical aberration imposes a significant limit on resolution, but it can be eliminated by using an off-axis parabolic collimating mirror in place of a spherical mirror. Any spherical aberration present in the system produces a symmetric broadening of spectral lines.

Coma is an important aberration in a Littrow system because of the necessity of working off-axis when using mirror optics and because the coma for the two reflections from the mirror is additive in an asymmetric system. Figure 6.3A is an end-on view of the monochromator looking from the slits into the collimator. The optic axis, which intersects the center of the collimator, is perpendicular to the figure. We assume for the present that straight slits are used and that they are very close together, so the separation can be ignored. With an off-axis angle α at the collimator, and with the slit ends defining an angle β from the optic axis, the field angle subtended by the ends of the slit is

$$\omega = \frac{h/2}{f \sin \beta}$$

The comatic patterns, sketched in the figure, have a length given by Eq. (5.12A) multiplied by 2 to allow for double use of the mirror. The projection of this length in a direction normal to the slit is thus (Fig. 6.3B), for a spherical collimator,

$$l'_c = \frac{3MD^2}{16f} \cdot \frac{h}{2f} \cdot \frac{\cos \beta}{\sin \beta}$$

$$= \frac{3MD^2}{16f} \cdot \frac{h}{2f} \cdot \frac{f \sin \alpha}{h/2} = \frac{3MD^2 \sin \alpha}{16f} \tag{6.1}$$

and this is approximately the width of the comatic image. Recall from Section 5.III that the distance of the aperture stop from the center curvature of the mirror is $2fM$. It is seen that l'_c is independent of slit height to this approximation, although it is clear from the figure that the broadening of the slit image is actually slightly greater at the ends of the slit. The comatic image of the slit is asymmetric (Fig. 6.3C). The image of a line object is sharp on the side closest to the optic axis and is de-

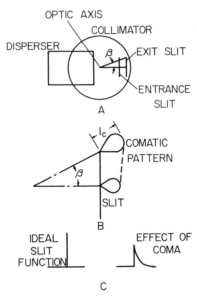

Fig. 6.3 Coma in an in-plane Littrow monochromator: (**A**) monochromator end view; (**B**) comatic pattern; (**C**) slit function.

graded on the other side. In an infrared Littrow monochromator of the type shown in Fig. 6.1A or B, the degradation occurs on the short-wavelength side of spectral lines.

The off-axis angle of an in-plane Littrow system with $M = 0.5$, that is, with the disperser one focal length from the collimator, must be at least

$$\alpha \geqslant \sin^{-1} \frac{D}{4f} \tag{6.2}$$

if we are willing to tolerate double dispersion. Therefore, from Eq. (6.1), the comatic image broadening for a system of this type is at least

$$l_c' \geqslant \frac{3 \times 0.5 D^2}{16 f} \cdot \frac{D}{4f} \approx 0.025 \frac{D^3}{f^2} = 0.025 \frac{f}{F^3}$$

where F is the f/number of the monochromator. For practical reasons, the off-axis angle must be somewhat larger than $\sin^{-1}(D/4f)$, making the comatic aberration proportionally larger.

We recall from Section 5.III that for the same off-axis angle, comatic aberration is twice as great when spherical aberration is removed by using a parabolic collimator. However, it is important to realize that when an off-axis paraboloid is used in a Littrow monochromator, the off-axis angle is not the same as it is when the same monochromator has a spherical collimator. The axis of the paraboloid bisects the line of separation between the entrance and exit slits. Thus the maximum off-axis angle is determined by the hypotenuse of the triangle formed by half the slit height and half the slit separation. Moreover, the comatic pattern lies along the slit length for small slit separation.

It is fortunate that spectrophotometer optics deal with a slit-like object, so only the aberrations that degrade the image in a direction perpendicular to the slit are important. It is customary to focus the monochromator with the appropriate astigmatic line image lying along the length of the slit. In the in-plane Littrow system the tangential image is chosen. This image lies on a curved surface in front of the paraxial image plane. With a straight slit and a spherical collimator the tangential image has a length given by Eq. (5.12c) multiplied by 2 to account for double use of the mirror. The component in the direction of the slit width at the extreme ends of the slits is given by (see Fig. 6.4A)

$$l'_{A,T} = 2M^2 D \sin^2 \omega \sin \beta \approx \frac{M^2 Dh}{f} \sin \alpha \qquad (6.3)$$

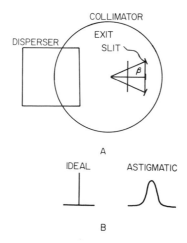

A

IDEAL   ASTIGMATIC

B

Fig. 6.4 Astigmatism in an in-plane Littrow monochromator: **(A)** monochromator end view; **(B)** slit function.

With the off-axis angle α restricted according to Eq. (6.2), at the extreme ends of the slits the astigmatic broadening takes values at least

$$l'_{A,T} \geq \frac{M^2 h D^2}{4f^2} = \frac{h D^2}{16f^2} \qquad (6.3)$$

Define the *l-number* of a monochromator as the ratio of slit height to focal length

$$l = h/f$$

This definition will be a useful one in several respects. For the moment, we use it to express Eq. (6.3) in the form

$$l'_{A,T} \geq \frac{lf}{16}\left(\frac{D}{f}\right)^2 = \frac{lf}{16F^2}$$

The astigmatic degradation of the slit image is decidedly dependent on the slit height. The *l*-number is generally no larger than about 1/20 in an in-plane Littrow system. At the center of the slit there is no astigmatic broadening. If a parabolic mirror is substituted to eliminate spherical aberration, the astigmatism also vanishes. The astigmatic broadening is symmetric, as shown in Fig. 6.4B.

The tangential astigmatic line lies on a curved surface of radius

$$R_{A,T} = \frac{f}{2(1 - 3M^2)}$$

after two reflections from the mirror. A point at the end of the slit is at least $\frac{1}{2}(h^2 + D^2)^{1/2}$ from the optic axis, so the tangential image of a point at the end of the slit falls a distance

$$z_{A,T} \geq (h^2 + D^2)(1 - 3M^2)/4f \qquad (6.4)$$

closer to the mirror from the Gaussian image plane. The beam is converging at a rate determined by the *f*/number of the system as it approaches the image. Therefore, the diameter of the broadened image of a point at the slit end due to astigmatic field curvature is at least

$$l'_{A,C} \geq \frac{D}{f} z_{A,T} = D \frac{(h^2 + D^2)(1 - 3M^2)}{4f^2}$$

$$= \frac{f}{4F}(1 - 3M^2)\left(l^2 + \frac{1}{F^2}\right)$$

This aberration also produces a symmetrically broadened image. There is no broadening at the center of the slit. The broadening by astigmatic field curvature can be avoided in principle by curving the plane of the slit. The contribution from the term containing *l* is usually not large, but it begins to dominate if *l* becomes larger than D/f, and it increases rapidly because *l* appears with exponent 2.

Notice that the tangential astigmatic image falls on a plane surface when $M = 3^{-1/2} = 0.58$ [see Eq. (6.4)]. Hence, the broadening of the slit

image due to astigmatic field curvature in an in-plane Littrow monochromator vanishes if the aperture stop is 0·84 focal lengths from the
collimator. This has been recognized by Mielenz (4), Sassa (5), and
Khrshanovskii (6). On the other hand, if a parabolic mirror is substituted in order to eliminate spherical aberration, the astigmatism
vanishes for $M = 0.5$. Astigmatism might also be reduced by the use of
an aspheric reflecting element, as, for example, a Newtonian mirror or
a Littrow mirror with a cylindrical surface or a collimator with a toroidal contour.

The collimator of a Littrow monochromator produces some image
distortion that could be easily compensated for by curving the slits.
However, the prism or grating introduces a strong image distortion that
is more serious because it varies with wavelength. The origin and
magnitude of this aberration was discussed in Sections 5.IV and 5.V.

The magnitude of the slit image broadening is easily written

$$l_{sc} = \frac{h^2}{8}\left(\frac{1}{r_{cD}} + \frac{1}{r_{cM}} - \frac{1}{r_c^0}\right)$$

where r_{cD} is the radius of curvature found from Eq. (5.30a) for a prism
or from Eq. (5.54a) for a grating, r_{cM} is the radius of curvature due to
the mirror, and r_c^0 is the excess radius of curvature of the exit slit over
that of the entrance slit. At wavelengths far from the wavelength chosen
for proper slit curvature, the slit image broadening from this distortion
can be by far the most serious of the aberrations present. It very definite-
ly limits the maximum useful l-number. In some monochromators, for
critical work at some wavelengths, it is desirable to mask the slit height
to reduce the magnitude of l_{sc}.

We have so far discussed the third-order aberrations independently.
There are, however, certain interactions that should be pointed out.
It is common practice to focus a monochromator having a spherical
collimator so that the slit plane is between the focal planes for the
paraxial and the marginal rays. The midway point, corresponding to
minimum circle of confusion, is not the best focusing position, but
rather a position which minimizes the spread of the energy density in
the focusing plane. The image broadening due to spherical aberration
is thus reduced to a value somewhere between the diameter of the
marginal ray pattern in the Gaussian image plane and the diameter of
the minimum circle of confusion. Moreover, this shifting of the slit
plane also tends to reduce the broadening of the image of the slit ends
by astigmatic field curvature.

If the monochromator is designed to place short wavelengths farther from the disperser, then the slit image curvature produced by a prism is away from the disperser and that introduced by the grating is toward the disperser, as shown in Fig. 6.5. The tangential astigmatic lines are seen to lie more or less along the curved slit in the grating monochromator, thus reducing the broadening from astigmatism; but in the prism monochromator the broadening is increased.

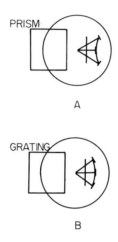

A

B

Fig. 6.5   Astigmatism and slit image curvature in an in-plane Littrow monochromator.

There is a variant of the Littrow system in which the separation of entrance and exit slits is achieved by displacing them along their length in a direction normal to the principal plane of the system, as shown in Fig. 6.6 in an end-on view. This is equivalent to using a common pair of slit jaws for entrance and exit slits, with one-half of the resulting long slit used for entrance and the other half for exit. This arrangement permits the off-axis angle in the principal plane to be minimized, but on the other hand, the slit ends are twice as far off-axis for equivalent slit heights.

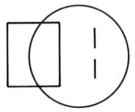

Fig. 6.6   Variation of an in-plane Littrow monochromator.

B. Off-Plane Littrow Monochromator

A ray-trace analysis of the off-plane Littrow monochromator with paraboloid mirror and prism or grating has been given by Yoshinaga, *et al.* (2). The off-plane Littrow system has entrance and exit slits placed close together on the same side of the optic axis, but above the disperser rather than beside it. This is illustrated in the end-on and side views of Fig. 6.7. An important property of this system is that the spectrum is dispersed above the disperser. There is no possibility of the disperser intercepting any of the spectrum and scattering or reflecting it back to the collimator as in the in-plane Littrow monochromator. Nevertheless, because the entrance and exit beams are nearly coincident on the mirror, radiation scattered from the mirror surface can proceed directly to the exit slit, just as in the in-plane Littrow monochromator.

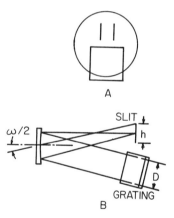

Fig. 6.7   Off-plane Littrow monochromator: (**A**) end view; (**B**) side view.

The spherical aberration is the same as in the in-plane Littrow system. Coma in the off-plane system is a little different because the axis of the comatic pattern is nearly parallel to the slit (Fig. 6.8) and because the size of the off-axis angle is determined differently. Coma and the other aberrations are additive for the two reflections from the mirror because of the asymmetry of the system.

The field angle for a point at the extreme end of the slit, again ignoring the small separation between the slits, is at least

$$\omega \geqslant \sin^{-1}\frac{D+h}{2f} \approx \frac{D+h}{2f} \qquad \text{with} \qquad M = 0.5 \qquad (6.5)$$

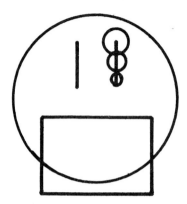

Fig. 6.8   Coma in an off-plane Littrow monochromator.

For this field angle, the comatic pattern at the slit end, after two reflections from the mirror, is, from Eqs. (5.26) and (6.5),

$$l'_c \geq \frac{D^2(D+h)}{16f^2} = 0.06\frac{f}{F^3}(1+lF)$$

This is perhaps 50% larger in size than the comatic broadening in a similar in-plane Littrow system, but in the off-plane system the broadening is symmetric.

    If the sagittal astigmatic line is made to fall on the slit, there is no astigmatic broadening, again ignoring the small separation between entrance and exit slits. However, the sagittal image surface is curved because the aperture stop is not coincident with the mirror. The radius of curvature of the sagittal surface is

$$R_{A,S} = \frac{-f}{2(M^2-1)}$$

after two reflections from the mirror.

    With the slit placed parallel to a tangent to this surface, the maximum focusing error is

$$Z_{AS} = \frac{h^2(M^2-1)}{2f}$$

and the image broadening is

$$l'_{AC} = \frac{h^2(M^2-1)}{2fF} = \frac{f}{2}(M^2-1)\frac{l^2}{F}$$

This broadening is about the same in magnitude as in a similar in-plane Littrow system when $M = 0.5$, but cannot be made to vanish for any convenient choice of M. Of course, we are not free to place the slit in a position which minimizes one aberration without considering the affect on others.

Moreover, slit image distortion introduced by the disperser causes the sagittal astigmatic line to no longer fall along the slit. This is illustrated in Fig. 6.9 for a straight entrance slit and a curved exit slit. Note that the exit slit is not only curved but tilted to make it conform to the distorted image.

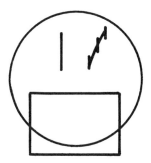

Fig. 6.9 Astigmatism and slit image distortion in an off-plane Littrow monochromator.

C. In-Plane Fastie–Ebert Monochromator

The Ebert spectrograph is very old, dating from 1889 (7), but it has attained widespread use as a monochromator only in the past decade or so since Fastie (8) described its advantages. More recently the optical properties of the in-plane Fastie–Ebert system have been discussed by Yoshinaga et al. (2), Kudo (9, 10), Megill and Droppleman (11), Leo (12), Sassa (5), Rosendahl (13), Mielenz (4, 14, 15), and Kaiser et al. (16). The in-plane Fastie–Ebert system comprises entrance and exit slits placed symmetrically on opposite sides of the disperser, which is usually a diffraction grating (see Fig. 6.10). The mirror, serving as both collimator and condenser, is much larger than in a Littrow system because the difference between the angles of incidence and diffraction is large. It follows that the Fastie–Ebert monochromator is somewhat more costly than the Littrow, and also somewhat less compact. Because the entrance and exit beams do not overlap on the mirror, it is impossible for stray radiation to be scattered from the entrance beam directly through the exit slit to the detector. The in-plane Fastie–Ebert monochromator has the same problem as the in-plane Littrow regarding dispersed radiation striking the grating and reappearing as stray radiation. Unlike the Littrow, double dispersion cannot be

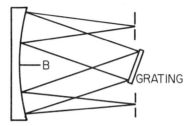

Fig. 6.10   Fastie–Ebert monochromator.

avoided in the Fastie–Ebert by increasing the off-axis angle because rays from the grating to the central region of the mirror are reflected back to the grating. A baffle is often placed on the mirror (B in Fig. 6.10) to help reduce this effect. An in-plane Fastie–Ebert monochromator using a spherical collimating mirror suffers the same spherical aberration as a Littrow monochromator of the same focal length and f/number. However, as we have seen, spherical aberration is relatively small in a monochromator of modest aperture. A parabolic mirror can be used in a Fastie–Ebert system. An off-axis paraboloid is used with the axis passing through the center of the grating. The slits are placed symmetrically off the axis of the parabola.

One of the great virtues of the Fastie–Ebert monochromator is its nearly complete freedom from coma. If the grating were replaced by a plane mirror, as in Fig. 6.11, it is evident that the comatic wavefront deformation introduced by the first reflection from the spherical mirror is exactly canceled at the second reflection. This is readily understood by considering that if the radiation entered the exit slit, the deformed wavefront (dashed line) would be the exact opposite. However, with the grating installed as in Fig. 6.10, the width of the wavefront after diffraction is not the same as before, causing a further distortion of the wavefront not exactly canceled upon the second reflection. The

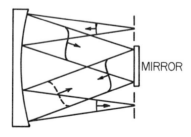

Fig. 6.11   Comatic aberration cancellation in the Fastie–Ebert monochromator.

magnitude of this *subsidiary coma* is dependent on the relative values of the angles of incidence and diffraction. This has been discussed by Fastie (8), Czerny and Turner (17), Megill and Droppleman (11), and Ford *et al.* (18). If the width of the grating is G, we can write $D = G \cos \theta_i$ for the incident beam and $D' = G \cos \theta_d$ for the diffracted beam. The comatic pattern has a net width of approximately

$$l_c = \frac{3M\omega}{16f}(D^2 - D'^2) = \frac{3M\omega G^2}{16f}(\cos^2 \theta_i - \cos^2 \theta_d) \qquad (6.6)$$

Equation (6.6) can be written in terms of the angle θ between the grating normal and the optic axis and the angle α between incident and diffracted rays

$$l_c = \frac{3M\omega G^2}{16f} \sin 2\theta \sin \alpha$$

Typically, $\sin \alpha$ might be $\frac{1}{5}$ while 2θ ranges up to perhaps 60°. Hence the subsidiary coma of a Fastie–Ebert monochromator can amount to as much as about 0.2 of the coma of a comparable Littrow monochromator. It is least when the angle of diffraction is closest to equaling the angle of incidence, that is, when the grating presents the greatest area as seen from the mirror.

Astigmatism is not canceled in the Fastie–Ebert system, but a trick permits it to be rendered unimportant. The trick is to use circularly curved entrance and exit slits, centered about the optic axis, with some

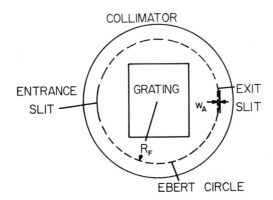

Fig. 6.12   Astigmatism in the Fastie–Ebert monochromator.

convenient radius R_F, as shown in Fig. 6.12. The slits lie on the so-called Ebert circle. The tangential astigmatic image of the entrance slit is thus tangential to the exit slit, and aside from the small sagittal distance,

$$w_A = \frac{l_{A,T}^2}{2R_F} = \frac{2M^4 R_F^3}{f^2 F^2}$$

the slit image is not broadened. Moreover, the astigmatic field curvature is the same everywhere on the Ebert circle. However, the plane of the tangential image may not coincide with the optimum focal plane for minimizing spherical aberration. Therefore it is sometimes desirable to introduce additional optical elements to compensate for astigmatism. For example, Newtonian mirrors with concave spherical or cylindrical surfaces can be used as in Fig. 6.13. If a paraboloid is used as the collimator and the grating is placed one focal length from the mirror ($M = 0.5$), there is no astigmatism.

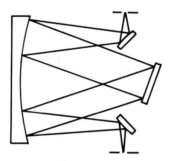

Fig. 6.13   Newtonian correction mirrors in the Fastie–Ebert monochromator.

An additional very fortunate circumstance occurs in a grating Fastie–Ebert monochromator. Slit curvature is correct for all wavelengths when the entrance and exit slits lie on the Ebert circle ([19, 20]). This can be demonstrated readily by first writing the diffraction grating equation (5.52) in the form

$$\frac{n\lambda}{2a} = \cos \alpha \cos \epsilon \sin \phi \qquad (6.7)$$

Now, with reference to Fig. 6.14. ϕ is the angle between the grating normal and the optic axis. A point on the entrance slit is defined by the angle ρ. Then for radiation from this point (dashed line) the half-angle α between the incident and diffracted rays is given by

$$\alpha = \sin^{-1}\left(\frac{R_F \cos \rho}{2f(1-M)}\right) \qquad (6.8)$$

where $2f(1-M)$ is the distance from the collimator to the grating and

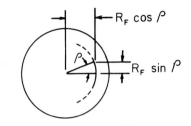

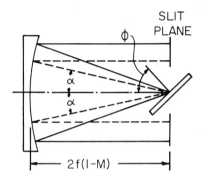

Fig. 6.14   Slit image distortion and the Ebert circle.

R_F is the radius of the Ebert circle. The off-axis angle ϵ is given by

$$\epsilon = \sin^{-1}\!\left(\frac{R_F \sin\rho}{2f(1-M)}\right) \tag{6.9}$$

From Eqs. (6.8) and (6.9) we can write

$$\cos\alpha\cos\epsilon = \left[1-\left(\frac{R_F\cos\rho}{2f(1-M)}\right)^2\right]^{1/2}\left[1-\left(\frac{R_F\sin\rho}{2f(1-M)}\right)^2\right]^{1/2}$$

$$= \left[1-\left(\frac{R_F}{2f(1-M)}\right)^2 + \left(\frac{R_F}{2f(1-M)}\right)^4\sin^4\rho\cos^4\rho\right]^{1/2} \tag{6.10}$$

Typically, R_F/f is of the same magnitude as the reciprocal of the f/number and $2(1-M)$ is close to unity. Therefore the last term in the bracket of Eq. (6.10) is very small and the product $\cos\alpha\cos\epsilon$ is not significantly dependent on ρ. It follows from Eq. (6.7) that the same diffracted wavelength falls on every point of the Ebert circle for a given choice of ϕ. Furthermore, image distortion by the collimator is immaterial because all points on the slit are equidistant from the optic axis.

D. Off-Plane Ebert Monochromator

A treatment of the off-plane Ebert system has been given by Yoshinaga et al. (2), and Mielenz (14). In an off-plane Ebert system, the entrance and exit slits are placed symmetrically above and below the grating, as shown in Fig. 6.15. The off-plane Ebert monochromator shares with the in-plane Fastie–Ebert monochromator the property of freedom of stray radiation from scatter at the collimator. Furthermore, it shares with the off-plane Littrow system freedom from stray due to dispersed radiation falling on the grating.

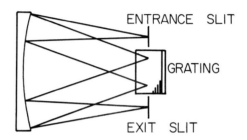

Fig. 6.15   Off-plane Ebert monochromator.

The symmetry of the off-plane Ebert system provides a cancellation of coma. Furthermore, there is no subsidiary coma because, unlike the in-plane system, the collimated beam has the same width both before and after diffraction by the grating.

The off-plane Ebert system does suffer from slit image distortion (Fig. 6.16) that varies with wavelength. Also, the sagittal astigmatic image lies on a curved surface, thus afflicting the monochromator with astigmatic field curvature, as in the off-plane Littrow system. If the

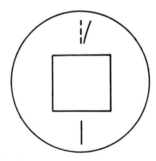

Fig. 6.16   Slit image distortion in the off-plane Ebert monochromator.

entrance slit is curved and the exit slit is straight as in Fig. 6.16, the sagittal astigmatic image lies along the length of the exit slit, and the astigmatic broadening is minimized. An off-plane Ebert monochromator with a paraboloid collimator and with a grating placed one focal length from the collimator has no astigmatism. This has been discussed by Welford (*21*), Baker (*22*), and Filler (*23*).

E. Czerny–Turner Monochromator

The Czerny–Turner monochromator was described by Czerny and Turner (*17*) in 1930 and by Czerny and Plettig (*24*). The optical properties of the Czerny–Turner monochromator have been discussed by Yoshinaga *et al.* (*2*), Kudo (*9, 10*), Rosendahl (*13*), Shafer, *et al.* (*25*), Mielenz (*14*), and Namioka and Seya (*26*). The Wadsworth prism monochromator, in which the mirror following the prism is set at such an angle that only a single pass is made through the prism may be considered as a special case of the Czerny–Turner. The theory of the Wadsworth monochromator has been discussed by Seya and Namioka (*27*). The Czerny–Turner monochromator (Fig. 6.17) is much like the in-plane Fastie–Ebert, except that separate collimator and condensing

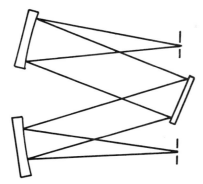

Fig. 6.17   Czerny–Turner monochromator.

mirrors are used. The Czerny–Turner has the advantage that two small mirrors are sometimes less expensive than one large one. The mirrors can be moved far enough apart to satisfy the condition for no double dispersion (*36, 37, 38*), namely, grating and mirror are on opposite sides of a radius to the extreme edge of the mirror as in Fig. 6.17. However, grating efficiency is sacrificed when the angles of incidence and diffraction are greatly different and, furthermore, subsidiary coma is increased.

If the centers of curvature of the two mirrors coincide, the Czerny–Turner is optically identical to the Ebert. Similarly, an off-axis Czerny–Turner is equivalent to the off-axis Fastie–Ebert.

Focusing of the separate mirrors of a Czerny–Turner monochromator is, if anything, easier than focusing the single mirror of a Fastie–Ebert monochromator. However, it might be difficult to ensure that the centers of curvature of the two mirrors coincide. If the centers of curvature do not coincide, either by accident or design, then the Ebert circle becomes an ellipse. In spite of this, there are some advantages to be had by using an asymmetrical Czerny–Turner system with non-coincidental centers of curvature. The subsidiary coma can be reduced to an amount less than is present in the symmetric form. Further modifications of the asymmetric Czerny–Turner monochromator utilize mirrors of different radii of curvature (25) or different angles of incidence on the two mirrors (28).

F. Pfund Monochromator

The optical properties of the Pfund monochromator are discussed by Kudo (9). Variations of the Pfund monochromator are shown in Figs. 6.18A, B, and C, in which the radiation from the slits is introduced through a hole in the grating, or with a plane transfer mirror, or with Newtonians that obscure part of the grating surface. A single small mirror is used. The Pfund arrangement minimizes the off-axis angle. Coma is compensated and subsidiary coma and astigmatism are minimized. However, the radius of the Ebert circle is very small, so

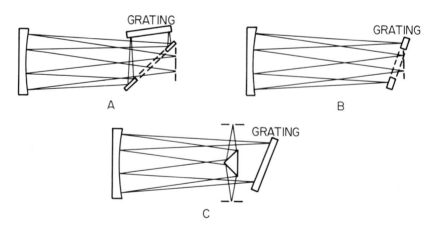

Fig. 6.18   Variations of the Pfund monochromator.

only short slits can be used if advantage is to be taken of the slit-image-distortion compensation offered by the symmetric monochromators. Moreover, the Pfund arrangement suffers from a very important defect. The dispersed spectrum is spread across the grating and consequently stray radiation is a problem.

G. Double Monochromators, Double-Disperser Monochromators, and Double-Passed Monochromators

A double monochromator has two monochromators arranged in tandem, as in Fig. 6.19. The output of the first monochromator enters the second and is further dispersed. The two monochromators usually share a common slit, the same slit acting as exit slit for the first mono-chromator and as entrance slit for the second.

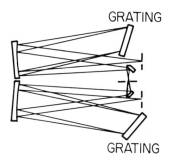

Fig. 6.19   Double monochromator.

A double-disperser monochromator uses two dispersing elements in a single monochromator. They can be arranged symmetrically or asymmetrically, as shown in Fig. 6.20.

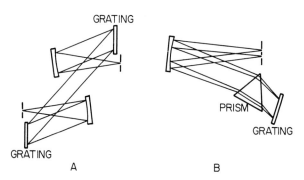

Fig. 6.20   Double-disperser monochromators: (**A**) two gratings; (**B**) grating and prism.

There are two forms of double-passed monochromators, the in-plane form or Walsh (29) system of Fig. 6.21A and the off-plane form of Fig. 6.21B. In the in-plane version, the dispersed radiation is intercepted by a pair of plane mirrors mounted close to a slit image. These mirrors return the beam to the collimator for a second pass at the disperser. Note that if a single plane mirror were used in place of the corner mirror, the dispersion produced by the first pass would be canceled by the second. A chopper is inserted at the focus of the singly dispersed radiation. Therefore, only that part of the singly dispersed radiation which returns to the disperser to be doubly dispersed is

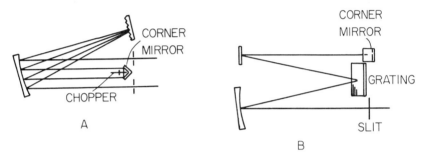

Fig. 6.21   Double-passed monochromators: (**A**) in-plane; (**B**) off-plane.

modulated. Singly passed, unmodulated radiation that passes through the exit slit is not detected. Only the double-passed modulated radiation is detected. Alternately, in the off-plane system the corner mirror and intermediate slit image are above the principal plane of the monochromator. There is no need to chop the radiation at the corner mirror because single-passed radiation cannot reach the slit. The optics of multipassed spectrometers are discussed by Shimomura (30).

There are two reasons for building one of these various kinds of double instruments. One is to increase the dispersion obtainable from a single instrument, and thereby increase both resolution and, as we shall see, the radiant power at the detector. This objective can be met by double monochromators, double-disperser monochromators, and double-passed monochromators.

The second reason is to improve the spectral purity of the radiation finally reaching the detector. Stray radiation originating in the first monochromator of a double monochromator is not of the proper wavelength to pass through the exit slit of the second monochromator. If, for example, a single monochromator has a stray radiation level of 3%, a double monochromator should have stray radiation of only $(0.03)^2 =$ 0.09%. Often a grating monochromator is used with a prism monochromator whose function is to prevent the occurrence of superposed grating orders. The prism monochromator also serves to reduce the general stray radiation level and, in addition, it adds a certain amount of dispersion. When the prism precedes the grating monochromator, it is termed a foreprism. It might just as well follow the grating monochromator. A double grating monochromator has increased dispersion and decreased stray radiation, but unless the gratings have different spacing or are arranged with different geometry (namely the angle α), the overlapping order problem will still be present.

The double-passed monochromator assists greatly in controlling stray radiation, but perhaps not as efficiently as the double monochromator because the corner mirror of the double-passed instrument, which plays the same role as the intermediate slit of the double monochromator, is usually somewhat wider than the slit. The double-disperser monochromator does little to improve stray radiation.

The comatic aberrations of the two monochromator sections cancel each other in the symmetrical arrangement of Fig. 6.19, provided the intermediate slit is wide enough to transmit the entire comatic image of the entrance slit. There will not even be any subsidiary coma if the two dispersers are equivalent. If the intermediate slit has the same width as the entrance and exit slits, some rays are stopped and coma is not canceled. The second monochromator makes a fresh start, so to speak, and although some energy is lost at the intermediate slit, only the coma of the second monochromator appears at the exit slit. This would be the case, for example, if the first monochromator served only as a filter, with all of the resolution being performed by the second monochromator. On the other hand, spherical aberration and astigmatism are not canceled when a wide intermediate slit is used, but are additive, whereas only single-monochromator aberrations are present when a narrow intermediate slit is used. Slit image distortion is not canceled in a double prism or double grating monochromator, except by the obviously unsatisfactory procedure of canceling dispersion as well. However, slit image distortion in a prism–grating double monochromator is partially canceled, because the distortions produced by prisms and gratings are in opposite directions.

The symmetrically arranged double-disperser monochromator of Fig. 6.20A has its coma canceled, but neither the form shown in Fig. 6.20B nor the double-passed monochromators of Figs. 6.21A and B can achieve any aberration compensation.

III. SOURCE OPTICS

The function of the source optics of a spectrophotometer is to collect radiation emitted by the source and introduce it into the monochromator in such a way as to flood the aperture of the monochromator and simultaneously illuminate the entire slit. It is important to recognize that accomplishing one of these objectives does not necessarily guarantee fulfillment of the other. The neophyte often adjusts his source optics by illuminating the slit without properly illuminating the collimator and disperser (as in Fig. 6.22). Similarly, the aperture can be filled when only part of the slit is illuminated.

Fig. 6.22   Improper illumination of a monochromator.

We have already discussed in Chapter 3 the factors involved in determining the required size of source optics. One must assure that the extreme rays passing through opposite corners of the entrance pupil and the slit are indeed intercepted by the source optics.

In double-beam instruments the source optics precede, or may be part of, the photometer optics. The source optics are then required to form an image of the source at some intermediate point, such as the attenuator comb (see Fig. 6.23). In any case, the source mirror forms an image of the source, which is usually a long cylindrical object, and eventually the image is transferred to the entrance slit of the monochromator. In single-beam instruments it is more important to introduce the maximum amount of radiation into the monochromator

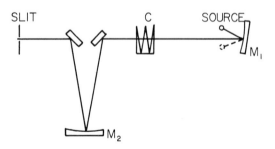

Fig. 6.23   Source optics of a double-beam system.

than it is to produce a high-quality image of the source on the slit jaw. Aberrations are therefore not too serious in source optics. Usually the tangential astigmatic image is selected to fall on the slit, and thus the focal length of the source mirror is taken to be $f_T = f \cos \omega$, following the discussion of Sect. 3.X.E. The angle ω is of course the off-axis angle of the source mirror and f is its Gaussian focal length. In double-beam optical null instruments, the edges of the teeth of the attenuator comb should be sharply focused on the slit. This means that the sagittal astigmatic image of the comb is focused on the slit in instruments whose optic axis remains in the same plane. In order to accomplish this, while at the same time imaging the edges of the source sharply on both the comb and the slit (tangential image), some instruments use a toroidal mirror to correct for astigmatism. Mirror M_2 in Fig. 6.23 can be a toroid. Both the image of the source on the comb and the image of the comb on the slit will be comatic, with the comatic pattern extending along the teeth of the comb. However, the image of the source on the slit is at least partly compensated for coma in the arrangement of Fig. 6.23. If the source is on the opposite side of the optic axis (dashed lines), coma is additive.

The shape of the source is not always compatible with the shape of the entrance slit. We have seen that the slit is usually curved in order to compensate for the slit image distortion introduced by the disperser. In the case of the Ebert–Fastie or Czerny–Turner mono-chromator, the slits can be quite long and highly curved. The source, on the other hand, is generally a straight cylinder. Image distortion can be deliberately introduced in the source optics to produce a dis-torted source image to fit the shape of the slit. An elegant device was used by Benesch and Strong (31) to illuminate an Ebert–Fastie mono-chromator. The optical principle is illustrated in Fig. 6.24. A mirror

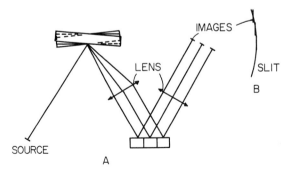

Fig. 6.24   Benesch–Strong image slicer.

is sliced into segments and the segments are angularly displaced relative to one another. This mirror forms a sequence of laterally displaced images of the source. These images are formed on the surface of another mirror that is also sliced into segments, so each source image is reflected in a different direction. They are reimaged (as seen in the figure) as a sequence of images displaced laterally and longitudinally in such a fashion as to approximate the shape of the entrance slit. If the sliced mirrors are properly figured, they can perform the duties of the focusing elements also. Such devices can be constructed with refracting elements as well as reflecting elements. The actual design of such a device is shown in Fig. 6.25. It is an adaptation of an element devised by Bowen for use in astronomical

Fig. 6.25 Refracting image slicer.

spectroscopy in imaging a star onto a spectrograph slit. A similar device is used in the Applied Physics Corporation Model 81 Raman spectrometer to image a round source (the end of the cell) onto the slit. These devices split a round image into displaced segments (Fig. 6.26) and then shift them to form a straight or a curved line compatible with the slit.

In some experiments it is necessary to rotate an image such that a horizontal line is imaged on a vertical slit. This might occur in studying the emission from a selected section of a flame a specified distance from the burner, as in Fig. 6.27A. It is wasteful to merely image the

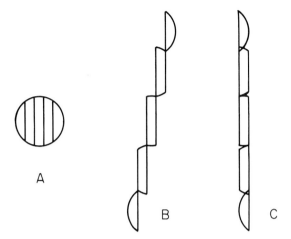

Fig. 6.26 Imaging of a round object on a slit aperture.

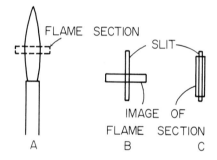

Fig. 6.27 Imaging of a flame on a monochromator slit.

Fig. 6.28 Dove prism for rotating an image.

section on a portion of the slit (Fig. 6.27B). It is desired to rotate the image to fill the slit (Fig. 6.27C). A Dove prism (Fig. 6.28) can accomplish this without deviating the direction of the beam. The Dove prism has the property of rotating the image when the prism is rotated about its axis. The image rotates at twice the rate of prism rotation. An equivalent reflecting device (*32*) can be constructed with three mirrors (Fig. 6.29). If the direction of travel of the beam need not be undeviated, two mirrors can accomplish an image rotation of 90°, as shown in Fig. 6.30.

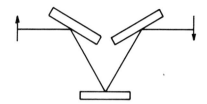

Fig. 6.29   Reflecting image rotation system.

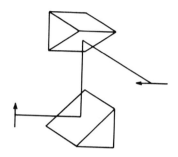

Fig. 6.30   Reflecting image rotation system.

Now let us discuss an error very commonly made by the neophyte in optics. The fallacy seems trivial, but in more subtle cases even the expert can stumble. The argument runs like this: We use a mirror (or lens as drawn in Fig. 6.31) to gather radiation from the source and focus it on the slit. We know, of course, that the beam approaching the slit should converge at a rate equal to the f/number of the mono-chromator. If it is less convergent, the collimator will not be filled completely, whereas if it is more convergent, the collimator will not receive all of the radiation. But why should we not use a shorter focal length lens, move the source closer, and thus gather a larger bundle of radiation? The answer is clear. A large bundle of rays is indeed gathered, in the universe ratio of the source-to-lens distances. But the size of the image of the source formed on the slit is also increased

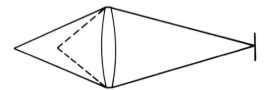

Fig. 6.31   Imaging the source on the slit.

in the same ratio. Equivalently, the image of the slit on the source is reduced in size in the direct ratio of the source-to-lens distances. What we gain in greater angle we lose by utilizing a smaller area of the source. A little reflection will convince the reader that this is just an illustration of Lagrange's law [Eq. (3.38)].

IV. PHOTOMETER OPTICS

The optical system used in comparing a sample with some reference is the photometer system of the spectrophotometer. In a double-beam instrument, the radiation beam from the source is split into two beams. One passes through the sample cell and the other serves as a reference and may pass through a cell filled with solvent. The beams are then recombined and allowed to continue to the monochromator. We shall illustrate with several examples.

Figure 6.32 shows the so-called box type of photometer optics as used in the Beckman IR-4-type instruments which are typical optical null spectrophotometers. Radiation from the source G is collected by the source mirror M_1. It is split into two beams by a rotating sector mirror M_2; M_2 is a semicircular mirror mounted in a rigid frame, as shown in Fig. 6.32B. The semicircular mirror, M_2, is rotated about a shaft at a rate of 11 cps. It alternately passes radiation into the sample beam S and reflects it into the reference beam R. A similar mirror M_5 is rotated 180° out of phase with M_2. Mirror M_5 either reflects the sample beam to M_6 and blocks the reference beam, or passes the reference beam to M_6 and passes the sample beam to the

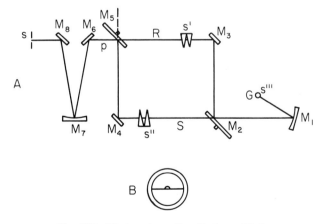

Fig. 6.32 Photometer optics: Beckman IR-4.

back of the instrument where it is blocked. From M_6 the combined beams are sent to M_7 which refocuses the radiation onto the entrance slit of the monochromator. The sample is placed between mirrors M_2 and M_4 and the reference is between M_3 and M_5. Mirror M_7 forms an image of the slit at s', where the reference beam attenuator comb is placed, and at s'', where an adjustable auxiliary comb is available for equalizing the empty sample and reference beams, that is, for adjusting 100% transmittance. A slit image also is formed on the source at s'' by M_1. Mirror M_7 forms an image of the aperture stop of the monochromator at p. An aperture is placed at p to block radiation which would overfill the monochromator and possibly contribute to stray radiation.

Notice that radiation traversing the sample and reference beams have identical path lengths and undergo the same number of reflections at the same angles of incidence. This is a necessary requirement of photometer optics if atmospheric absorptions are to be compensated for and if the I_0 (100% line with no sample or reference cell) is to be a straight line.

Another requirement of infrared photometer optics is that chopping elements are placed before and not after the entrance slit, unless certain unusual precautions are taken. This is because an infrared detector responds to thermal radiation from all objects in its field of view. If a beam-switching photometer system is inserted between the exit slit and the detector, the detector will sense the differences in thermal radiation emitted over all wavelengths by objects in the sample and the reference beams. This can be many orders of magnitude greater than the small signal corresponding to the wavelength increment passed by the monochromator. By placing the photometer optics before the entrance slit, only thermal radiation of the wavelength passed by the monochromator is seen by the detector. Normally this is a very small fraction of the radiation originating at the source, and does not contribute a measurable error. However, if the sample is quite hot or quite cold, significant errors can occur. This is discussed in greater detail in a later chapter, but for now we remark that certain modifications can be made in the photometer optics to avoid such errors. In the IR-4 system, for example, the second rotating mirror M_5 can be replaced by a nonchopping beam-recombining element. Only radiation originating at the source is chopped by mirror M_2. Thermal radiation from the sample or reference is not chopped, and therefore not detected by the detector. The nonchopping element might be in the form of the strip mirror described in Section 5.III.

An image of the aperture stop falls on M_5, so the strip mirror samples radiation in bands across the aperture. Alternatively, rotating mirror M_5 can be stopped and locked in the half-way position, so one-half of the aperture is illuminated through the reference beam. An inherent feature of any nonchopping beam combiner is the loss of at least half of the useful radiation. Therefore the use of these devices is normally limited to cases in which it is necessary to avoid errors introduced by extraneous emission, such as when studying very hot or very cold samples, or when doing quantitative work in the far infrared. There is an alternate electronic method of eliminating error due to sample radiation in optical null instruments. It too imposes a 50% energy loss. It is discussed further in a later section.

Spectrophotometers used in the visible and ultraviolet spectral regions have their photometer optics following the slit in order to avoid illuminating the sample with intense ultraviolet radiation that can cause fluorescence and photochemical decomposition. This is permissible because the detectors of such instruments are insensitive to infrared thermal radiation. Sometimes it is necessary to avoid exposing a sample to the initial beam in an infrared spectrophotometer. Such might be the case when a high-powered microbeam illuminator (infrared microscope) is used to condense the incident radiation onto a small area of the sample. The condensed beam from a Nernst glower can scorch paper or melt sulfur. It is possible, though usually in-convenient, to mount the sample following the exit slit of a single-beam spectrophotometer or a double-beam instrument operating in the single-beam mode. It is even possible to retain double-beam operation. The sample and reference beams are masked at the points where slit images occur. The masks are positioned so that the top half of the slit is illuminated through one beam and the bottom half through the other beam. The sample is placed following the monochromator close to the slit, or placed at a slit image in such a way as to cover only the half of the slit illuminated through the sample beam. As the beam switchers rotate, the detector sees radiation first through one-half of the slit and the sample, and then through the other half of the slit and the reference. If the sample or cell is bulky, it is necessary to construct some optics to separate, in space, the images of the slit halves. If aperture stop images are available in the sample and reference beams, it might be better to work with them rather than slit images in order to retain the full width of the reference beam attenuator.

A symmetrical photometer system, shown in Fig. 6.33, is used in Perkin–Elmer Model-21-type instruments. The source at G is viewed

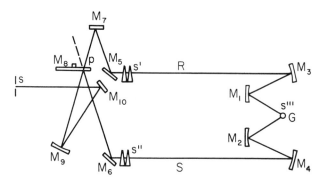

Fig. 6.33   Photometer optics: Perkin–Elmer Model 21.

from different sides by collecting mirrors M_1 and M_2 and sent through the sample and reference beams by M_3 and M_4. The sample beam is deflected by M_6 to the rotating mirror M_8 where it is alternately reflected or allowed to pass out of the system. The reference beam, after reflections from M_5 and M_7, is alternately blocked by M_8 or allowed to pass. Mirror M_8 rotates at 13 cps in the Perkin–Elmer instruments. After passing M_8, the combined beams are focused onto the slit by a toroidal mirror M_9, which is toroidal in order to compensate for astigmatism. A slit image is formed by M_9 at s' and s'' where the reference and sample beam attenuators are placed. The slit is also imaged on the source. An image of the aperture stop is formed at p.

A symmetrical photometer system is also used in the Beckman IR-5-type instrument (Fig. 6.34). In this system the slit is imaged on the reference and sample beam attenuators at s' and s'' and the aperture stop is imaged at p and p'. The sample and reference cells are then placed between slit and aperture stop images. This is a decided advantage because it minimizes the size of the beam through the sample and reference areas, thus permitting use of small-sized cells with minimum sample volume. This is made clear in Fig. 6.35. Rays through the corners of the aperture and the slit, or their images, define the size of the beam. Clearly the beam is smallest between the aperture and slit. The IR-5 system is unusual in that the source is imaged not on the slit, but on the prism, that is, the aperture stop.

The photometer optics of the Baird spectrophotometer are particularly simple (see Fig. 6.36). Radiation from the source G is collected by two spherical mirrors M_1 and M_1' and sent through the sample and the reference. The reference beam attenuator is at C. The beams are re-

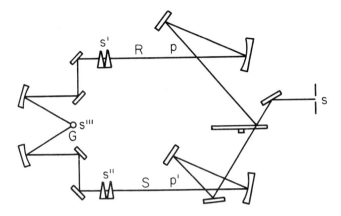

Fig. 6.34   Photometer optics: Beckman IR-5.

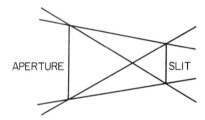

Fig. 6.35   Envelope of beam between slit and aperture images.

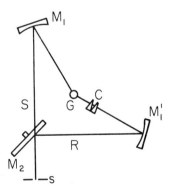

Fig. 6.36   Photometer optics: Baird.

combined with a rotating mirror M_2 that transmits the sample beam and blocks the reference beam half of the time, and reflects the reference beam and blocks the sample beam the other half. The mirror rotates at 10 cps. Notice that the reference beam undergoes one reflection more than the sample beam. A variation in reflectivity with wavelength will affect the flatness of the 100% line.

The Perkin–Elmer Infracord photometer optics are sketched in Fig. 6.37. This is an example of a photometer system in which path lengths and the number of reflections in the two beams are the same, but in which the angles of incidence on the mirrors are not all the same. This is of no importance in an instrument of this type, but it could become significant in certain types of studies using polarized radiation because of the different ellipticity of radiation reflected at different angles from metallic surfaces.

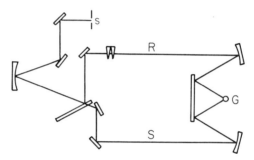

Fig. 6.37   Photometer optics: Perkin–Elmer Infracord.

The photometer optics of ratio recording instruments are necessarily somewhat different from optical null systems because of the way in which choppers are used. A Halford–Savitsky system, as used in the Perkin–Elmer Model 301 far-infrared spectrophotometer, is shown in Fig. 6.38. This arrangement is somewhat more complicated than the usual Halford–Savitsky system because extra reflections are introduced to provide optical filtering. There are two elements of interest. The double chopper, C, interrupts sample and reference beams thirteen times per second, 90° out of phase. The electronic system constructs the ratio of sample to reference energy. The mirrors at M act as the beam combiner. An image of the aperture appears at M. The top half of the aperture is filled with radiation from one beam, and the bottom half with radiation from the other beam, much as in the IR-4 system with the second rotating mirror stopped.

The Cary Model 90 is a ratio recording spectrophotometer which

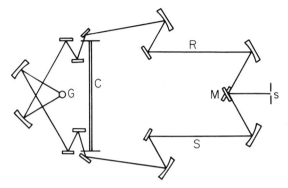

Fig. 6.38   Photometer optics: Halford–Savitzky system of Perkin–Elmer Model 301.

discriminates sample and reference beam radiation by chopping at different frequencies. The photometer optics system is quite complex (see Fig. 6.39.) Radiation from the source G is collected by the mirror M_1 and directed to the multifaceted beam splitter M_2; M_2 divides the beam into two parts, reflected above and below the plane of the figure. A set of flat mirrors at M_3 redirects the beams to the sample and

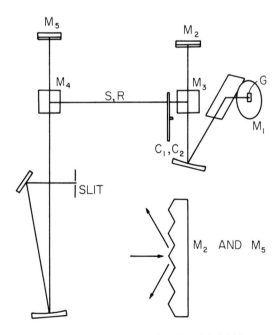

Fig. 6.39   Photometer optics: Cary Model 90.

reference, S and R. At this point the beams are parallel, one above the other. A similar set of mirrors, M_4, and a beam splitter, M_5, recombine the beams into a single wave front. The choppers, C_1 and C_2, chop the sample beam at a rate of $13\frac{1}{3}$ cps and the reference beam at $26\frac{2}{3}$ cps.

V. DETECTOR OPTICS

The beam of radiation leaving the exit slit of the monochromator must be collected and placed on the detector, which transduces it into an electrical signal. We see in a later chapter that the response of a thermal detector varies inversely as the square root of its area. Consequently, the radiation receiver of a detector is quite small: typically 1 m wide and several millimeters tall. This requires that the size of the radiation beam be reduced considerably. Usually the slit is imaged on the detector, though occasionally the aperture stop is placed there. A tenfold demagnification of the slit image is usually required.

This presents the optical designer with a problem. Demagnification of this order implies that an optical element must be placed close to the detector. If he images with mirror optics (Fig. 6.40), he can either use a large off-axis angle where the attendant astigmatism will cause some of the rays to miss the detector, or he can place the detector on the axis where it will obscure some of the incident radiation. Fortunately, some

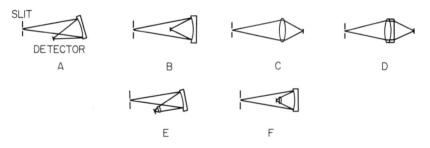

Fig. 6.40   Detector optics.

detectors, including most thermocouples and bolometers, can be made small. But others, such as Golay cells and some cooled detectors, are fairly large. The designer can reduce some of the mirror aberrations by using aspheric surfaces. He might minimize the astigmatism of an off-axis mirror by using a toroid; or he can minimize the spherical aberration by using an ellipsoid having one focus at the center of the slit and the other on the detector. Of course, the aberrations of the images of the slit ends can then become even more severe. The designer might prefer to use a lens system. He will either have to tolerate the accompanying chromatic aberration or he will have to design an achromat, which is not too easy for the infrared. Finally, the designer can compromise on a system comprising both a mirror and a lens.

Beckman spectrophotometers of the IR-4 type use a spherical mirror on-axis to attain a fivefold reduction of image size and a lens for an additional twofold reduction, thereby giving a net tenfold reduction. The problem of lens and mirror design was discussed in Chapter 3. IR-5-type instruments use a similar system, but with the mirror mounted off-axis. Perkin–Elmer designers use an ellipsoid to achieve all of the image reduction with an on-axis mirror. Baird Atomic uses a single on-axis spherical mirror, but the bolometer detector used in the Baird instruments has a larger area, so the image reduction is not as great. The Cary instrument uses an off-axis toroidal mirror followed by an on-axis condensing mirror.

Detectors that are mounted in bulky bodies present more of a problem. Such a detector is the Golay cell used in far-infrared instruments. Some solutions are shown in Fig. 6.41. If the rate of divergence from the exit slit is not too great (i.e., a large f/number monochromator), a spherical mirror can be used off-axis and the aberrations tolerated (Fig. 6.41A). This is the case in the Beckman IR-11 and the Cameca

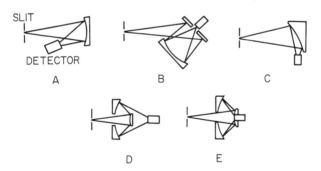

Fig. 6.41   Detector optics for a bulky detector.

SI 36. In the Unicam SP 100, 130, 200, and 200G (Fig. 6.41B), the spherical condensing mirror is kept on-axis. A plane mirror with a central hole obscures some of the incident radiation, but not as much as the detector body would if it were in the beam. The Perkin–Elmer Models 301 and 125 make use of a 90° off-axis ellipsoid (Fig. 6.41C). The Hitachi far-infrared instrument uses a 60° off-axis ellipsoid. Cassegrain reducing optics (Fig. 6.41D) are adaptable to bulky detectors and have been used by some workers. Greenler (*34*) devised a system using two concentric mirrors, one positive and one negative (Fig. 6.41E). He points out that spherical aberration is corrected when the curvatures of the mirrors' surfaces are equal. Robinson (*33*) constructed a far-infrared Ebert spectrometer using Greenler's (*34*) image reducer. In addition, because the slits of the monochromator are quite long, he inserted an image slicer of the type devised by Benesch and Strong (*31*). The Jarrell–Ash far-infrared instrument uses a similar arrangement (*35*).

Detectors operating at reduced temperature present additional problems because not only are they likely to be quite bulky, but also the receiver usually is buried deep within an enclosure and preceded by cold filters and baffles. Sometimes lenses are mounted inside the cold chamber, as they can also serve as cold filters. Especially in the far infrared, light pipes and cones can be used to transfer radiation into the enclosure and condense it onto the detector.

VI. Transfer Optics

The discussion of optical systems associated directly with samples and sample handling is postponed to later chapters. Such devices as microbeam illuminators and microscopes, folded path gas cells, reflectance apparatus, and so on, are included in that category. However, it is often required to transfer the radiation beam from its usual path in the spectrophotometer to a region where a particular experiment is underway, and then return it to the original path without disturbing the focus or apertures. Such might be the case if one is studying the reflections from a sample in a magnetic field or the transmission of a gas in a very hot chamber. Or, one may wish to transfer radiation from an external source to the monochromator, or from a monochromator to an external detector. In this section we present briefly a few of the optical tricks used to accomplish these tasks.

An image of some element of the spectrophotometer, the slit for

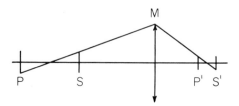

Fig. 6.42   Transfer optics.

example, can, in principle, be formed anywhere and have any reasonable size by proper choice and position of a single imaging element. In practice, however, we find that the more remote the image, the larger must be our imaging element. This is illustrated in Fig. 6.42 where it is seen that the required size of mirror M is determined by the extreme ray through the edges of the slit and the pupil and the distance from the slit. Not only are large optical elements costly, but quite often there is insufficient room in existing apparatus to accommodate an outsized optical element.

The size of the beam can be reduced to a more comfortable diameter by use of a field element, as in Fig. 6.43. The field element M_1 is placed

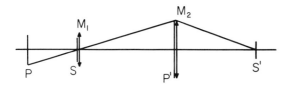

Fig. 6.43   Field element in transfer optics.

close to the slit and forms an image of the pupil on the surface of element M_2. M_2 images S at S' just as before, but now its size is determined by the size of the image P'. If large distances are involved, this process can be repeated, as in Fig. 6.44. Periscopes are constructed in this way to transmit a beam down a long narrow tube without loss of field angle.

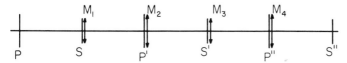

Fig. 6.44   Multiple field elements.

All of these schemes distort the original beam in that the sizes of pupil and slit images and their separation are not changed in the same ratio, although Lagrange's law is of course obeyed. It is often desired to image two objects (slit and pupil for example) the same distance apart in image space as they were in object space and also in the same size. This can be done by using *two identical elements separated by twice their focal length*. We prove this by referring to Fig. 6.45 which shows two optical elements a distance D apart. It is convenient to use the Newtonian form of the image equations. An object at a distance x from

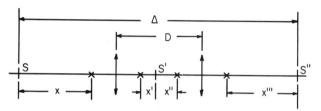

Fig. 6.45 Transfer optics with identical elements separated by twice their focal length.

the focal point of the first lens is imaged at $x' = f^2/x$. This image acts as an object for the second element with object distance

$$x'' = D - 2f - x' = D - 2f - f^2/x$$

The final image is formed at

$$x''' = \frac{f^2}{D - 2f - f^2/x} = \frac{f^2 x}{Dx - 2fx - f^2} \tag{6.11}$$

The requirement that any two objects a given distance apart are imaged the same distance apart is equivalent to requiring that

$$\frac{dx'''}{dx} = \frac{f^4}{[(D - 2f)x - f^2]^2} = -1$$

which is satisfied by $D = 2f$, as we set out to prove.

Moreover, with $D = 2f$, Eq. (6.11) becomes $x''' = -x$ and, consequently, the magnitude of S'' is the same as S. Notice that with the object S fixed, a motion of y of the two-element assembly produces a motion of $2y$ of the image with no change in its size. The distance introduced between object and image is

$$\Delta = 4f + x + x''' = 4f$$

An illustration of a system based on this principle is shown in Fig. 6.46. The beam is intercepted at a slit image, for example, by a plane mirror that deflects it to a pair of focusing mirrors. A second plane

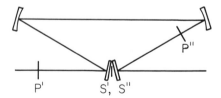

Fig. 6.46 Transfer optics with no disturbance of original images.

mirror returns the beam to its original direction. The focal points of the spherical mirrors coincide midway between them and also fall on the plane mirrors. The shift in the image position is $\Delta = 4f$ and the slit is reimaged in practically its original position. Thus, the beam is effectively undisturbed as far as the rest of the system is concerned, but a section of collimated beam is made available between the two spherical mirrors.

VII. Luminosity of a Spectrophotometer

There is a great deal of confusion about the importance of various parameters in determining the optical efficiency of a spectrophotometer. Let us discuss the amount of radiation reaching the detector for a given amount leaving the source. This might be called the transmittance of the spectrometer, although this term seems to be associated only with losses due to absorption and reflection. It is also called luminosity and optical throughput. None of these terms are especially appealing, but in this book we use the term luminosity.

Consider the spectrometer shown schematically in Fig. 6.47. A source emits a continuum of J_0 watts of radiant power per unit area into a unit of solid angle over a unit wavelength interval. The source is imaged on the entrance slit of the monochromator with unity magnification. The area of the source intercepted by the slit is the product of the slit width w_e and the slit height h. The solid angle is the area of the entrance pupil of the monochromator divided by the square of its distance from the entrance slit. This is essentially the area of the prism or grating A divided by the square of the focal length f of the collimator. Nothing is changed if the source is imaged on the detector with some magnification other than unity because the area of the source image and the solid angle of collected radiation change in opposite directions by amounts which just compensate each other. Thus the radiant power entering the monochromator over a unit wavelength interval is

$$J_m = \frac{J_0 A}{f^2} w_e h \tag{6.12}$$

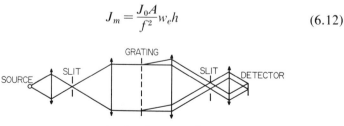

Fig. 6.47   Luminosity of a spectrophotometer.

The function $Aw_e h/f^2$ is just the Lagrange constant of the optical system.

The grating or prism has an angular dispersion $D_a = d\psi/d\lambda$ determined by the properties of the disperser, that is, the spacing of the rulings on the grating or the apex angle and dispersion of the refractive index of the prism. The dispersed radiation is spread along the plane of the exit slit. The linear dispersion D_l is proportional to D_a and to the focal length of the condensing mirror, which we assume is also f,

$$\frac{dx}{d\lambda} = D_l = D_a f$$

The range of wavelengths passed through an exit slit of width w_x is

$$\Delta\lambda = w_x D_l^{-1} = w_x \frac{D_a^{-1}}{f} \tag{6.13}$$

The radiant power falling on the detector is therefore

$$J_d = J_m \Delta\lambda T = J_0 T A h w_e w_x D_l^{-1}/f^2 \tag{6.14}$$

The factor T has been included to take into account the reflectance and transmittance of the optical elements in the system. Usually the entrance and exit slits have the same width, so Eq. (6.14) can be written

$$J_d = J_0 T A l w^2 D_l^{-1}/f \tag{6.15}$$

where l is the l-number of the monochromator, defined by Strong as the slit-height to focal-length ratio. Thus the power falling on the detector increases as the *square* of the slit width.

Equation (6.15) can be written in terms of the spectral bandwidth of the monochromator and the angular dispersion, using Eq. (6.13),

$$J_d = \frac{J_0 T A l (\Delta\lambda)^2}{D_l^{-1} f} = J_0 T A l (\Delta\lambda)^2 D_a$$

The luminosity of a monochromator is therefore proportional to the area of the disperser and the l-number. Note that the f/number of the monochromator plays no role in determining its luminosity! Equation (6.15) shows very clearly why the change to diffraction gratings during the past several years has done so much to improve the performance of spectrophotometers: For a given resolution the luminosity is proportional to the angular dispersion.

If we are considering an emission spectrograph or a spectrometer used to record radiation from very narrow emission lines (narrow with respect to the resolution of the spectrometer), then $\Delta\lambda$ is no longer to be considered. The detected radiation, in this case, is the same as J_m in Eq. (6.12). In this case the f/number is a significant parameter and the equation for detected radiation can be written

$$J_d = \frac{J_0 T w_e l}{F}$$

VIII. Optical Alignment and Focusing

Occasionally the optical elements of a spectrophotometer need to be focused or aligned. This is usually the case after a source or a detector has been replaced. The installation of some optical accessories always requires a certain amount of alignment. Several useful techniques are described in this section.

A visual alignment procedure is easier if a bright light is used. The usual spectrophotometer source is not bright enough and a more powerful incandescent lamp or a mercury arc can be used. White file cards are very useful as targets in observing images at various places in the optical path and in determining whether an optical element is properly filled.

Alignment is always carried out with respect to the monochromator. It is essential to ensure that the slit and the aperture of the monochromator are correctly illuminated. The monochromator must be set to transmit visible radiation. This is done in different ways, depending on the monochromator. Sometimes a special stop or cam position is provided. Often a cork can be placed between the cam and the follower to obtain visible light.

Detector optics are adjusted by illuminating the entrance slit of the monochromator and observing the light beam leaving the exit slit. The detector should be protected from the intense beam by placing a card before it, and the various optical elements are aligned in succession until a good patch of light is observed on the card at the proper place. The brightness of the source should be reduced before the beam is permitted to fall on the detector. Final adjustment of the detector optics can be carried out by maximizing the signal from the detector with the spectrophotometer operating in the usual way with its own source. If the slit is imaged on the detector, the adjustment should be made with a wide slit opening.

The source and photometer optics are also adjusted by illuminating the slit and aperture of the monochromator with a strong visible light. In this case, however, the exit slit should be illuminated and the beam from the entrance slit adjusted element by element until it finally falls on the source. Source optics can rarely be aligned satisfactorily by working from the source to the slit. The photometer optics of a double-beam system pose a special problem because not only must both sample and reference beams form an image of the slit (usually) on the source, but the magnification must be the same. Moreover, the images should coincide in systems of the type exemplified by the Beckman IR-4. It is useful to observe both the slit and the aperture images in the combined beams in the vicinity of the source while rotating the beam-switching mirrors. Optical elements should be adjusted until no perceptible dancing of the images takes place. In Perkin–Elmer Model-21-type systems, care is required to be sure the source is not warped or, if it is, to ensure that both sample and reference beams are imaged completely on the source.

If an optical accessory is placed in the sample beam but not in the reference beam, it might be impossible to match the two beams at the source. In this case it is necessary to compromise, and the quality of the I_0 or 100% line will suffer.

The optical adjustment of a monochromator should probably not be attempted by the amateur. However, we outline some of the procedures here for the interested reader. First of all, the grating or prism must be correctly aligned in its mount. This is usually done by the manufacturer with a special alignment fixture and an autocollimator before the mount is fastened to the base of the monochromator. The rulings of a grating must be parallel to the axis of rotation, and the axis of rotation, in turn, must be perpendicular to the optic axis of the monochromator. Otherwise, the diffracted image of the entrance slit will translate along the length of the exit slit as the grating is rotated.

The collimator (and telescope) is adjusted to flood the grating or prism and to image the entrance slit sharply on the exit slit. If a parabola is used, its axis must be correctly defined. Critical focusing is done with a modified Foucalt test. The monochromator is illuminated with light from a source having a narrow emission line as, for example, the green line from a mercury arc. The line is scanned across the exit slit by slowly rotating the disperser, while the collimator is observed through the exit slit with the aid of a telescope. In a properly focused monochromator, the mirror becomes bright and then dark in a uniform fashion. Spherical aberration causes the outer and inner zones to become bright separately and a doughnut pattern is observed. In a poorly focused monochromator, the collimator first becomes bright at one edge, and a patch of light travels across the mirror as the spectral line is scanned.

REFERENCES

1. W. T. Welford, *Progress in Optics*, Vol. IV, E. Wolf, ed., Wiley, New York, 1965.
2. H. Yoshinaga, B. Okazaki, and S. Tatsuoka, *J. Opt. Soc. Am.*, **50**, 437 (1960).
3. K. Kudo, *Sci. of Light*, **9**, 65 (1960).
4. K. D. Mielenz, *Optik*, **20**, 28 (1963).
5. N. Sassa, *Sci. of Light*, **10**, 53 (1961).
6. S. A. Khrshanovskii, *Opt. Spectry*, **9**, 207 (1960).
7. H. Ebert, *Wiedemann's Ann.* **38**, 489 (1889).
8. W. G. Fastie, *J. Opt. Soc. Am.*, **42**, 641, 647 (1952).
9. K. Kudo, *Sci. of Light*, **9**, 1 (1960).
10. K. Kudo, *J. Opt. Soc. Am.*, **55**, 150 (1965).
11. L. A. Megill and L. Droppleman, *J. Opt. Soc. Am.*, **52**, 258 (1962).
12. W. Leo, *Z. Angew. Physik*, **8**, 196 (1956).
13. G. R. Rosendahl, *J. Opt. Soc. Am.*, **51**, 1 (1961); **52**, 408, 412 (1962).
14. K. D. Mielenz, *J. Res. NBS*, **68C**, 195, 201, 205 (1964).
15. K. D. Mielenz, *Optik*, **19**, 510 (1962).
16. H. Kaiser, K. D. Mielenz, and F. Rosendahl, *Z. Instr. Kde.*, **67**, 269 (1959).
17. M. Czerny and A. F. Turner, *Z. Physik*, **61**, 792 (1930).
18. M. A. Ford, W. C. Price, and G. R. Wilkenson, *J. Sci. Inst.*, **35**, 55 (1958).
19. W. G. Fastie, *J. Opt. Soc. Am.*, **43**, 1174 (1953).
20. C. S. Rupert, *J. Opt. Soc. Am.*, **42**, 779 (1952).
21. W. T. Welford, *J. Opt. Soc. Am.*, **53**, 766 (1963).
22. S. C. Baker, *J. Opt. Soc. Am.*, **54**, 271 (1964).
23. A. S. Filler, *J. Opt. Soc. Am.*, **54**, 424 (1964).
24. M. Czerny and V. Plettig, *Z. Physik*, **63**, 590 (1930).
25. A. B. Shafer, L. R. Megill, and L. Droppleman, *J. Opt. Soc. Am.*, **54**, 879 (1964).
26. T. Namioka and M. Seya, *Sci. of Light*, **16**, 140 (1967).
27. M. Seya and T. Namioka, *Sci. of Light*, **16**, 129 (1967).
28. C. D. Allemand, *J. Opt. Soc. Am.*, **58**, 159 (1968).
29. A. Walsh, *J. Opt. Soc. Am.*, **42**, 94 (1952).
30. T. Shimomura, *Japan. J. Appl. Phys.*, **3**, 459 (1964).
31. W. Benesch and J. Strong, *J. Opt. Soc. Am.*, **41**, 252 (1951).
32. C. J. Cremers and E. R. F. Winter, *Appl. Spectry*, **20**, 421 (1966).
33. D. W. Robinson, *J. Opt. Soc. Am.*, **49**, 966 (1959).
34. R. G. Greenler, *J. Opt. Soc. Am.*, **46**, 433 (1956).
35. T. M. Hard and R. C. Lord, *Appl. Opt.*, **7**, 589 (1968).
36. J. J. Mitteldorf and D. O. Landon, *Appl. Opt.*, **7**, 1431 (1968).
37. A. B. Shafer and D. O. Landon, *Appl. Opt.*, **8**, 1063 (1969).
38. R. C. Hawes, *Appl. Opt.*, **8**, 1063 (1969).

CHAPTER 7

Slit Functions and Spectral Modulation Transfer Functions of Monochromators

In this chapter we treat the nature of the narrow band of spectral frequencies transmitted by a monochromator. In particular, we discuss the shape and width of this band and its effect on the resolving power of the spectrophotometer.

I. THE SLIT FUNCTION

The mathematical function that has come to be called the slit function of a monochromator can be defined and interpreted in two distinct ways. Let us discuss these separately, and see how they are related.

Consider first a monochromator whose entrance slit is illuminated with monochromatic radiation. Somewhere in the plane of the exit slit will appear a monochromatic image of the entrance slit. By adjusting the angle of the grating or Littrow mirror, this image can be swept across the exit slit. We can record the power passing through the exit slit as a function of the angle of the grating or mirror, as a function of the indicated wavenumber of the monochromator, as a function of the coordinate of the center (or edge perhaps) of the image, or as a function of time. Such a plot is the *slit function* of the monochromator. If aberrations and diffraction can be ignored, and if the image of the

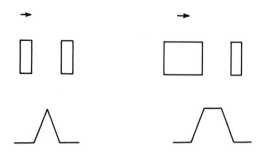

Fig. 7.1 Slit function.

entrance slit is of the same size as the exit slit, as is usually the case, then the slit function is clearly triangular, as in Fig. 7.1, with the base of the triangle equal to twice the width of the slit. In this illustration we have elected to plot the image in terms of the coordinate of the slit image. If one slit is wider than the other, then the slit function is trapezoidal. The base of the trapezoid is equal to the sum of the slit widths, and the width of the top is the difference of the slit widths. Now let us generalize these results on a somewhat more satisfactory mathematical basis.

Represent the transmission through the opening between the exit slit jaws by $u_x(x)$

$$u_x(x) = \begin{cases} 1 & \frac{1}{2}(-w_x) \leqslant x \leqslant \frac{1}{2}(+w_x) \\ 0 & \text{otherwise} \end{cases}$$

where w_x is the width of the slit and x is a coordinate measured from the center of the slit. Also represent the irradiance of the entrance-slit image by $u_n(x - x')$. With parallel slit jaws and negligible diffraction and aberrations,

$$u_n(x - x') = \begin{cases} 1 & \frac{1}{2}(-w_n) \leqslant x - x' \leqslant \frac{1}{2}(+w_n) \\ 0 & \text{otherwise} \end{cases} \tag{7.1}$$

where w_n is the width of the entrance slit and x' is the displacement of the image of the center of the entrance slit measured from the center of the exit slit. The radiation passing through the exit slit as a function of x' is written

$$\sigma(x') = \int_{-\infty}^{\infty} u_x(x) u_n(x - x') \, dx \tag{7.2}$$

where the infinite limits of integration are only formal, as the integration needs to extend only across the entrance-slit opening. The integral in Eq. (7.2) is a convolution integral such as we have already encountered in Section 4.VII.

Now let us reinterpret the function $\sigma(x')$. We do this by defining the displacement x' in terms of the frequency of the monochromatic radiation v', the frequency to which the monochromator is set v, and the linear dispersion $D = (dx'/dv')$. The frequency of the monochromator is determined by the geometry of the monochromator and its disperser. It is defined to make $v - v'$ vanish when the image of the entrance slit "best" coincides with the exit slit. Thus x' can be written

$$x' = D(v - v') \tag{7.3}$$

and the function $\sigma(x')$ can be written in terms of frequency from Eqs. (7.2) and (7.3),

$$\sigma(v - v') = \int_{-\infty}^{\infty} u_x(x) u_n[x - D(v - v')] \, dx$$

The function $\sigma(v - v')$ is the *slit function*. It is the radiant power transmitted by a monochromator when it is set to pass radiation of frequency v and the incident radiation is monochromatic and has frequency v'.

Now consider the entrance slit to be irradiated with polychromatic radiation having some distribution $J_{in}(v')$ representing the radiant power in a frequency interval between v' and $v' + dv'$. For each frequency an image of the entrance slit appears at a different place in the plane of the exit slit. That is, for a given setting of the monochromator, there is a function $\sigma(v - v')$ for each frequency in the incident radiation, and the radiant power passed through the exit slit is clearly just the integral of these weighted by the distribution $J_{in}(v')$, or

$$J_{out}(v) = \int_{-\infty}^{\infty} J_{in}(v') \sigma(v - v') \, dv'$$

$$= \int_{-\infty}^{\infty} J_{in}(v - v') \sigma(v') \, dv' \tag{7.4}$$

The negative frequencies implicit in Eq. (7.4) are formal, but we retain them for reasons of convenience, as will be seen later. Equation (7.4) describes the way in which a monochromator alters the apparent shape of an input distribution as it forms and presents a spectrum.

If the input radiation has but a single frequency. v_0, it can be represented as a delta function

$$J_{in}(v') = \delta(v' - v_0) = \begin{cases} 1 & v' = v_0 \\ 0 & \text{otherwise} \end{cases}$$

With the introduction of this input, Eq. (7.4) becomes

$$J_{out}(v) = \int_{-\infty}^{\infty} \delta(v' - v_0) \sigma(v - v') \, dv' = \sigma(v - v_0) \tag{7.5}$$

Now the magnitude of a distribution J can be sampled at any frequency v_0 by forming the product $J(v)\,\delta(v-v_0)$. In particular, if $J(v)$ is a uniform distribution over some frequency interval,

$$J(v) = 1 \quad \text{and} \quad J(v)\,\delta(v-v_0) = 1$$

Hence Eq. (7.5) provides us with an alternative definition of slit function: A monochromator set at frequency v_0 and irradiated with radiation having a uniform frequency distribution transmits an amount $\sigma(v-v_0)$ of radiant power of frequency v. Thus the slit function $\sigma(v-v_0)$ has been defined in two different ways, but the mathematical form of the function is the same in both cases.

In the discussion in Section 4.VII we showed that an image $i(x')$ can be expressed as the convolution of the irradiance distribution in the object $o(x)$ and the line spread function $s(x-x')$. Thus the image of the entrance slit can be written in the form

$$u_n(x-x') = \int u_{no}(x'')s(x-x'-x'')\,dx''$$

and Eq. (7.2) can be written as a convolution of three factors

$$\sigma(x') = \int \int u_x(x)u_{no}(x'')s(x-x'-x'')\,dx\,dx''$$

The contribution to the line spread function by aberrations can be found, for example, from a detailed ray trace of the monochromator, and the contribution from diffraction can be written from the equations of Sections 4.VI and 4.VII.

A. The Geometric Slit Function

In a monochromator having equivalent entrance and exit slits and negligible aberrations and diffraction, the slit function is triangular and can be represented by

$$\sigma_T(v-v') = \begin{cases} 1 - \dfrac{|v-v'|}{\Delta v_T} & |v-v'| < \Delta v_T \\ 0 & |v-v'| \geq \Delta v_T \end{cases} \qquad (7.6)$$

Δv_T is the effective bandwidth (EBW) of the monochromator and is given by

$$\Delta v_T = D^{-1}w$$

where D^{-1} is the reciprocal linear dispersion of the monochromator and w is the width of the slit; Δv_T is half the width of the base of the triangle or the width at half its height. Defined in this way, the effective bandwidth can be extended to other forms of slit function differing somewhat from the triangular.

Aberrations, including slit image distortion, and diffraction are generally negligible when a monochromator is used with wide slits, and so the triangular function is a good approximation to the actual slit function. A very narrow spectral line scanned with a triangular slit function is recorded with a triangular form of the same width as the slit function, according to Eq. (7.5).

Two very narrow spectral lines lying close together are recorded as the sum of two triangular functions. This is illustrated in Fig. 7.2 for two lines of equal peak height. There is no confusion in the two lines until the bases begin to overlap (Fig. 7.2B). Two recognizable peaks are visible for separations greater than half the width of the base of the triangle (Fig. 7.2C), but when the lines are separated by half the base width (Fig. 7.2D) the minimum between the peaks disappears and a trapezoidal structure is seen. Closer lines (Fig. 7.2E) overlap to form a single peak. The situation of Fig. 7.2D is said to represent

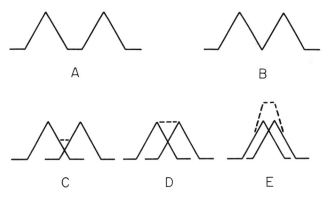

Fig. 7.2   Overlapping lines.

two just-resolved lines. This definition of resolution is Sparrow's criterion, namely, two lines are just resolved when the central minimum between them vanishes. Thus the effective bandwidth of the triangular slit function is the resolution attainable by the monochromator, in the sense of Sparrow's criterion and for very narrow lines of equal intensity.

A spectral band, or line of finite width, when scanned with a triangular slit function (or any symmetric slit function), is distorted in two ways: its peak height is reduced and its width is increased. The area of the band, that is, the integral over frequency, is not altered by the slit function provided the slit function is normalized.

$$\int \sigma(\nu - \nu') \, d\nu' = 1$$

for example, by adjusting the amplifier gain. This is simply a consequence of conservation of energy, but it is instructive to consider the integral of the recorded band, from Eq. (7.4),

$$\int_0^\infty J_{\text{out}}(\nu) \, d\nu = \int_0^\infty \int_0^\infty J_{\text{in}}(\nu') \sigma(\nu - \nu') \, d\nu'$$

$$= \int_0^\infty J_{\text{in}}(\nu') \left[\int_0^\infty \sigma(\nu - \nu') \, d\nu \right] d\nu' = \int_0^\infty J_{\text{in}}(\nu') \, d\nu'$$

This is just the area of the band itself and is independent of the functional form of the slit function.

The problem of distortion of a band by the slit function of a monochromator has been discussed many times in the past (1–34). Various assumptions have been made about the shape of the slit function and the band.

We now show how a particular band shape is distorted by a triangular slit function. Let us assume a band with a Gaussian contour arbitrarily centered about the origin $\nu = 0$,

$$J_{\text{in}}(\nu) = \frac{1}{\sqrt{2\pi}a} \exp\left(\frac{-\nu^2}{2a^2}\right) \qquad (7.7)$$

This represents an emission band, but the modification to absorption is trivial. The parameter a^2 is the variance or second moment of the distribution and is a measure of width. In statistical terms, a is the standard deviation of the Gaussian distribution. The variance of a function $f(x)$ is defined by the relationship.

$$\text{var} f(x) = \langle (x - \bar{x})^2 \rangle = \int_{-\infty}^\infty (x - \bar{x})^2 f(x) \, dx$$

where $\bar{x}$ is the mean,

$$\text{mean} f(x) = \langle x \rangle = \int_{-\infty}^\infty x f(x) \, dx$$

The variance is therefore the mean-square bandwidth. In our Gaussian distribution we have set the mean equal to zero by centering at the origin.

A more usual measure of bandwidth in spectroscopy is the half-bandwidth (HBW or $\Delta_{1/2}\nu$), which is defined as the width of a band at half its maximum intensity (Fig. 7.3). For a Gaussian band, we have the relationship

$$(\text{HBW})_{\text{Gauss}} = 2a(2\ln 2)^{1/2} = 2.35a \tag{7.8}$$

HALF—BAND—WIDTH

Fig. 7.3 Half-bandwidth.

The factor $1/\sqrt{2\pi}a$ in Eq. (7.7) is a normalizing factor to maintain unit area as a is varied.

There is no physical basis for assuming a Gaussian band shape beyond the observation that often bands resemble a Gaussian contour. The main reason for working with a Gaussian is that it is a convenient mathematical form. The triangular slit function is given by Eq. (7.6). The distorted band shape presented by the monochromator is, according to Eq. (7.4),

$$J_{\text{out}}(\nu') = \frac{1}{\sqrt{2\pi}a} \int_{\nu'-\Delta\nu}^{\nu'+\Delta\nu} \exp\left(\frac{-\nu^2}{2a^2}\right)\left[1 - \frac{|\nu-\nu'|}{\Delta\nu}\right] d\nu$$

$$= \frac{1}{\sqrt{2\pi}a}\Bigg\{\int_{\nu'-\Delta\nu}^{\nu'+\Delta\nu} \exp\left(\frac{-\nu^2}{2a^2}\right) d\nu - \frac{1}{\Delta\nu}\int_{\nu'}^{\nu'+\Delta\nu} \exp\left(\frac{-\nu^2}{2a^2}\right)(\nu-\nu')\, d\nu$$

$$- \frac{1}{\Delta\nu}\int_{\nu'-\Delta\nu}^{-\nu'} \exp\left(\frac{-\nu^2}{2a^2}\right)(\nu'-\nu')\, d\nu\Bigg\} \tag{7.9}$$

The integrations in Eq. (7.9) can be expressed in terms of the error function

$$\text{erf}\, x = \int_0^x e^{-\xi^2}\, d\xi$$

whose values may be found in standard tables. The result of the integration is

$$
\begin{aligned}
J_{out}(\nu') &= \frac{1}{\sqrt{2\pi}a}\Bigg\{ a\sqrt{\frac{\pi}{2}}\Bigg[\left(1+\frac{\nu'}{\Delta\nu}\right)\operatorname{erf}\frac{\nu'+\Delta\nu}{2^{1/2}a} - \frac{2\nu'}{\Delta\nu}\operatorname{erf}\frac{\nu'}{2^{1/2}a} \\
&\quad - \left(1-\frac{\nu'}{\Delta\nu}\right)\operatorname{erf}\frac{\nu'-\Delta\nu}{2^{1/2}a}\Bigg] + \frac{a^2}{\Delta\nu}\Bigg[\exp\left(\frac{-(\nu'+\Delta\nu)^2}{2a^2}\right) \\
&\quad - 2\exp\left(\frac{-\nu'^2}{2a^2}\right) + \exp\left(\frac{-(\nu'+\Delta\nu)^2}{2a^2}\right)\Bigg]\Bigg\} \\[6pt]
&= \frac{1}{\sqrt{2\pi}a}\Bigg\{ a\sqrt{\frac{\pi}{2}}\Bigg[\left(1+\frac{\nu'}{\Delta\nu}\right)\operatorname{erf}\frac{\nu'+\Delta\nu}{2^{1/2}a} - \frac{2\nu'}{\Delta\nu}\operatorname{erf}\frac{\nu'}{2^{1/2}a} \\
&\quad - \left(1-\frac{\nu'}{\Delta\nu}\right)\operatorname{erf}\frac{\nu'-\Delta\nu}{2^{1/2}a}\Bigg] + \frac{2a^2}{\Delta\nu}\exp\left(\frac{-\nu'^2}{2a^2}\right) \\
&\quad \cdot \Bigg[\exp\left(\frac{-\Delta\nu^2}{2a^2}\right)\cosh\frac{\nu'\,\Delta\nu}{a^2} - 1\Bigg]\Bigg\}
\end{aligned}
$$

This equation is plotted in Fig. 7.4 for a band of unity variance and a series of effective bandwidths. Evidently the distortion takes the form of a reduction in peak height and an increase in bandwidth.

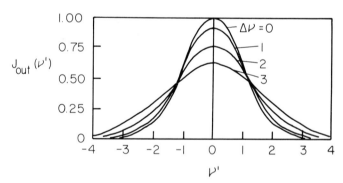

Fig. 7.4   Gaussian band of unity variance distorted by scanning with triangular slit functions of various effective bandwidths.

The peak height of a distorted Gaussian band is plotted against monochromator effective bandwidth in Fig. 7.5. Also plotted in this figure is the peak height of a band having a Cauchy or Lorentz shape of unit area.

$$
J_{in}(\nu) = \frac{1}{\pi(1+(\nu-b)^2)} \tag{7.10}
$$

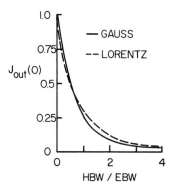

Fig. 7.5 Peak height of Gauss- and Lorentz-shaped bands scanned with triangular slit function for various ratios of half-bandwidth (HBW) to effective bandwidth (EBW).

The variance of a Cauchy distribution does not exist, but the half-bandwidth is given by

$$(\text{HBW})_{\text{Lorentz}} = 2(b + (1 + 2b^2)^{1/2})$$

The curves of Fig. 7.5 are based on Gaussian and Cauchy bands of equal half-bandwidths. It is evident from the figure that the sensitivity of the peak height to changes in the effective bandwidth of the monochromator depends on the shape of the band. As a practical working rule, the effective bandwidth of the monochromator should be less than about one-fifth the half-bandwidth of the band under study in order to keep the peak height error less than about 2 or 3%.

It is common practice in spectroscopic laboratories to state that the observed bandwidth (OBW) of an emission band distorted by passage through a monochromator is given by the sum of the effective bandwidth of the monochromator and the half-bandwidth of the band

$$\text{OBW} \approx \text{EBW} + \text{HBW} \qquad (7.11)$$

There is in fact no mathematical basis for such a statement except as a very rough approximation. A rigorous relationship between the monochromator slit width and the bandwidth exists in terms of the variance:

$$\text{var}(J_{\text{out}}) = \text{var}(J_{\text{in}}) + \text{var}(\sigma)$$

The derivation of this relationship can most easily be carried out by using the spectral modulation transfer function that is introduced in Section II. As the variance is the mean-square bandwidth, we might expect an approximation to hold

$$(\text{OBW})^2 \approx (\text{HBW})^2 + (\text{EBW})^2 \qquad (7.12)$$

and, to first order in a power series expansion, Eq. (7.11) is an approximation of Eq. (7.12).

The variance of a triangular function centered at $\nu = 0$ is calculated from the integral

$$\langle \nu^2 \rangle_T = \frac{\int_{-\Delta\nu}^{0} \nu^2 (1 + \nu/\Delta\nu)\, d\nu + \int_{0}^{\Delta\nu} \nu^2 (1 - \nu/\Delta\nu)\, d\nu}{\int_{-\Delta\nu}^{0} (1 + \nu/\Delta\nu)\, d\nu + \int_{0}^{\Delta\nu} (1 - \nu/\Delta\nu)\, d\nu} \tag{7.13}$$

The denominator is a normalizing factor. The variance is clearly the same if the function is centered at any value of ν, but the integration is simplified when the mean of the function is zero. The evaluation of Eq. (7.13) is straightforward and results in

$$\langle \nu^2 \rangle_T = \tfrac{1}{6}(\Delta\nu)^2 = \tfrac{1}{6}(\text{EBW})^2 \tag{7.14}$$

We already know from Eq. (7.8) that the variance of a Gaussian function is $(1/2.35)^2$ times the square of its half-bandwidth. Hence the variance of a Gaussian band scanned with a triangular slit function is

$$(\nu^2) = 0.187(\text{HBW})^2 + 0.167(\text{EBW})^2$$

and to the approximation that the distorted band has nearly a Gaussian shape

$$0.187(\text{OBW})^2 \approx 0.187(\text{HBW})^2 + 0.167(\text{EBW})^2 \tag{7.15}$$

This is recognized as an approximation to Eq. (7.12).

We mention in passing that squares of the half-bandwidths of two convolved Gaussian functions are added and that the half-bandwidths of two Cauchy functions are added.

B. The Slit Function with Diffraction

We know from the discussion of Section 4.VI that a point image is actually not a point, but is broadened even in a perfect optical system by diffraction. We also know, from Section 4.VII, how this broadening affects the image of an extended object and how to describe it with the spread function. The image of the entrance slit of a monochromator is also broadened by diffraction. So in place of Eq. (7.1) for the irradiance of the image of the entrance slit formed in the plane of the exit slit, we must write

$$u_n(x - x') = \int_{\xi} s_d(\xi) u_n(x - x' - \xi)\, d\xi$$

where $s_d(\xi)$ is the spread function due to diffraction and is given by

Eq. (4.42) for a rectangular aperture and incoherent radiation. We rewrite Eq. (4.42) in the form

$$s_d(\xi) = \frac{\sin^2 \pi\xi/F\lambda}{(\pi\xi/F\lambda)^2}$$

where F is the f/number of the monochromator. The radiation passing through the exit slit is now written, in place of Eq. (7.2),

$$\sigma(x') = \int_\xi \int_x u_x(x) s_d(\xi) u_n(x - x' - \xi)\, dx\, d\xi \qquad (7.16)$$

We suppose that u_x and u_n represent rectangular slit openings of equal width. The order of integration in Eq. (7.16) is immaterial. If we choose first to integrate over u_x and u_n, we find a triangular function just as before, and this must then be convolved with the diffraction spread function. The diffraction spread function can be rewritten in terms of frequency

$$s_d(v') = \frac{\sin^2 \pi Dv_0 v'/F}{(\pi Dv_0 v'/F)^2} \qquad (7.17)$$

where D is the linear dispersion of the monochromator, v_0 is the center frequency or nominal frequency passed by the monochromator, and v' represents small variations about v_0. Then, with Eq. (7.6) for the triangular slit function with no diffraction, we have an expression for the diffraction-broadened triangular slit function centered at $v = 0$

$$\sigma(v) = \int_{-\Delta v}^{\Delta v} \left(1 - \frac{|v'|}{\Delta v}\right) \frac{\sin^2 \pi Dv_0(v - v')/F}{(\pi Dv_0(v - v')/F)^2}\, dv' \qquad (7.18)$$

The meaning of Eq. (7.18) is this (see Fig. 7.6): In order to calculate $\sigma(v)$, we center a diffraction pattern at v', giving it a central height

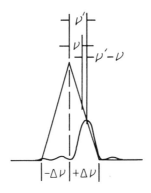

Fig. 7.6   Diffraction-broadened triangular slit function.

equal to the value of the triangular function at ν'. We then determine the value of this diffraction pattern at ν from $s_d(\nu-\nu')$. This process is repeated for all ν's in the range $-\Delta\nu$ to $+\Delta\nu$ and the results are summed by the integration of Eq. (7.18) to find $\sigma(\nu)$. This integral can be evaluated, but not in closed form, and the computation is tedious, though straightforward.

The variance of the diffraction pattern, Eq. (7.17), is not defined because the integral

$$\int_{-\infty}^{\infty} \nu^2 s_d(\nu)\, d\nu$$

diverges. However, the predominant effect of diffraction originates in the central lobe, with the role of the subsidiary maxima remaining relatively minor. Thus, we can determine an effective variance of the diffraction pattern in the slit plane from

$$\langle x^2 \rangle_d = \left[\int_{-F\lambda}^{F\lambda} x^2 \frac{(\sin^2 \pi x/F\lambda)}{(\pi x/F\lambda)^2}\, dx\right] \Big/ \left[\int_{-F\lambda}^{F\lambda} \frac{(\sin^2 \pi x/F\lambda)}{(\pi x/F\lambda)^2}\, dx\right] \quad (7.19)$$

where the integration is carried out between the first zeros of the function. The value of Eq. (7.19) is

$$\langle x^2 \rangle_d = 0.224(F/\nu)^2$$

We really desire the variance in terms of ν, so with the introduction of the reciprocal dispersion D^{-1} we have

$$\langle \nu^2 \rangle_d = 0.224\left(\frac{FD^{-1}}{\nu}\right)^2 \quad (7.20)$$

The variance of the diffraction-broadened triangular slit function is thus, from Eqs. (7.14) and (7.20),

$$\langle \nu^2 \rangle_{T+d} = 0.167(\text{EBW})^2 + 0.224\left(\frac{FD^{-1}}{\nu}\right)^2$$

$$= 0.167(D^{-1}w)^2 + 0.224\left(\frac{FD^{-1}}{\nu}\right)^2$$

$$\approx 0.2(D^{-1})^2\left(w^2 + \left(\frac{F}{\nu}\right)^2\right)$$

with w the physical slit width. Analogous to Eq. (7.15), we now can write for the variance of a Gaussian band distorted by both the geometric triangular slit function and diffraction,

$$0.187(\text{OBW})^2 \approx \langle \nu^2 \rangle = 0.187(\text{HBW})^2 + 0.167(\text{EBW})^2 + 0.224\left(\frac{FD^{-1}}{\nu}\right)^2$$

or, approximately,

$$(OBW)^2 \approx (HBW)^2 + (EBW)^2 + \left(\frac{FD^{-1}}{\nu}\right)^2$$

$$= (HBW)^2 + (EBW)^2 + (DBW)^2$$

where we use DBW to stand for diffraction bandwidth.

Let us consider a monochromator having an f/number of $F = 10$ and a reciprocal dispersion $D^{-1} = 20 \text{ cm}^{-1}/\text{cm}$, and let us consider $[\langle \nu^2 \rangle_{T+d}]^{1/2}$ as a function of slit width at a frequency of 1000 cm^{-1}, as shown in Fig. 7.7. Notice how the diffraction contribution dominates for slit widths less than about 0.01 cm. There is little to be gained in

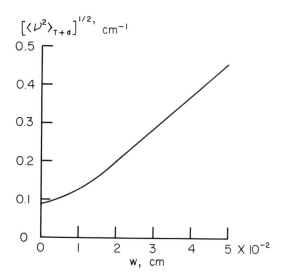

Fig. 7.7 Total slit function variance (diffraction and geometric) vs. slit width for an $f/10$ monochromator with reciprocal dispersion $20 \text{ cm}^{-1}/\text{cm}$ at 1000 cm^{-1}.

reduced slit function variance by narrowing the slits beyond this value and, in general, the narrowest practical slit width is

$$w_d \approx F/\nu = F\lambda \tag{7.21}$$

which is called the *diffraction-limited slit width* or merely the *diffraction slit*. Equation (7.21) is an important relationship which should be kept in mind by the spectroscopist as he selects an operating slit width for his monochromator.

The entrance pupil of a monochromator is irradiated with incoherent radiation if a thermal source is placed at the slit or imaged on the slit with an optical element whose aperture is much larger than the collimator of the monochromator. On the other hand, the radiation is coherent if a laser is used or if a thermal source is located a great distance from the slit (e.g. a star) or is imaged on the slit with an element having a very small aperture. Intermediate cases of partial coherence are obtained with thermal sources using optical elements whose apertures are comparable to the collimator. The effect of coherence on the diffraction slit function has been discussed by van Cittert (4) and more recently by Sakai (35) and Roseler (34, 36). Sakai notes that for geometric slit widths of half the diffraction width, or less, or greater than about four times the diffraction width, the state of coherence is not important. At intermediate geometric slit widths, resolution is somewhat greater with coherent radiation than with incoherent.

C. The Slit Function with Aberrations

When aberrations are present in a monochromator, they exert an influence on the shape of the slit function. The final form of the slit function is found by convolving the diffraction-broadened slit function with the spread function due to the aberrations. Of course, the order in which the convolutions are carried out does not matter and the aberration spread function can be introduced first. Moreover, the spread functions of the individual third-order aberrations can be introduced separately and successively.

The symmetric aberrations – spherical aberration, astigmatism, field curvature, and defocusing – broaden the slit function by adding their own variances. The asymmetric aberrations – coma and distortion, including slit image distortion – broaden the slit function and also skew it into an unsymmetric shape. Let us consider some simple cases in order to get an idea of the magnitude of the effect of aberrations on the slit function.

A defocused optical system images a point object as a disk in a reference plane. Hence, the point spread function of a defocused system is a pillbox function of diameter l which can be written

$$l = \xi/F \qquad (7.22)$$

where ξ is the distance of the reference plane from the Gaussian image plane and F is the f/number of the system. The line spread function is found by integration across a chord of the pillbox, as shown in Fig. 7.8.

$$s_D(x) = \int_{-(l^2-x^2)^{1/2}}^{(l^2-x^2)^{1/2}} dy = 2(l^2 - x^2)^{1/2} \qquad x \leqslant l \qquad (7.23)$$

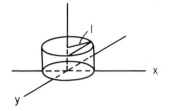

Fig. 7.8 Point spread function of a defocused system.

Upon substitution of Eq. (7.22) and the reciprocal dispersion, Eq. (7.23) becomes

$$s_D(\Delta\nu) = 2\left[\frac{\xi^2(D^{-1})^2}{F^2} - (\Delta\nu)^2\right]^{1/2}$$

The form of this equation is sketched in Fig. 7.9. The width of the line spread function in units of optical frequency is

$$\Delta\nu = D^{-1}\xi/F \qquad (7.24)$$

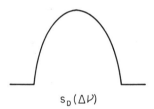

$$s_D(\Delta\nu)$$

Fig. 7.9 Line spread function of a defocused system.

and the variance of the spread function is easily calculated:

$$\langle\nu^2\rangle = \left(\frac{D^{-1}\xi}{2F}\right)^2$$

The variance of the defocusing spread function thus equals the variance of the central lobe of the diffraction spread function [Eq. (7.20)] when

$$\xi = \frac{0.224\,F^2}{0.25\,\nu} \approx \frac{F^2}{\nu}$$

If this is taken as a measure of the tolerance in defocusing, we see that an $f/10$ monochromator working at $1000\ \mathrm{cm}^{-1}$ must be properly focused within $0.1\ \mathrm{cm}$.

Let us approximate the line spread function associated with de-

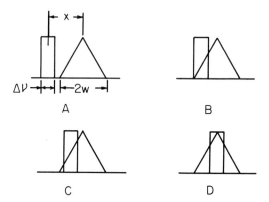

Fig: 7.10   Convolution of triangle and rectangle.

focusing with a rectangle of width $\Delta\nu$ given by Eq. (7.24). The convolution of the rectangle with the triangular geometric slit function can be written with reference to Fig. 7.10:

$$s(x) = 0 \qquad\qquad\qquad x \geqslant w + \tfrac{1}{2}(\Delta\nu)$$

$$s(x) = \frac{(w + \Delta\nu/2 - x)^2}{2w} \qquad w - \tfrac{1}{2}(\Delta\nu) \leqslant x \leqslant w + \tfrac{1}{2}(\Delta\nu)$$

$$s(x) = \frac{\Delta\nu}{w}(w - x) \qquad\qquad \tfrac{1}{2}(\Delta\nu) \leqslant x \leqslant w - \tfrac{1}{2}(\Delta\nu)$$

$$s(x) = \left(\Delta\nu - \frac{(\Delta\nu)^2}{4w}\right) - \frac{x^2}{w} \qquad x \leqslant \tfrac{1}{2}(\Delta\nu)$$

The resulting function, shown in Fig. 7.11, consists of parabolic sections at top and bottom connected by straight line segments. The straight line portion is not present, of course, when $\Delta\nu \geqslant w$. Notice that the function of Fig. 7.11 resembles a Gaussian function, particu-

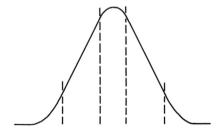

Fig. 7.11   Convolute of triangle and rectangle.

larly when the straight line segments are short. We shall return to this point a little later. For the moment, let us approximate the function with a trapezoid whose sides have the same slope as the triangle and whose top is as wide as the rectangle. The peak height of a Gauss-shaped band [Eq. (7.7)] and a Lorentz-shaped band [Eq. (7.10)] were calculated for a variety of trapezoidal slit functions. The results are shown in Fig. 7.12. The abscissa is one-half the ratio of the HBW to the EBW of the basic triangular slit function. The ordinate is the peak height. The parameter y is one-half the ratio of the broadening (i.e., the width of the top of the trapezoid or the width of the defocusing spread function) to the EBW of the basic triangular slit function. It was remarked earlier (Fig. 7.5) that the peak height of a Lorentz-shaped band is somewhat more sensitive to the action of a triangular slit function. It is not surprising to observe from Fig. 7.12 that the Lorentz-shaped band is also more sensitive to defocusing.

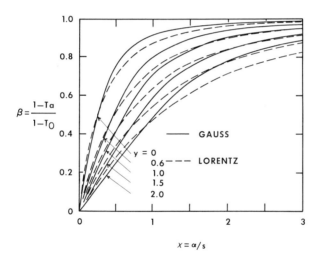

Fig. 7.12   Peak height of Gauss and Lorentz bands scanned with defocused mono-chromator.

Spherical aberration causes radiation from a zone of radius ρ in the aperture of an optical system to appear as a circle of radius r in the Gaussian image plane. Let us assume that the aperture is uniformly illuminated with one unit of radiant power per unit area and define an illumination density function $J_s(r)$ as the radiant power intercepted by the Gaussian image plane in a ring of radius r to $r + dr$. We equate the power passing through the aperture with the power intercepted by the image plane by writing

$$\int_0^\rho \pi\rho \, d\rho = \int_0^r J_s(r) \, dr$$

Integrate the left-hand side and substitute the relationship between r and ρ from Section 3.X.A.

$$\frac{\pi\rho^2}{2} = \frac{\pi}{2}\left(\frac{r}{a_{11}R}\right)^{2/3} = \int_0^r J_s(r) \, dr \tag{7.25}$$

$J_s(r)$, found by differentiation of Eq. (7.25), is of course the point spread function

$$J_s(r) = s_s(r) = \begin{cases} \dfrac{\pi}{3(a_{11}R)^{2/3}r^{1/3}} & 0 \leqslant r \leqslant r_s \\ 0 & r > r_s \end{cases}$$

where r_s is the radius of the circle originating in the outermost ring of the aperture. J_s is the point spread function of a system with spherical aberration. The line spread function is found by integration of J_s along a chord, as sketched in Fig. 7.13,

$$s_s(x) = 2\int_0^{y_{\lim}} s_s(x, y) \, dy = \frac{2\pi}{3(a_{11}R)^{2/3}} \int_0^{[r_s^2 - x^2]^{1/2}} \frac{dy}{(x^2 + y^2)^{1/6}} \qquad x \leqslant r_s \tag{7.26}$$

The integration of Eq. (7.26) can be carried out numerically, but we shall not do it here. A well-adjusted monochromator in which spherical aberration is the greatest contributor to geometric image degradation will not be focused with the slit on the Gaussian image plane anyway, but will have the slit plane somewhere between the paraxial and marginal focal planes.

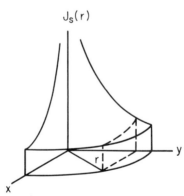

$J_s(r)$

Fig. 7.13 Point spread function of a system with spherical aberration.

In a system with pure astigmatism, the line spread function is a delta function if the sagittal astigmatic image falls along the direction of the slit in the Gaussian image plane, and it is a rectangle of width equal to the length of the sagittal image if the latter is normal to the slit.

The line spread function associated with slit image distortion is given by

$$s_D(x) = 2 \cos \theta \qquad |y| \leq y_m$$

where y_m is half the slit height (see Fig. 7.14). Because the distorted slit image has a parabolic shape, we can write from Eq. (5.30) or (5.54),

$$x = Ky^2 \quad \text{or} \quad \Delta x = 2Ky\Delta y$$

and

$$\cos \theta = \frac{\Delta y}{[(\Delta y)^2 + (\Delta x)^2]^{1/2}} = [1 + 4K^2y^2]^{-1/2}$$

so

$$s_D(x) = 2/(1 + 4Kx)^{1/2} \qquad 0 \leq x \leq Ky_m^2$$

or

$$s_D(x) = 2/(1 + 4K\nu/D^{-1})^{1/2}$$

The form of $s_D(\nu)$ is sketched in Fig. 7.15. We can consider $s_D(\nu)$ as roughly a rectangle with one edge at ν_0 and width equal to the sagittal distance of the curved slit image, expressed in spectral frequency units.

The line spread function for comatic aberration has a symmetric shape, sketched in Fig. 7.16, when the tail of the comatic pattern extends along the slit, and an asymmetric shape when the tail lies across the slit. We shall not calculate the shapes of these functions, but the procedure is straightforward.

It has been noted in Section 6.II that the individual aberrations encountered in typical infrared monochromator systems are of roughly the same magnitude as diffraction in the mid-infrared region. With a

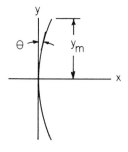

Fig. 7.14   Slit image distortion.

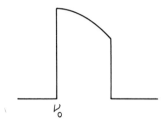

Fig. 7.15   Slit function with slit image distortion.

narrow geometric slit width, the overall slit function thus involves a multiple convolution of several functions having roughly the same variance. We have noted above that the distribution obtained by convolution of a triangular function with a rectangular function resembles a Gaussian function. There is a theorem in statistical mathematics, the central limit theorem, which states that the multiple convolutions of n functions, when they fulfill certain requirements, approach a Gaussian function as n approaches infinity. The sufficient set of requirements on the component functions are:

(1) the functions vanish for arguments greater than an arbitrary infinite constant; and

(2) the variances of the functions are all greater than a constant.

Our aberration functions and the geometric slit function satisfy these requirements and so does the diffraction function if we approximate it by making it vanish for large arguments. Now we certainly are not dealing with an infinite convolution process, but nevertheless it is known that often only a few convolutions are required to closely approximate a Gaussian function, and we have seen that a single convolution of a triangle and a rectangle gives a good approximation. With this justification we make the following generalizations about slit functions of infrared monochromators:

(1) At longer wavelengths, with very narrow slits the diffraction spread function is dominant and a diffraction slit function occurs.

Fig. 7.16   Slit functions with coma along and across the slit length.

(2) With wide slits the diffraction and aberration contributions are negligible and the slit function is triangular.

(3) When the slit width is of the same magnitude as the width of the aberrations and diffraction functions, and there are no dominant asymmetric aberrations (slit image distortion and coma across the slit), the slit function is nearly Gaussian.

(4) When the slit width, aberrations, and diffraction are of the same magnitude and strong asymmetric aberrations are present, the slit function takes on a somewhat skewed shape.

II. THE SPECTRAL MODULATION TRANSFER FUNCTION

The Fourier transform of the spread function is the optical transfer function. This has been discussed in Section 4.VII, where we pointed out that the optical transfer function describes the ability of the optical system to transmit information about an object. More specifically, it describes the capability of the system to transmit amplitude and phase information about a periodic object with a specified spatial periodicity, and therefore about any object that can be considered to be a superposition of many spatially periodic objects.

It is natural to define a function $S(\tau)$, which we shall call the spectral modulation transfer function, or simply the spectral transfer function, to be the Fourier transform of the monochromator slit function. The spectral transfer function describes the ability of a monochromator to transmit information about a spectrum. If, for example, channel spectra are being studied (i.e., spectra whose intensity varies periodically with wavenumber, such as the interference fringes often seen in spectra of thin films or empty cells), the spectral transfer function describes the amplitude and the phase shift of the recorded spectra as a function of periodicity. Let us call the periodicity of a channel spectrum the spectral modulation frequency, τ, in analogy with the spatial modulation frequency introduced in Section 4.VIII.

We must not be confused by the various usages of the word frequency. We might, for example, study a specimen with electromagnetic radiation having a wavenumber frequency of 1000 cm^{-1}. The molecules in the specimen are vibrating with a vibrational frequency of 3×10^{13} cps. If the spectrum of the specimen is a channel spectrum with a maximum every 10 cm^{-1}, it has a spectral modulation frequency of $1/10$ cycles cm^{-1}. If the spectrum is dispersed across the slit plane of a monochromator having a reciprocal dispersion of $1 \text{ cm}^{-1}/\text{mm}$, it is said to be an image with a spatial periodicity of 1 cm, or a spatial modulation frequency of 1 cm^{-1}. In addition, there is a chopper in the spectro-photometer system which is modulating the radiation at a time rate of perhaps 10 cps. When we discuss in a later chapter the electrical pro-perties of the spectrophotometer, and again when we discuss inter-ferometry, we encounter some additional kinds of frequency.

A spectrum can be considered as a superposition of channel spectra of various amplitudes, phases, and spectral modulation frequencies in just the same way as we considered a luminous object to be repre-sented as a superposition of many sinusoidal light distributions in Section 4.VII. The Fourier transform of the spectrum then represents the relative proportions of each channel spectrum.

The action of the slit function in forming an output distribution from the input spectrum was described by the superposition or convolution integral of Eq. (7.4). We recall that the Fourier transform of a convolu-tion integral is the product of the transforms of the two convolved functions. Hence, the Fourier transform of Eq. (7.4) is

$$Y(\tau) = S(\tau)X(\tau) \tag{7.27}$$

where $S(\tau)$, $X(\tau)$, and $Y(\tau)$ are the Fourier transforms of $\sigma(\nu)$, $J_{in}(\nu)$, and $J_{out}(\nu)$, respectively. Moreover, as the slit function is itself the multiple convolution of the entrance and exit slits (which give us a triangular slit function), the diffraction distribution function, and the various aberration distribution functions, we can write $S(\tau)$ in terms of its various components

$$S(\tau) = S_T S_D S_{SA} S_C \cdots$$

where the subscripts refer to triangle function, diffraction, spherical aberration, coma, and so on.

The contribution to the spectral transfer function by diffraction at a rectangular aperture is found by finding the Fourier transform of the diffraction spread function or, alternatively, by calculating the self-convolution of the intensity distribution in the exit pupil.

When this is done for a rectangular aperture, we find

$$S_D(\tau) = \begin{cases} \Delta\nu_D\left(1 - \dfrac{|\tau|\,\Delta\nu_D}{2\pi}\right) & |\tau| < \dfrac{2\pi}{\Delta\nu_D} \\[2ex] 0 & |\tau| \geq \dfrac{2\pi}{\Delta\nu_D} \end{cases} \tag{7.28}$$

where $\Delta\nu_D$ is given in terms of the f/number and reciprocal dispersion of the monochromator, and the frequency of the radiation by

$$\Delta\nu_D = \frac{FD^{-1}}{\nu}$$

Equation (7.28) is a triangular function that is identically zero for spectral modulation frequencies greater than

$$\tau_D = 2\pi/\Delta\nu_D \qquad (7.29)$$

That is, no information can be transmitted by the monochromator at spectral modulation frequencies greater than $2\pi/\Delta\nu_D$; $\Delta\nu_D$ is the diffraction-limited resolution of the monochromator, which we have already defined from the first zero of the diffraction pattern.

The spectral transfer function component arising from the triangular geometric slit function is found by Fourier transforming Eq. (7.6) for the triangular slit function. The result is

$$S_T(\tau) = \Delta\nu_T \left(\frac{\sin \tau \, \Delta\nu_T/2}{\tau \, \Delta\nu_T/2}\right)^2 \qquad (7.30)$$

where $\Delta\nu_T$ is the geometric resolution found from the reciprocal dispersion and the geometric slit width, $\Delta\nu_T = D^{-1}w$. Notice the interesting reciprocity: The diffraction slit function has a $\sin^2 \theta/\theta^2$ form and the diffraction spectral transfer function is a triangle, whereas the geometric slit function is a triangle and the geometric spectral transfer function is a $\sin^2 \theta/\theta^2$ function.

The first zero of $S_T(\tau)$ occurs where

$$\tau_T = 2\pi/\Delta\nu_T \qquad (7.31)$$

and in this case we also have a correspondence between a vanishing of the spectral transfer function and the resolution defined by the width of the slit function. It is important to recognize that information *can* be transmitted by the monochromator at spectral modulation frequencies greater than τ_T, provided $\tau_T < \tau_D$ and excepting the discrete points where $S_T(\tau)$ vanishes. We discuss the significance of this observation in the next section.

The contribution to the spectral transfer function from aberrations can be found by Fourier transforming the aberration line spread functions. We must bear in mind that it is permissible to convolve the diffraction and aberration spread functions, or, equivalently, to multiply the transfer functions, only when the aberrations are sufficiently large relative to diffraction. Otherwise, the aberration and diffraction spread functions are found as a single function, and so is the transfer function. That is, we must apply the diffraction theory of aberrations. Even when the aberrations are small, however, we can consider a separate aberration spread function calculated from geometric optics as an indication of the magnitude of the aberration. Of course, we lose the fine detail introduced by diffraction.

Optical transfer functions for an optical system with a square aperture and various types and amounts of aberrations are given in a review by Miyamoto (*37*). He shows the aberration transfer functions calculated from diffraction theory and from geometric optics, the latter by expanding the function in wavelength and retaining only the wavelength-independent term. In general, if the aberration distortion of the wave front in the exit pupil is two or more wavelengths, the approximation of separable diffraction and aberration is very good. We shall usually proceed as though this were so.

A. Limiting Spectral Resolution

We have seen that the optical transfer function, to the approximation we are considering, is the product of contributions from diffraction, aberrations, and the geometric slit function. The diffraction transfer function, for a uniformly illuminated rectangular aperture, is a straight line intercepting the spectral modulation frequency axis at a value τ_D [Eq. (7.29)] and remaining zero for larger values. The value of τ_D defines the diffraction-limited resolution of the monochromator. No information can be transmitted at higher spectral modulation frequencies if the system is diffraction limited. Information is transmitted at lower spectral modulation frequencies with amplitude reduced according to the shape of the transfer function. The suppression of some spectral modulation frequencies causes the shape of a spectral band to be distorted by the monochromator.

The limiting resolution of the monochromator, from Eq. (7.29), is found from the limiting spectral modulation frequency,

$$\Delta \nu = 2\pi/\tau$$

Remember that this resolution refers to a channel spectrum with maxima separated by $\Delta \nu$. We have seen in Section 4.VII that this is equivalent to Rayleigh's criterion for the resolution of two line objects.

Now there is usually a source of noise in the system, contributed perhaps by the detector or by fluctuations in the source. Information transmitted at those spectral modulation frequencies for which the amplitude is small in comparison with the noise amplitude will be so badly corrupted by this noise that it will not be usable. Hence, the effective limiting spectral modulation frequency τ_D' is reduced somewhat in the presence of noise. This is illustrated in Fig. 7.17. Infrared spectrophotometers are almost always *noise limited* rather than diffraction limited.

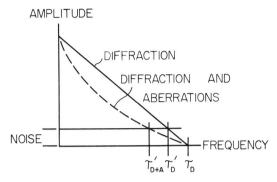

Fig. 7.17 Spectral modulation transfer function showing limiting effects of noise, diffraction, and aberrations.

The presence of significant aberrations causes the spectral transfer function to be reduced in magnitude at all spectral modulation frequencies, and this produces a further distortion of the spectrum. The limiting spectral modulation frequency τ_D is generally not affected, but the effective limiting spectral modulation frequency τ'_{D+A} is further reduced in the presence of noise.

Now we must multiply the product of the diffraction and aberration spectral transfer functions by the contribution from the geometric slit function. This contribution has a $\sin^2 \theta/\theta^2$ form for a triangular slit function [Eq. (7.30)], which has zeros at a sequence of values of spectral modulation frequency (Fig. 7.18), the first zero falling at τ_T. As the slit opening is made narrower, the value of τ_T becomes larger. When

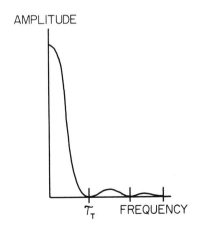

Fig. 7.18 Spectral modulation transfer function for a triangular slit function.

the slit is made so small that its width equals the diffraction slit width, τ_T coincides with τ_D. At still narrower slit widths, multiplication of $S_D S_A$ by S_T does not change the shape of $S_D S_A$ to any great extent. Once again, we have arrived at the spectroscopist's general rule: *resolution is not improved by narrowing the slit width much beyond the diffraction limit.* This is true with regard to the value of τ where the spectral transfer function becomes zero. However, we should also consider the effect of distortion of the spectrum by attenuation of the higher modulation frequencies. Clearly, these frequencies are less attenuated when we narrow the slit substantially beyond the diffraction limit. In the presence of noise this affects not only the distortion of the recorded spectrum but also the effective limiting spectral modulation frequency. However, because infrared spectrophotometers are almost always noise limited, it is rarely feasible to operate with slits narrower than the diffraction width, and we usually have $\tau_T \lesssim \tau_{D+A}$. The properties of noise and noise filters are discussed in more detail in a later chapter. We should point out here that the limiting spectral modulation frequency τ_T is inversely proportional to the slit width. The noise amplitude relative to the signal passed by the monochromator is inversely proportional to the slit width squared because the luminosity of the monochromator varies as the slit width squared. Thus, under some conditions, the effective limiting spectral modulation frequency τ' might be reduced by noise at a faster rate than τ_T increases with narrowing slits.

Now let us consider once more the variance of the slit function and how it can be determined from the spectral transfer function. There is a theorem on Fourier transforms (see the Appendix) which states that the nth moment of a function is found from the nth derivative of its Fourier transform (if it exists) evaluated at the origin. For our purposes, we take the mean of the function to be zero, but this is not a real restriction. Specifically,

$$\langle x^n \rangle_{f(x)} = \frac{1}{(-j)^n} \frac{d^n F(\omega)}{d\omega^n} \bigg|_{\omega=0}$$

As the spectral transfer function $S(\tau)$ is the Fourier transform of the slit function $\sigma(\nu)$, we can write for the variances, or second moment, of the slit function,

$$\langle \nu^2 \rangle_\sigma = -\frac{d^2 S(\tau)}{d\tau^2} \bigg|_{\tau=0}$$

Furthermore, from this theorem it is easily shown (see the Appendix) that the variance of a function which is the convolution of several functions is the sum of the variances of the convolved functions. This result has already been used in the previous discussion.

B. *Spurious Spectral Resolution*

In connection with Eq. (7.31) for the zeros of the spectral transfer function associated with a triangular slit function, we pointed out that information can be transmitted at modulation frequencies higher than τ_T. This is because the transfer function does not remain zero except at modulation frequencies greater than the diffraction limit τ_D. There is an interesting consequence of this. The spectral modulation transfer function $S_T(\tau)$ tells us how the amplitude of a channel spectrum depends on slit width. In infrared spectroscopy, spectra are often encountered which resemble channel spectra over limited regions. The interference fringes observed with an empty cell or a thin film or the rotational structure in the resolved absorption bands of some light symmetrical molecules are examples of spectra which have periodically repeating peaks of more or less the same height.

The periodicity of the spectrum is defined in terms of the separation of peaks $\Delta\nu_s$ in the spectrum,

$$\tau_s = 2\pi/\Delta\nu_s$$

When the spectrometer slit is widened, the amplitude of the peaks decreases according to $S_T(\tau)$ and vanishes when τ_T equals τ_s. Resolution of the peaks ceases at this slit width. If the slit is now widened further, the spectrum reappears, because the amplitude takes on the values given by the second lobe of $S_T(\tau)$. There is no phase shift or frequency error if the slit function is truly triangular. The resolution of the peaks in the spectrum is real, but we call the effect *spurious spectral resolution* (*38, 39*) in analogy with the spurious resolution of optical images. Spurious spectral resolution is illustrated by the rotational structure of the carbon monoxide spectrum in Fig. 7.19. This band is centered at about 2150 cm^{-1}. It consists of two branches, and is typical of vibration–rotation absorption bands of diatomic molecules. The lines in the low-frequency edge of the P branch are separated by about 5 cm^{-1}, and the spacing decreases gradually to about 3 cm^{-1} at the high-frequency edge of the R branch. The band is scanned repeatedly with successively narrower slits. The lines on the R branch become unresolved when the slit width is 1.7 mm (EBW $= wD^{-1} = 3.6$ cm^{-1}) *and the P* branch is unresolved with 2.5-mm slits (EBW $= 5.3$ cm^{-1}). Notice that resolution of the R branch returns with wider slits; this

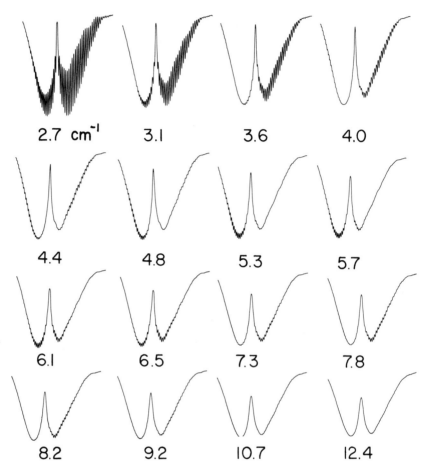

Fig. 7.19 Spurious spectral resolution: CO absorption at 2150 cm⁻¹ for various effective bandwidths.

is the phenomenon of spurious spectral resolution. It is lost again with 3.5-mm slit width (EBW = 7.3 cm⁻¹). The lines in the P branch are also resolved at wider slits, disappearing again with a 5.1-mm slit (EBW = 10.7 cm⁻¹). Spurious spectral resolution of the rotational structure of the carbon dioxide spectrum near 640 cm⁻¹ is shown in Fig. 7.20. In this example, resolution of the line is lost with a 1.3-mm slit (EBW = 1.6 cm⁻¹), returns (spurious resolution) with a 1.9-mm slit (2.4 cm⁻¹) is lost again at 2.6 mm (3.3 cm⁻¹), and returns a second time with a 3.2-mm slit (4.0 cm⁻¹) as the periodicity of the spectrum coincides with the second lobe of the spectral transfer function.

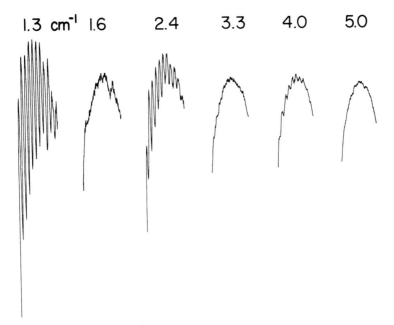

Fig: 7.20 Spurious spectral resolution: CO_2 absorption at 640 cm^{-1} for various effective bandwidths.

These rotation–vibration bands are only approximations to channel spectra. A much better channel spectrum can be generated with an interferometer. A Fabry–Perot etalon with uncoated sodium chloride windows produces interference fringes of modest amplitude that are very nearly sinusoidal. An interferometer of this type was constructed to give fringes separated by 4.8 cm^{-1}. The amplitude of the fringes was measured near 1200 cm^{-1} with a grating monochromator as a function of slit width; the results are plotted in Fig. 7.21. The solid line

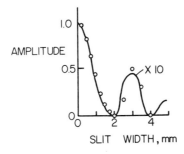

Fig. 7.21 Measured amplitudes of Fabry–Perot fringes for various slit widths.

is calculated from the spectral transfer function of a triangular slit function. The diffraction spread function and the aberration effects are negligible at slit widths greater than a few tenths of a millimeter. The measured amplitudes (normalized to unity at zero slit width) fall very close to the theoretical curve.

III. OPTIMUM SLIT FUNCTION AND SPECTRAL MODULATION TRANSFER FUNCTION

In the discussion of slit functions up to now we have considered only the triangular geometric slit function because this is the form encountered with entrance and exit slits of equal widths and with straight, parallel slit jaws. We have also considered only a $\sin^2 \theta/\theta^2$ form of diffraction spread function because this is the form contributed by a uniformly illuminated rectangular exit pupil. We have considered, in a general sort of way, the contribution of aberrations to the slit function and implied at least that all aberrations are bad and should be reduced to a minimum. It is time now to ask whether there are other forms of realizable slit functions that might not in some way improve the performance of the monochromator. We consider three problems: In the first, a mathematical theory for the optimum slit function is sketched; secondly, the question of alteration of the exit-pupil illumination is discussed; and thirdly the properties of a monochromator with several slits are treated.

A. The Optimum Slit Function

We know that the variances of convolved functions are added to find the variance of the generated function. The increase in variance can be taken as a useful measure of the broadening of a recorded spectral line and, therefore, of the resolution of the instrument. We should, therefore, seek a slit function having a small variance. Of the class of functions having unit area and a value of unity at its maximum, the rectangle has the least variance, $\langle \nu^2 \rangle = 0.083$. But a rectangular slit function is not physically realizable. A triangle has a variance of 0.167 and a Gauss function has a somewhat smaller variance of 0.159. The variance of a Cauchy function does not exist.

The effective width of a slit function might also be taken as the width at half its maximum value. In this case, the width of a rectangle or a triangle of unit area and height is 1.00, the width of the Gauss function is 0.94, and the width of a Cauchy function is 0.64.

Now let us consider as a criterion of resolution not the broadening of a recorded spectral line, but the ability to distinguish two nearby lines. This is, after all, the reason for the term "resolution." Consider two very narrow, equally intense, spectral lines (delta functions) separated by a spectral frequency interval δ. Let us define resolution by stating that the lines are resolved by the monochromator if the intensity at the valley between the two peaks is 81.1% of the intensity at the peaks. This is equivalent to Rayleigh's criterion. The reciprocal of the minimum value of δ for which resolution is achieved is proportional to the resolving power. On this basis we find for slit functions of rectangular, Cauchy, Gauss, and triangular shape and unit area and height resolving powers of $1/\delta = 0.50, 0.55, 0.58$, and 0.84, respectively.

Thus, we see that the choice of best slit function depends on the criterion we use to judge resolution. Of the four slit functions considered in these examples, the triangular function is worst when we speak of broadening of a recorded line, and best when we consider distinguishing two nearby narrow lines.

A criterion that is frequently used for judging the performance of an electrical system is the integral-squared-error criterion. If an input $y(t)$ into a system appears as an output $x(t)$, the difference between the input and output is the error, and the integral-squared, error is

$$\epsilon = \int [y(t) - x(t)]^2 \, dt \qquad (7.32)$$

The integral is squared in order that positive and negative errors will not cancel each other. This could be accomplished by integrating $|y(t) - x(t)|$, but not as conveniently. Moreover, Eq. (7.32) leads to some particularly useful relationships when the input is statistical in nature, as in the case of noise. The shape of the output function is said to be best matched to the shape of the input when ϵ is minimized. Linfoot (40) proposed a similar integral for the difference between the illumination of an optical image and its object,

$$\Phi_0 = 1 - \frac{\iint [O(x, y) - I(x, y)]^2 \, dx \, dy}{\iint [O(x, y)]^2 \, dx \, dy}$$

The integration is carried out over the image plane. The term Φ_0 is called the *image fidelity* by Linfoot. Its value is unity when the image is perfect and zero when the image is so bad that I vanishes over the image plane.

We can consider a similar least-squared-error criterion for spectra by defining a *spectrum fidelity* function

$$\Phi_s = 1 - \frac{\int_{-\infty}^{\infty} [J_{\text{out}}(\nu) - J_{\text{in}}(\nu)]^2 \, d\nu}{\int_{-\infty}^{\infty} [J_{\text{in}}(\nu)]^2 \, d\nu}$$

The limits of integration extend, at least formally, from $-\infty$ to $+\infty$. We wish to select a slit function to maximize the spectrum fidelity function. This is equivalent to minimizing the function

$$I' = \int [J_{\text{in}}(\nu) - J_{\text{out}}(\nu)]^2 \, d\nu$$

We shall generalize somewhat by writing instead the integral

$$I = \int [J_{\text{in}}(\nu'; \kappa, \nu_0) - J_{\text{out}}(\nu')]^2 \, d\nu' \qquad (7.33)$$

where the designation $J_{\text{in}}(\nu'; \kappa, \nu_0)$ represents the desired form of the output spectrum. It could, for example, stand for $J_{\text{in}}[K\nu' + \nu_0(1 - K)]$, meaning that $J_{\text{in}}(\nu')$ is expanded by the substitution of $K\nu'$ for ν', and then the expanded distribution is shifted to bring its peak back into coincidence with the peak of the original J_{in} at ν_0. In this way, we can permit a certain amount of broadening of the output by comparing the output with a broadened modification of the input.

Now, substitute Eq. (7.4) into Eq. (7.33) and square the integrand

$$I = \int_{\nu'} \left[J_{\text{in}}(\nu'; \kappa, \nu_0) - \int_{\nu} \sigma(\nu) J_{\text{in}}(\nu' - \nu) \, d\nu \right]^2 \, d\nu'$$

$$= \int_{\nu'} \int_{\nu} \int_{\nu''} \sigma(\nu) \sigma(\nu'') J_{\text{in}}(\nu' - \nu) J_{\text{in}}(\nu' - \nu'') \, d\nu'' \, d\nu \, d\nu'$$

$$- 2 \int_{\nu'} \int_{\nu} \sigma(\nu) J_{\text{in}}(\nu' - \nu) J_{\text{in}}(\nu'; \kappa, \nu_0) \, d\nu \, d\nu' + \int_{\nu'} J_{\text{in}}^2(\nu'; \kappa, \nu_0) \, d\nu'$$

We consider the optimum slit function to be the one that minimizes I. We now follow the usual procedure of elementary calculus of variations and substitute the function $\sigma + \epsilon \sigma_v$ in place of σ, where σ_v is a variation function which is arbitrary except that it vanishes at the limits of integration ($-\infty$ and $+\infty$). The parameter ϵ is a small quantity. This substitution provides us with the function

$$I = \int_{\nu''} \int_{\nu'} \int_{\nu} [\sigma(\nu) \sigma(\nu'') + \epsilon \sigma_v(\nu) \sigma(\nu'') + \epsilon \sigma(\nu) \sigma_v(\nu'')$$

$$+ \epsilon^2 \sigma_v(\nu) \sigma(\nu'')] J_{\text{in}}(\nu' - \nu) J_{\text{in}}(\nu' - \nu'') \, d\nu \, d\nu' \, d\nu''$$

$$-2\int_{\nu'}\int_{\nu}[\sigma(\nu)+\epsilon\sigma_v(\nu)]J_{\text{in}}(\nu'-\nu)J_{\text{in}}(\nu';\kappa,\nu_0)\,d\nu\,d\nu'$$

$$+\int_{\nu'}J_{\text{in}}^2(\nu';\kappa,\nu_0)\,d\nu'$$

If I is to be an extremal (i.e., either a minimum or a maximum), it must satisfy

$$\lim_{\epsilon\to 0}\frac{\partial I}{\partial\epsilon}=0$$

or

$$0=\int_{\nu''}\int_{\nu'}\int_{\nu}[\sigma_v(\nu)\sigma(\nu'')+\sigma(\nu)\sigma_v(\nu'')]J_{\text{in}}(\nu'-\nu)J_{\text{in}}(\nu'-\nu'')$$

$$\cdot d\nu\,d\nu'\,d\nu''-\int_{\nu'}\int_{\nu}\sigma_v(\nu)J_{\text{in}}(\nu'-\nu)J_{\text{in}}(\nu';\kappa,\nu_0)\,d\nu\,d\nu'$$

$$=2\int_{\nu''}\int_{\nu'}\int_{\nu}\sigma_v(\nu)\sigma(\nu'')J_{\text{in}}(\nu'-\nu)J_{\text{in}}(\nu'-\nu'')\,d\nu\,d\nu'\,d\nu''$$

$$-2\int_{\nu'}\int_{\nu}\sigma_v(\nu)J_{\text{in}}(\nu'-\nu)J_{\text{in}}(\nu';\kappa,\nu_0)\,d\nu\,d\nu'$$

$$=2\int_{\nu}\sigma_v(\nu)\left[\int_{\nu''}\int_{\nu'}\sigma(\nu'')J_{\text{in}}(\nu'-\nu)J_{\text{in}}(\nu'-\nu'')\,d\nu'\,d\nu''\right.$$

$$\left.-\int_{\nu'}J_{\text{in}}(\nu'-\nu)J_{\text{in}}(\nu';\kappa,\nu_0)\,d\nu'\right]d\nu \qquad (7.34)$$

But, because the variation function σ_v is arbitrary, in order for the integral of Eq. (7.34) to vanish, the function inside the brackets must vanish. Equation (7.34) thus reduces to

$$\int_{\nu''}\int_{\nu'}\sigma(\nu'')J_{\text{in}}(\nu'-\nu)J_{\text{in}}(\nu'-\nu'')\,d\nu'\,d\nu''$$

$$-\int_{\nu'}J_{\text{in}}(\nu'-\nu)J_{\text{in}}(\nu';\kappa,\nu_0)\,d\nu'=0 \qquad (7.35)$$

Define the following functions

$$r_{ii}(\nu-\nu'')=\int_{\nu'}J_{\text{in}}(\nu'-\nu)J_{\text{in}}(\nu'-\nu'')\,d\nu'$$

$$r_{id}(\nu)=\int_{\nu'}J_{\text{in}}(\nu'-\nu)J_{\text{in}}(\nu';\kappa,\nu_0)\,d\nu'$$

These are, respectively, the autocorrelation function of the input and the cross-correlation function of the input with the desired output. In terms of these functions, Eq. (7.35) is written

$$\int_{\nu''}\sigma(\nu'')r_{ii}(\nu-\nu'')\,d\nu''-r_{id}(\nu)=0 \qquad (7.36)$$

This is an integral equation for determining the form of the slit function which best reproduces a spectrum with an allowable distortion of its lines. This equation is very similar to the Wiener–Hopf equation of communications theory (41). There is a significant difference however: the Wiener–Hopf equation treats a communication signal as a function of time. Since the response of a signal cannot precede the excitation there is a restricted range of validity for the time parameter equivalent to our ν. In our monochromator theory there is no such restriction. Consequently, unlike the Wiener–Hopf equation, we can formally solve Eq. (7.36) very simply by Fourier transforming it.

$$S(\tau)R_{ii}(\tau) = R_{id}(\tau)$$

There are numerous variations possible in this approach to optimization of the slit function. We can, for example, specify additional constraining conditions. If we wish the monochromator to transmit a specified amount of radiant power from a given source of uniform spectral distribution, we might append a term

$$\mu\left[\int \sigma(\nu)\,d\nu\right]^2$$

to Eq. (7.33). The factor μ is a Lagrange multiplier. With this addition, Eq. (7.36) becomes

$$\int_{\nu'}\sigma(\nu'')r_{ii}(\nu-\nu'')\,d\nu'' = r_{id}(\nu)+\mu'$$

There are other criteria for optimum slit functions which can be used in place of the minimum-squared-error criterion. We might ask that the quantity

$$\int\left[J_{\text{out}}(\nu)\right]^2 d\nu$$

be a maximum. This is analogous to Linfoot's relative structure content.

The mathematical problem of the optimum slit function is left as unfinished business. Assuming that a satisfactory solution can be found, we are then left with the practical problem of how to realize the optimum slit function. Certainly some very desirable slit functions are not realizable. A delta function would exactly reproduce an input spectrum, but is not realizable. It would seem that we have three approaches open to us in an attempt to create a realizable slit function which is in some sense optimum. The first of these is in the area of optical design and implies a reduction of all aberrations to an acceptable level. This is accomplished in the original choice of monochromator system, its parameters, such as off-axis angle and slit height, and the decision to use spherical or aspherical optical components. Aberrations cannot be eliminated completely, but they can be balanced. We have seen, for example, that a certain amount of defocusing has a beneficial effect on spherical aberration. We have seen also that often a great deal of astigmatism can be tolerated, provided the astigmatic line image lies along the slit and not across it.

Next, the shape of the geometric slit function can be altered by curving the edge of the slit jaw. This is the way slit image distortion is compensated. A Gaussian slit function, for example, might be approximated by making the entrance and exit slit jaws in the form of Gauss-shaped openings. Remember that a Gauss function convolved with another Gauss function is still Gaussian. A more elaborate treatment of the slit shapes is discussed in Section III.C.

Finally, the diffraction contribution to the slit function can be modified by altering the form of the exit pupil of the monochromator. This technique is discussed in the next section.

B. Apodization

The contribution of diffraction to the spectral transfer function is the auto correlation function of the exit-pupil irradiation function. The diffraction spread function is the inverse Fourier transform of the diffraction spectral transfer function. The effect of diffraction on the spread function, therefore, can be varied by controlling the shape of the pupil with an aperture, or the illumination of the pupil with an absorbing filter of appropriate transmittance varying over the aperture, or the phase of the radiation leaving the pupil with a phase filter that introduces a retardation varying over the aperture. The latter two approaches have been used effectively with image-forming devices such as cameras and enlargers which can be used with monochromatic radiation, or at least over a fairly narrow spectral range. It is possible to introduce corrections for aberrations and defocusing in this way and thus to sharpen diffuse images. However, when a wide range of wavelengths are to be studied, as in an infrared spectrophotometer, we are practically limited (at present) to changing the shape of the aperture. The future may see variable phase filters or spectrally neutral filters for the infrared, which can change this situation of course.

We have seen that the diffraction pattern (line spread function) from a rectangular pupil consists of a central maximum and a series of side lobes of diminishing intensity. The first lobe has a maximum about 5% as high as the central peak, the next lobe has a value only about 2% of the central peak, and so on. These side lobes can be serious in some spectroscopic problems, as when a weak line lies very close to a strong one. Treatment of the exit pupil to suppress these side lobes has been termed *apodization* (from the Greek meaning "to take away the feet") by the French opticist Jacquinot. Jacquinot and Roizen-Dossier have recently given a review of the theory of apodization (*42*).

We must stress that apodization cannot increase the ultimate resolution of the instrument because the limiting spectral modulation frequency is always determined by the maximum width of the aperture. Moreover, the presence of an apodizing aperture must entail the loss of a certain amount of energy, and usually the central lobe of the diffraction pattern is widened somewhat. Nevertheless, there are certain circumstances where it is desirable to apodize, such as the case of the weak satellite line mentioned above.

Let us discuss a few apodizing apertures for purposes of illustration. A diamond-shaped aperture, as in Fig. 7.22, has the spectral transfer and line spread functions plotted in the figure. The corresponding functions for a rectangular aperture are shown for comparison (dashed lines). The diffraction pattern has the form

$$\left(\frac{\sin \theta/2}{\theta/2}\right)^4$$

for a diamond aperture, and therefore the first side lobe occurs twice as far from the central maximum as it does for a rectangular aperture. Moreover, it is weaker by more than a factor of 10. However, the central lobe of the diamond aperture is wider by roughly half at the half-maximum-intensity point and the transmission is only half that of the rectangle.

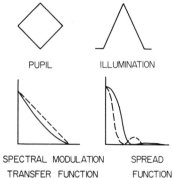

Fig. 7.22 Spectral modulation transfer function and spread function for a diamond pupil.

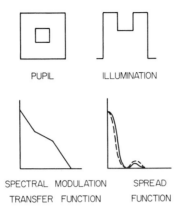

PUPIL   ILLUMINATION

SPECTRAL MODULATION   SPREAD
TRANSFER FUNCTION   FUNCTION

Fig. 7.23 Spectral modulation transfer function and spread function for a pupil with a central obscuration.

An aperture with a central obstruction is interesting because it occurs naturally in some optical systems, such as a Cassegrain condenser or a Pfund monochromator. The spectral modulation transfer function can be plotted at once from the autocorrelation of the pupil (Fig. 7.23), assuming a rectangular pupil and obstruction. The effect of this pupil is to suppress the middle spectral modulation frequencies and accentuate the high frequencies. The line spread function has its central lobe somewhat broadened relative to that of the uniformly illuminated rectangular pupil.

C. Monochromators with Multiple Slits or Grills

The spectroscopist is constantly faced with a compromise between resolution and radiant power. He can gain resolution by narrowing his slits, but only at the expense of lost radiant power, and therefore poorer signal-to-noise ratio. In the two previous sections we have indicated some techniques that might be used to optimize resolution (in some sense) without suffering too greatly in lost power. But at best these techniques offer only slight improvement, and the infrared spectroscopist wants orders of magnitude. Drastic measures are required.

Resolution improves with narrowed slits, and radiant power increases with wider slits. Why not have both by using an array of narrow slits instead of a single wide slit? Let us see how this might be done and what some of the attendant problems are.

Consider first a pair of parallel entrance slits (1 and 2 in Fig. 7.24) and a corresponding pair of exit slits (3 and 4) in an otherwise standard

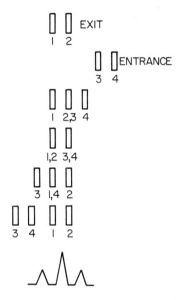

Fig. 7.24 Slit function for double entrance and exit slits.

monochromator. The slits are separated by a distance s, and D^{-1} is the reciprocal dispersion of the monochromator. We can derive the geometric slit function by convolving the entrance slits with the exit slits. In other words, we imagine what will happen if the entrance slits are flooded with monochromator radiation and their images scanned across the exit slits by rotation of the disperser. As the image of entrance slit 1 crosses exit slit 4, a triangular function is generated. As the image of slit 1 crosses slit 3 and the image of slit 2 simultaneously crosses slit 4, a second triangular function is generated, twice as tall as the first one and separated from it by the distance separating the slits. Finally, as the image of slit 2 crosses slit 3, a third triangle is generated, as tall as the first triangle. The central triangle represents a geometric slit function as narrow as the slit function generated by a single entrance and exit slit, but with *twice the optical power*. With polychromatic radiation illuminating the entrance slits we have, for a given setting of the disperser, radiation of frequency, ν_0 centered at the central triangle, and half of this amount of radiation of frequency $\nu_0 + D^{-1}s$ centered on one of the side triangles and radiation of frequency $\nu_0 - D^{-1}s$ centered on the other triangle. The system is no longer a monochromator in the strict sense, but we shall retain this term for this and other multislit instruments.

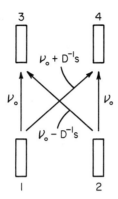

Fig. 7.25 Generation of slit function for a double-slit system.

The generation of the slit function with polychromatic radiation can be considered in another way (see Fig. 7.25). Radiation of frequency ν_0 from entrance slit 1 passes through exit slit 3 and from entrance slit 2 through exit slit 4. Radiation of frequency $\nu_0 + D^{-1}s$ from entrance slit 1 passes through exit slit 4. Radiation of frequency $\nu_0 - D^{-1}s$ from entrance slit 2 passes through exit slit 3.

Three parallel entrance slits and a corresponding set of exit slits produce the slit function sketched in Fig. 7.26. There are five triangles with heights in the ratio $1 : 2 : 3 : 2 : 1$. A set of n entrance slits and n exit slits produces a slit function consisting of $2n - 1$ triangles with heights in the ratio $1 : 2 : \cdots : n-1 : n : n-1 : \cdots : 2 : 1$.

As far as the central triangle is concerned, the luminosity is increased very greatly with no sacrifice in resolution. In fact, the radiant power in the central triangle is proportional to the number of slits. Unfortunately, the other triangles are also present and together they form a slit function that is not very useful at all. What is needed is a means of eliminating all but the central triangle or, alternatively, a means of discriminating between the central triangle and the others. We now discuss several possibilities.

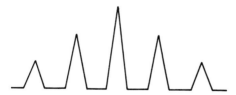

Fig. 7.26 Slit function for triple entrance and exit slits.

Suppose we have available a bandpass filter with a bandwidth narrower than the spectral interval between the first two side lobes of the multitriangle slit function. This filter permits us to retain the central triangle and reject the others. This is not particularly useful unless the filter can be tuned over a reasonable range. Such filters are available in the form of interference filters formed on circular substrates. We mentioned these in Chapter 5. The frequency of the passband is tuned by rotating the filter. This approach to multislit monochromator construction has not yet been utilized in a working instrument.

The concept of frequency discrimination might be applied as a means of preferentially detecting the spectral frequencies in the central triangle. Golay (43) proposed this method long ago, but did not actually use it. Each slit of the entrance array is chopped at a different frequency, the kth slit being chopped at frequency $f_n + k\,\Delta f$. Each slit of the exit array is also chopped at a different frequency, the kth slit at the rate of $f_x + k\,\Delta f$. Now a signal that is simultaneously modulated at two different frequencies, f_1 and f_2, gives rise to signals modulated with the difference $f_1 - f_2$ and with the sum $f_1 + f_2$ of the modulation frequencies. This is easily seen from the trigonometric identities:

$$\cos (f_1 t + f_2 t) = \cos (f_1 t) \cos (f_2 t) - \sin (f_1 t) \sin (f_2 t)$$

and

$$\cos (f_1 t - f_2 t) = \cos (f_1 t) \cos (f_2 t) + \sin (f_1 t) \sin (f_2 t)$$

which combine to give

$$\cos (f_1 t) \cos (f_2 t) = \cos (f_1 + f_2) t + \cos (f_1 - f_2) t$$

The radiation that we wish to detect enters an entrance slit and leaves the corresponding exit slit, that is, it enters the mth entrance slit and leaves the mth exit slit. This radiation will be modulated at the frequency $f_n - f_x$ for any corresponding pair of entrance and exit slits. Radiation entering the mth entrance slit and leaving the nth exit slit ($n \neq m$) is modulated at the frequency $f_n - f_x + (m - n)\,\Delta f$. An amplifier tuned to the frequency $f_n - f_x$ will therefore detect only the desired radiation, provided that none of the individual slit modulation frequencies, $f_n + k\,\Delta f$ or $f_x + k\,\Delta f$, and none of the sum frequencies coincide with $f_n - f_x$.

Notice that in multislit spectrometers a great deal of unused radiation falls on the detector in addition to that used to generate the signal. This is a feature common to all multislit systems and is of no consequence when a thermal detector (thermocouple, bolometer, or Golay cell) is used, because in such detectors noise is independent of signal level. But a background of radiation on a photoconducting or photo-emitting detector results in a noise level proportional to the square root of the signal level, making the multislit method less advantageous with such detectors. Stated differently, the signal-to-noise ratio increases in proportion to the number of slits when a thermal detector is used, but only in proportion to the square root of the number of slits when a quantum detector is used.

The possibility of using phase discrimination has apparently not been explored. All slits, entrance and exit, are chopped at the same frequency but with a selected phase. The nth entrance slit is chopped with the phase $n\delta$ and the nth exit slit with the phase $-n\delta$. Then, radiation entering the nth entrance slit and leaving the mth slit is modulated with

$$\cos(ft+n\delta)\cos(ft-m\delta)$$

$$=\cos(2ft+(n-m)\delta)+\cos(n+m)\delta \qquad (7.37)$$

The second term on the right-hand side of Eq. (7.37) is independent of time and represents a dc term that is ignored by the amplifier. The first term on the right-hand side indicates that the exit radiation is modulated at frequency $2f$, and that only radiation passing through corresponding entrance and exit slits ($n = m$) has zero phase. If a phase-sensitive electronic system with sufficient sensitivity is available, radiation through noncorresponding entrance and exit slits ($n \neq m$) can be discarded.

Another approach, one which Golay (43) actually used to construct a working instrument, is based on a binary modulation scheme. It can best be explained by giving the example for six slits from Golay's article. The basic modulation period is divided into eight parts and each of the slits is either blocked or unblocked during each of these parts according to the following scheme, where 0 means blocked and 1 means not blocked.

	Entrance slits		Exit slits	
Slit	1st half-period	2nd half-period	1st half-period	2nd half-period
1	0 1 0 1	0 1 0 1	1 0 1 0	0 1 0 1
2	0 0 1 1	0 0 1 1	1 1 0 0	0 0 1 1
3	0 1 1 0	0 1 1 0	1 0 0 1	0 1 1 0
4	0 1 0 1	1 0 1 0	1 0 1 0	1 0 1 0
5	0 0 1 1	1 1 0 0	1 1 0 0	1 1 0 0
6	0 1 1 0	1 0 0 1	1 0 0 1	1 0 0 1

Radiation is not transmitted through corresponding entrance and exit slits during the first half-period, and is transmitted for half of the time during the second half-period. Thus the desired radiation is modulated with a period equal to the basic modulation period. Undesired radiation is either not transmitted at all during either half-period (e.g., entrance slit 1 and exit slit 4) or it is transmitted for one-fourth of the time during both the first and second half-periods and, therefore, is not modulated with the basic modulation period (e.g., entrance slit 1 and exit slit 2). Therefore, only the desired radiation is modulated at a frequency that is amplified by the tuned amplifier. The radiation transmitted by the system is increased over a single-slit instrument by a factor equal to one-half of the number of entrance or exit slits used; one-half because desired radiation is transmitted during only half of the second half-period. The multislit instrument has an entrance port of width W and individual slit openings of width w. Thus it has a resolution determined by w and a luminosity determined by W. By comparing the luminosity of the multislit instrument with that of a single-slit instrument of equivalent resolving power, assuming the height of the openings are the same in both instruments, we find

$$\text{luminosity gain} = W/2w \qquad (7.38)$$

The factor $\frac{1}{2}$ occurs because one-half of the area of the ports is blocked.

In Golay's instrument, slits were cut in rotating disks that served both as slit and chopper, one disk serving as entrance and another as exit slits. The preparation of the disks was done with a computer-controlled machine that generated the binary modulation scheme and cut the disks. With a 64-slit instrument Golay was able to demonstrate a tenfold increase in signal level, with resolution equivalent to the width of the individual slits. Losses in the system associated mostly with its mechanical construction account for the failure to attain the theoretical factor of 32 for the increase in energy. Golay termed this method *dynamic multislit spectrometry*.

Later, Golay (44) devised a method called *static multislit spectrometry*. The entrance port is divided into two parts, A and B (Fig. 7.27). Each half contains an array of many parallel slits, all having the same height and width but positioned in a random way in a manner described by Golay. The exit port is also divided into two parts, C and D. One-half (C) is the image of the corresponding part of the entrance port (A). The other half (D) is the negative of the corresponding part of the entrance port (B). Consequently, all radiation of the desired wavelength, for which the entrance port is imaged on the exit port, is transmitted by

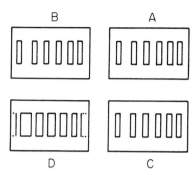

Fig. 7.27   Multislit array of Golay.

both the A half of the entrance port and the C half of the exit port. But radiation of the same wavelength which is transmitted by the B half of the entrance port is blocked by the corresponding D half of the exit port. On the other hand, because of the randomness of the distribution of slits, radiation of undesired wavelengths is transmitted equally well by both halves of the exit port. Therefore, the convolution of the first pair (A and C) of corresponding entrance and exit halves appears something like Fig. 7.28A, and that of the second pair (B and D) as in Fig. 7.28B. The difference of the two functions was recorded by using separate detectors for the two halves of the exit port. The difference appears as in Fig. 7.28C. This is the apparatus function (slit function) of the system.

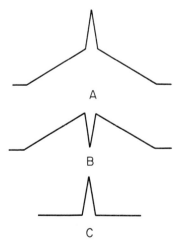

Fig. 7.28   Slit function of multislit system.

In spite of the potential of the method, multislit spectrometry remained dormant until Girard's (*45, 46, 47*) work began to appear in 1959. The Girard grill spectrometer is similar to Golay's static multislit spectrometer, except that the openings in the entrance and exit ports, or grills, are arranged in a regular pattern. In an early form, the grill was made as shown in Fig. 7.29. This is, in essence, a one-dimensional Fresnel zone plate. The convolution of this pattern with itself, minus the convolution of the pattern with its negative, gives the apparatus

Fig. 7.29   Girard's grill array.

function, just as in Golay's work. The function shown was unsatisfactory because of the excessive amplitude of the side lobes of the apparatus function. Girard later considered an array of openings, as shown in Fig. 7.30A, consisting of several series of openings equally spaced within a series, but each series having different spacing. In the limit of a very large number of series, the grill becomes the hyperbolic grill of Fig. 7.30B. Girard was able to modify the shape of the apparatus function by shaping the contour of the grill, much as apodization of the exit pupil modifies the diffraction pattern. A square aperture of hyperbolic openings gives very strong side lobes, but a diamond-shaped aperture greatly suppresses them.

The resolution of a grill spectrometer is determined by the smallest opening of the grill pattern, obeying the same relationship between the width of the opening and the reciprocal dispersion that determines the geometric resolution of a spectrometer with a single slit. In addition, diffraction broadening and aberrations must be considered. Both Golay and Girard chose to use in-plane Littrow monochromator systems with paraboloidal collimators because of the freedom of this system from astigmatism. The luminosity gain of the Girard instrument is, in prin-

A   B

Fig. 7.30   Generation of Girard's hyperbolic array.

ciple, the same as Golay's [Eq. (7.38)], reduced by any losses imposed by the apodizing aperture.

The mock interferometer is another form of multislit spectrometer. It was invented by Mertz (48). More recently, the theory was discussed by Selby (49) and a working instrument was constructed by Ring and Selby (50). The plane of the entrance and exit ports contains a large grid (a Ronchi ruling) having coarse, uniformly spaced lines. The grid can be rotated about a point, as shown in Fig. 7.31. Images of the entrance port appear in the plane of the exit port, with images of different wavelength displaced by distances determined by the dispersion of the monochromator. When the grid is rotated, the images are modulated at a rate that depends on the distance of the image from the center of

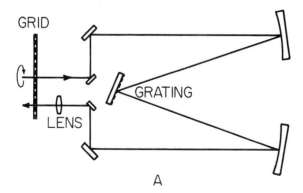

A

Fig. 7.31A   Mock interferometer.

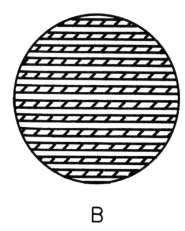

B

Fig. 7.31B   Mock interferometer.

rotation. Therefore, different wavelengths are modulated at different rates and can be discriminated with a tuned amplifier or by the Fourier techniques discussed in the next chapter. The resolution of the mock interferometer is determined by the width of the rulings, and the luminosity by the transmitting area of the entrance port.

IV. Measurement of the Resolution of a Monochromator

The resolution of a monochromator can be defined in different ways, as we have seen, and these different definitions lead to different methods of measurement. Three general methods are: (A) demonstration that two spectral lines of known separation can be distinguished, thereby establishing an upper limit on the two-point resolution; (B) measurement of the width of a very narrow spectral line as it is recorded by the monochromator. This is in effect the width of the slit function; (C) measurement of the spectral modulation transfer function. Let us discuss these in a little greater detail.

A. *Two-Point Resolution*

This really comes closest to what we intuitively understand by resolution. Two lines might be said to be resolved if they are just distinguishable by the occurrence of an inflection point between them; this is Sparrow's criterion. Or we might insist that a definite minimum exists between the peaks, with an intensity of about 80% of the peak values;

this is Rayleigh's criterion. Unfortunately, we do not always have available spectral lines separated by an amount appropriate for the monochromator in question and occurring throughout the spectral region of interest. Nevertheless, this is a common way of comparing competitive instruments and a convenient routine test for checking the performance of a monochromator. Polystyrene is particularly useful for prism instruments. It is readily available and in a convenient form. It provides a pattern of bands in the 3000-cm^{-1} region, where the performance of a NaCl prism is at its worst. The appearance of this spectrum under high resolution is shown in Fig. 7.32. The spectrum of ammonia vapor near 1000 cm^{-1} is another traditional test of instrument resolution; a section of this spectrum is shown in Fig. 7.33. A word of caution to the spectroscopist is necessary here. The separation of resolved lines must be known from measurements made with a spectrophotometer of higher resolving power than the one being tested. A measurement of the separation of lines with an instrument which is just resolving them is likely to be erroneous and, in fact, to indicate better resolution (i.e., smaller separation) than is actually the case. To better understand this, consider two lines of unity intensity, separated

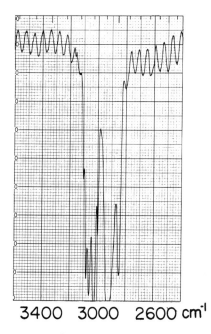

3400   3000   2600 cm^{-1}

Fig. 7.32   Polystyrene spectrum near 3000 cm^{-1}.

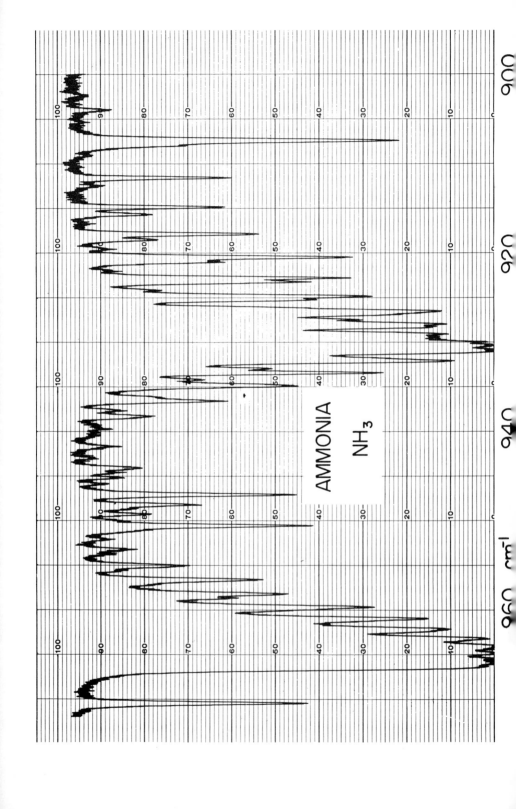

AMMONIA
NH₃

by a small amount 2δ, and suppose that the contours of the individual lines are Gaussian with variance β^2. This is the case of two very sharp lines scanned with a spectrometer having a Gaussian slit function. The recorded spectrum is

$$y(\nu) = \exp\left[-\frac{(\nu-\delta)^2}{2\beta^2}\right] + \exp\left[-\frac{(\nu+\delta)^2}{2\beta^2}\right]$$

where, for convenience, we have centered the pattern at $\nu = 0$. The frequencies of maximum intensity in the spectrum are found by setting

$$\frac{\partial y(\nu)}{\partial \nu} = -\frac{(\nu-\delta)}{\beta^2}\exp\left[-\frac{(\nu-\delta)^2}{2\beta^2}\right] - \frac{(\nu+\delta)}{\beta^2}\exp\left[-\frac{(\nu+\delta)^2}{2\beta^2}\right]$$

$$= -\frac{(\nu-\delta)}{\beta^2}\exp\left[-\frac{(\nu^2+\delta^2)}{2\beta^2}\right] \cdot \exp\left(\frac{2\nu\delta}{2\beta^2}\right)$$

$$-\frac{(\nu+\delta)}{\beta^2}\exp\left[-\frac{(\nu^2+\delta^2)}{2\beta^2}\right] \cdot \exp\left(-\frac{2\nu\delta}{2\beta^2}\right)$$

$$= 0$$

From this equation we have

$$\frac{\delta-\nu}{\delta+\nu} = \exp\left(-\frac{2\nu\delta}{\beta^2}\right)$$

which is the desired equation for the frequencies of maximum intensity. Note that $\nu = 0$ is a solution to this equation, but this defines the point of minimum intensity between the peaks. This equation can be solved numerically and the results are plotted in Fig. 7.34, where we show the apparent separation of the lines versus the actual separation in units of half-bandwidths of the Gaussian function. When the actual separation is greater than about 1.5 bandwidths, the apparent separation is correct, but for smaller actual separations, the apparent separation is considerably in error. For example, the apparent separation is 0.4 bandwidth when the lines are really separated by 0.9 bandwidth. The resolution of the instrument is overestimated by a factor of 2!

A more complicated example of this effect is shown in Fig. 7.35, where a section of the spectrum of methane near 1240 cm^{-1} is recorded with successively poorer resolution. The resolution shown on the figure is the geometric effective bandwidth computed from the product of the slit width and the reciprocal dispersion. The spacing between neighboring pairs of lines, from left to right (high to low frequency), is about 0.60, 0.63, and 0.95 cm^{-1}. At a resolution of 0.72 cm^{-1}, the two

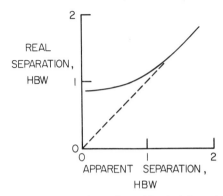

Fig. 7.34   Real and apparent separations of two identical lines, in terms of half-band-widths.

higher frequency lines are no longer resolved. As the slits are widened further through 0.92, 1.04, and 1.16 cm^{-1}, the unresolved pair appear to shift progressively closer to the other two lines and to increase in relative intensity. At 1.16 cm^{-1} resolution, the band has the appearance of a partially resolved triplet of lines spaced about 0.55 and 0.95 cm^{-1} apart. With further widening of the slits to 1.44 cm^{-1}, the band resembles a doublet with a low-frequency shoulder, and at 1.6 cm^{-1} the apparent doublet structure persists with a measured separation of less than 0.6 cm^{-1}. If the resolution of the instrument were to be estimated from the measured separation of the "doublet" components, the result would be optimistic by a factor of more than 2!

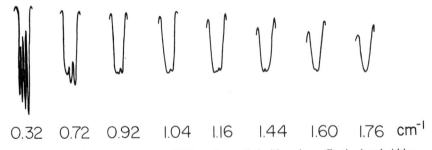

0.32   0.72   0.92   1.04   1.16   1.44   1.60   1.76   cm^{-1}

Fig. 7.35   Methane spectrum near 1240 cm^{-1} recorded with various effective bandwidths.

B. Measured Width of Narrow Lines

If an emission line is very narrow, relative to the resolution of a monochromator, it can be approximated by a delta function:

$$I\delta(\nu - \nu_0) = \begin{cases} I & \nu = \nu_0 \\ 0 & \text{otherwise} \end{cases}$$

The recorded spectrum is then the slit function itself:

$$y(\nu') = I\int \delta(\nu - \nu_0)\sigma(\nu' - \nu)\, d\nu = I\sigma(\nu' - \nu_0)$$

The half-width of the recorded spectral line is the half-width of the slit function. Similarly, if an absorption line can be represented by

$$1 - I\delta(\nu - \nu_0) = \begin{cases} 1 - I & \nu = \nu_0 \\ 1 & \text{otherwise} \end{cases}$$

then the recorded spectrum is a constant minus the slit function:

$$y(\nu') = \int [1 - I\delta(\nu - \nu_0)]\sigma(\nu' - \nu)\, d\nu$$

$$= \int \sigma(\nu' - \nu)\, d\nu - I\sigma(\nu' - \nu_0)$$

$$= \text{const} - I\sigma(\nu' - \nu_0)$$

The constant is the 100% T level of the spectrophotometer and can be set equal to 1 in the initial adjustment of the instrument. Narrow absorption lines are available in the gas-phase spectra of many substances. If the molecules of the substance are small, with light atoms, the rotational structure of the spectrum is resolvable. The pressure of the sample should be kept low to avoid pressure broadening. Lines with multiple structure should not be used. The rotational spectra of molecules with high symmetry are most likely to be simple. Just as in the measurement of two-point resolution, a record from a spectrophotometer of very high resolving power should be available to ensure that the lines are really narrow. The rotation–vibration bands of methane at $3000\ cm^{-1}$, carbon monoxide at $2200\ cm^{-1}$, hydrogen chloride at $2900\ cm^{-1}$, and carbon dioxide at $2300\ cm^{-1}$ are especially useful. These substances have lines that are only a fraction of a wavenumber wide. Some lines of water vapor and ammonia can also be used. In the far infrared, the pure rotational spectra of hydrogen chloride, hydrogen fluoride, water vapor, and ammonia are useful.

The half-width of the slit function is only an incomplete indication of the resolution of a monochromator. Consider the slit functions sketched in Fig. 7.36. All have the same width at half-maximum height, but the wide wings of B certainly degrade resolution; C shows the appearance of recorded spectral lines; and D can introduce spurious structure.

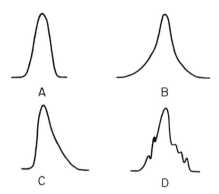

Fig. 7.36   Several hypothetical slit functions.

These measurements provide not only the width of the slit function, but its shape as well. Unfortunately, it is difficult to find pure, narrow lines not overlapped to some extent, so the shape of the slit function can usually not be determined in this way far from its center, that is, into the *wings* of the slit function. Nevertheless, it is interesting to compare slit functions of different classes of monochromators in this way. The shapes of narrow lines measured with a Fastie–Ebert monochromator are usually nicely symmetrical, for example, while the same lines measured with a Littrow monochromator are generally somewhat asymmetrical due to the effects of slit image distortion.

Narrow emission lines are not as plentiful as absorption lines in the infrared. Emission lines generated in a flame, for example, are apt to be broadened. Many laser transitions are known to occur throughout the infrared. The helium–neon laser, for example, can provide a very nice line of high intensity and extremely narrow width at about 3.5 μ. Moreover, there are no nearby lines to overlap it. This transition has been proposed by some workers as a means of measuring the slit function. One would recognize that the radiation from a laser is highly coherent, so the contribution of diffraction to the slit function is different from the case in which incoherent radiation is used. Furthermore, the illumination of the pupil of the instrument can be affected somewhat by diffraction at the entrance slit, and this will be different in the two cases.

C. Spectral Modulation Transfer Function

Both the two-point resolution and the slit function measurements are limited by the availability of naturally occurring spectral lines. The spectral modulation transfer function, as we have seen, is a natural and complete way of defining resolving power. In order to measure it we require a spectrum in which the intensity varies sinusoidally with frequency, with a dc or unmodulated component, of course, because the intensity cannot be negative.

There are several possible ways of generating such channel spectra, using interferometers. Two-beam interferometers, such as the Michelson, might be used. Sakai and Vanasse (*51*) and Hoffman and Vanasse (*52*) have used a lamellar grating interferometer for this purpose (see Chapter 8 on interferometry). A multipath interferometer in the form of a Fabry–Perot etalon was used by Roseler (*53*), Coates and Hausdorff (*39*), and Vasilev (*54*). However, fringes are of low amplitude unless the plates are partially aluminized, in which case the fringes are no longer sinusoidal. Stewart (*55*) used a polarization interferometer in the visible region and, in principle, the instrument and methods are applicable to the infrared.

V. COMPUTATIONAL CORRECTION OF OPTICALLY DISTORTED SPECTRA

A recorded spectrum is necessarily distorted to some extent by the monochromator. The question of how to extract the true line shape from the recorded one has occupied spectroscopists for a long time.

If it is known that the slit function has a particular form, for example, triangular, with a specified bandwidth, and that the band has a well-characterized shape, such as Gaussian or Lorentzian, then the error in the recorded peak height and bandwidth can be obtained from formulas and figures discussed earlier and which appear in the literature (*1–34*). Usually, however, these serve at best to indicate the approximate size of error to be expected.

If the shape of the actual spectral line is unknown *a priori* it is possible, in principle, to find it from the recorded line and the slit function, assuming that the slit function is known, by solving the integral equation

$$y(\nu') = \int \sigma(\nu' - \nu)x(\nu)\,d\nu \qquad (7.39)$$

for the unknown $x(\nu)$. This can be done in various ways. Numerical deconvolution of the integral by means of techniques known to applied mathematicians and computer programmers is one way (*56–60*). Khidir and Decius (*61*) expressed the equation (7.39) as a matrix equation, with spectral and slit function information given at discrete points separated by finite increments,

$$|Y| = |S|\,|X|$$

Matrix multiplication on the left by $|S|^{-1}$, the inverse matrix to $|S|$, gives

$$|X| = |S|^{-1}|Y|$$

for the values of $x(\nu)$ at discrete points. The convergence of the method was not completely satisfactory, but the approach has some potential.

The spectral modulation transfer function formulation of Eq. (7.27), $Y = SX$, lends itself naturally to such calculations (62–66). The recorded spectral line is Fourier transformed by numerical means and then divided by the spectral transfer function. The inverse transformation of the quotient function is the original spectral line. In this formulation it is easy to see the origin of some potential trouble. Both the spectral modulation transfer function and the transform of the recorded spectrum are corrupted somewhat by noise. This particularly introduces uncertainties at the higher spectral modulation frequencies where the values of S and Y are small. After dividing one small uncertain number by another small uncertain number to find X, it is clear that X is very uncertain indeed at the higher frequencies.

Recall that the spectral modulation transfer function associated with a triangular slit function vanishes at certain values of the modulation frequency. But between these zeros it has nonvanishing values. Thus it is possible in principle to deduce detail in a spectrum which is finer than the geometric resolution limit given by the product of the reciprocal linear dispersion and the geometric slit width. This is related to the spurious resolution that was discussed earlier. It should be remembered that the practical implementation of this possibility is limited by the corrupting presence of noise, by the failure of the system to transmit information at those modulation frequencies at which the transfer function vanishes, and by the vanishing of the transfer function at all modulation frequencies greater than the limit imposed by diffraction. The possibility of analytically extracting unresolved detail from nonspectroscopic images has been discussed by Harris (67), for example, but no application has been made as yet in spectroscopy.

A somewhat different method of correcting for slit function distortion was devised by Burger and van Cittert (5). It is called pseudo-deconvolution by Jones et al. (68). A recorded spectrum is further distorted by convolving it again with the monochromator slit function. The difference between the recorded spectrum and the convolute is taken as a measure of the distortion in the spectrum itself and an appropriate correction is made. The process is repeated as required until it is judged that correction is complete.

The successful use of computers in removing the distorting effect of the monochromator in simple spectra has led several workers to develop methods for treating more complex spectra (*69–73*). These techniques will become more useful in the future and ultimately infrared spectrophotometers will be coupled routinely to computers.

REFERENCES

1. J. B. Willis, *Aust. J. Sci. Res.*, **A4**, 172 (1951).
2. D. A. Ramsey, *J. Am. Chem. Soc.*, **74**, 72 (1952).
3. V. Z. Williams, *Rev. Sci. Inst.*, **19**, 135 (1948).
4. P. H. van Cittert, *Z. Physik*, **65**, 547 (1930); **69**, 298 (1931).
5. H. C. Burger and P. H. van Cittert, *Z. Physik*, **79**, 722 (1932); **81**, 428 (1933).
6. S. Brodersen, *J. Opt. Soc. Am.*, **43**, 877 (1953); **44**, 22 (1954).
7. J. R. Nielsen, V. Thornton, and E. B. Dale, *Rev. Mod. Phys.*, **16**, 307 (1944).
8. A. R. Philpotts, W. Thain, and P. G. Smith, *Anal. Chem.*, **23**, 268 (1951).
9. D. M. Dennison, *Phys. Rev.*, **31**, 503 (1928).
10. H. J. Babrov, *J. Opt. Soc. Am.*, **51**, 171 (1961); **52**, 831 (1962).
11. V. P. Koslov and E. O. Federova, *Opt. and Spectry.*, **10**, 348 (1961).
12. A. Lempicki, H. Samelson, and A. Brown, *J. Opt. Soc. Am.*, **51**, 35 (1961).
13. H. J. Kostkowski and A. M. Bass, *J. Opt. Soc. Am.*, **46**, 1060 (1956).
14. J. E. Stewart, *Appl. Opt.*, **2**, 1303 (1963).
15. K. S. Seshadri and R. N. Jones, *Spectrochim. Acta*, **19**, 1013 (1963).
16. S. G. Rautian, *Sov. Phys. Usp.*, **66**, 245 (1958).
17. G. G. Petrash, *Opt. and Spectry.*, **6**, 516 (1959).
18. G. G. Petrash and S. G. Rautian, *Inzh-Fiz. Zh.*, **1**, 61, 80 (1958).
19. G. G. Petrash and S. G. Rautian, "Materialy x vsesoiuz," *Soveshch. Spek.*, **1** (Mol. Spekt.), 102, 107 (1957).
20. G. W. King and A. G. Emslie, *J. Opt. Soc. Am.*, **41**, 405 (1951); **43**, 664 (1953).
21. A. G. Emslie and G. W. King, *J. Opt. Soc. Am.*, **43**, 658 (1953).
22. A. Schuster, *Astrophys. J.*, **21**, 197 (1905).
23. R. A. Russell and H. W. Thompson, *Spectrochim. Acta*, **9**, 133 (1957).
24. K. D. Mielenz, *Optik*, **13**, 437 (1956).
25. A. Lohmann, *Opt. Acta*, **6**, 173 (1959).
26. V. P. Kozlov, *Opt. and Spectry.*, **16**, 271 (1964).
27. F. Moser and F. Urbach, *Phys. Rev.*, **102**, 1519 (1956).
28. C. R. De Prima and S. S. Penner, *J. Chem. Phys.*, **23**, 757 (1955).
29. S. S. Penner, *Quantitative Molecular Spectroscopy and Gas Emissivities*, Addison-Wesley, Reading, Mass., 1959.
30. J. T. Davies and J. M. Vaughan, *Astrophys. J.*, **137**, 1302 (1963).
31. H. J. Bernstein and G. Allen, *J. Opt. Soc. Am.*, **45**, 237 (1955).
32. G. Pirlot, *Bull. Soc. Chim. Belg.*, **59**, 352 (1950).
33. T. G. Kyle and J. O. Green, *J. Opt. Soc. Am.*, **55**, 895 (1965).

34. A. Roseler, *Infrared Phys.*, **5**, 37, 51 (1965); **6**, 111 (1966).

35. H. Sakai, *Japan. J. Appl. Phys.*, **4**, Suppl. 1,, 308 (1965).

36. A. Roseler, *Monatsh. Deut. Akad. Wiss. Berl.*, **9**, 237 (1967).

37. K. Miyamoto, *Progress in Optics*, Vol. I, E. Wolf, ed., Wiley, New York, 1961, pp. 31–66.

38. J. E. Stewart, *Appl. Opt.*, **4**, 609 (1965).

39. V. J. Coates and H. Hausdorff, *J. Opt. Soc. Am.*, **45**, 425 (1955).

40. E. H. Linfoot, *J. Opt. Soc. Am.*, **46**, 740 (1956).

41. Y. W. Lee, *Statistical Theory of Communication*, Wiley, New York, 1960.

42. P. Jacquinot and B. Roizen-Dossier "Apodization," *Progress in Optics*, Vol. III, Wiley, New York, 1964, pp. 29–186.

43. M. J. E. Golay, *J. Opt. Soc. Am.*, **39**, 437 (1949).

44. M. J. E. Golay, *J. Opt. Soc. Am.*, **41**, 468 (1951).

45. A. Girard, R. Lenfant, and N. Louisnard, *Rech. Aeron.*, **72**, 35 (1959).

46. A. Girard, *Opt. Acta*, **7**, 81 (1960); *Appl. Opt.*, **2**, 79 (1963); *J. Phys.*, **24**, 139 (1963); *Rech. Aeron.*, **99**, 27 (1964).

47. A. Girard, "Etude d'un spectrometre a modulation selective," *Office National d'Etudes et de Recherches Aerospatiales*, Publ. No. 117, 1967.

48. L. Mertz, *Transformations in Optics*, Wiley, New York, 1965.

49. M. J. Selby, *Infrared Phys.*, **6**, 21 (1966).

50. J. Ring and M. J. Selby, *Infrared Phys.*, **6**, 33 (1966).

51. H. Sakai and G. A. Vanasse, *J. Opt. Soc. Am.*, **56**, 357 (1966).

52. J. E. Hoffman and G. A. Vanasse, *Appl. Opt.*, **5**, 1167 (1966).

53. A. Roseler, *Infrared Phys.*, **5**, 81 (1965); *Monatsh. Deut. Akad. Wiss. Berl.*, **9**, 380 (1967).

54. A. F. Vasilev, *Opt. and Spectry.*, **14**, 74 (1963).

55. J. E. Stewart, *Appl. Opt.*, **6**, 1523 (1967).

56. W. F. Herget, W. E. Deeds, N. M. Gailar, R. J. Lovell, and A. H. Nielsen, *J. Opt. Soc. Am.*, **52**, 1113 (1962).

57. J. M. Render and R. J. Lovell, *J. Opt. Soc. Am.*, **56**, 51 (1966).

58. D. Montegny, *Spectrochim. Acta*, **20**, 1373 (1964).

59. R. N. Bracewell, *J. Opt. Soc. Am.*, **45**, 873 (1955).

60. V. P. Sachenko, *Bull. Acad. Sci. USSR, Phys. Ser.* **25**, 1043 (1961).

61. A. L. Khidir and J. C. Decius, *Spectrochim. Acta*, **18**, 1629 (1962).

62. A. C. Hardy and F. M. Young, *J. Opt. Soc. Am.*, **39**, 265 (1949).

63. A. F. Bondarev, *Opt. and Spectry.*, **12**, 282 (1962).

64. B. N. Grechushnikov, *Opt. and Spectry.*, **12**, 70 (1962).

65. J. S. Rollett and L. A. Higgs, *Proc. Phys. Soc.*, **79**, 87 (1962).

66. H. Stone, *J. Opt. Soc. Am.*, **52**, 998 (1962).

67. J. L. Harris, *J. Opt. Soc. Am.*, **54**, 931 (1964).

68. R. N. Jones, R. Venkataraghavan, and J. W. Hopkins, *Spectrochim. Acta*, **23A**, 925, 941 (1967).

69. J. Pitha and R. N. Jones, *Can. J. Chem.*, **44**, 3031 (1966).

70. R. N. Jones, K. S. Seshadri, N. B. W. Jonathan, and J. W. Hopkins, *Can. J. Chem.*, **41**, 750 (1963).

71. R. N. Jones, *Appl. Opt.*, **8**, 597 (1969).

72. P. A. Jansson, R. H. Hunt, and E. K. Plyler, *J. Opt. Soc. Am.*, **58**, 1665 (1968).

73. D. W. Berreman, *Appl. Opt.*, **7**, 1447 (1968).

CHAPTER 8

Interference Spectroscopy

I. INTRODUCTION

Progress in infrared spectrophotometric instrumentation has proceeded along two paths, with high resolution and high luminosity and signal-to-noise ratio as the goals. These paths are not necessarily independent, and indeed the goals are in a very real sense mutually exclusive. We have seen that in a conventional spectrophotometer, high resolution is obtained by narrowing the slits, at least up to the point of the limits imposed by diffraction or aberrations. On the other hand, the luminosity of the instrument decreases with narrowing slits, and therefore so does the signal-to-noise ratio.

In the previous chapter we described some ingenious devices for increasing luminosity without sacrifice of resolution: the multislit or grill monochromators. These devices are conventional monochromators, with gratings used to disperse the radiation. Their novelty lies in the nature of the entrance and exit "slits" and the means used to discriminate among overlapping "slit functions." In the present chapter we discuss the use of interferometers in infrared spectroscopy to increase both luminosity and resolution.

Interferometers appear to have not even a superficial resemblance to monochromators; but in a more fundamental sense, they have many of the same properties, as we shall recognize. In fact, we must concede that a diffraction grating *is* an interferometer that sorts out narrow wavelength intervals by means of interference. A case can be made for classifying a prism, or any refracting or reflecting optical element for that matter, as an interferometer.

We begin this chapter with a discussion of the Fabry–Perot interferometer, followed by a treatment of several two-beam interferometers. An excellent book on interferometry in general has been written by Francon (*1*) and a useful review of interference spectroscopy has been given by Jacquinot (*2*).

II. Multibeam Interferometry: The Fabry–Perot Etalon

The basic properties of the Fabry–Perot etalon have been discussed in Section 4.V. We saw that a monochromatic extended source is imaged as a set of concentric circular rings or fringes, with angular separation given by

$$\cos \theta_k - \cos \theta_{k+1} = \lambda/2bn$$

The angular separation decreases with increasing plate separation. In order to use the etalon as a spectrometer, a diaphragm is placed at the focal plane of the lens to permit only a portion of the radiation to pass (Fig. 8.1). Usually a simple circular opening is used centered on the fringe pattern. From Eq. (4.34), the relationship between the angular position of a fringe of order k and wavelength λ is

$$\lambda = \frac{2bn \cos \theta}{k} \qquad k = 1, 2, 3, \ldots \qquad (8.1)$$

The wavelength λ_k appearing at the center of the diaphragm is

$$\lambda_k = 2bn/k \qquad (8.2)$$

If the angle defined by the edge of the diaphragm is θ', then the wave-

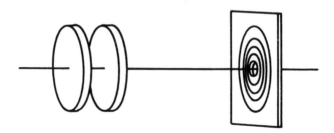

Fig. 8.1   Fabry–Perot interferometer with diaphragm.

length of the same order which is just passed by the diaphragm is

$$\lambda'_k = \lambda_k \cos \theta$$

Therefore, the range of wavelengths passed by the diaphragm is, for order k,

$$\delta \lambda = \lambda_k - \lambda'_k = \lambda_k (1 - \cos \theta) \tag{8.3}$$

Simultaneously, ranges of wavelengths of other orders are also passed by the diaphragm. Thus the apparatus function, analogous to the geometric slit function of a monochromator, appears as in Fig. 8.2A.

It is necessary to provide some means of separating one of the peaks of Fig. 8.2 corresponding to only one order. In infrared applications this can be done by combining the etalon with a monochromator. The monochromator provides a band of wavelengths narrower than the wavelength interval $\Delta\lambda$ between maxima of adjacent orders ($\Delta\lambda = \lambda_k - \lambda_{k-1}$), and the etalon with its diaphragm isolates a narrow band $\delta\lambda$ ($\delta\lambda = \lambda'_k - \lambda_k$) and presents it to the detector. The band isolated by the etalon may be scanned by changing the optical path length bn. We temporarily defer a discussion of methods for doing this. The wavelength interval $\Delta\lambda$ between maxima is termed the *free spectral range* of the interferometer. The width of the fringes $\delta\lambda$, or the width at half-maximum intensity in the general case, defines the resolution of the interferometer in the same way as the effective bandwidth of a monochromator. The *finesse* of a Fabry–Perot interferometer was defined as the ratio of $\Delta\phi$ to $\delta\phi$ in Chapter 4. Clearly, this is equal to the ratio of $\Delta\lambda$ to $\delta\lambda$, or the equivalent in terms of frequency,

$$\mathscr{F} = \Delta\lambda/\delta\lambda = \Delta\nu/\delta\nu$$

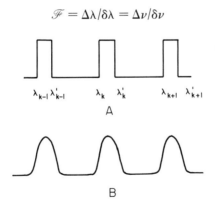

A

B

Fig. 8.2   Wavelength passed by Fabry–Perot interferometer.

In other words, the finesse is the number of resolvable spectral elements per free spectral range. The resolving power, or resolvance, is defined as usual

$$R = \lambda/\delta\lambda = \nu/\delta\nu \tag{8.4}$$

We can calculate $\Delta\lambda$ from Eq. (8.2),

$$\Delta\lambda = \lambda_k - \lambda_{k+1} = 2bn\left(\frac{1}{k} - \frac{1}{k+1}\right)$$

$$= \frac{2bn}{k(k+1)} = \frac{\lambda_{k+1}}{k} \tag{8.5}$$

A relationship between the finesse, the resolving power, and the order of interference follows:

$$R = \frac{\lambda}{\Delta\lambda} \cdot \frac{\Delta\lambda}{\delta\lambda} = k\mathscr{F} \tag{8.6}$$

The resolution of an etalon can thus be increased by increasing the spacing between the plates, which increases the order k.

The luminosity of a Fabry–Perot etalon is determined, just as for any optical instrument, from the product of the transmission efficiency, the area of the pupil of the system (normally the interferometer plates), and the solid angle subtended by the diaphragm. Hence

$$L = TA\Omega = TA(f\sin\theta)^2 \pi/f^2$$

$$= TA\pi \sin^2\theta \tag{8.7}$$

The focal length of the lens, f, drops out of this equation.

If θ is not too large, Eq. (8.7) can be written in terms of resolving power with the aid of Eqs. (8.3) and (8.4),

$$L = \pi AT(1 - \cos^2\theta) \approx 2\pi AT(1 - \cos\theta) = 2\pi AT/R$$

Thus the luminosity of an etalon varies inversely with the resolving power, not inversely with the square as in a monochromator. Moreover, the solid angle subtended by the diaphragm is considerably larger than that of the narrow slit of a monochromator, which provides a significant advantage in luminosity for the etalon that amounts to perhaps the order of hundreds.

Now let us return to some of the experimental methods used with the Fabry–Perot etalon. Obviously, the plates must be fabricated of a material that is transparent in the infrared. Quartz and fused silica are excellent materials for plates, but unfortunately are limited to wavelengths shorter than about 3.5 μ. Fluorite and rock salt have been used in infrared etalons, and the other alkali halides can of course be used in appropriate spectral regions. The plates must be quite flat, but working in the infrared has its advantages in this regard. It is not too difficult to polish infrared materials to within 1/20 of a wavelength of *visible* radiation. At 5 μ this is 1/200 of a wavelength, which is as good as the best quartz plates available for the visible region.

The plates must be coated with a highly reflecting film to provide an adequate finesse. Thin films of aluminum or silver are used, but metals attenuate the radiation by absorption. Multilayer coatings of dielectric materials having alternately high and low refractive indices offer high reflectivity and low absorption. In the near infrared ZnS–MgF$_2$ coatings have been used. At longer wavelengths semiconducting materials such as germanium, silicon, tellurium, or selenium are used as high-index materials. Alkali halides such as NaCl and KBr are useful low-index materials. The reflectivity of multilayer coatings is somewhat wavelength selective, thus limiting the usefulness of a given coating to a specific spectral region.

We mentioned the necessity of isolating one spectral order from the sequence of peaks presented by the etalon. Combining an etalon with a monochromator is one way of doing this, and very effective spectroscopy has been done in this way. However, the limited luminosity of the monochromator prevents us from taking full advantage of the luminosity of the etalon. An alternative procedure is to use two etalons with different free spectral range, as illustrated in Fig. 8.3. The etalon F_1 has a narrower spacing between plates than F_2, and thus it is used in lower order [Eq. (8.5)] and has a greater free spectral range. On the other hand, it has lower resolution [Eq. (8.6)]. The combination of the two etalons provides the free spectral range of F_1 and the resolution of F_2. Further filtering is required to eliminate the remaining orders, and additional etalons can be used. The selectivity of the dielectric reflecting coatings may be considered advantageous in that they provide additional filtering. Greenler's (3) infrared interferometer consists of only two etalons, with the filtering provided by KBr–Te coatings.

Scanning of wavelength with a Fabry–Perot spectrometer can be accomplished by changing the value of the optical path length, bn [Eq. (8.2)]. It can also be done by tilting the etalon to change the angle of incidence of the radiation, but this is usually unsatisfactory because the fringes become elliptical and eventually their appearance approaches parallel lines. A slit-shaped aperture is required, rather than a circle, and a loss in luminosity is suffered. The product bn can be

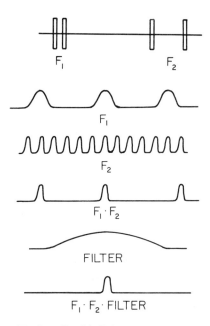

Fig. 8.3   Double Fabry–Perot system.

varied by changing either the plate separation b or the refractive index n. If one chooses to vary b, one plate of the etalon can be mounted on a carriage so that it can be displaced relative to the other plate. Alternatively, the spacer can be made of a piezoelectric material which changes in size when an electric field is imposed on it. The refractive index of the path between the plates can be varied by changing the pressure of the gaseous medium. A change in pressure which is linear in time can be generated by programming or servocontrolling the motion of a piston or by permitting the gas to escape through an orifice at supersonic velocity. The fractional wavelength change for a change in air pressure of 1 atm. is about 3×10^{-4}.

Large changes in optical path length with refractive index are not possible unless very high gas pressures are used. On the other hand, mechanically displacing one plate over large distances without introducing a tilt is difficult. Fortunately, it is only necessary to change the spacing by $\frac{1}{2}\lambda$ to scan over a complete free spectral range. The procedure for scanning a spectrum is illustrated in Fig. 8.4. The grating monochromator isolates a spectral range of width equal to (or less than) the free spectral range $\Delta\lambda$ of the etalon. The etalon separates out a very narrow wavelength increment $\delta\lambda$ of order k which is passed through the diaphragm. The optical path length is increased smoothly,

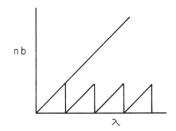

Fig. 8.4 Scanning method for Fabry–Perot spectrometer.

and the etalon scans across the free spectral range. The monochromator is then advanced $\Delta\lambda$, the optical path length is reduced to its original value and is once more smoothly increased, this time separating out a narrow wavelength increment of order $k-1$. In practice the monochromator need not be advanced stepwise, but can be scanned smoothly so the maximum of its slit function coincides with the wavelength passed by the etalon. Synchronization of monochromator and etalon can present a difficult problem.

Notice that if two etalons of different thickness are used in tandem and they are to be scanned by changing the plate spacing, then the rate of change must be different for the two etalons if they are to pass the same wavelength. Indeed, from Eq. (8.1)

$$\Delta b_1 = \frac{b_1}{\lambda}\Delta\lambda \quad \text{and} \quad \Delta b_2 = \frac{b_2}{\lambda}\Delta\lambda \qquad b_1 \neq b_2$$

On the other hand, if scanning is done by varying n, then

$$\Delta n = \frac{n}{\lambda}\Delta\lambda$$

is independent of b. This means that two or more etalons can be placed in a common enclosure in tandem, and when wavelength is scanned by varying n, the etalons are automatically synchronized.

III. Two-Beam Interferometry: The Michelson Interferometer and Periodic Modulation

We now consider a different type of interferometer, the two-beam interferometer typified by the Michelson. In Section 4.V we discussed the basic properties of this device. We considered the radiation from a monochromatic source imaged on a narrow opening, collimated, and introduced into the interferometer. We saw that the radiation is modulated by constructive and destructive interference as the length of one beam is varied uniformly while the length of the other beam remains fixed. The modulation frequency is given by

$$f = \frac{2}{\lambda}\frac{d\xi}{dt} = 2\nu\frac{d\xi}{dt}\text{ cycles per unit time}$$

or

$$\omega = \frac{4\pi}{\lambda}\frac{d\xi}{dt} = 4\pi\nu\frac{d\xi}{dt}\text{ radians per unit time} \qquad (8.8)$$

where $d\xi/dt$ is the velocity with which the mirror is moved. The path difference, of course, changes at twice this rate,

$$\frac{d\Delta}{dt} = 2\frac{d\xi}{dt} = 2v \qquad (8.9)$$

Equation (8.8) informs us that the frequency of modulation depends on the rate of travel of the movable mirror and the wavelength of the radiation. For example, a mirror speed of 0.05 mm/sec results in modulation of 10-μ radiation at a rate of 10 cps. If our source is polychromatic, each constituent wavelength is modulated at a different rate. A spectrometer can be constructed on this basis, and indeed several devices have been made in this way with somewhat different interferometer configurations (4,5). An ac amplifier that is tuned to the modulation frequency of the desired wavelength is used. Only the intensity at that wavelength and a narrow band around it is recorded. The more highly tuned the amplifier is (the higher its Q), the narrower will be the band of wavelengths. A different wavelength can be chosen by changing the tuning frequency of the amplifier. A better approach, particularly with slowly responding infrared detectors, is to fix the amplifier but change the rate of movement of the interferometer mirror.

It appears that the spectral resolving power depends only on the electrical properties of the amplifier. This is not entirely true, as the resolution also depends on the extent of travel of the mirror. Let us see how this surprising property comes about. The output of the interferometer, as received by the detector, is a function of time. This can be expressed by introducing Eq. (8.9) into Eq. (4.32) and dropping the dc term

$$S(t) = \frac{1}{2}\int_0^\infty P(\nu)\cos\left(4\pi\frac{d\xi}{dt}t\nu\right)d\nu$$

$$= \frac{1}{2}\int_0^\infty P(\nu)\cos 4\pi\nu v t\, d\nu$$

Implicit in this formulation, of course, is the constancy of v and the definition of zero time corresponding with zero path difference. If we consider a perfectly monochromatic source represented by a delta function, $P(\nu) = \delta(\nu - \nu_0)$, then the output is

$$S(t) = \tfrac{1}{2} \cos 4\pi\nu_0 vt = \tfrac{1}{2} \cos \omega t \qquad (8.10)$$

when the mirror is moving with uniform velocity v. This statement is, of course, in agreement with Eq. (8.8). It states that the output signal is a periodically varying signal of frequency ω, and only ω. This is in accordance with the statement that the Fourier transform of $S(t)$ is

$$s(\omega') = \tfrac{1}{2}\pi[\delta(\omega - \omega') + \delta(\omega + \omega')]$$

(We include negative frequencies for formal completeness.) However, this is dependent on the uniformity of the velocity v. In fact, v is not uniform, because the mirror must be stopped when its maximum displacement is reached. Its actual motion is a sawtooth, as depicted in Fig. 8.5A. In each segment of motion the output is periodic, as in Eq. (8.10), but at the corners there is an abrupt phase change, as shown in Fig. 8.5B. This destroys the delta-function character of the frequency content of the output. The center segment of Fig. 8.5B is the product of the cosine function [Eq. (8.10)] and a square or boxcar function of the type

$$u_T(t) = \begin{cases} 1 & -T < t < T \\ 0 & t < -T \quad t > T \end{cases} \qquad (8.11)$$

The Fourier transform of the product is

$$s_0(\omega') = \frac{1}{2} \int_{-T}^{T} \cos \omega t \, \exp(-j\omega' t) \, dt$$

$$= \frac{1}{2} \left\{ \frac{\sin(\omega' - \omega)T}{\omega' - \omega} + \frac{\sin(\omega' + \omega)T}{\omega' + \omega} \right\}$$

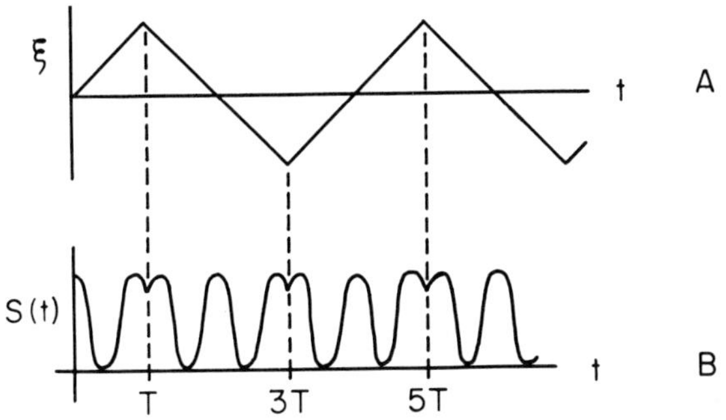

Fig. 8.5   Sawtooth scanning.

The other segments of Fig. 8.5B are identical with the center segment, but shifted by multiples of $2T$. The transforms of these segments are thus

$$s_n(\omega') = \frac{1}{2}\left\{\frac{\sin(\omega'-\omega)T}{\omega'-\omega} + \frac{\sin(\omega'+\omega)T}{\omega'+\omega}\right\}\exp(-j\cdot 2nT\omega')$$

$$n = \pm 1, \pm 2, \ldots$$

The complete Fourier transform of the output is the sum of all of these, or

$$s(\omega') = \sum_{n=0}^{\infty} s_n(\omega')$$

$$= \left\{\frac{\sin(\omega'-\omega)T}{\omega'-\omega} + \frac{\sin(\omega'+\omega)T}{\omega'+\omega}\right\}\sum_{n=0}^{\infty}\cos 2n\omega'T \quad (8.12)$$

Thus, as we vary the tuning frequency of the amplifier (or, equivalently, change the velocity of the mirror), a monochromatic source is recorded as the function of Eq. (8.12) (see Fig. 8.6). The amplifier is swept over only the region about $+\omega'$. The frequency patterns are repeated at higher frequencies, $n\omega'$, but infrared detectors have decidedly limited frequency response, and no additional information could be obtained from these higher frequencies anyway. Consequently, the significant output is merely

$$s(\omega') = \frac{\sin(\omega'-\omega)T}{\omega'-\omega} \qquad \omega \approx \omega' \qquad (8.12a)$$

Figure 8.6B and Eq. (8.12a) represent the apparatus function of the interferometer. It is analogous to the slit function of a monochromator.

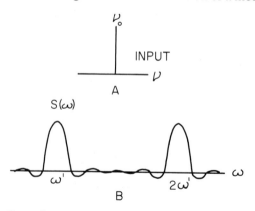

Fig. 8.6 Recording of a monochromatic line with periodic modulation.

We can adopt the half-width of the function between the first zeros as a measure of the resolving power of the interferometer, just as we did with the monochromator slit function. Hence

$$\Delta\omega = \pi/T$$

This result is related to spectral resolving power through Eq. (8.8) and the maximum path difference

$$|x_{max}| = 2|\xi_{max}| = 2vT$$

by writing

$$\Delta\nu = \frac{1}{4\pi v}\Delta\omega = \frac{1}{4vT} = \frac{1}{4|\xi_{max}|} = \frac{1}{2|x_{max}|} \qquad (8.13)$$

Equation (8.13) is the desired relationship between resolution and maximum path difference.

The apparatus function given by Eq. (8.12) has very strong side lobes, considerably stronger than the side lobes of the diffraction pattern of a rectangular aperture. Hence, it is very desirable to introduce some sort of apodization. This could be done by attenuating the signal, either electrically or optically, as a function of mirror travel. In this way, for example, the boxcar function of Eq. (8.11) might be replaced with a triangular function, and the apparatus function would then have the $(\sin^2\phi)/\phi^2$ form of the diffraction pattern of a rectangular aperture.

We have considered the Michelson interferometer as though it were illuminated through a pinhole of infinitesimal size. However, the opening must be finite in order to transmit a finite amount of radiation. The luminosity is proportional to the solid angle defined by the opening. On the other hand, it must be small enough to isolate a small portion of an interference ring. Just as with the Fabry–Perot etalon, a compromise between resolution and luminosity is required. The size of the fringe pattern depends on the path difference in the two beams. When the path lengths are very nearly equal, the central fringe is very large and covers the entire field, but as the path difference increases, the rings become smaller. The analysis is very much as in the case of the etalon, and so we do not repeat it. For a given diaphragm size the fringe contrast therefore lessens as the path-length difference increases. This amounts to a built-in apodizing function. The Michelson interferometer shares with the etalon a considerable advantage in luminosity over the conventional monochromator.

In constructing a Michelson interferometer, one of the mirrors must be mounted in such a way that it can be moved periodically over distances up to centimeters or so at a uniform rate. The motion must not permit a change in angle of the mirror. The use of parallelograms and V grooves are two possible arrangements. The mirror can be driven with a lead screw or by a cam. It is necessary to stop the mirror abruptly at the end of its travel and immediately start it moving in the opposite direction.

The beam splitter of a Michelson interferometer is of crucial importance. Alkali halide crystals can serve as supports for thin metallic films for use in the mid-infrared region. Multilayer dielectric reflecting films are less wasteful of energy, but are useful over a more limited wavelength region. A single layer of a semiconducting material such as germanium or selenium provides good reflectivity and low absorption over a wide portion of the infrared spectrum. In the very far infrared, thin stretched films of a polymer such as Mylar are used as substrates for beam splitters. They can be coated with metal or germanium films, but usually they are used without any coating, relying on the reflectivity of the plastic itself. Metal screens, grids, or mesh are also useful beam splitters in the far infrared.

When a thin-film beam splitter is used, its efficiency is found to vary significantly with wavelength. This is shown in Fig. 8.7. There is a general lowering of efficiency at longer wavelengths because the dielectric constant is falling and therefore so is the reflectivity. In addition, there is a periodic variation in the efficiency because of multiple reflections within the beam splitter. This can be described in a fashion similar to the discussion leading to Airy's formula for a Fabry–Perot interferometer. From Fig. 8.8 we see that the amplitude (electric vector) of the beam transmitted by the beam splitter is

$$E_T = tt' + tt'r'^2e^{-j\phi} + tt'r'^4e^{-j\phi} + \cdots$$
$$= \frac{tt'}{1 - r'^2e^{-j\phi}}$$

where r and t refer to external reflection and transmission of the electric

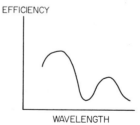

Fig. 8.7   Efficiency of a thin-film beam splitter.

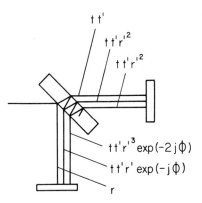

Fig. 8.8   Beam splitter.

vector at each surface, and r' and t' refer to internal reflection and transmission. Also, ϕ represents the phase difference of successive rays due to different optical path lengths through the film and also any phase change occurring at the reflecting surfaces. The internal reflection coefficient r' may be seen, from the Fresnel equations for a dielectric, to be numerically equal to the external reflection coefficient, but opposite in phase. That is,

$$r' = -r$$

Also, a relationship between t, t', and r can be derived by considering the imaginary experiment of Fig. 8.9. An incident beam with unit electric vector is partly reflected and partly refracted. The refracted and reflected beams are now returned, with their phases carefully maintained by mirrors placed at nodes of the waves. The beam reflected back into the medium at the surface has amplitude

$$r't + rt = 0$$

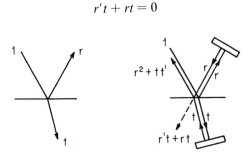

Fig. 8.9   Internal reflection.

The beam returned to the source has amplitude

$$r^2 + tt' = 1$$

and this is the desired relationship. The amplitude of the ray transmitted through the beam splitter can now be written

$$E_T = \frac{1 - r^2}{1 - r^2 e^{-j\phi}}$$

The transmitted radiant power is, therefore,

$$I_T = E_T E_T^* = \frac{(1 - r^2)^2}{(1 - r^2)^2 + 4r^2 \sin^2 \phi/2}$$

$$= \frac{(1 - R)^2}{(1 - R)^2 + 4R \sin^2 \phi/2}$$

The amplitude of the reflected beam is

$$E_R = r + tt'r'e^{-j\phi}(1 + r'^2 e^{-j\phi} + r'^4 e^{-2j\phi} + \cdots)$$

$$= r + \frac{tt'r'e^{-j\phi}}{1 - r'^2 e^{-j\phi}}$$

$$= \frac{r(1 - e^{-j\phi})}{1 - r^2 e^{-j\phi}}$$

and the radiant power in this beam is

$$I_R = E_R E_R^* = \frac{4r^2 \sin^2 \phi/2}{(1 - r^2)^2 + 4r^2 \sin^2 \phi/2}$$

$$= \frac{4R \sin^2 \phi/2}{(1 - R)^2 + 4R \sin^2 \phi/2}$$

The radiation ultimately reaching the detector through either beam has been subjected to one reflection and one transmission. Its power is therefore

$$I = I_R I_T = \frac{4R(1 - R)^2 \sin^2 \phi/2}{[(1 - R)^2 + 4R \sin^2 \phi/2]^2} \qquad (8.14)$$

The power reaching the detector through both beams is equal to twice $I_R I_T$, modulated by the usual fine Michelson interferometer fringes.

The values attained by $2I_R I_T$ are plotted in Fig. 8.10 as a function of the reflectance of the beam splitter material. Even with rather low values of R, the radiation reaching the detector is a substantial fraction of the incident radiation. This is the reason that uncoated plastic films can be used as beam splitters.

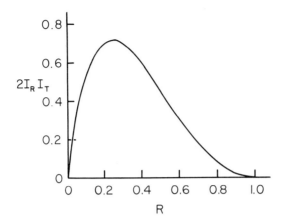

Fig. 8.10 Power reaching detector as a function of beam splitter reflectivity.

Now let us continue to discuss the variation in beam splitter efficiency displayed in Fig. 8.7. This is produced by the $\sin^2 \frac{1}{2}\phi$ terms in Eq. (8.14) for $I_R I_T$. If the angle of incidence on the beam splitter is 45°, the angle of the refracted ray is

$$\theta = \sin^{-1}\left(\frac{1}{\sqrt{2}n}\right)$$

or

$$\cos \theta = \frac{(2n^2 - 1)^{1/2}}{\sqrt{2}n}$$

The phase difference between successive rays due to two traversals through the film is given by

$$\phi = \frac{4\pi bn \cos \theta}{\lambda}$$

So

$$\frac{\phi}{2} = \sqrt{2}\pi b \frac{(2n^2 - 1)^{1/2}}{\lambda} = 2\pi b\nu(2n^2 - 1)^{1/2}$$

We should add the contribution to the phase shift from reflections, but this is either 0 or π for the two planes of polarization, assuming a dielectric beam splitter, and therefore will not affect the results.

From the equation for $I = I_R I_T$, it is clear that minima (in fact, zeros) in the beam splitter efficiency occur when $\frac{1}{2}\phi = k\pi$, $k = 0, 1, 2, \ldots$, or at frequencies

$$\nu_{min} = \frac{k}{\sqrt{2}b(2n^2 - 1)^{1/2}}$$

Frequencies of maximum power occur midway between the zeros. Metallic mesh reflectors used as beam splitters display similar peaked behavior in their efficiencies even more pronounced than dielectric beam splitters. This property of Michelson interferometers means that beam splitters must be changed according to the spectral region being studied. This is not necessarily a disadvantage in the far infrared, because the beam splitters thereby also serve as filters to remove radiation of higher frequencies.

IV. FOURIER TRANSFORM SPECTROSCOPY

In the previous section we described a method for extracting a spectrum from the output of a two-beam interferometer. The method involves the use of a tuned amplifier or filter which samples the signal piecemeal, one band of frequencies at a time. Thus *at any given moment*, most of the information in the output is discarded, with only the information contained in one narrow band of spectral frequencies actually being used. This is wasteful in exactly the same way that the conventional monochromator wastes most of the information in the beam of radiation falling on the slit jaws, except for the narrow band passed by the slit opening. Of course, the interferometer has an advantage in luminosity over the conventional monochromator and can better afford wasteful habits, but nevertheless it would be very nice if information were not wasted at all. As it turns out, this is quite feasible provided the spectroscopist is willing to accept a little inconvenience in obtaining a spectrum. The method of Fourier spectroscopy, which we describe briefly in this section, has been reviewed in detail by Vanasse and Sakai (6) and by Gebbie and Twiss (7). The reader may also consult Mertz's (8) book. Fourier spectroscopy is proving to be a particularly valuable tool in the far infrared where the signal-to-noise ratio with conventional spectrometers is very poor.

In Section 4.V we described an *interferogram*, which is the output of an interferometer as a function of optical path difference. The interferogram, we said, is the Fourier transform of the spectrum of the input radiation plus a constant that is the integral of the input

$$S(\xi) = \tfrac{1}{2}P_0 + \tfrac{1}{2}\int_0^\infty P(\nu) \cos 4\pi\xi\nu \, d\nu$$

The spectrum itself can be found by computing the inverse Fourier transform of the interferogram,

$$P(\nu) = \tfrac{1}{2}\pi \int_{-\infty}^\infty F(\xi) \cos 4\pi\xi\nu \, d\xi \tag{8.15}$$

This inversion process is exactly what is done by the tuned amplifier of the previous section. The tuned amplifier extracts the spectrum one frequency at a time when the interferometer scans periodically. We now see that the entire interferogram can be recorded as the interferometer makes a single scan, and the spectrum extracted by computation at our leisure. The importance of this is that all information about the spectrum is under observation at all times and none is wasted. This advantage, usually called the *Fellgett advantage* because it was first recognized by Fellgett (9), can be discussed quantitatively, but let us defer that briefly.

The computation of the spectrum from the interferogram can be carried out, in principle, from Eq. (8.15). In practice, the equation is modified in two important respects. In the first place, the infinite limits of integration are not realistic, but must be replaced by the finite limit of mirror movement. In the second place, the integration is replaced by a sum over finite increments in order that it can be done numerically with a computer. We thus arrive at the form

$$P(v) \approx \tfrac{1}{2}\pi \sum_{n=-N}^{N} F(n\Delta\xi) \cos(4\pi n\Delta\xi v)\, \Delta\xi$$

$$= \pi \sum_{n=0}^{N} F(n\Delta\xi) \cos(4\pi n\Delta\xi v)\, \Delta\xi \qquad (8.16)$$

where advantage has been taken of the symmetry of $F(n\,\Delta\xi)$.

Let us decide how far apart the sampling points in the interferogram can be; that is, what is the maximum value of $\Delta\xi$? The highest frequency which we shall encounter in the cosine term corresponds to the highest spectral frequency in the spectrum, v_{max}. The cosine should be sampled at least twice per period. This can be shown by the methods of information theory, but we can show the plausibility of this contention by referring to Fig. 8.11. Sampling once per period does not characterize the cosine at all. If we accept the twice-per-cycle criterion, we can write at once from the argument of the cosine

$$\Delta\xi = 1/4v_{max}$$

Thus the frequency of sampling is determined by the highest spectral frequency present in the spectrum.

Not only is the calculation of the spectrum from Eq. (8.16) performed from discrete points in the interferogram, but if it is done on a digital computer, the calculations are made at a discrete set of spectral frequencies. The final plotted spectrum is constructed from the calculated

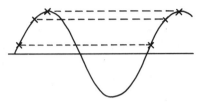

Fig. 8.11   Sampling frequency in Fourier spectroscopy.

points, with intermediate points obtained by interpolation. Just as the cosine of the interferogram is completely characterized by sampling twice per period, the spectrum is completely characterized by sampling twice per spectral element $\Delta\nu$, where $\Delta\nu$ is the resolution limit of the instrument. The resolution, as we know [Eq. (8.13)], is determined by the maximum path difference x_{max} used in recording the interferogram. Hence, the spectrum is calculated at frequency intervals

$$\Delta\nu_{sampling} = \tfrac{1}{2}\Delta\nu = 1/4x_{max}$$

Now we return to the Fellgett advantage. The spectrum presented by a conventional spectrometer is also completely characterized by points taken at $\frac{1}{2}\Delta\nu$ intervals. The argument is the same whether the instrument scans uniformly or stepwise from one point to the next. If a time Δt is required to register the information from each point, then the time necessary to scan a spectral region of width $\nu_{max} - \nu_{min}$ is

$$T = 2\frac{(\nu_{max} - \nu_{min})}{\Delta\nu}\Delta t = 2N\Delta t$$

where

$$N = \frac{(\nu_{max} - \nu_{min})}{\Delta\nu}$$

is the number of spectral resolution elements in the spectrum. An interferometer, on the other hand, records all spectral elements simultaneously. Each spectral element is observed for only half the time because of the destructive interference implicit in interferometry. Therefore, aside from computation time, the interferometer required only the time $2\Delta t$ to gather all the necessary information to reproduce the spectrum. This is a speed advantage of N in favor of the interferometer — which is not an altogether fair statement since the scanning rate is really limited by the response time constant of the detector–amplifier–recorder system.

However, it is fair to consider recording an interferogram and a spectrum in the same total time. In this case, there is available a factor of $\frac{1}{2}N$ more time for obtaining information about each spectral element in the interferogram than in the spectrometer record. From statistics or information theory this means an advantage of $(\frac{1}{2}N)^{1/2}$ in signal-to-noise ratio with the interferometer. This is equivalent to the same increase in effective luminosity of the instrument and is sometimes called the

$$\text{Fellgett factor} = (\tfrac{1}{2}N)^{1/2}$$

The statistical Fellgett factor is in addition to the luminosity advantage provided by the geometry of the interferometer.

It should be stressed that the $(\frac{1}{2}N)^{1/2}$ advantage in signal to noise is enjoyed only when using a detector whose noise level is independent of the signal level. This is true of most thermal detectors used in the mid and far infrared, such as the thermocouple, bolometer, or Golay detector. But it is not true of quantum detectors, such as photoemitters and photoconductors. We discuss detectors in some detail in Chapter 11. For the moment, we point out that the shot noise generated by a quantum detector is proportional to the square root of the total radiant power falling on it. Because this is on the average a factor of $(\frac{1}{2}N)^{1/2}$ greater in the interferometer than in the spectrometer, it exactly cancels the Fellgett factor.

The apparatus function and apodization requirements are the same in Fourier spectroscopy as in periodic interference modulation. However, it is possible to apodize *a posteriori* by multiplying the interferogram by an apodizing function just before computations of the spectrum.

It would take us rather far afield to discuss in detail the procedure for computing the spectrum from the interferogram (*10, 11, 12*). It is necessary to extract information from the interferometer at regular well-defined path lengths. The path-length points can be fixed by using a precision lead screw and nut or a stepping motor to advance the interferometer, by producing a subsidiary fringe pattern with the primary interferometer or another coupled interferometer using a monochromatic source, or by using a Moire fringe system. The radiant power measured at the points defined in this way are converted to digital form. They can be punched on cards, or stored on tape and later read into a general-purpose computer which computes the spectrum (*13*). Or the values can be retained and treated by a special-purpose computer built into the interferometer (*14*).

V. The Lamellar Grating Interferometer

A major problem is encountered in the Michelson interferometer because of the properties of the beam splitter. The beam splitter is called upon to divide the amplitude of a wave front and channel it into two separate beams. If the division is not equal, the contrast of the fringes does not suffer, but the optical power reaching the detector does. An alternate approach is to divide the wave front itself into two parts. One instrument for doing this is the lamellar grating sketched in Fig. 8.12. It consists of two interleaving sets of bars with polished and aluminized reflecting surfaces. One of the two sets can be moved with respect to the other. The lamellar grating is mounted in what appears to be a conventional Littrow arrangement (Fig. 8.13). The plane wave front approaching the grating is divided into two parts, half reflected by one set of bars and half by the other. A phase difference of

$$\Delta\phi = 2\pi \cdot 2\xi/\lambda$$

is thus introduced between the two halves, where ξ is the depth of separation between the bars. When the two sections of wave front are recombined, they interfere with varying degrees of destructivity as ξ is changed. The device is seen to function much as a Michelson interferometer does. In fact, we can imagine it to be equivalent to the idealized amplitude-dividing interferometer illustrated in Fig. 8.14 in which the fixed mirror A is 50% reflecting and 50% transmitting and the movable mirror B is 100% reflecting. Note that the two sets of bars of the lamellar grating can be made coplanar and ξ can be made either positive or negative. The mirrors of the idealized Michelson interferometer, on the other hand, cannot be made to pass through each other, and they cannot even be made coplanar. Of course, the

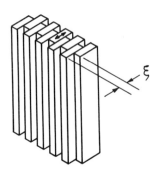

Fig. 8.12 Lamellar grating.

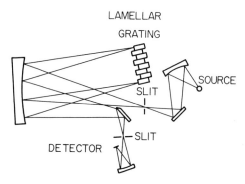

Fig. 8.13 Lamellar grating interferometer.

mirrors of the actual Michelson interferometer can, in effect, pass through each other.

There is an important difference between the Michelson interferometer and the lamellar grating in addition to the efficiency of beam splitting. The Michelson interferometer forms circular fringes of successive orders of interference. A circular diaphragm is used to select the central fringe. The lamellar interferometer is after all a grating, and it forms straight fringes of successive orders of diffraction. A straight slit is used to select the central fringe, or zero-order diffracted beam. The Michelson interferometer has greater inherent luminosity than the lamellar grating because of the rotational symmetry of its diaphragm. On the other hand, the acceptance angle and area of the detector often determine the limiting aperture, and in this case, the greater beam splitting efficiency of the lamellar grating interferometer makes it superior. This is especially likely at very long wavelengths of the very far infrared ($\lambda > 200\,\mu$).

Let us discuss the mathematics of the lamellar grating in a little more detail, as it does not obey quite the same diffraction relationship as the ordinary ruled grating. Refer to Fig. 8.15: The width of each bar is a and the depth of one set with respect to the other is ξ. Two rays

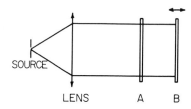

Fig. 8.14 Michelson interferometer equivalent to lamellar grating interferometer.

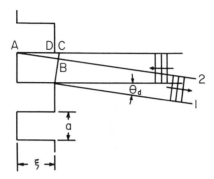

Fig. 8.15 Diffraction by a lamellar grating.

are incident normally on adjoining bars a distance a apart, and are diffracted at angle θ_d. Ray 2 travels a distance $\xi + \overline{AB}$ farther than ray 1. Now

$$\overline{AB} = \overline{AC} \cos \theta_d \quad \overline{AC} = \xi + \overline{DC} \quad \overline{DC} = a \tan \theta_d$$

So, in order for rays 1 and 2 to constructively interfere, we must have

$$n\lambda = a \sin \theta_d + \xi(\cos \theta_d + 1) \qquad n = 0, \pm 1, \pm 2, \ldots \qquad (8.17)$$

When the bars are coplanar, $\xi = 0$ and Eq. (8.17) reduces to the usual diffraction grating equation. For radiation diffracted normal to the grating, the equation becomes

$$n\lambda = 2\xi \qquad (8.18)$$

which is the usual two-beam interferometer equation. Notice that n and ξ have the same sign. For angles θ_d not too large, Eq. (8.17) can be written approximately,

$$n\lambda \approx a \sin \theta_d + 2\xi \qquad (8.19)$$

Now Eq. (8.18) gives us the depth ξ corresponding to a certain order of interference of a specific wavelength in the central order. The same wavelength occurs with order $n + k$ at an angle found from Eq. (8.19)

$$(n+k)\lambda - 2\xi = (n+k)\lambda - n\lambda = a \sin \theta_d$$

or

$$k\lambda = a \sin \theta_d \qquad k = \pm 1, \pm 2, \ldots$$

It also occurs with order $n + k + \frac{1}{2}$ (i.e., with destructive interference) when

$$(k+\tfrac{1}{2})\lambda = a \sin \theta_d$$

Similarly, when λ suffers destructive interference in the central order

$$(n+\tfrac{1}{2})\lambda = 2\xi$$

and in the side orders

$$(n+\tfrac{1}{2})\lambda \approx a \sin \theta_d + 2\xi$$

We can show that destructive interference occurs simultaneously at

$$\sin \theta_d = k\lambda/a$$

and constructive interference at

$$\sin \theta_d = \frac{(k+1/2)\lambda}{a}$$

If a lamellar grating interferometer is illuminated with monochromatic radiation and ξ is uniformly changed, the distribution of diffracted radiation will change alternately from the pattern of Fig. 8.16A to that of Fig. 8.16B. This is no different from the behavior of the circular fringes of the Michelson interferometer. A more rigorous theory, taking into account the effects of diffraction on the intensity distribution, would yield patterns more like those sketched in Fig. 8.17A and B.

The same relationship between maximum path difference and resolution holds as in the case of the Michelson interferometer, and so do the considerations of sampling frequency.

VI. THE SISAM: SPECTROMETRE INTERFERENTIAL A SELECTION PAR L'AMPLITUDE DE MODULATION

In 1957, P. Connes (*10*) disclosed an instrument which makes use of both the dispersion properties of diffraction gratings and the inter-

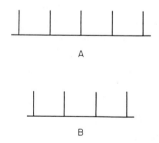

Fig. 8.16 Diffraction orders of a lamellar grating.

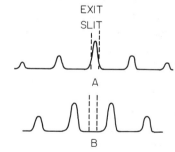

Fig. 8.17 Diffraction orders with slit effects.

ference modulation properties of a two-beam interferometer. The instrument, sketched in Fig. 8.18, is a Michelson interferometer. Its mirrors are replaced with identical diffraction gratings, or alternatively, with prisms aluminized on the back surface. The gratings are rotated in unison by means of a steel tape which couples their motion. As the gratings are rotated, one of them is moved along the axis of the interferometer to vary the path length of one beam. In practice this motion would present severe mechanical difficulties, so Connes provided an equivalent translational motion by oscillating a compensating plate in one beam.

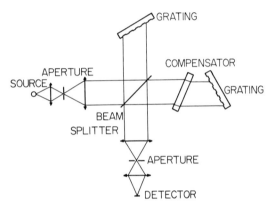

Fig. 8.18   Connes' SISAM spectrometer.

Each grating disperses a spectrum across the exit aperture. The angle of incidence on the grating is the same, so each grating causes the same wavelength to fall on the center of the aperture. But the orientation of the gratings is such that the spectra are dispersed in the opposite sense. Therefore, when the path length is varied, the radiation at the center of the aperture is strongly modulated, but the modulation falls off very rapidly for radiation passing through the aperture offcenter. Thus the exit aperture can be dispensed with altogether. Radiation of shorter wavelength from higher diffraction orders is also modulated, of course, but at a higher modulation frequency, so it is easily filtered by the electronic circuitry.

Thus we see that the SISAM produces a spectrum directly, as does the periodically modulated Michelson interferometer, with no computer needed to reduce the data. Consequently, the SISAM does not have the advantage of the Fellgett factor. It does, however, have the same luminosity advantage as the Michelson interferometer. For equivalent aperture size and amplifier filtering characteristics, it achieves higher resolution than the Michelson because of the selective modulation of the coincident wavelengths from the two gratings. It is as though we had a grating monochromator operating with very wide slits and, therefore, with high optical power, but somehow with the resolution associated with very narrow slits.

The apparatus function (slit function) of the SISAM has the form $\sin \theta/\theta$ which, as we have seen, has very strong side lobes. Therefore, apodization is necessary. This is done optically by placing appropriately shaped masks at the exit pupil. A diamond-shaped pupil function gives the SISAM; the same $(\sin \theta/\theta)^2$ apparatus function as a diffraction-limited monochromator with rectangular aperture. Some loss in resolution is suffered, of course.

REFERENCES

1. M. Francon, *Optical Interferometry*, Academic Press, New York, 1966.
2. P. Jacquinot, "New developments in interference spectroscopy," *Rept. Progr. Phys.*, **23**, 267 (1960).
3. R. G. Greenler, *J. Phys. and Radium*, **19**, 375 (1958).
4. J. Strong and G. Vanasse, *J. Phys. and Radium*, **19**, 192 (1958).
5. L. Genzel and R. Weber, *Z. Angew. Physik*, **10**, 127, 195 (1958).
6. G. A. Vanasse and H. Sakai, "Fourier spectroscopy," *Progress in Optics*, Vol. VI, E. Wolf, ed., Wiley, New York, 1967.
7. H. A. Gebbie and R. Q. Twiss, "Two-beam interferometric spectroscopy," *Rep. Progr. Phys.*, **29**, Pt. II, 729–756 (1966).
8. L. Mertz, *Transformations in Optics*, Wiley, New York, 1965.
9. P. Fellgett, *J. Phys. and Radium*, **19**, 187 (1958).
10. J. W. Cooley and J. W. Tukey, *Math. Comp.*, **19**, 297 (1965).
11. M. L. Forman, *J. Opt. Soc. Am.*, **56**, 778 (1966).
12. H. L. Buijs, *Appl. Opt.*, **8**, 211 (1969).
13. H. Sakai, G. A. Vanasse, and M. L. Forman, *J. Opt. Soc. Am.*, **58**, 84 (1968).
14. J. E. Hoffman, *Appl. Opt.*, **8**, 323 (1969).
15. P. Connes, *J. Phys. and Radium*, **19**, 215 (1958); *Opt. Acta* **4**, 136 (1957).

CHAPTER 9

Mechanics of Infrared Spectrophotometers

I. INTRODUCTION

In the preceding six chapters we have discussed the elements of geometric and physical optics, optical components, and optical systems. We now turn to the methods used in mechanically mounting, adjusting, and moving these optical components in an infrared spectrophotometer.

II. THE KINEMATIC MOUNT

The principle of kinematic mounting is of fundamental importance in the construction of optical instruments. The objective is to mount an object, such as a mirror or base plate, in such a way that: (1) it has only one equilibrium position; (2) it is stable with respect to small displacements; that is, if displaced slightly, it returns to its original position; (3) it can be removed and returned to its original position; and (4) its orientation can be adjusted with respect to certain movements, as, for example, rotations about two axes and a translation in a direction perpendicular to the two axes (see Fig. 9.1). Let us consider these requirements separately.

The requirement of a unique equilibrium position is best met by

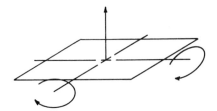

Fig. 9.1   Rotation about two axes and translation in one direction.

mounting on three points with the center of pressure or gravity well inside the perimeter of the three points; for example, a four-legged chair rocks, but not a three-legged stool. In addition to preventing rocking, however, we must also fix the positions of the three points. Let us see how this can be done by using balls, cones, slots, and flats. Let us consider that object A is to be mounted with respect to plane B (Fig. 9.2). The supports for the object are in the form of balls. A conical hole is milled into B at the point where ball 1 is to be fixed (Fig. 9.3).

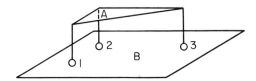

Fig. 9.2   Mounting object A relative to plane B.

The ball contacts the wall of the cone in a circle. The center of the ball is fixed, but the ball and the object are still free to rotate. This rotation might be prevented by providing a cone for another of the balls to rest in, but the mechanical tolerances would be extreme: the distance between the centers of the cones would have to equal exactly the distance between balls. This difficulty is easily avoided by positioning

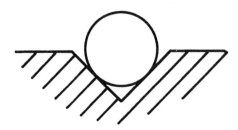

Fig. 9.3   Ball and cone.

ball 2 in a V-shaped slot in surface *B*. The axis of the slot lies more or less in the direction of the cone, as in Fig. 9.4, but the tolerance is not strict. The ball contacts the walls of the slot at two points. Clearly, the possibility of rotation about ball 1 has now been removed and there is only one equilibrium position. Ball 3 contacts *B* at a single point. The surface of *B* is polished in the vicinity of the ball, but otherwise it can be rough except for the walls of the cone and the slot.

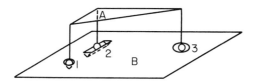

Fig. 9.4   Cone, slot, and flat kinematic mounts.

The second condition regarding stability to small displacements is also satisfied, because object *A* can be moved only by allowing ball 1 to climb partly out of its cone or ball 2 out of its slot. Gravity or a restoring spring promptly pushes the balls back in place, provided there is no undue amount of friction to overcome. The balls and milled surfaces are smooth to reduce friction, and they are made of a hard material (e.g., steel). If *B* is constructed of a soft material or one which is difficult to machine, steel pads with cones, slots, and flats can be cemented in place on *B*. The walls of the cone and slot are steep to provide a large restoring force for a small displacement—45° from the axis is a good choice.

The third condition is also evidently met, because object *A* can be lifted from surface *B* and replaced in exactly the same equilibrium position. Finally, the fourth condition on adjustability can be satisfied by mounting the balls on threaded shafts which are screwed through threaded holes in *A* (Fig. 9.5). Rotation about an axis through balls 1 and 2 can be produced by turning the screw on which ball 3 is mounted. Rotation about the axis through balls 2 and 3 is accomplished with screw 1. Translation of the entire object is produced by turning all three screws by the same amount. The screws generally have fine threads when fine adjustments are required. They must fit tightly into the threads of the mount. Very often the mount itself is not threaded, but threaded nuts are inserted into the mount. For most satisfactory performance, the balls are machined on the ends of the screws so that their centers lie on the axis of the screws. Separate balls can be cemented on the ends of the screws if some nutation of the center of the ball

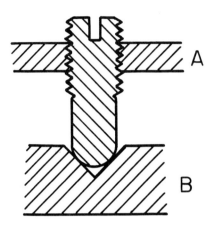

Fig. 9.5   Ball and cone with translation.

with screw rotation can be tolerated. Steel ball bearings and epoxy cement make a good combination.

The arrangement we have discussed is only one typical form of kinematic mount. Other forms can be used. In Fig. 9.6 we show some of the alternate arrangements. Form A is the one we have been describing. Forms B and C are obvious variations of A. Variations of form D are sometimes used. In this form, rotation about axis α is produced with screw 3, but rotation about axis β is inconvenient as it requires adjustment of both screws 1 and 2. Form E and its variations make use of three cones rather than a cone–slot–flat combination. In this form the mounted object is constrained from rotating by the parallel slots 1 and 2, but it is free to translate along their axes. Slot 3 completes the kinematic mount by preventing the translation.

There are many examples of kinematic mounts in infrared spectrophotometers. A large casting is used as the basic structure on which the instrument is built-up. However, the alignment of certain portions of the instrument, specifically the monochromator, is very critical. The basic casting suffers slight distortions that can vary from time to time, as when a spectator leans on the instrument or a large accessory is bolted to it. To avoid disturbing the monochromator when this happens, the components of the monochromator of a large instrument, including the slits, are frequently mounted on a separate smaller casting that is kinematically mounted on the large casting. Distortions of the large casting are not transmitted to the monochromator. The grating and mirrors are kinematically mounted on their holders, and these are described in somewhat more detail in later sections. In prism mono-

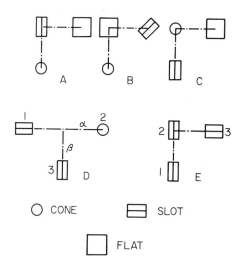

Fig. 9.6 Various kinematic mounting configurations.

chromators, the prism is clamped on a table that is kinematically mounted on the inner casting, thus making it possible to adjust the alignment of the prism and to replace it easily when other spectral regions are to be studied. Often, in more complex instruments, an interchange casting containing the grating or prism mount, scanning mechanism, cams, Newtonian mirrors, and so on, is kinematically mounted on the inner casting. The entire interchange casting is removed and replaced by a different one when other spectral regions are studied.

Some large spectrophotometers are evacuated to remove the disturbing influence of atmospheric water vapor and carbon dioxide. Obviously, the distortion of an evacuated chamber can be extreme, so the entire instrument is mounted kinematically inside the vacuum chamber.

III. MIRROR AND GRATING MOUNTS

Kinematic mounts of the types just described are used in mounting mirrors. They provide rotation about two axes for alignment and also translation for focusing adjustments. The cones and slots can be cut into the back surface of the glass mirror blank itself, or steel pads can be cemented to the back of the mirror, or the mirror can be attached to a metal plate that contains the cones and slots. Spectrophotometer

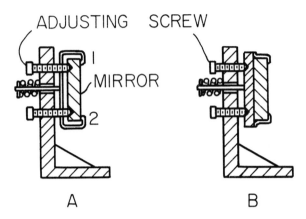

Fig. 9.7 Typical kinematic mirror mounts.

mirrors are usually mounted vertically, so a force must be provided to hold the balls in their seats. Usually a spring of some sort is used. Two typical mounts are sketched in Fig. 9.7. In Fig. 9.7A, the mirror fits loosely in a sheet-metal wraparound box. Protrusions at 1 and 2 press against the mirror opposite the adjusting screws. This is necessary to avoid distorting the mirror. The adjusting screws penetrate through holes in the sheet metal and the mirror is held against the screws with a spring. In Fig. 9.7B the mirror is attached to a metal plate, which is held against the screws by a spring. If the mirror needs to be adjusted by rotation about an axis perpendicular to its surface, as when an aspheric mirror is used, the plate containing the adjusting screws can be attached to the mirror mount with a large screw on which it can be rotated. More often, the slots and cones are milled into the mirror during manufacture so that the axis is sufficiently well defined to eliminate the need for rotation adjustments.

A diffraction grating in a spectrophotometer presents a special problem. It must be adjustable about two axes lying in its surface in order to properly return the radiation beam to the collimator. But, in addition, the rulings must be parallel to the axis of rotation of the grating. A typical arrangement for accomplishing this is sketched in Fig. 9.8. Screw 1 is seated in a cone and is used to tilt the face of the grating. Ball 3 and screw 2 are seated in slots. Ball 3 is attached to a spring. Screw 2 serves to adjust the rulings parallel to the axis of rotation. The grating is rotated about the two large balls 4 and 5, which are seated in cones.

The wavelength calibration of a prism monochromator changes with

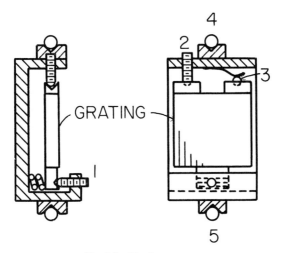

Fig. 9.8   Grating mount.

temperature because of changes in the refractive index of the prism material. The temperature of infrared monochromators is often regulated to minimize wavelength errors (and also to control the relative humidity when hygroscopic optical materials are used).

In addition, temperature compensation is sometimes used. A monochromator mirror, usually the Littrow or Newtonian, is mounted so that its angle of incidence is controlled by thermal expansion of a compound bar or spring. A proper choice of material and geometry permits good compensation of wavelength shift over a range of many degrees.

IV. SPECTRAL SCANNING METHODS

In order to scan a spectrum with a monochromator, it is necessary to rotate the grating or the Littrow mirror. Early spectroscopists did this manually by reading the angular setting from a graduated circle or drum and referring to a calibration table to determine wavelength. With a grating, the wavelength can be calculated easily from the spacing of the grating, the geometry of the instrument, and the measured angle of incidence. In a prism instrument the calculation is less direct, and standardization relative to lines of known wavelength is the usual procedure. When the early instruments were adapted for use with pen-and-ink recorders, the grating or mirror was rotated at a uniform rate, usually with a synchronous motor and worm gear arrangement. A microswitch on the drive system was made to close in synchronization with the rotation, thus discharging a capacitor and causing a series of pips to appear on the spectrum. These served as markers from which the wavelengths were eventually determined. Many special instruments continue to be operated in this way even today, but the necessity of referring to a calibration chart makes this method much too slow and tedious for routine use.

In modern commercial instruments the calibration is built into the instrument. The grating or Littrow mirror is rotated by a cam that is fabricated in such a way that it provides a spectrum linear in either wavelength or wavenumber. The contour of the cam is computed from the geometry of the instrument and the refractive index of the prism material or line spacing of the diffraction grating. A master cam is carefully machined and finished by hand until it reproduces the positions of known spectral lines within the required tolerances. Production cams are compared with the master.

The calculation of a cam contour is not as straightforward as one might expect. It is relatively easy to calculate the rise required for a given rotation of the cam. But the cam follower itself introduces complications. The cam follower is usually a ball or a roller that rides on the edge of the cam under a light spring force. In Fig. 9.9 we see that the line from the center of rotation of the cam to the point of contact OC is not necessarily coincident with the line from the center of the follower to the point of contact FC. It is the locus of the center of the follower whose rise is calculated for given cam rotation, and the relationship between this rise and cam shape is complicated. Some grating instru-

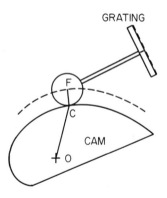

Fig. 9.9 Cam arrangement for rotating a grating.

ments use a grating in two or more orders. The same cam can be used for all orders, giving a scale expansion in the ratio of the order numbers. Alternatively, a separate cam lobe can be used for each order. If more than one grating is used, a separate cam lobe must be made for each grating. In addition, provision must be made for positioning the proper grating in the beam. This is usually done by kinematically mounting each grating on a turret that can be rotated to expose the desired grating. The entire turret is rotated by the wavelength-scanning mechanism. In double monochromators the dispersing elements must be properly coupled so that the same wavelength is selected by each monochromator. A prism–grating monochromator has a separate prism and grating cam coupled together and rotated by the same motor.

It is possible to take advantage of the form of the diffraction equation for gratings in constructing a special kind of cam known as the sine-bar linkage. The equation, we recall, is

$$n\lambda = 2a \sin \theta_i \cos \alpha \quad \text{or} \quad \lambda \propto \sin \theta_i \qquad (9.1)$$

If the grating is mounted with reference to a triangle in such a way that the side opposite θ_i is changed uniformly while the hypotenuse is held fixed, then $\sin \theta_i$ and λ will change uniformly. A typical arrangement for doing this is shown in Fig. 9.10. A nut N is advanced uniformly by rotation of a lead screw S. The bar AB is carried along by riding on another bar D attached to the nut, and as it moves it rotates the grating about A. The length of the bar is fixed, and as the nut moves the length of line AC is changed uniformly. The wavelength λ is thus scanned uniformly. Note that the bar BD is actually a straight cam and that the end of bar BA is a cam follower. The tolerances on BD are every bit as strict as the tolerances on the cams discussed earlier.

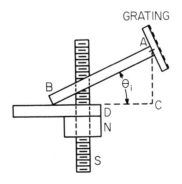

Fig. 9.10 Sine-bar linkage for wavelength scanning.

The sine-bar linkage of Fig. 9.10 offers a means of scanning wavelength uniformly. A slight modification permits scanning wavenumber uniformly, which is the more usual requirement for infrared grating instruments. Equation (9.1) is rewritten in the form

$$n/\nu = 2a \sin \theta_i \cos \alpha \quad \text{or} \quad \nu \propto \csc \theta_i$$

Hence, in order to scan wavenumber uniformly, we wish to hold the side opposite θ_i fixed and vary the hypotenuse. The cosecant-bar linkage sketched in Fig. 9.11 accomplishes this. The mechanical limit of either the lobed cam or the sine-bar linkage lies in eccentricities of gears or lead screws and the stick–slip action of sliding contacts.

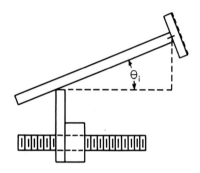

Fig. 9.11   Cosecant-bar linkage for wavenumber scanning.

A very elegant wavelength-scanning method has been devised by Shearer and Wiggins (*1*) for use in a high-resolution monochromator used in conjunction with a Fabry–Perot etalon. It is based on an optical scanning system used in range finders; its principle is illustrated in Fig. 9.12A. Rays of neighboring wavelength are intercepted by a prism of small wedge angle and diverted to the exit slit S_X. With the grating

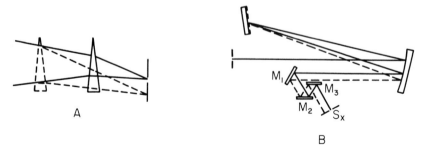

Fig. 9.12   Scanning method for a high-resolution monochromator.

fixed, the prism is translated in the direction shown, with the result that different wavelengths are selected at the slit. The grating is rotated stepwise and the motion of the prism is repeated. The advantage of the device is that a large motion of the prism introduces a small motion across the slit. Therefore, a very smooth scanning action is available over a small wavelength increment. A later variation (2) is shown in Fig. 9.12B. A set of mirrors is substituted for the prism, making the device achromatic.

V. SLITS

We usually think of monochromator slits as though they are cut in an infinitesimally thin plane surface. This is an ideal, of course. Actually slit jaws are fabricated of sheet material of finite thickness and the edges are beveled sharp. The correct and incorrect arrangements of the bevel are shown in Figs. 9.13A and B, respectively. When the entrance slit is arranged as in Fig. 9.13B, radiation from the beveled edge is reflected into the monochromator where it can add to stray

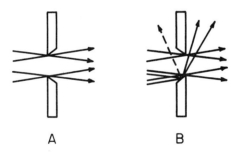

A   B

Fig. 9.13   Correct (**A**) and incorrect (**B**) illumination of slit jaws.

radiation. If the exit slit is arranged this way, radiation outside the desired wavelength interval can be reflected through the slit and detected, or it can be scattered back into the monochromator. The edge of the slit jaw must be curved, as in Fig. 9.14, to match the slit image distortion at some compromise wavelength. Moreover, the slit jaw might be bent to compensate for field curvature, though this is only rarely done.

Slit jaws are usually fabricated from some readily machined metal. Brass is a good choice. A clever method of machining many slit jaws at once is to clamp a number of metal rectangles tightly together but

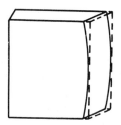

Fig. 9.14   Curved slit jaw.

tilted, as shown in Fig. 9.15. The assembly is cut and ground as shown by the dashed line. When separated, each piece is one-half of a slit jaw pair with convex curvature. The mating jaw, with concave curvature, is made with the same tool, but with the pieces tilted in the opposite sense. Notice that this operation gives an elliptical shape to the slit edges. This is an approximation to the parabolic shape called for by the theory, which is after all an approximate theory.

The slit jaws must be mounted in such a way that their separation is uniform along their length and can be changed easily and reproducibly. The motion of the jaws is usually bilateral, that is, the jaws are moved the same amount in opposite directions. Otherwise, the center of the slit opening would shift, thereby affecting the wavelength calibration of the monochromator. Many mechanisms for mounting slit jaws have been invented.

Our discussion is restricted to a few examples. In Fig. 9.16 is sketched the parallelogram mechanism (3) used, for example, in the Beckman IR-4 series of infrared monochromators. This mechanism is also suited to other monochromator arrangements, such as the Czerny–Turner, in which entrance and exit slits are mounted with their jaws parallel. Two parallel bars, B_1 and B_2, form two sides of a parallelogram; the other two sides are defined by two spring steel members, S_1 and S_2.

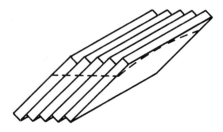

Fig. 9.15   Generation of curved slit jaws.

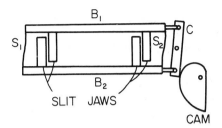

Fig. 9.16 Parallelogram slit mount.

One entrance slit jaw and one exit slit jaw are mounted on bar B_1, while the other two jaws are mounted on B_2. The bars are coupled through the crank C so that they move in opposite directions by the same amount when C is rotated. The slit jaw motion is thus bilateral and the opening of the entrance and exit slits is the same.

The slit mechanism (4) developed for the Perkin–Elmer Model 12 and 21 series is sketched in Fig. 9.17. In this monochromator the entrance and exit slits are physically close to one another, but not parallel. The two jaws of the exit slit are mounted on concentric shafts, one jaw on each shaft. One shaft is rotated by an attached arm, thus causing the edge of the slit jaw to move away from its original position. This shaft is coupled to the corresponding shaft of the entrance slit with a steel tape, so the shafts move simultaneously by the same amount. Similarly, the other two corresponding shafts are coupled with tape. In addition, a diagonal reversing tape couples one shaft of the exit slit to the noncorresponding shaft of the entrance slit so that they must rotate equal amounts in opposite directions. Consequently, when the driven shaft of the exit slit is rotated, the entrance and exit slit jaws move bilaterally in unison.

The Unicam instrument of Tarbet and Daly (5) uses a magnetic slit mount. Opposing slit jaws are mounted on separate loudspeaker-type coils. The coils are wired in series so that the slit jaws move the same amount in opposite directions.

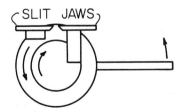

Fig. 9.17 Concentric shaft slit mount.

When an extended spectral region is to be recorded, it is necessary to adjust the slit width continuously to maintain a more-or-less constant level of radiant power falling on the detector. The slit width required at a given wavelength is determined by such factors as source output, monochromator dispersion, grating efficiency or prism absorption, overall optical efficiency, and detector sensitivity. In early Perkin–Elmer Model 12 spectrometers, the wavelength drive was coupled to a nonlinear cam mounted on the slit adjusting shaft. The cam was a tapered helix cut to program the slit width and the coupling was made with a string. Later instruments used lobed cams mounted on the wavelength drive shaft and some present-day inexpensive instruments continue to control the slits in this way. The trouble with a mechanical cam is its inflexibility. It is possible to introduce multiplying factors to allow variation of the resolving power, but the general shape of the slit program function is fixed. Therefore, the function gives at best an approximation to constant optical power, and variations in sources, detectors, etc., cannot be compensated. Accordingly, high-performance instruments usually use electrical programmers to control the slits; these are considerably more versatile. The discussion of the electrical details of this method is continued in Chapter 12. An electrical signal derived from the programmer can be used to drive a servomotor which moves the slits through a gear train or a mechanical cam. In the Unicam electromagnetic slit mechanism, no motor is required, as the coils can be excited directly from the programming signal.

VI. Attenuators, Baffles, and Shutters

Optical null spectrophotometers perform by equalizing the radiant power in two beams. This is done by automatic adjustment of an attenuator in the reference beam. A similar, but less precise, attenuator is manually adjusted to trim the sample beam and set the 100% T level. Numerous optical schemes can be devised for use as attenuators, but mechanical attenuators offer the advantages of simplicity, ease of coupling to a servomechanism, and, most important of all, attenuation nearly independent of wavelength. Mechanical attenuators are of two general types: apertures which modify the actual transmittance of a beam, by changing its area for example, and choppers which control the average transmittance over a time interval.

An aperture-type attenuator is used to vary the area of some aperture of the spectrophotometer, as viewed through one beam. The only two possibilities are the slit and the exit pupil of the monochromator, and the attenuator is placed at an image of one of these in the reference beam. We do not wish to modify the width of either the exit pupil or the slit, as this will affect the resolving power of the instrument. It will not do for one beam of a double-beam instrument to reproduce spectral lines more sharply than the other. Hence, we are left with changing the effective height of the slit or the disperser. One method for accomplishing this is by means of a blade moving along the length of the image, as illustrated in Fig. 9.18A. A better arrangement uses two blades moving bilaterally, as shown in Fig. 9.18B. An almost equivalent scheme uses a wedge opening moving laterally, as shown in Fig. 9.18C. There is a change in the rate of attenuation with attenuator travel when the point of the V reaches the edge of the radiation beam. The slit image is narrower than the disperser image so a wedged attenuator is best placed at a slit image. This point is discussed again in Chapter 14 in connection with photometric accuracy.

The single-opening attenuators of the types sketched in Figs. 9.18A, B, and C are usually not suitable for high-performance instruments for several reasons. The off-axis aberrations are affected when either the slit height or the disperser height is changed, and this affects the slit function for one beam. The beam may not be uniformly illuminated due to inhomogeneities in samples and cells or other optical elements, variations in source brightness, or faulty optical alignment. The attenuation will, therefore, not be a linear function of position, and photometric accuracy will suffer. Moreover, the slit is also usually imaged

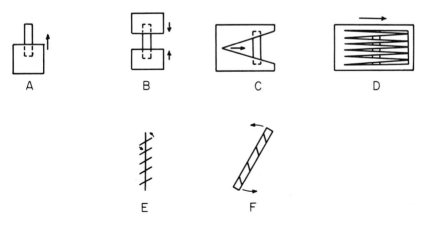

Fig. 9.18 Attenuators.

on the detector, and the sensitivity of the detector may be nonuniform. To avoid these problems, the attenuator is formed as a comb, shown in Fig. 9.18D. This gives a better sampling along the length of the slit image.

The comb can be photoetched from very thin metal sheet. It is mounted on a frame that travels laterally on a track, moved by a servo-motor through a gear reduction system. The comb is very delicate and should be protected against damage. Nicks in the edges of the teeth will show up on the spectral record when narrow slits are used. Even dust particles on the teeth can be observed, but these can be blown off with care.

An attenuator similar to the comb can be constructed in the form of a venetian blind. The attenuation (Fig. 9.18E) depends on the sine of the angle of rotation of the slats, but it can be linearized by driving the slats from a cam or a sine bar if desired. Such an attenuator has not been used as a servo-driven reference beam attenuator, but a simple form (Fig. 9.18F) is a very convenient accessory attenuator for use in the reference beam to expand the transmittance scale when a highly absorbing sample is in the sample beam. In this form, the slats are mounted rigidly in a frame and the frame is rotated for adjustment.

Transmittance-controlling attenuators based on polarization can be imagined. They would modify the transmittance of a beam without changing its area. This approach has been used in the visible region, but the inefficiency of infrared polarizers makes it less attractive for infrared spectrophotometers. Similarly, a wedge-shaped neutral filter, whose transmittance is independent of wavelength, might be inserted into the beam. However, substances whose transmittance is strictly constant over very large wavelength regions are unavailable. Finally, we mention in passing that the optical properties of semiconductors depend on the density of change carriers, though in a wavelength-dependent way. Attenuators (actually modulators) have been made of germanium, and the transmittance is varied with the applied voltage.

Attenuators of the second general type are rotating choppers or sector disks whose average transmittance is determined by the relative extent of open and closed portions during a complete rotation (see Fig. 9.19)

$$T = \frac{\theta_o}{\theta_o + \theta_c} \qquad (9.2)$$

If the disk is rotated fast enough, flicker disappears and the trans-mittance appears to the eye to be the average transmittance given by Eq. (9.2). The eye integrates and averages the light passed by the

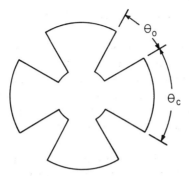

Fig. 9.19 Sector disk.

sector. This principle is called *Talbot's law*. Talbot's law can be generalized to detectors other than the eye. The question of how fast the disk must be rotated depends on the nature of the detector and its response time and that of the associated electrical and mechanical systems. This is discussed in greater detail in Chapter 14, but for the moment we merely state that it is easy to achieve the necessary chopping rate. An adequate rate of 360 chops per second is provided by a sector with twelve openings rotated at 1800 rpm.

A simple fixed sector of the type shown in Fig. 9.19 is, of course, limited to a single transmittance value. Two general methods for varying the transmittance without stopping the sector have been used. The first of these makes use of a star-shaped sector, shown in Fig. 9.20, mounted on a carriage in such a way that it can be moved into or out of the beam while it is rotating. The transmittance depends on how far

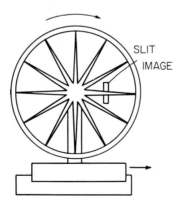

Fig. 9.20 Star wheel attenuator.

into the beam the sector is inserted. This sector attenuator has a significant advantage over the comb of being a nondiscrete attenuator, so local variations in beam illumination, detector sensitivity, or source brightness are of no importance. The linearity of the device still depends on the straightness of the edges of the sector, but it is difficult to imagine a dust particle clinging to an edge. Unicam (5) uses a sector attenuator of this type as the reference beam attenuator in their infrared spectrophotometers, with a similar sector used for sample beam trimming.

The second method for varying the transmittance of a sector is illustrated in Fig. 9.21. Two identical sectors are mounted and rotated concentrically. Each sector alone transmits 50%, but by varying the phase of one with respect to the other, the transmittance can be made to vary from 50 to 0%. The relative position of the two disks can be controlled by driving them at the same speed but with separate motors whose phases can be changed. Mechanical methods can also be used to vary the phases of the disks. Sectors of this sort have been used (6), or at least proposed, as optical attenuators for infrared spectrophotometers. The author has used a sector with disks adjustable through a differential gear train as a standard for establishing photometric linearity of infrared spectrophotometers, and as a manually adjusted reference beam attenuator for use in precise photometric measurements. Double disk sectors are limited to 50% transmittance. Measurements at higher transmittance can be made, of course, by attenuating the other beam of the spectrophotometer, but this is wasteful of energy. Variations of double disk sectors have been constructed with much lower loss. These are built on the order of a collapsible fan.

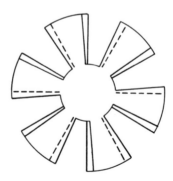

Fig. 9.21　Variable attenuation sector disk.

Baffles are used to confine radiation to desired regions. They control stray radiation by preventing radiation from reaching surfaces from which it can be scattered. Baffles in infrared spectrophotometers are usually apertures constructed of sheet metal or plastic and painted black. It must be remembered that no painted surface is absolutely black and that a surface which appears black to visible radiation might be a good reflector or transmitter of infrared. Black plastic tape, the type used for electrical insulation, is an example of a material that is excellent for constructing experimental baffles for use in the visible region, but should always be checked for transmission when used in the infrared. Even the best material scatters to some extent, and it is best to baffle a monochromator at an aperture image rather than inside the monochromator whenever possible.

Shutters are used to interrupt the radiation beam temporarily, as in establishing the zero transmittance level. Shutters can be constructed of metal or plastic (beware of transmittance in the infrared) and hinged on the instrument case or inserted in the sample compartment. Sometimes shutters are made of a material that transmits shorter wavelengths than the region under study. Thus, stray radiation, which is generally of shorter wavelengths, is not interrupted, and the zero level is compensated for stray radiation. Such shutters can be made of glass or an appropriate alkali halide.

VII. Radiation Choppers

Choppers in single-beam instruments are most easily made of sheet metal with radial openings. The chopper is mounted on a shaft and rotated with a synchronous motor, either directly or through gears or a timing belt. Contacts for mechanical or optical demodulators may be mounted on the same shaft. Sometimes, especially in far-infrared spectroscopy, the chopper blade is made of a material transparent to shorter wavelengths. Radiation of shorter wavelengths is, therefore, not chopped, and signals from stray radiation are thereby reduced.

When space is limited, as, for example, in double-passed mono-chromators, the chopper can be constructed as sketched in Fig. 9.22A by rotating a sheet-metal blade about its surface. A variation is the barrel chopper of Fig. 9.22B. Compact choppers have been constructed by using electrically driven tuning forks (7, 8).

In most double-beam instruments, the chopper is really a beam switcher and consists of a rotating segmented mirror. The optical

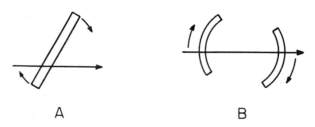

A   B

Fig. 9.22   Choppers.

alignment of this element is more critical than the ordinary chopper.

The rotation of choppers and mirrors necessarily introduces some vibration into the base of the spectrophotometer. Infrared detectors are usually sensitive to vibration; that is, they are *microphonic*. It is desirable to reduce the vibration to a minimum. In addition, the detector and also the preamplifier, if it is mounted on the instrument base, should be held in vibration damping mounts or pads.

REFERENCES

1. J. N. Shearer and T. A. Wiggins, *J. Opt. Soc. Am.*, **45**, 133 (1955).
2. K. N. Rao, C. J. Humphreys, and D. H. Rank, *Wavelength Standards in the Infrared*, Academic Press, New York, 1966, p. 152.
3. R. V. Jones, *J. Sci. Inst.*, **29**, 345 (1952).
4. J. U. White and M. D. Liston, *J. Opt. Soc. Am.*, **40**, 29 (1950).
5. C. S. C. Tarbet and E. F. Daly, *J. Opt. Soc. Am.*, **49**, 603 (1959).
6. J. E. Stewart, *Appl. Opt.*, **1**, 75 (1962).
7. H. G. Lipson and J. R. Littler, *Appl. Opt.*, **5**, 472 (1966).
8. F. Dostal, *Opt. Spectra*, **2**, No. 2, 29 (1968).

CHAPTER 10

Elements of Electronics

I. Introduction

We know that infrared detectors of various kinds respond to incoming optical signals by producing electrical outputs. The electrical signal is generally very feeble and must be amplified considerably before it can be recorded. Early in the process of amplification, the signal is embedded in strong noise. Care is necessary to ensure that the signal is extracted from the noise and sent along to the next stage of processing without introducing additional noise. Farther along in the processing, additional steps are taken to reject the noise as much as possible. When the signal is sufficiently robust, it is further treated by analog computation, in double-beam instruments, and transformed into a record of the ratio of sample beam power to reference beam power. All of this is done by the methods of modern electronics and servomechanisms.

Meanwhile, other electronic systems are busy at other tasks: controlling slit width to maintain constant reference beam power, regulating source brightness, monitoring the temperature of the instrument, and in some sophisticated spectrophotometers, adapting to the requirements imposed by rapidly changing signals by controlling the scanning speed or response time.

In this chapter we review some of the basic principles of electronics and control system theory. In the next two chapters, we specialize these concepts to the detectors and electronic systems used in modern infrared spectrophotometers. In Chapter 13, the effects of all these, especially the effect of the noise filter, in distorting the recorded spectra are discussed.

Components of electronic systems may be roughly classified as passive or active, depending on whether they require the addition of outside energy in order to act on an impressed electrical signal. In some cases, the distinction is not altogether clear. Resistors, capacitors and inductors are clearly passive because they modify an input signal only by attenuating it. There is no amplification, and no energy is added to the component in addition to the signal. A transformer amplifies a voltage, but no external energy is supplied, so it is considered to be passive. Switches and vacuum-tube rectifiers do require external energy, but we often consider them to be passive anyhow. Vacuum-tube and transistor amplifiers, on the other hand, are clearly active components.

II. Passive Components and Networks

A resistor acts as an impediment to the flow of electrons. A resistor can be constructed of a coiled length of wire, a metallic film, or a cylinder of compressed carbon particles bonded together. A potentiometer is a wire resistor with a movable contactor or wiper which can pick off a portion of a voltage applied across the two ends of the wire.

The relationship between the electrical current i (which by convention is in a direction opposite to the flow of electrons), the potential difference between the two terminals of the resistor v, and resistance R, is expressed by Ohm's law

$$v = iR$$

In the practical system of units v is expressed in volts, i in amperes, and R in ohms. The electrical power in watts dissipated by a resistor is

$$P = vi = i^2R = v^2/R$$

The resistance of several resistors in series is

$$R_{series} = \sum_i R_i$$

and of several resistors in parallel

$$R_{\text{parallel}} = \frac{1}{\sum (1/R_i)}$$

A pure resistor treats alternating current in the same way as it does direct current. That is, it reduces the potential according to Ohm's law, but it does not alter the functional form of a time-varying signal.

A rectifier is a device having a high resistance to current flowing in one direction and a low or negligible resistance to current flowing in the opposite direction. It has the ability to convert an alternating current to a pulsating direct current. A rectifier can be a vacuum-tube diode, a semiconductor diode, or a selenium or cuprous oxide device.

Capacitors serve as storage reservoirs for electrical charge. They are constructed of metal plates or foil separated by a dielectric material. The relationship between the stored charge q, the potential difference across the plate of the capacitor, v, and the capacitance C, is

$$v = q/C \qquad (10.1)$$

The unit of charge is the coulomb and capacitance is expressed in farads. Electrical current is the time rate of flow of charge

$$i = dq/dt$$

so Eq. (10.1) can be rewritten in the form

$$v(t) = \frac{1}{C} \int_{-\infty}^{t} \frac{dq}{dt'} \, dt' = \frac{1}{C} \int_{-\infty}^{t} i(t') \, dt' \qquad (10.2)$$

If the capacitor is originally uncharged (at $t = -\infty$), Eq. (10.2) can be written in an equivalent form by differentiation

$$\frac{dv(t)}{dt} = \frac{1}{C} i(t) \qquad (10.3)$$

The electrical energy in joules stored in a capacitor is

$$E = \tfrac{1}{2} vq = \tfrac{1}{2} C v^2$$

The capacitance of several capacitors in series is

$$C_{\text{series}} = \frac{1}{\sum_i (1/C_i)}$$

and in parallel

$$C_{\text{parallel}} = \sum C_i$$

If an alternating potential of frequency ω radians per second is impressed on a capacitor,

$$v(t) = V_0 \sin \omega t$$

then from Eq. (10.3), the current is

$$i(t) = C \frac{dv(t)}{dt} = CV_0\omega \cos \omega t = CV_0\omega \sin(\omega t + \tfrac{1}{2}\pi) \qquad (10.4)$$

Two important deductions can be made from Eq. (10.4). First, the *current leads the applied potential* by 90°. Secondly, the relationship between the current and potential can be written in a form resembling Ohm's law

$$v = i\left(\frac{-j}{\omega C}\right) = iX_C \qquad (10.5)$$

The quantity $X_C = -j/\omega C$ is called the capacitive reactance. The insertion of $j = \sqrt{-1}$ is a convenient way to describe the 90° phase difference between potential and current. The capacitive reactance is greatest for lower frequencies.

An inductor is a coil of wire surrounding either air or a magnetic material. When a time-varying current is passed through the wire, a potential is induced according to the relationship

$$v(t) = L \frac{di(t)}{dt} \qquad (10.6)$$

where L is the inductance in henrys. The energy stored in the magnetic field of the coil is given by

$$E = \tfrac{1}{2} L i^2$$

The inductance of inductors in series or parallel obey the same equations as resistors.

An alternating current of frequency ω

$$i(t) = I_0 \sin \omega t$$

produces an induced voltage, from Eq. (10.6),

$$v(t) = I_0\omega L \cos \omega t = I_0\omega L \sin(\omega t + \tfrac{1}{2}\pi)$$

Hence, in an inductor the *potential leads the current* by 90°. Note the contrast with the case of the capacitor. The relationship between the amplitude of current and potential is

$$v = i(j\omega L) = i \cdot X_L \qquad (10.7)$$

where $X_L = j\omega L$ is the inductive reactance.

A transformer consists of two independent coils wound about the same core. A change in the current in one induces a potential across the other and, therefore, a current through it. The first coil is the primary and the second one is the secondary. The ratio of the number of turns in the secondary to the number of turns in the primary is the turns ratio n. The ratio of ac voltage induced in the secondary to voltage impressed on the primary is

$$v_s/v_p = n$$

On the other hand, the ratio of currents is

$$i_s/i_p = 1/n$$

A generalized Ohm's law for an element containing resistance, capacitance, and reactance can be written by defining the *impedance Z*

$$v = Zi$$

with

$$Z = R + X_C + X_L = R - \frac{j}{\omega C} + j\omega L$$

The absolute value of the impedance is

$$|Z| = (Z\,Z^*)^{1/2} = [R^2 + (X_L - X_C)^2]^{1/2}$$

The impedance of components in series and parallel is found from the same equation as for pure resistors. Sometimes the relationship between potential and current is written

$$i = Yv$$

where Y is the *admittance*.

The analysis of networks containing passive elements is carried out by applying Kirchhoff's laws, which state that the sum of currents entering and leaving a point in the circuit is zero:

$$\sum_k i_k = 0$$

and that the sum of potentials around a circuit loop, including any sources such as batteries or generators, is zero:

$$\sum_k v_k = 0 \quad \text{or} \quad \sum_{\text{sources}} E = \sum_k Z_k i_k$$

Let us apply these laws to some examples. Consider first the simple *RC* circuit of Fig. 10.1. It consists of a source, a resistor, and a capacitor in series. Let us suppose that the source is a battery of potential E_0 and that it is connected at time $t = 0$. We would like to

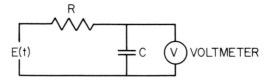

Fig. 10.1 Simple *RC* circuit.

know the voltage appearing across the capacitor as a function of time. An equation for the sum of the potentials around the loop is

$$E(t) = iR + \frac{1}{C} \int_0^t i \, dt \qquad (10.8)$$

or

$$\frac{dE(t)}{dt} = 0 = R\frac{di}{dt} + \frac{i}{C}$$

This differential equation is solved subject to the final condition

$$i = 0 \quad \text{at} \quad t = \infty$$

after the capacitor is fully charged. The solution is

$$i = \begin{cases} i_0 e^{-t/RC} & t > 0 \\ 0 & t < 0 \end{cases}$$

The potential measured across the capacitor is

$$v_c = \frac{1}{C} \int_0^t i \, dt = i_0 R (1 - e^{-t/RC})$$

$$= \begin{cases} E_0(1 - e^{-t/RC}) & t > 0 \\ 0 & t < 0 \end{cases} \qquad (10.9)$$

If we had started with the capacitor fully charged to potential E_0, and applied a short circuit across the battery at time $t = 0$, we would find

$$v_c = \begin{cases} E_0 e^{-t/RC} & t > 0 \\ E_0 & t < 0 \end{cases} \qquad (10.10)$$

In either case, the time required for the capacitor to arrive at $1/e$ (37%) of its final potential is

$$\tau = RC$$

The time τ is the *time constant* of the circuit.

If instead of a steady potential, we apply an impulse at time $t = 0$, given in terms of the delta function

$$E(t) = E_0 \delta(t)$$

the solution for the potential across C is

$$v_c/E_0 = \begin{cases} e^{-t/RC} & t > 0 \\ 0 & t < 0 \end{cases}$$

This is the impulse response of the circuit. Finally, if the applied potential is an alternating voltage applied at time $t = 0$

$$E(t) = \begin{cases} E_0 \sin \omega t & t > 0 \\ 0 & t < 0 \end{cases}$$

the solution for the potential on the capacitor is

$$\frac{v_c}{E_0} = \frac{\omega e^{-t/RC}}{\omega^2 + (1/RC)^2} + \frac{1}{[(1/RC)^2 + \omega^2]^{1/2}} \sin(\omega t - \phi) \qquad (10.11)$$

where $\phi = \tan^{-1} \omega RC$.

The term in $e^{-t/RC}$ is the transient contribution, which eventually dies out. The second term represents the steady-state behavior. We see at once that a sinusoidal input results in a sinusoidal output, but with a phase lag depending on the frequency and on the parameters of the circuit. Moreover, the amplitude of the output signal is frequency dependent. Low-frequency signals are transmitted with little loss, but high-frequency signals are greatly attenuated. The circuit is thus a low-pass filter and is widely used as a noise filter.

For the final example in this section, consider the circuit of Fig. 10.2 which contains a resistor, a capacitor, and an inductor. This is a resonant circuit for reasons that we shall see presently. The input signal is taken to be a sinusoidally varying potential and we shall not be concerned with the transient term. We need not set up and solve the differential equation for the circuit because we can write immediately,

$$E(t) = iR - \frac{ji}{\omega C} + jiL\omega \qquad (10.12)$$

or

$$i = \frac{E}{\{R^2 + [\omega L - (1/\omega C)^2]\}^{1/2}}$$

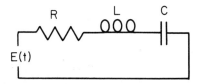

Fig. 10.2   Series *RLC* circuit.

The current vanishes for frequencies approaching zero or infinity. When the frequency is

$$\omega_R = \frac{1}{\sqrt{LC}}$$

the denominator of Eq. (10.12) is smallest and the current attains a maximum value. The frequency ω_R is the resonant frequency, and at ω_R the current is

$$i = E/R$$

which does not depend on the value of L or C.

III. ACTIVE COMPONENTS AND AMPLIFIERS

Active components are used to amplify or otherwise modify input signals. The vacuum tube has been the traditional active device, but in the past decade it has been supplanted in many applications by the semiconductor solid-state device. In this section we very briefly treat the properties of these components.

Good introductory treatments of electronics for nonspecialists have been given by Malmstadt et al. (1), by Phillips (2), and in a book edited by Cannon (3).

A. Vacuum Tubes

The simplest form of vacuum-tube device is the diode rectifier. It consists of a cathode which emits a stream of electrons and an anode which collects them. They are mounted inside an evacuated enclosure, usually a glass envelope. Remember that the current, as defined in engineering practice, flows in the opposite sense to the electron flow. Electrons are negatively charged particles, so the current is equivalent to a flow of positive charge. The cathode can be an electrically heated metallic filament, or an oxide material heated by an auxiliary heater. The impedance to electron flow from cathode to anode is low, but the impedance to a flow in the reverse direction is quite high. Thus the device effectively permits current to flow in only one direction and it converts an alternating current to a pulsating or half-wave rectified direct current as sketched in Fig. 10.3.

The relationship between voltage applied across the vacuum diode and the current flowing is typically of the form shown in Fig. 10.4. It is linear over much of the curve. The nonlinear region at low potential

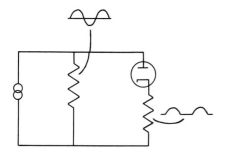

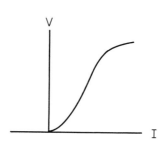

Fig. 10.3 Half-wave rectification with a diode.

Fig.10.4 Voltage–current relationship for a vacuum diode.

is due to space charge. At high currents a saturation condition is reached.

A third element can be introduced between the anode and cathode to form a *triode*. The element is a wire screen and is called a *grid*. It is used to control the flow of current and it makes possible the amplification of an input potential. The anode in a triode vacuum tube is called the plate. A primitive triode amplifier is sketched in Fig. 10.5. The cathode is shown as being at ground potential, but this is not necessary. When the voltage on the grid is zero with respect to the cathode, the tube behaves as a diode rectifier. If the grid is made more negative than the cathode, it repels the electrons and impedes the flow of current. If the grid is positive with respect to the cathode, the flow of current is enhanced. The relationships between output current (plate current) and grid potential for various values of plate voltage (or B^+ voltage) are called the triode mutual characteristics. They

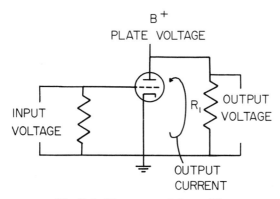

Fig. 10.5 Elementary triode amplifier.

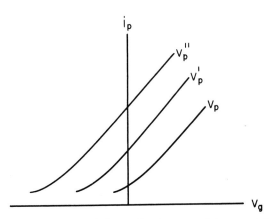

Fig. 10.6 Triode mutual characteristics: grid voltage vs. plate current for several plate voltages.

appear as in Fig. 10.6. A small change in grid potential can provide a large change in plate current; or, equivalently, a large change in potential across the load resistor R_l. Thus, the triode is an amplifier. Over an appropriate region of a characteristic, the relationship between input and output is very nearly linear.

The characteristic curves and other properties of the vacuum tube are modified and improved by the introduction of additional grids to form tetrodes, pentodes, and so on.

The amplification factor or gain μ provided by a vacuum tube is defined as the rate of change of plate voltage with respect to grid voltage at constant plate current

$$\mu = -\left[\frac{\partial v_p}{\partial v_g}\right]_{i_p}$$

Typically, μ is of the order of tens to hundreds. Other parameters commonly used to characterize tubes are the plate resistance ρ and anode conductance g_a defined by

$$g_a = \frac{1}{\rho} = \left[\frac{\partial i_p}{\partial v_p}\right]_{v_g}$$

and the mutual conductance or transconductance g_m

$$g_m = \left[\frac{\partial i_p}{\partial v_g}\right]_{v_a}$$

These parameters are related by the equation

$$\mu = g_m \rho$$

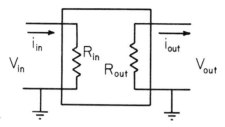

Fig. 10.7 Equivalent amplifiers.

Amplifiers are constructed with two or more tubes cascaded, the output of one tube serving as the input for the next one. In this book we are not concerned with the details of amplifier design. We shall consider the amplifier to be a "black box," as sketched in Fig. 10.7. An input potential is applied at one end and an input current flows across the equivalent input impedance. At the other end an output current flows through an equivalent output impedance across which an output potential appears. There is generally a phase difference between the input and output signals.

B. Transistors

The atomic structure of a metal is such that many electron orbitals overlap. Electrons are freed from ownership by individual atoms and are able to wander throughout the lattice. Under the influence of an applied electric field they drift readily across the metal, and we say that a current flows. The metal is thus a conductor. In insulators, on the other hand, electrons are tightly bound to individual atoms and electrical conductivity is very poor. Semiconductors form an intermediate case. Electrons must be promoted to a conducting state (or a conduction band of states) by absorption of energy. Commonly, a semiconductor material is produced by adding small controlled amounts of impurities to a lattice of a substance such as germanium or silicon. Some impurities are donors, supplying electrons which are readily excited into the conduction band, thus forming n-type (negative) semiconductors. Other impurities are acceptors which remove electrons from a filled band, leaving behind a vacancy or hole which drifts through the material under the influence of a field in much the same way as electrons do but in the opposite direction. The current in this case consists of positively charged holes, and the material is a p-type semiconductor.

A semiconductor diode consists of a p-type material in contact with

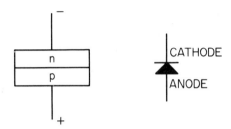

Fig. 10.8   Semiconductor diode.

an *n*-type material. A potential applied as in Fig. 10.8, in the forward-biased direction, causes both electrons and holes to move toward the boundary between the *n* and *p* regions. There they combine; that is, the surplus electrons from the *n* region are incorporated into the vacancies in the band of the *p* region. Meanwhile, electrons from the cathode are introduced into the *n* region, the *p* region releases electrons into the anode, and a net current flows through the diode. Only a feeble current can flow in the reverse direction if the polarity of the potential is reversed (reverse biased) unless it is made so large that the material breaks down. The relationship between current and potential, analogous to Fig. 10.4 for triodes, is shown in Fig. 10.9. The semiconductor diode is a rectifier, just as the vacuum-tube diode.

   The transistor consists of two layers of similar material, *n* or *p* type, separated by a thin layer of the opposite type. Transistors can be *pnp* type (and they usually are) or they can be *npn* type. They are shown schematically in Fig. 10.10. The center layer is called the *base* of the transistor and it plays a role analogous to the grid of a vacuum-tube triode. The outer layers are the *emitter* and the *collector*. They are analogous to the cathode and the plate. (This analogy should not be

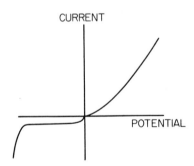

Fig. 10.9   Voltage–current relationship for a semiconductor.

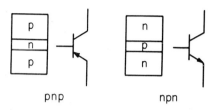

Fig. 10.10 Transistors.

pushed too far.) A simplified *pnp* transistor amplifier is shown schematically in Fig. 10.11. When a negative potential is applied to the base, the emitter–base combination acts as a forward-biased diode and positive holes migrate across the boundary. Some of them combine with electrons from the base, but because the base is very thin, many of them drift right through it and into the collector. The base–collector combination acts as a reverse-biased diode, so it cannot contribute significant current. The number of positive holes which migrate into the base region is controlled by the number of free electrons in the base. Hence, the current through the transistor is increased by making the base more negative. A current of electrons must be injected into the base to compensate for the electrons lost by recombination with the holes from the emitter. The transistor is a current amplifier, in contrast with the vacuum-tube triode which is a voltage amplifier and whose grid draws no significant current.

The current gain β (for small signals) of a transistor is defined as the rate of change of collector current with respect to base current, at constant collector-to-emitter potential

$$\beta = \left[\frac{\partial i_c}{\partial i_b}\right]_{v_{ce}}$$

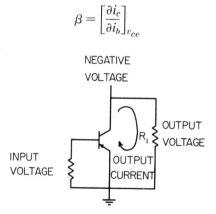

Fig. 10.11 Elementary transistor amplifier.

Values of β range from tens to hundreds. Other parameters used to characterize transistors are the forward current transfer ratio α

$$\alpha = \left[\frac{\partial i_c}{\partial i_e}\right]_{v_{cb}} = \frac{\beta}{1+\beta}$$

and the base input resistance

$$r_{be} = \left[\frac{\partial v_{be}}{\partial i_b}\right]_{v_{ce}}$$

The significance of the subscripts is obvious.

IV. IMPEDANCE CONSIDERATIONS

When two electronic systems are to be coupled together, it is very important to take into account the impedances of the two devices. Let us consider for example, a voltmeter placed across the output terminals of an amplifier, as shown in Fig. 10.12. The amplifier is considered to be a voltage generator with output impedance Z_a. The voltmeter has an input impedance Z_m. The current flowing from amplifier to meter is

$$i = \frac{V_0}{Z_a + Z_m}$$

and the voltage appearing at the input terminals of the meter is

$$V_m = V_0 \frac{Z_m}{Z_a + Z_m}$$

Unless Z_a is very much less than Z_m, the potential seen by the meter will be substantially less than the actual output voltage and an erroneous reading will be made. This holds for an input–output pair of

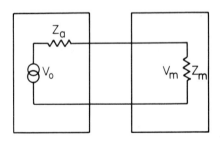

Fig. 10.12   Impedance of coupled systems.

devices, of course, and a general rule is as follows:

The maximum input voltage (or an input voltage closest to the output voltage) is obtained if the input impedance is much larger than the output impedance.

If the objective is not to obtain a high (or accurate) input voltage but to transfer the greatest amount of electrical power, the impedance requirements are quite different. The power dissipated in the input resistance of the system sketched in Fig. 10.12 is

$$P = iV_m = \frac{V_0{}^2 Z_m}{(Z_a + Z_m)^2}$$

The maximum power is found by setting

$$\frac{\partial P}{\partial Z_m} = V_0{}^2 \cdot \frac{Z_a{}^2 - Z_m{}^2}{(Z_a + Z_m)^4} = 0$$

or

$$Z_a = Z_m$$

That is, for maximum power transfer, the input impedance should equal the output impedance.

Vacuum-tube circuits of the type sketched in Fig. 10.5 are called common cathode amplifiers. The input impedance is very high (e.g., 1 MΩ), which is to say that the grid conducts very little current. This is why a vacuum-tube voltmeter is used to measure the potential across a high resistance. Moving coil voltmeters have input impedances of perhaps 20 kΩ/V at best and will give erroneous results. The output impedance of a common cathode amplifier is determined by the load resistor R_l which, in a typical case, might be ten or a hundred kilohms. The load resistor is in parallel with the resistance of the tube itself, which is quite high for current in the plate-to-cathode direction. The corresponding transistor circuit of Fig. 10.11 is called a common emitter amplifier. Its input impedance is not as high as the vacuum-tube circuit because the base of a transistor necessarily conducts some current. A typical input impedance is in the order of thousands of ohms. The output impedance is determined primarily by the load resistor and is usually several kilohms.

Often it is desired to perform an impedance conversion in an electronic system. For example, we may wish to measure the output voltage across a moderately high load resistor and record it with a recorder having a fairly low input impedance. We need a circuit having a high input impedance and a low output impedance which we can insert

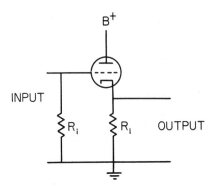

Fig. 10.13   Cathode follower.

between the amplifier and the recorder. Such a device is the *cathode follower* shown in Fig. 10.13. The input impedance is the value of the resistor R_i and can be megohms or larger. The output impedance for this circuit is determined by the load resistor in parallel with the tube, but in this case the impedance of the tube is very low in the cathode-to-plate direction. Thus, the output impedance of a cathode follower is very low, typically a few hundred ohms. The cathode follower does not provide an amplification of voltage—indeed the voltage gain is nearly unity or slightly less. Because of the relationship between input and output impedance, the current gain is high. The transistor analog of the cathode follower, shown in Fig. 10.14, is the *emitter follower*. Its properties are similar to those of the cathode follower, but the input impedance is generally not as high and the output impedance is somewhat lower.

Sometimes it is desired to present a low input impedance to a source and a high output impedance to the elements following the source. The vacuum-tube circuit of Fig. 10.15 does just this. It is called a *grounded*

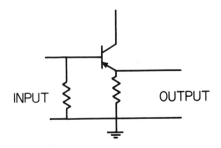

Fig. 10.14   Emitter follower.

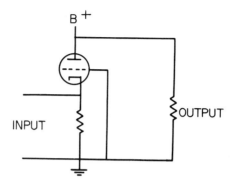

Fig. 10.15 Grounded grid amplifier.

grid amplifier. The corresponding transistor circuit, the *grounded base amplifier*, is sketched in Fig. 10.16.

The transformer has important impedance-converting properties, in addition to its ability to transform potential and current. An impedance Z_i in the primary circuit acts as an equivalent impedance

$$Z_0' = Z_i n^2$$

in the secondary circuit, where n, as before, is the ratio of the number of turns in the secondary coil to the number in the primary.

V. FEEDBACK

The principle of negative feedback is of utmost importance in modern electronics and in the area of automatic control systems. Consider the system shown schematically in Fig. 10.17. A signal V_i, is introduced. This is not necessarily a voltage or even an electrical signal in the general case. It might be hydraulic pressure, for example, or irradiance by a light beam. The signal is amplified by an amplifier with gain K. We

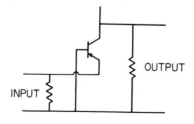

Fig. 10.16 Grounded base amplifier.

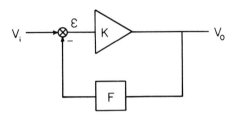

Fig. 10.17 Feedback system.

ignore for the present the complications of phase shift or frequency-dependent gain or attenuation. A fraction F of the signal V_0 delivered by the amplifier is fed back to the input end of the amplifier and subtracted from the input signal. Hence the magnitude of the signal entering the amplifier is really

$$\epsilon = V_i - FV_0$$

But V_0 is given by

$$V_0 = K\epsilon$$

These equations can be combined to give

$$V_0 = \frac{KV_i}{1 + KF}$$

If the feedback were removed by setting $F = 0$, we would have

$$V_0 = KV_i$$

So the effect of negative feedback is first of all to reduce the amplitude of the output signal. The term K is the open-loop gain of the system and $K/(1 + KF)$ is the closed-loop gain, or simply loop gain.

The output impedance of an amplifier with negative feedback is reduced by the same factor as the gain, namely, $1/(1 + KF)$. The input impedance is increased by the reciprocal of this factor.

The application of negative feedback has a stabilizing effect on the output of an amplifier. To understand this consider two amplifiers, one with gain K' and no feedback and the other with higher open-loop gain K, but with feedback reducing the closed-loop gain to an amount equaling K', that is,

$$K' = \frac{K}{1 + KF}$$

The effect on the output of a variation in gain of the first amplifier is

easily shown by differentiation to be

$$\frac{\delta V_0}{V_0} = \frac{\delta K'}{K'}$$

The effect on the output of the second amplifier, however, is

$$\frac{\delta V_0}{V_0} = \frac{1}{1+KF}\frac{\delta K}{K}$$

The variation in output is less in the second case for the same fractional change in amplifier gain. Thus, in the interest of stability, it is advantageous to use a high-gain amplifier with negative feedback.

The impedance of many circuits depends on frequency. We have discussed, for example, the resonant circuit of Fig. 10.2, which preferentially transmits signals in a certain frequency band. Other circuits preferentially reject signals in a specific band of frequencies. An example is the important twin-T filter circuit. It is possible to obtain a bandpass effect from a band-rejection circuit by placing the circuit in the feedback loop of the system. Signals of frequencies different from the rejection frequency are fed back nearly undiminished, and by adjusting the fraction fed back they can be made to cancel the input.

VI. Automatic Control Systems

The principle of negative feedback is utilized in the design of automatic control systems. As an illustration, let us consider the system sketched in Fig. 10.18, representing a positioning servo device. A dc signal voltage R is introduced into the system. The difference between R and the fed-back voltage is detected. This difference, or *error signal*, ϵ, is amplified to a level $K\epsilon$, which is sufficient to drive a dc motor. The angular position of the shaft of the motor is denoted by θ. The wiper

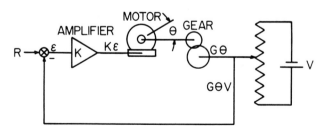

Fig. 10.18 Servo positioning device.

shaft of a potentiometer is coupled to the motor shaft through a reduction gear train so the angular position of the wiper is given by $G\theta$. The potentiometer has a fixed dc voltage V across its length and the wiper picks off a fraction of this voltage, $G\theta V$. The wiper voltage is fed back to the beginning of the system and subtracted from R to complete the loop. The angular velocity of the motor is proportional to the voltage $K\epsilon$, that is,

$$\frac{d\theta}{dt} = MK\epsilon$$

If ϵ is positive, the motor turns in a direction to increase $G\theta V$, thereby reducing ϵ until it is zero. On the other hand, if ϵ is negative, the motor turns in the other direction, decreasing $G\theta V$, and thereby making ϵ less negative until it is zero. Questions concerning how rapidly ϵ approaches zero or whether the system oscillates about $\epsilon = 0$ are part of the theory of automatic control systems. We return to these questions in later sections.

We have seen that our elementary positioning servo automatically responds to an input R by making

$$G\theta V = R \quad \text{or} \quad \theta = R/GV$$

If we attach a pen to the motor shaft in such a way that its position is proportional to θ, the device serves as an automatic potentiometric recorder. This is indeed the way in which most recorders operate.

The nature of the subtracting device at the input has not yet been discussed. There are very many ways to perform the subtraction operation. One method commonly used for recorders is sketched in Fig. 10.19. The dc input signal and the dc signal from the potentiometer

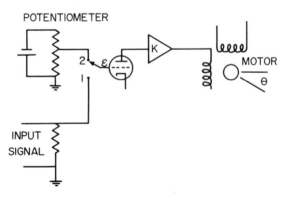

Fig. 10.19   Signal subtracting device.

wiper are alternately sampled by means of a switch driven synchronously from the main power lines. The difference between these signals
is thus converted to an ac error signal. After amplification, the error
signal is used to excite one winding of an ac motor. The other winding
is excited from the main power line and therefore is in synchronization
with the sampling switch. The speed of the motor is governed, as
before, by the magnitude of the amplified error signal. The direction of
rotation depends on the relative phase of the currents in the two
windings. If the original dc error signal before chopping is positive, the
phase will have one sense; and if the dc signal is negative, the phase
will have the opposite sense. This is clear from Fig. 10.20. In Fig.
10.20A, the potential at the potentiometer (contact 2 on the switch) is
greater than the input signal (contact 1); in Fig. 10.20B, the reverse is
true.

There are numerous good treatments of automatic control systems
theory. Among these are books by Clark (4) and by Truxal (5).

VII. TRANSFER FUNCTIONS

We have seen that the behavior of circuits containing resistors,
capacitors, and inductors is described by linear differential equations.
This is readily extended to circuits containing amplifiers by considering
the gain of the amplifier and an equivalent passive network to account
for the phase shift. A particularly convenient mathematical device for
solving linear differential equations is the Laplace transform, defined as

$$\text{Laplace transform of } f(t) = L\{f(t)\} = F(s) = \int_0^\infty f(t)\, e^{st}\, dt$$

The properties of the Laplace transform are surveyed in the Appendix. Let us write the relationships between current and potential in a

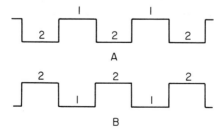

A

B

Fig. 10.20 Phases of difference signals.

resistor, a capacitor, and an inductor, together with the Laplace transforms of these equations.

For a resistor

$$v(t) = i(t)R \quad \text{and} \quad V(s) = I(s)R$$

For a capacitor

$$\frac{dv(t)}{dt} = \frac{1}{C}i(t) \quad \text{and} \quad sV(s) = v(0^+) = \frac{1}{C}I(s) \tag{10.13}$$

where

$$v(0^+) = \lim_{t \to +0} v(t)$$

or alternatively,

$$v(t) = \frac{1}{C}\int_0^t i(t)\, dt \quad \text{and} \quad V(s) = \frac{1}{sC}I(s)$$

For an inductor

$$v(t) = L\frac{di(t)}{dt} \quad \text{and} \quad V(s) = LsI(s) - Li(0^+) \tag{10.14}$$

Now let us reconsider the simple RC circuit of Fig. 10.1, but with a
general potential $v(t)$ turned on at time $t = 0$. The equation for the
circuit is Eq. (10.8),

$$v(t) = i(t)R + \frac{1}{C}\int_0^t i(t)\, dt$$

and the Laplace-transformed equation is

$$V(s) = I(s)R + \frac{1}{Cs}I(s) \quad \text{or} \quad I(s) = \frac{V(s)Cs}{1 + RCs}$$

The factor $Cs/(1+RCs)$ relating the Laplace transform of the current
to the Laplace transform of the applied potential is the transfer function of the network. In block diagram form it is usually drawn as in
Fig. 10.21. The potential appearing on the capacitor is

$$v_c(t) = \frac{1}{C}\int_0^t i(t)\, dt$$

and the Laplace transform is

$$V_c(s) = \frac{1}{Cs}I(s) = \frac{V(s)}{1 + RCs} \tag{10.15}$$

The factor relating input and output potentials, $1/(1+RCs)$, is also a

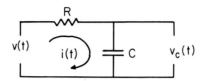

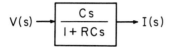

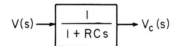

Fig. 10.21   Transfer functions.

transfer function. One needs to be careful in defining which transfer function is being considered.

If the applied potential is a constant voltage suddenly applied at time $t = 0$

$$v(t) = \begin{cases} 0 & t < 0 \\ E_0 & t > 0 \end{cases}$$

the Laplace transform is

$$V(s) = E_0/s$$

Then from Eq. (10.15),

$$V_c(s) = \frac{E_0}{s(1+RCs)} = \frac{E_0/RC}{s(1/RC+s)} \qquad (10.16)$$

and the inverse transform of Eq. (10.16) gives us, for the potential on the capacitor,

$$v_c(t) = \begin{cases} E_0(1-e^{-t/RC}) & t > 0 \\ 0 & t < 0 \end{cases}$$

which is the same as Eq. (10.9), of course.

If the applied voltage is in the form of a very sharp pulse at time zero, that is,

$$v(t) = E_0 \delta(t)$$

the Laplace transform is simply

$$V(s) = E_0$$

The Laplace transform of the voltage on the capacitor is

$$V_c(s) = \frac{E_0}{1 + RCs}$$

and the voltage itself, from the inverse transform, is

$$v_c(t) = \begin{cases} E_0 e^{-t/RC} & t > 0 \\ 0 & t < 0 \end{cases}$$

and this is the same as Eq. (10.10). Thus, given the transfer function of a circuit, we can determine the time behavior of the output for any input whose Laplace transform can be written. This is, of course, not limited to electrical circuits but holds as well for any system with dynamics describable by linear differential equations.

The general relationship between the Laplace transform of input $C(s)$ and output $R(s)$ and the transfer function of a linear system $H(s)$ is

$$R(s) = C(s)H(s) \tag{10.17}$$

As the Laplace transform of an impulse (delta function) is unity, it is evident that the transfer function is the Laplace transform of the impulse response. Moreover, the inverse transform of Eq. (10.17) is

$$r(t) = \int c(t')h(t-t')\, dt'$$

That is, the output is determined by the convolution of the input and the impulse response. We have seen similar relationships involving the optical spread function and optical modulation transfer function and the monochromator slit function and spectral modulation transfer function.

The Laplace variable is a complex variable whose imaginary component is the frequency ω. We can often determine the frequency response of a system directly from its transfer function by substituting

$$s = j\omega$$

Thus, the transfer function of a capacitor, Eqs. (10.13) with $v(o^+) = 0$, gives us

$$\frac{V(\omega)}{I(\omega)} = \frac{1}{j\omega c} = \frac{-j}{\omega C} = X_c(\omega) \tag{10.13a}$$

in agreement with Eq. (10.5).

For an inductor, from Eqs. (10.14),

$$\frac{V(\omega)}{I(\omega)} = j\omega L = X_L(\omega) \tag{10.14a}$$

in agreement with Eq. (10.7). The frequency response of the RC network of Fig. 10.21, from Eq. (10.15) for the voltage transfer function, is given by

$$H(\omega) = \frac{V_C(\omega)}{V(\omega)} = \frac{1}{1+j\omega RC} = \frac{1}{1+j\omega\tau} \qquad (10.15a)$$

from which we find the modulus of $H(\omega)$

$$|H(\omega)| = [H(\omega)H^*(\omega)]^{1/2} = \left[\frac{1}{1+\tau^2\omega^2}\right]^{1/2} \qquad |10.15b)$$

and the phase is given by

$$\tan\phi = \frac{\text{imaginary part of } H(\omega)}{\text{real part of } H(\omega)} = -\omega\tau \qquad (10.15c)$$

This agrees with the solution for a sinusoidal input, Eq. (10.11).

A transfer function or signal level can be expressed in decibels,

$$H(\omega) = 20 \log\frac{H(\omega)}{H(0)}$$

A graph of H in decibels versus the logarithm of frequency is called a Bode plot. Bode plots have particularly simple forms, typified by the graph of Eqs. (10.15b and c) shown in Fig. 10.22. The response is seen to drop to -3 dB at a frequency given by $\omega_C = 1/\tau$, and then to fall off at a uniform rate of 20 dB/decade. The *bandwidth* of the system is ω_c. The two asymptotes intersect at ω_C. The phase shift is $-45°$ at ω_C and approaches $-90°$ at large frequencies. The asymptotic behavior of more complicated systems can be constructed readily on Bode plots.

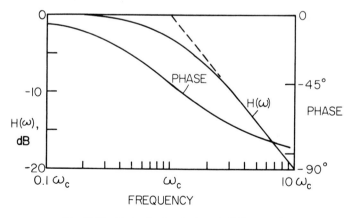

Fig. 10.22 Bode plot for single-pole RC network.

Control systems engineers discuss the response and stability of systems by studying the behavior of the transfer function in the s plane. We restrict our treatment in this section to a discussion of a system having a transfer function of the form

$$H(s) = \frac{ab}{(s+a)(s+b)} \tag{10.18}$$

This system is said to be *overdamped*. The impulse response of a system is determined in general by the location of the poles and zeros of the transfer function, that is, the values of s for which the denominator and numerator, respectively, vanish. $H(s)$ has two poles, one at $s = -a$ and one at $s = -b$. The inverse transform of Eq. (10.18), with a and b real, is

$$h(t) = \frac{ab}{b-a}(e^{-at} - e^{-bt}) \tag{10.19}$$

If either a or b is negative, the corresponding pole is positive and Eq. (10.19) diverges as $t \to \infty$. The system, in this case, is clearly unstable. This is also the case if a and b are complex quantities with negative real parts. Hence, for stability, the poles of $H(s)$ must not lie in the right half of the complex plane. This is a general rule for all linear systems. If a and b are pure imaginary numbers, such that

$$a = j\omega \quad b = -j\omega$$

Eq. (10.19) becomes

$$h(t) = \omega \sin \omega t$$

and the system oscillates at frequency ω. Incidentally, if one pole of a transfer function has an imaginary component, there must be another pole which is the complex conjugate.

In the case of two coincident real poles, $a = b$, the impulse response is

$$h(t) = a^2 t e^{-at}$$

which can be obtained from Eq. (10.19) by letting a approach b in the limit. This system is said to be *critically damped*.

When a and b are complex conjugates with positive real parts,

$$a = \alpha + j\beta \quad \text{and} \quad b = \alpha - j\beta$$

the transfer function is written

$$H(s) = \frac{\omega_n{}^2}{s^2 + 2\zeta\omega_n s + \omega_n{}^2}$$

with

$$\omega_n = (\alpha^2 + \beta^2)^{1/2} \quad \text{and} \quad \zeta = \alpha/\omega_n$$

The impulse response is

$$h(t) = \frac{\omega_n^2}{\beta} \sin \beta t e^{-\alpha t}$$

This signal is seen to decay with a time constant $1/\alpha$ determined by the real part of the poles, but it *overshoots* and *rings*, that is, undergoes a damped oscillation with frequency β given by the imaginary part of the poles. This system is said to be *underdamped*. The amount of overshoot is determined by ζ, which is called the *damping ratio*.

The transfer function of a feedback system can be derived in the same way as the gain equations of Section V. Thus, the transfer function of the simple feedback system of Fig. 10.23 is written in terms of the transfer function $F(s)$ of the feed-forward portion and the transfer function $G(s)$ of the feedback portion of the loop.

$$H(s) = \frac{F(s)}{1 + F(s)G(s)}$$

VIII. ELECTRICAL NOISE

Random fluctuations in the electrical output of a system are called *noise*. Noise corrupts and obscures the desired signal. The level of noise present in a system ultimately sets the limit on the minimum detectable signal level. In the absence of noise exceedingly feeble signals might be amplified to a respectable level, but signals much less than the magnitude of the noise are difficult to extract from the noise.

Some kinds of noise arise in an essential manner in the components of the electronic system or in the detector itself. Consequently, this noise cannot be eliminated, and it determines the fundamental limit of sensitivity. Other kinds of noise occur because of defects in components,

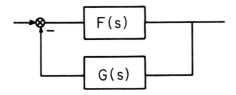

Fig. 10.23   Feedback system.

construction, or adjustment of the system. Let us discuss some of the important sources of noise in an infrared spectrophotometer.

Thermal Noise. The electrons in a conductor are caused to drift under the influence of an electric potential, and this drift is the electric current. Superposed on this drifting motion is a random motion due to thermal agitation, much like the Brownian motion observed in small particles suspended in a gas. The random motion is present whether or not there is a net current flowing. During the course of the fluctuation, there will be moments when there is a slight preponderance of electrons at one end of the conductor, and moments when the other end is richer in charge. Consequently, a fluctuating potential difference exists across the conductor. Nyquist (6) derived an equation for the mean-square potential

$$\langle v^2 \rangle = 4kTR\,\Delta f \qquad (10.20)$$

where k is Boltzmann's constant, T is the absolute temperature, R is the resistance of the conductor, and Δf is the width of the band of frequencies considered. The fluctuations are often called *white noise*, meaning that all frequencies are present to the same extent, although this is certainly not true for infinitely high frequencies. The Nyquist formula for thermal noise was verified by Johnson (7) and so thermal noise is also called Johnson noise.

Thermal noise is the most important source of noise in thermal detectors (thermocouples, bolometers, etc.). In a properly designed system with such a detector, the thermal noise originating in the detector is the fundamental limitation to sensitivity.

Shot Noise. Shot noise occurs because of the discrete nature of an electric current as a stream of individual electrons. Statistical arguments show that the extent of the fluctuations in such a stream is proportional to the average value of the stream. This leads to Schottky's formula (8) for the fluctuations in the current, or shot noise.

$$\langle i^2 \rangle = 2qi\,\Delta f \qquad (10.21)$$

where i is the average value of the current and q is the charge on an electron. As before, Δf is the width of the band of frequencies under observation. The right-hand side of Eq. (10.21) must be multiplied by another factor when space-charge effects are significant. Shot noise is also essentially white noise for practical frequencies.

Shot noise is an essential feature of devices that make use of a stream of electrons. These include photocells and photomultipliers, which are not important as infrared detectors, and photoconductors, which are. These also include vacuum tubes and transistors.

It is important to recognize a fundamental difference in the characteristics of thermal noise and shot noise, which is evident in comparing Eq. (10.20) and (10.21). Thermal noise is independent of the signal applied, whereas the mean-square amplitude of shot noise is proportional to the signal current.

Current Noise and Flicker Noise. These sources are considered together because of their similar functional dependence on signal and frequency. The mechanism of generating such noise is quite different. Current noise is observed as a fluctuation in the current carried by a conductor. It is distinct from thermal noise. Current noise is observed mostly in carbon resistors, semiconductors, and metallic films, but not in wire-wound resistors. It occurs because of a statistical fluctuation in the conductivity. The mean-square noise voltage, at frequencies from 0.001 cps up to tens of kilocycles per second, obeys a law of the form

$$\langle v^2 \rangle \propto \frac{i^2}{f^2} \Delta f \tag{10.22}$$

Thus, the root-mean-square noise is proportional to the current and to the reciprocal of the frequency. Current noise is also called excess noise or $1/f$ noise.

Flicker noise is observed in vacuum tubes and seems to be associated with statistical fluctuations in the emitting properties of the cathode. Flicker noise also follows Eq. (10.22), modified by the effects of space charge.

Because of the $1/f$ dependence of the root-mean-square amplitude of current noise and flicker noise, it is advantageous to deal with ac signals of high frequency whenever possible.

Component Parameter Fluctuations. Random variations in the circuit parameters can appear as noise in the output signal. Defective resistors, capacitors, and soldered connections can have random or intermittent changes in impedance. Vibration transmitted to tubes or detectors can result in random changes in parameters. This is generally called microphonics. Variations in tube gain can originate in power supply fluctuations affecting plate voltage or filament current.

Pickup. Extraneous signals can be picked up by an electronic system in various ways. Surges or ac signals from power lines can be transmitted into the input of an amplifier by inductive or capacitive coupling. Electromagnetic signals can also be picked up and amplified. Amplifiers with high input impedance are particularly susceptible because a given induced current produces a higher potential across a high resistance. The effects of pickup can be minimized by paying attention to the shielding of leads and their location and length.

Instability. Under certain conditions of loop gain, a servosystem can become unstable and oscillate. It is important for the operator of an infrared spectrophotometer to realize this and know how to adjust his instrument properly. We discuss this in considerably greater detail in later chapters. Degradation of components also can make a control system unstable.

REFERENCES

1. H. V. Malmstadt, C. G. Enke, and E. C. Toren, Jr., *Electronics for Scientists*, Benjamin, New York, 1963.
2. L. F. Phillips, *Electronics for Experimenters*, Wiley, New York, 1966.
3. G. C. Cannon, ed., *Electronics for Spectroscopists*, Wiley-Interscience, New York, 1960.
4. R. N. Clark, *Introduction to Automatic Control Systems* Wiley, New York, 1962.
5. J. Truxal, *Automatic Feedback Control System Synthesis*, McGraw-Hill, New York, 1955.
6. H. Nyquist, *Phys. Rev.*, **32**, 110 (1928).
7. J. B. Johnson, *Phys. Rev.*, **32**, 97 (1928).
8. W. Schottky, *Ann. Phys.*, **57**, 541 (1918).

CHAPTER 11

Infrared Detectors

I. INTRODUCTION

In the previous chapter we laid the groundwork for a discussion of the electronic properties of the infrared spectrophotometer. We are now ready to discuss the electronic systems in greater detail. This chapter is devoted to the infrared detector. A detailed treatment of infrared detectors has been given by Smith *et al.* (*1*). Useful reviews have been published by Putley (*2*) and by Levinstein (*3*).

The infrared detector is a transducer. Its purpose is to intercept a beam of infrared radiation and convert it into the form of an electrical signal. The *responsivity* of a detector is a measure of its sensitivity. For example, the responsivity of a radiation thermocouple might be expressed as volts of output signal per watt of radiant power input signal. The responsivity of a detector depends on such factors as the wavelength of the radiation and the temperature of the detector. It is important to understand the conditions under which a specified responsivity is attained.

The responsivity of a detector is not necessarily its most important characteristic, because practical detectors also generate noise and the noise power limits the ability of the detector to detect very small signals. The *noise-equivalent power* W_n is defined as the radiation input signal which, when converted to an output signal by the detector, produces a signal equal to the noise generated by the detector. For example, if a detector has a responsivity of E volts per watt and it generates a thermal noise of N volts, the noise-equivalent power is

$$W_n = N/E$$

Noise-equivalent power is often abbreviated NEP, and it is also called minimum detectable power. Signal levels less than W_n can be detected under proper conditions, but W_n is nevertheless a good indication of minimum detectable power.

The value of W_n depends on circumstances for a given type of detector. The wavelength and temperature are factors, as we indicated above. The area of a detector affects W_n in a predictable way. Most infrared spectrophotometers make use of ac (chopped) radiation, and as the response time of many infrared detectors is rather slow, the NEP depends on the frequency of chopping. The NEP is also greatly affected by electrical filtering. The signal is concentrated at the chopping frequency while the noise covers a very broad range of frequencies. Hence, a filter that suppresses frequencies other than the chopping frequency, preferentially suppresses the noise.

It is common practice to specify noise-equivalent power for a detector of 1-cm^2 area used with a noise filter having a 1-cps bandwidth, with the chopping frequency, detector temperature, wavelength, etc., specified. Sometimes the total polychromatic radiation from a blackbody source is used, and in this case the temperature of the source is specified.

Notice that the smaller W_n is, the better is the detector. Hence a parameter reciprocally related to W_n is often used: the specific detectivity or D-star parameter,

$$D^* = 1/W_n$$

Infrared detectors are of two general classes: thermal and quantum. Thermal detectors absorb infrared radiation and convert it to heat, which produces a change in temperature. Some temperature-dependent property of the detector is then measured. In quantum detectors the absorbed photons exite electrons of the detector into conduction states or cause them to be emitted from the surface, and the resulting electric currents are measured.

II. NOISE AND FLUCTUATIONS

The noise generated by an infrared detector is an example of a random or stochastic process. Several samples of noise $x(t)$ recorded as a function of time, generated by the same detector under the same circumstances, will be completely different in detail. However, certain statistical properties of the noise samples will tend to be the same, such as the mean values, the variances, and so on. The excursions of the noise record are characterized by a probability density function $p(x)$, where

$p(x)\Delta x =$ probability that the noise has a value between x and $x + \Delta x$ at any instant.

A random process is said to be stationary if its probability function does not change with time. Many important random processes, including thermal and shot noise, are Gaussian, meaning that the probability density function takes the form

$$p(x) = \frac{1}{(2\pi)^{1/2}\sigma} \exp\left[\frac{-(x-\bar{x})^2}{2\sigma^2}\right]$$

where $\bar{x}$ is the mean value

$$\bar{x} = \int_{-\infty}^{\infty} xp(x)\,dx$$

and σ is the variance, or mean-square value of the fluctuation of $x(t)$ about the mean value,

$$\sigma = \int_{-\infty}^{\infty} (x-\bar{x})^2 p(x)\,dx$$

An important statistical concept is the *autocorrelation function* of a random function.[*]

$$R(\tau) = \lim_{T \to \infty} \frac{1}{2T} \int_{-T}^{T} x(t+\tau)x(\tau)\,dt \qquad (11.1)$$

Note that $R(0)$ is the mean-square fluctuation of $x(t)$. The Fourier transform of the autocorrelation function is the *power spectral density function*

$$S(\omega) = \int_{-\infty}^{\infty} R(\tau)e^{-j\omega\tau}\,d\tau$$

This is an important consequence of communication theory. Its meaning in our discussion is this: The noise is a superposition of a large

---

[*]This assumes that we are dealing with a real ergodic random process, which we shall not discuss further.

number of randomly oscillating voltages or currents of various frequencies. The mean power associated with those oscillations having frequencies between ω and $\omega + \Delta\omega$ rad/sec is $S(\omega)\Delta\omega$. Noise for which $S(\omega)$ is a constant (at least up to some limiting frequency) is called white noise. If something other than voltage or current is fluctuating, $S(\omega)$ may not have the units of power, but it has become traditional to call it power spectral density anyhow; $S(\omega)$ is also called the *Wiener spectrum.*

If a random function is introduced into the input of a linear system, the output is also a random function, modified by the properties of the system. If the time behavior of a sample of the input is known, then the time behavior of the output can be calculated from the impulse response of the system,

$$y(t) = \int h(t - t')x(t')\,dt'$$

However it is seldom desired to know the detailed time behavior of a sample of the output. Rather, the statistical functions $R(\tau)$ and $S(\omega)$ are used to characterize the input and output. From the definition of the autocorrelation function [Eq. (11.1)], it is easy to show that the autocorrelation function of the output $R_y(\tau)$ is given by a double convolution of the autocorrelation function of the input $R_x(\tau)$ with the impulse response, or

$$R_y(\tau) = \int_{-\infty}^{\infty} \int_{-\infty}^{\infty} h(t')h(t'')R_x(\tau + t' - t'')\,dt'\,dt''$$

The spectral density is given by

$$S_y(\omega) = H(\omega)H^*(\omega)S_x(\omega) = |H(\omega)|^2 S_x(\omega) \qquad (11.2)$$

where $H(\omega)$ is the transfer function (in this case the Fourier transform of the impulse response of the system).

Let us now consider a body, such as an infrared detector, in thermal equilibrium with its surroundings. We know from elementary thermodynamics that the average temperature of the body is constant, but we also know that the temperature undergoes small fluctuations about the average which are associated with fluctuations in the rates of receipt and loss of radiant power. The mean-square magnitude of the fluctuation in energy of a system in equilibrium with its surroundings can be shown from statistical thermodynamics to be

$$\overline{\Delta E^2} = kT^2 C_T \qquad (11.3)$$

in terms of the absolute temperature T, the heat capacity C_T, and

Boltzmann's constant k. A change in energy is related to a change in temperature by

$$\Delta E = C_T \Delta T \qquad (11.4)$$

So, the mean-square fluctuation in the temperature is given by

$$\overline{\Delta T^2} = \frac{1}{C_T^2} \overline{\Delta E^2} = \frac{kT^2}{C_T} \qquad (11.5)$$

Now the rate of flow of energy into or from a body is proportional to the small temperature difference between the body and its surroundings and also to the rate of change of its temperature. Thus, a differential equation is written

$$C_T \frac{d\Delta T}{dt} + \frac{\Delta T}{R_T} = W(t) \qquad (11.6)$$

R_T is analogous to electrical resistance, and is called the *thermal resistance*. It is discussed in greater detail a bit later. This equation and its solution are very much like those of the RC electrical circuit of the previous chapter with heat capacity playing the role of electrical capacitance. The relationship between the Fourier transforms of input $W(\omega)$ and output $\Delta T(\omega)$ and the transfer function $H(\omega)$ is

$$\Delta T(\omega) = H(\omega) W(\omega) = \frac{1/C_T}{(1/R_T C_T) + j\omega} W(\omega)$$

From Eq. (11.2), the spectral densities of W and ΔT are related by

$$S_{\Delta T}(\omega) = \frac{(1/C_T)^2}{(1/R_T C_T)^2 + \omega^2} S_W(\omega) \qquad (11.7)$$

The spectral density $S_W(\omega)$ is not frequency dependent, at least over the region where the factor multiplying it in Eq. (11.7) is significant. We write, for a bandwidth $\Delta\omega$

$$S_W(\omega) = K\Delta\omega \qquad (11.8)$$

and integrate Eq. (11.7) over $0 \leq \omega \leq \infty$ to get

$$\int_0^\infty S_{\Delta T}(\omega) \, d\omega = \frac{\pi K R_T \Delta\omega}{2C_T} \qquad (11.9)$$

But the left-hand side of Eq. (11.9) is just the mean-square amplitude of the temperature fluctuations, $\overline{\Delta T^2}$, and so by comparison with Eq. (11.5) we can evaluate the constant K in Eq. (11.8),

$$K\Delta\omega = \frac{2kT^2}{\pi R_T} = S_W(\omega)$$

The autocorrelation function $R_W(\tau)$ is the inverse Fourier transform of $S_W(\omega)$. Thus

$$R_W(\tau) = \frac{2kT^2\Delta\omega}{\pi R_T}\delta(\tau)$$

where $\delta(\tau)$ is the delta function. Thus the variance, or mean-square fluctuations, of $W(t)$ in a frequency band $\Delta\omega$ rad/sec are

$$\overline{\Delta W^2} = R_W(0) = \frac{2kT^2}{\pi R_T}\Delta\omega$$

or, following the common practice of considering a band of width Δf cps,

$$\overline{\Delta W^2} = \frac{4kT^2\Delta f}{R_T} \tag{11.10}$$

A somewhat more general form of this equation will be given below.

Now we turn to another form of noise associated with thermal fluctuations. This is the Johnson noise introduced in the previous chapter. The derivation of the Nyquist equation follows very closely the derivation just completed for radiation fluctuations. Consider the circuit of Fig. 11.1. The thermal noise voltage v generated by the

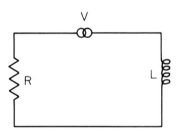

Fig. 11.1 Equivalent circuit for generation of Johnson noise.

resistor is represented by a generator. Energy is stored in the magnetic field of the inductor according to the discussion of Section 10.11, and the fluctuation in this energy is given by thermodynamics

$$\overline{\Delta E^2} = \tfrac{1}{2}L\overline{(i^2)} = \tfrac{1}{2}(kT) \tag{11.11}$$

Equation (11.11) is analogous to Eqs. (11.3)–(11.5). The differential equation for the circuit, its solution, and an equation for the transfer function are easily written. They can in fact be obtained from the RLC circuit discussed in Chapter 10 by setting $C \to \infty$. Thus the transfer function is

$$H(\omega) = \frac{i(\omega)}{v(\omega)} = \frac{1/L}{R/L + j\omega}$$

The spectral density of the voltage is independent of frequency, at least for frequencies in the range we are considering, so

$$S_v(\omega) = K\Delta\omega$$

and the spectral density of the current is therefore

$$S_i(\omega) = \frac{K/L^2}{(R/L)^2 + \omega^2} \qquad (11.12)$$

The value of the constant K is determined by integrating Eq. (11.12) over the range $0 \leqslant \omega \leqslant \infty$ and comparing the results with Eq. (11.11), analogous to the treatment of Eq. (11.9). This procedure gives us

$$K\Delta\omega = 2kTR\Delta\omega = S_v(\omega)\Delta\omega$$

The autocorrelation function of the fluctuating voltage follows,

$$R_v(\tau) = 2kTR\Delta\omega\,\delta(\tau)$$

and the mean-square voltage fluctuation is

$$\overline{\Delta v^2} = R_v(0) = (2/\pi)kTR\Delta\omega = 4kTR\Delta f \qquad (11.13)$$

which is the Nyquist formula given in Section 10.VIII, Eq. (10.20). The inductance L does not appear in this equation. It was introduced in the first place only as a convenient place to store energy. A capacitor would have served as well.

This same procedure can be used to study other kinds of thermal noise, provided a thermodynamic expression for a fluctuation about equilibrium is available and an expression for a spectral density is known or can be assumed. But now let us turn to a different class of noise, typified by shot noise. We again assume a frequency-independent spectral density, but instead of a thermodynamic fluctuation, we invoke a result from statistics regarding the fluctuation in a statistical average. We start with a discussion of shot noise. A "steady" stream of electrons is observed by counting the number of electrons n_τ arriving at an anode during an interval of time of length τ_c. The average count is $\bar{n}_\tau$. The average rate of arrival is

$$n = \bar{n}_\tau/\tau_c$$

The current is defined for each counting interval

$$i_\tau = en_\tau/\tau_c$$

and the average current is

$$i_0 = e\bar{n}_\tau/\tau_c = en$$

where e is the charge of an individual electron. The variance of the electron counts about the mean value is given by statistics, and is just equal to the mean value itself,

$$\overline{\Delta n_\tau^2} = \overline{(n_\tau - \bar{n}_\tau)^2} = \bar{n}_\tau \qquad (11.14)$$

Hence, the variance of the current is

$$\overline{\Delta i_\tau^2} = \overline{(i_\tau - i_0)^2} = \frac{e^2}{\tau_c^2}\overline{(n_\tau - \bar{n}_\tau)^2}$$

$$= \frac{e^2\bar{n}_\tau}{\tau_c^2} = \frac{e^2 n}{\tau_c} = \frac{ei_0}{\tau_c}$$

If this current flows through a resistor in series with the anode, a fluctuating potential of mean value $v_0 = i_0 R$ and variance

$$\overline{\Delta v^2} = R^2 ei_0/\tau_c = Rev_0/\tau_c \qquad (11.15)$$

is observed.

The density spectrum of shot noise is constant for practical frequencies,

$$S_i(\omega) = K\Delta\omega \qquad (11.16)$$

Let us suppose that instead of counting electron arrivals, we use an integrating capacitor in parallel with the anode resistor, as in Fig. 11.2.

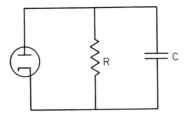

Fig. 11.2   Integrating capacitor.

The transfer function (Fourier transform) of this configuration is readily calculated:

$$H(\omega) = \frac{1}{1 + j\omega RC}$$

The spectral density of the voltage fluctuations on the capacitor is therefore

$$S_v(\omega) = \frac{R^2 K \Delta \omega}{1 + \omega^2 R^2 C^2}$$

and the total voltage fluctuation is determined by integrating this equation over $0 \leqslant \omega \leqslant \infty$

$$\overline{\Delta v^2} = RK/4C \qquad (11.17)$$

Now the use of an integrating capacitor is equivalent to counting electron arrivals. We merely state, without an attempt at proof, that the equivalent counting interval is equal to two time constants of the RC network, that is,

$$\tau_c = 2RC$$

With this assumption, we can compare Eqs. (11.17) and (11.15) to evaluate K,

$$K = \frac{2ev_0}{R} = 2ei_0$$

The current fluctuation is found by inserting this value of K into Eq. (11.16) and finding the autocorrelation function. The result is

$$\overline{i_f^2} = 2ei_0 \Delta f$$

which is the Schottky formula given in Section 10.VII, Eq. (10.21).

Equation (11.14) for the fluctuations in a stream of electrons is not strictly correct. Electrons obey a somewhat different kind of statistics (Fermi–Dirac statistics), which requires a modification in the equation. However, for the electron densities we are concerned with, Eq. (11.14) is a satisfactory approximation. When we turn to consideration of the fluctuations in a radiation stream of photons, however, we find that the statistics of photons (Bose–Einstein statistics) requires a modification that we cannot ignore. We must write

$$\overline{\Delta n(v)^2} = \frac{n(v) \, dv}{1 - e^{-hv/kT}}$$

for the fluctuation in the number of photons with frequencies between v and $v + dv$ emanating from a body at temperature T.

By making use of this equation a more general expression than Eq. (11.10) can be derived for the fluctuations in radiant power exchanged between bodies. The requirement that the bodies be in thermal equilibrium can be relaxed; that is, they need not be at the same temperature. The general equation is

$$\overline{\Delta W^2} = 4k \left(\frac{T_1^2}{R_{T_1}} + \frac{T_2^2}{R_{T_2}} \right) \Delta f$$

where T_1 and T_2 represent the temperature of the body and its surroundings and R_{T_1} and R_{T_2} are the corresponding thermal resistances.

III. THERMAL DETECTORS

We have already written and made use of the differential equation (11.6) that describes the change in temperature of a body when it is exposed to and absorbs a beam of radiant power. The equation contains the heat capacity that we recognized as playing a role analogous to electrical capacitance. Heat that is delivered to the detector by a beam of radiation is partly stored in the detector, thus raising its temperature. The equation also contains the thermal resistance that we saw was analogous to electrical resistance. Excess heat that is not stored in the detector flows from it either by conduction through wires and mounting structures or the surrounding atmosphere, or by reradiation to surrounding objects. Together, the heat capacity and the thermal resistance form the complex thermal impedance Z_T, defined by

$$\frac{1}{Z_T} = \frac{1}{R_T} + jC_T\omega$$

Z_T appears in the solution of the differential equation in just the same way as the electrical impedance in an RC circuit.

The relationship between heat flow and thermal resistance is

$$dW/dt = 1/R_T \tag{11.18}$$

Now the total power radiated into space from the surface of a body of area A, emissivity ϵ, and temperature T is given by Planck's law

$$W = \frac{2\pi A h}{C^2} \int_0^\infty \frac{\epsilon \nu^3 d\nu}{e^{h\nu/kT} - 1} \tag{11.19}$$

From Eqs. (11.18) and (11.19) the thermal resistance associated with radiation is found to be

$$\frac{1}{R_{T.R}} = \frac{2\pi A h}{C^2} \int_0^\infty \nu^3 \left[\frac{(e^{h\nu/kT} - 1)(d\epsilon/dT) + \epsilon(h\nu/kT^2)e^{h\nu/kT}}{(e^{h\nu/kT} - 1)^2} \right] \cdot d\nu$$

In the likely event that the emissivity is independent of temperature, or of negligible dependence for small changes in temperature, the first term in the integrand can be dropped. If the emissivity is effectively independent of wavenumber as well, as it is likely to be in a thermal detector, the integration is readily carried out to give

$$R_{T.R} = \frac{1}{4\sigma\epsilon A T^3}$$

where σ is Stefan's constant. With this expression for thermal resistance due to radiation, Eq. (11.10) for radiation power fluctuation becomes

$$\overline{\Delta W^2} = 16kT^5\sigma\epsilon A\Delta f \qquad (11.20)$$

For a black body ($\epsilon = 1$) of area 1 mm² in equilibrium with its surroundings at 300°K only by radiation exchange, Eq. (11.20) tells us that

$$(\overline{\Delta W^2})^{1/2} = 5.5 \times 10^{-12}\text{ W} \quad \text{for} \quad \Delta f = 1\text{ cps}$$

This represents the noise-equivalent power of an ideal thermal detector at room temperature.

Now let us consider the heat escaping from the detector by conduction along a supporting member. The details of the problem are simplified by restricting it to one dimension. That is, we assume that the cross-sectional area of the support is constant (e.g., a wire) and that the temperature is uniform over any cross section. The rate of flow of heat in the x direction is proportional to the temperature gradient, the thermal conductivity K, and the cross-sectional area a

$$W = -Ka\frac{\partial T}{\partial x}$$

The differential equation satisfied by T is

$$\frac{\partial T}{\partial t} = k\frac{\partial^2 T}{\partial x^2} \qquad (11.21)$$

where the coefficient k (not to be confused with Boltzmann's constant) is the diffusivity defined by

$$k = K/C_w\rho \qquad (11.22)$$

in terms of the heat capacity C_w and the density ρ of the wire. Suppose one end of a wire of length l is at the temperature T_0 of the surroundings and the other end, attached to the detector, is raised to $T_0 + \Delta T$ with ΔT a constant. After equilibrium is established, $\partial T/\partial T = 0$ and Eq. (11.21) has as its solution

$$T = T_0 + (1 - x/l)\Delta T$$

The rate of heat flow is

$$\Delta W = \frac{Ka}{l}\Delta T$$

and therefore the thermal resistance due to conduction is given by

$$\frac{1}{R_{T.C}} = \frac{\Delta W}{\Delta T} = \frac{Ka}{l} = \frac{kC_w \rho a}{l} \tag{11.23}$$

Now suppose that ΔT is not a constant but a sinusoidal function of time. We must solve Eq. (11.21) subject to the boundary conditions

$$T(0, t) = T_0 + \Delta T \sin \omega T \quad \text{and} \quad T(l, t) = T_0$$

and the initial condition

$$T(x, 0) = T_0$$

The equations can be reformulated in terms of ΔT,

$$\frac{\partial \Delta T}{\partial t} = k \frac{\partial^2 \Delta T}{\partial x^2}$$
$$\Delta T(0, t) = \Delta T_0 \sin \omega t$$
$$\Delta T(l, t) = 0$$
$$\Delta T(x, 0) = 0$$

The solution of these equations [see Smith et al. (1)] is a complex function

$$\frac{\Delta T}{\Delta T_0} = \sin \omega t \frac{\sinh[(1+j)(l-x)/(2k/\omega)^{1/2}]}{\sinh[(1+j)l/(2k/\omega)^{1/2}]}$$

The complex impedance is found by calculating the rate of heat flow at the end of the wire

$$\Delta W = \frac{\partial \Delta T}{\partial x}\bigg|_{x=l}$$

and then finding $\Delta W / \Delta T_0$. The result for a very short wire is

$$\frac{1}{Z} = \frac{kC_w}{l}\left(1 + \frac{j\omega l^2}{3k}\right) \quad l \ll \left(\frac{2k}{\omega}\right)^{1/2}$$

from which

$$R_{T.C} = \frac{l}{kC_w} \quad \text{and} \quad C_T = \frac{C_w l}{3}$$

For long wires the real imaginary parts of the impedance are independent of the length of the wire, and

$$R_T = \frac{1}{kC_w}\left(\frac{2k}{\omega}\right)^{1/2} \quad \text{and} \quad C_T = C_w \left(\frac{k}{2\omega}\right)^{1/2}$$

A. Thermocouples

A thermocouple, as shown schematically in Fig. 11.3, consists of two dissimilar metals. If the two junctions are maintained at different

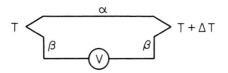

Fig. 11.3 Radiation thermocouple.

temperatures an emf appears across them, proportional to the temperature difference

$$E_T = P(\alpha, \beta) \cdot \Delta T$$

This is the *thermoelectric* or *Seebeck* effect; $P(\alpha, \beta)$ is called the thermoelectric power for materials α and β. The radiation thermocouple has one junction in thermal contact with a receiver whose purpose is to absorb the incident radiation and thereby raise the temperature of the junction relative to a reference junction.

The *Peltier* effect, related to the Seebeck effect, is a change in temperature of the junction occurring when a current flows through it. A current generated at a heated junction by the Seebeck effect reduces the temperature of the junction by the Peltier effect. The heat removed from the junction when a current i is flowing is given by

$$\Delta W = - P(\alpha, \beta) T i$$

where T is the temperature. One consequence of Peltier cooling is an increase in the effective electrical resistance of the thermocouple circuit by an amount

$$R_d = P^2 R_T T.$$

Let us assume that the thermal emf is to be measured open-circuit so that no current flows and we can ignore the Peltier effect. The temperature of the thermocouple target of absorptance ϵ, when exposed to a periodic beam of incident radiation $W e^{j\omega t}$, obeys the equation

$$C_T \frac{d\Delta T}{dt} + R_T^{-1} \Delta T = \epsilon W e^{j\omega t} \tag{11.24}$$

The solution for the thermal emf is

$$E = P\Delta T = \frac{P R_T \epsilon W e^{j\omega t}}{1 + j\omega R_T C_T}$$

and for the responsivity

$$r = \frac{(EE^*)^{1/2}}{W} = \frac{PR_T\epsilon}{(1 + \omega^2 R_T^2 C_T^2)^{1/2}} \tag{11.25}$$

Notice that the time constant for the temperature response is given by $\tau = R_T C_T$, just as in an electrical RC circuit. The responsivity decreases with increasing frequency in the same way as the output of the RC circuit. At a frequency

$$\omega_{3dB} = \frac{1}{\tau} \quad \text{or} \quad f_{3dB} = \frac{1}{2\pi\tau}$$

the responsivity is 3 dB less than the dc value and at higher frequencies it approaches a loss rate of 20 dB/decade of ω.

The responsivity of typical thermocouples lies in the range from 1 to 6 V/W for dc operation. Time constants of about 30 msec are common, necessitating chopping rates usually no faster than 15 or 20 cps.

Radiation thermocouples are usually mounted in evacuated chambers. This eliminates conduction through the surrounding atmosphere and restricts the thermal impedance to heat flow by radiation and by conduction along the wires. Notice that increasing R_T in this way increases the responsivity of the thermocouple, but it also lengthens the time constant.

In general, materials that are good electrical conductors also are good thermal conductors. In fact, the electrical conductivity σ and the thermal conductivity K are related by the Wiedemann–Franz law

$$K/\sigma T = L$$

where L is the Lorenz number

$$L \approx 2.4 \times 10^{-8} \ (V/°C)^2$$

for metals.

There are two essential sources of noise in a thermocouple: Johnson noise and the noise associated with fluctuations in the radiation exchanged with its surroundings. Johnson noise dominates in practical thermocouples. In order to reduce Johnson noise, a low value of electrical resistance is desired. But according to the Wiedemann–Franz relationship, this implies a low value for the thermal resistance as well, and this reduces the responsivity. It happens that the best compromise is reached by making the thermal resistance due to conduction equal to the thermal resistance due to radiation.

If we take the Johnson noise, from Eq. (11.13), as the largest noise source in a thermocouple and the dc responsivity given by Eq. (11.25), with ω set equal to zero, we have for the noise-equivalent power

$$W_m{}^2 = \frac{v_j{}^2}{r^2} = \frac{4kTR\Delta f}{(PR_T\epsilon)^2} \qquad (11.26)$$

Equation (11.23) is an equation for the thermal resistance due to conduction along a wire. The electrical resistance of the wire is

$$R = l/\sigma a.$$

Thus, a relationship between R and $R_{T,C}$ can be written

$$R = (K/\sigma)R_{T,C} = LTR_{T,C}.$$

If the thermal resistances due to conduction and to radiation have been made equal, we can set

$$R_T = R_{T,C} + R_{T,R} = 2R_{T,R}$$

Equation (11.26) can therefore be written

$$W_m{}^2 = \frac{4kT\,\Delta f}{(P\epsilon)^2} \cdot \frac{LTR_{T,R}}{(2R_{T,R})^2} = \frac{kT^2L\,\Delta f}{P^2\epsilon^2 R_{T,R}} = \frac{4kT^5LA\sigma\Delta f}{P^2\epsilon} \qquad (11.27)$$

Equation (11.27) gives us the important relationship

$$W_m \propto \sqrt{A}$$

for matched resistance and dc operation. The noise-equivalent power increases with the square root of the area of the detector. With this relationship and the dependence on f and $\sqrt{\Delta f}$, the noise-equivalent power of actual detectors can be reduced to a common basis of area, frequency, and bandwidth. A typical good thermocouple has a noise-equivalent power of about 10^{-10} W for an area of 1 mm², a bandwidth of 1 cps, and a modulation frequency in the neighborhood of 5–10 cps.

Thermocouples are commonly of either the pin type, the wire type, or the film type. The Schwarz thermocouple used in Perkin–Elmer instruments is typical of pin-type construction (Fig. 11.4). A thin gold foil acts as the radiation receiver. It is welded to cone-shaped supports at each end. One of the cones is constructed of a material having a positive thermal emf with respect to gold. This material is a composition of copper, silver, tellurium, selenium, and sulfur. The material of the other cone is a mixture of silver sulfide and silver selenide and it has a negative thermal emf relative to gold. The cones are attached to the ends of pins which provide electrical connection and mechanical support.

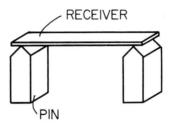

Fig. 11.4   Pin-type thermocouple.

Wire-type thermocouples are used in Beckman and Cary spectro-photometers. They consist of a gold foil with several wires welded to the edges (Fig. 11.5). The wires are 0.2 mm long and the foil is 2×10^{-4}

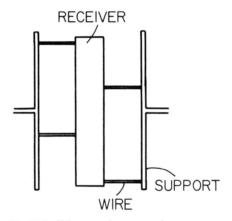

Fig. 11.5   Wire-type thermocouple.

mm thick and 2×0.3 mm in area. Alternate wires are bismuth–tin alloy.

Film-type thermocouples are fabricated by vacuum deposition of metals onto a substrate.

The cold or reference junction of a radiation thermocouple is usually the point of attachment of the fine thermoelectric wires to the relatively massive lead wires passing through the evacuated enclosure. The temperature of the reference junction thus remains more or less constant, while the temperature of the irradiated junction varies with incident power. However, the temperature of the reference junction does vary with changes in ambient temperature, and for this reason, it was necessary to control closely the temperature of the early dc spectrometers. Even so the detector had to be standardized regularly during the course of a measurement by inserting a shutter in the beam and reestablishing the zero level. With the use of chopped radiation, the standardization is accomplished automatically by the chopper. Only the ac signal is detected, representing the difference in temperature of the two junctions. There is another important reason for using chopped radiation, having to do with the coupling of the thermocouple to the amplifier. This is discussed in the next chapter.

Thermocouple receivers are coated with a material that absorbs the incident radiation. In modern detectors the absorber is usually gold black. It is deposited on the receiver by evaporation of gold under carefully controlled conditions of evaporation rate, atmosphere composition and pressure, and temperature of the receiver. Blacks can also be deposited electrolytically. The function of the absorber is to provide a dense array of tiny particles which serve to scatter and trap the radiation on the surface of the receiver so that eventually, after many reflections, the radiation is almost completely absorbed. Unfortunately, this process is not efficient for longer wavelengths, so blackened receivers are not useful in the far infrared.

A radiation thermocouple is a delicate device and must be protected from electrical, mechanical, and thermal shocks. When soldering to the lead wires of a thermocouple, the wires should be grasped between the soldering iron and the thermocouple with pliers to serve as a heat sink. A low-powered iron should be used, not a soldering gun. Local fields, including static fields generated by polishing the window or even removing a desiccant cap too rapidly, can damage or destroy the receiver. Attempts to measure the resistance of a thermocouple with an ohmmeter usually destroy the detector. Finally, the detector should be protected from sudden exposure to a strong radiation source.

B. Resistance Bolometers

A resistance bolometer is a resistance thermometer. The electrical resistivity of a conductor depends on its temperature. The dependence is expressed in terms of the temperature coefficient of resistance

$$\alpha = \frac{1}{R}\frac{dR}{dT} \qquad (11.28)$$

The coefficient is positive for metals and is approximately given by

$$\alpha \approx 1/T$$

so the percentage change in resistance equals the percentage change in temperature. For semiconductors α is negative and is approximately given by

$$\alpha \approx -K/T^2$$

The change in resistance accompanied by a *small* change in temperature is written, from Eq. (11.28),

$$\Delta R \approx \alpha R \, \Delta T$$

This small change in resistance is detected by passing a fixed current through the bolometer element and measuring the potential drop when the temperature changes, or by placing a fixed voltage across the element and measuring the current through it. A bridge circuit, as in Fig. 11.6, is commonly used, and the current measured is the out-of-

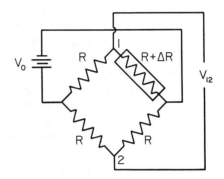

Fig. 11.6   Radiation bolometer in bridge circuit.

balance bridge current. In the symmetric arrangement shown, it is easy to show that the potential appearing across points 1 and 2 is

$$V_{12} = V_0 \frac{\Delta R}{2R + \Delta R} \approx \frac{V_0 \, \Delta R}{2R}$$

The heat-balance equation for the bolometer differs from that of the thermocouple in that we must include the heating effect of the biasing current W. That is, we have

$$C_T \frac{dT}{dt} + \frac{T - T_0}{R_T'} = \epsilon W e^{j\omega t} + W_h$$

in place of Eq. (11.24).

T_0 is the ambient temperature, and the actual mean temperature of the bolometer can differ significantly from ambient. Consequently, the thermal resistance R'_T also is different from that of a thermocouple. The contribution to $R'_T(T - T_0)$ due to radiation has the form $A\epsilon\sigma(T^4 - T_0^4)$. The heating power W_h is also dependent on temperature. Smith et al. (1) discuss the solution of this equation, including the effect of the temperature gradient along the bolometer element. They derive an equation for the responsivity (volts per watt) analogous to Eq. (11.25) for the thermocouple,

$$r = \frac{\epsilon i\alpha RR_e}{(1 + \omega^2 R_e^2 C_T^2)^{1/2}} = \epsilon\alpha R_e \left[\frac{R(T - T_0)}{R'_T(1 + \omega^2 R_e^2 C_T^2)}\right]^{1/2}$$

where i is the biasing current and R_e is an effective thermal resistance that involves the nature of the electrical circuit as well as the effects of radiation and heat conduction.

The responsivity increases with biasing current but so does the mean temperature, so there is an optimum value of the biasing current for best NEP. The responsivity increases with decreasing thermal capacity, so the element is made as thin as possible. The responsivity is optimized by making

$$C_T R_e = 1/\omega$$

and the effective thermal resistance is usually adjusted to satisfy this condition. One way to do this is by controlling the pressure of a thermally conducting gas in the bolometer enclosure. Bolometer elements are also sometimes mounted on a glass or quartz substrate to increase the thermal conductivity. The substrate can be the back surface of a condensing lens, on which an image of the monochromator slit or exit pupil is formed.

The noise-equivalent power of a bolometer is given by

$$W_m^2 = \frac{4KT^2}{R'_T}\Delta f\left[1 + \frac{\beta R'^2_T(1 + \omega^2 C_T^2 R_e^2)}{\alpha^2 R_e^2\epsilon^2 T(T - T_0)}\right]$$

where β is the ratio of R'_T to the thermal resistance for a small temperature difference from ambient. Again, we are quoting the results given by Smith et al. (1). The first term represents thermal fluctuation and the second term is due to Johnson noise. Because both the responsivity and Johnson noise are proportional to $\sqrt{R}$, the resistance does not effect the noise-equivalent power of a bolometer and can therefore be high, in contrast with the thermocouple.

There is another significant contribution to W_m^2 not included in this discussion: the current noise due to the fluctuations in a nominally steady current. It appears to be highest in semiconductors or very thin metallic films or wires and is likely to be the limiting noise source in a high-performance radiation bolometer.

Metallic resistance bolometers have been made of thin metal foils or evaporated metal films, properly blackened. In general, metal bolometers have not rivaled thermocouples in performance. This is probably because in order to satisfy the condition for optimum performance, $C_T R_e = 1/\omega$, the element must be very thin. But current noise becomes excessive in thin films, and in fact the electrical properties of bulk materials fail to be realized in thin films. For example, bismuth, which is a metallic conductor in ordinary thicknesses, behaves as a semiconductor with a negative coefficient of resistance when it is deposited as a very thin film.

Czerny et al. (4) have constructed self-absorbing bolometers of thin evaporated films of bismuth or antimony on collodion substrates. The bolometers are not blackened, but rely for absorption on an interesting property of metallic films discussed by Murmann (5) and Woltersdorff (6). The transmittance of a very thin film can be written approximately

$$T = \left(1 + \frac{2\pi}{c}\sigma d\right)^{-2}$$

in terms of its thickness d and electrical conductivity. The reflectivity of the film is approximately

$$R = \left(1 + \frac{c}{2\pi d\sigma}\right)^{-2}$$

The product σd is recognized as the resistance per square of the film. The absorption of the film, $1 - T - R$, reaches a maximum value of 50% when the thickness is adjusted to give a film resistance of 188.5 Ω/sq. With this resistance the film reflects 25% and transmits 25% of the incident radiation. A film of bismuth 0.01 μ thick has a resistance of 188.5 Ω/sq. Thin-film bolometers of the self-absorbing type would seem to be ideal for detection of the far infrared, but because of limiting current noise, they have not delivered performance equal to the Golay detector discussed further on in Section III.D.

Semiconductors are frequently used in making bolometers because their large temperature coefficient of resistance and high resistivity can provide large values of responsivity. Moreover, the high resistance of a semiconductor bolometer makes it easy to match to an amplifier without the use of special transformers. Thermistor bolometers are made from sintered nickel, manganese, and cobalt oxides.

There are definite advantages in low-temperature operation of bolometers. In the first place, the Johnson noise is reduced, and so is the noise due to fluctuations in background noise. At liquid helium temperature (4°K or less) the contribution from the background fluctuations consists of the nearly negligible radiation from the surroundings plus the radiation from the components of the spectrometer at room temperature which are seen through the entrance window of the detector. In order to reduce this to a minimum, the detector element is usually placed behind a cooled filter, which limits the band of wavelengths incident on the detector from outside the enclosure to those that are to be observed with the spectrometer. The field of view is limited by cooled baffles. Another less obvious advantage of low-temperature operation is that the thermal capacity of the detector becomes very small near absolute zero. Thick bolometer elements, therefore, can be used, and in the case of semiconductors used as far-infrared detectors, absorption of the radiation by free electrons can be utilized without any need for blackening the receiver. Semiconductor bolometers for low-temperature operation have been made of germanium doped with gallium or indium, and of gallium arsenide. NEP values of 10^{-11}–10^{-12} W have been obtained. Bolometers have also been made of carbon resistance elements constructed from ordinary carbon resistors, with NEP's of about 10^{-11} W.

Superconductors have been used as infrared detectors and NEP's of 10^{-12} W have been attained. It is necessary to control the temperature of superconducting bolometers very accurately in order to hold them close to the transition temperature for superconductivity. This necessitates control to perhaps 10^{-4}°K.

C. Dielectric Bolometers

Dielectric constants vary with temperature and therefore can serve as a means of detecting infrared radiation. The significant limiting noise source in thermocouples is Johnson noise. In resistance bolometers used at room temperature, both Johnson and current noises are important. Dielectric bolometers are free of both of these noise sources and should, in principle, provide good detectors with NEP approaching the limit set by temperature fluctuations.

The capacitance C of a dielectric bolometer plays the same role as resistance in a resistance bolometer. The responsivity of a dielectric bolometer is (7)

$$r = \frac{\epsilon i_b \alpha_D R_T}{\omega_b C(1 + \omega^2 R_T^2 C_T^2)^{1/2}}$$

where the bias current i_b is alternating current of frequency ω_b, not necessarily the same as the modulation frequency of the incident radiation ω. The temperature coefficient of dielectric constant is α_D. In order to have a large value of α_D, and therefore responsivity, a ferroelectric material is often chosen as the bolometer element, and in particular, a material whose Curie temperature for the ferroelectric–dielectric transition is slightly above ambient temperature. $BaTiO_3$ is a common choice. These bolometers are more properly called ferro-electric bolometers. The limiting noise source in ferroelectric bolo-meters appears to be somewhat analogous to the Barkhausen noise of ferromagnetic materials, and it seems to be large near the Curie temperature. Noise-equivalent powers of about 10^{-10} W have been obtained with these detectors, which is not much different from the performance of thermocouples and the best resistance bolometers.

Ferroelectric materials, in addition to having a large change in dielectric constant with temperature, are characterized by the appearance of a large spontaneous electric polarization below the Curie temperature. This is the pyroelectric effect. It too is strongly temperature dependent and has been used as the basis of infrared detectors (8). $BaTiO_3$, triglycine sulfate, and triglycine selenate have been used, with results comparable to other detectors. They seem to be particularly promising as detectors of far-infrared radiation. The responsivity of the pyroelectric bolometer detector is

$$r = \frac{\epsilon \omega R A q R_T}{(1 + \omega^2 R^2 C^2)^{1/2}(1 + \omega^2 \tau^2)^{1/2}}$$

where C is the capacitance of the element, R is the value of a resistor in parallel with it, q is the pyroelectric coefficient, A is the area of the element, and τ is a characteristic time constant associated with the pyroelectric effect. Note the presence of ω in the numerator of this equation. There is no dc response, and the responsivity *increases* with frequency up to a value determined by RC and τ.

D. Pneumatic Detectors

It is well known that matter expands when heated (except for a few substances in restricted temperature regions). Thermometers can be constructed by taking advantage of this, and so can infrared detectors. The thermal expansion of a solid has been used in a radiation detector of low sensitivity, and in fact the very first detector of infrared radiation (Herschel's) was an expanding-liquid-mercury thermometer. However, large temperature coefficients of expansion and small heat capacities make gases a better choice.

It is necessary to provide a mechanism for absorbing radiation, transferring the heat to the gas, and sensing the expansion. Golay (9) devised an ingenious solution to these problems and developed the detector bearing his name (see Fig. 11.7). A radiation-absorbing

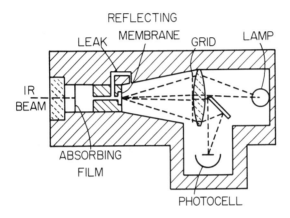

Fig. 11.7   Golay detector.

membrane is suspended in a cavity. This receiver could be a film blackened with gold black, for example, but Golay chose as his coating a thin film of metal of proper thickness to absorb the optimum amount of radiation. This is similar to the receiver used in Czerny's bolometer which was discussed in Section III.B. This feature makes the Golay detector useful in the far infrared where blackened receivers are no longer absorbing. Indeed, the Golay detector is the only room-temperature detector commonly used in the far infrared. The cavity containing the receiver is filled with a gas. Xenon is usually chosen because of its thermal capacity and heat conductivity. The pressure of the gas is adjusted for best balance between sensitivity and response speed. When radiation is introduced into the detector, the gas is heated by contact with the receiver and it expands. The back of the cavity is terminated with a thin membrane of metallized plastic. The expanding gas distorts this membrane and the amount of distortion is a measure of the temperature of the gas. A leakage path is provided around the membrane to establish dc equilibrium, but is small enough so that it does not interfere with ac response. The distortion of the membrane might be measured in a number of ways. For example, it could form one of a pair of capacitor plates. Golay devised an optical amplification device to take advantage of the properties of photocells. A grid is illuminated by a small light bulb. The light is reflected from the membrane and reimaged back on the grid, displaced just enough so that no light passes the grid. However, when the membrane is distorted, some light leaks through and is detected with a photocell.

The response of the Golay cell extends to the limit of the far infrared if a proper window is fitted to it; crystal quartz or diamond is the usual choice. The theoretical limiting noise source is the Brownian fluctuations of the gas atoms. The NEP usually attained is about the same as that of a good thermocouple in the mid infrared but is superior at the longer wavelengths where the black of the thermocouple becomes inefficient. The time constant is usually around 15 msec. Golay cells are inherently susceptible to microphonic disturbances and some care is needed in mounting them. They are easily damaged by exposure to intense light beams, and even with normal use the diaphragm seems to lose its elasticity, necessitating frequent replacement.

Special pneumatic detectors are commonly installed in nondispersive instruments used to monitor chemical processes. These detectors generally consist of cavities charged with a quantity of a sensitizing gas. When the detector is exposed to radiation from a suitable source, only those wavelengths are absorbed which correspond to absorption bands of the sensitizing gas. If a gaseous sample is inserted between the source and the detector, and the sample contains a quantity of the sensitizing gas, the amount of absorbed radiation is reduced. Other gases in the sample do not effect the absorption of radiation except by accidental overlap of the absorption bands of the sensitizing gas. The absorption is measured from the distortion of a thin membrane, usually as a change in capacitance as described above, but occasionally a Golay cell is modified to serve as a sensitized detector.

E. Other Thermal Detectors

In devising new thermal detectors the basic problem is to find a mechanism for transforming a temperature change into a change in some other measurable quantity. The main reason for the success of the Golay detector is that infrared absorption is made to produce a visible light signal. Advantage is then taken of the superior properties of photodetectors of visible light. Other detectors have been proposed or constructed to convert infrared to light signals in various ways. We shall mention a few of these very briefly.

Long ago Czerny (*10*) invented a method of producing a visible image of a heat-emitting object. An instrument based on Czerny's method has been marketed by Baird Atomic and called the Evaporograph (*11*). The device consists of a means of imaging an object on a film of volatile oil. using only the near-infrared radiation from the object. Where the image corresponds to a hotter portion of the object. the oil is raised to a higher temperature than in cooler portions. The hotter oil evaporates at a greater rate, and thereby a "relief" image is produced with a thinner oil film corresponding to regions of higher temperature in the object. Oil films assume colors by interference. and the color depends on the thickness of the film. Thus. a colored image of the thermal source can be observed or photographed. This instrument is a thermal image converter rather than a spectrometer detector.

The sniperscope. developed during World War II. is also an image converter. but not a thermal detector. It operates by stimulating the emission of phosphorescence radiation from a screen that is originally excited by light or ultraviolet radiation.

In 1954 Alkemade (*12*) suggested a novel detector. It is based on the change in thermal radiation emitted by a black element when its temperature is changed by absorption of radiation. The element absorbs infrared radiation, but its emission is observed in the visible or near infrared with a photoconductive detector. The steep edge of the Planck curve at wavelengths shorter than the peak is a very sensitive function of temperature. Also, the response of a photo-conductive detector is quite good, as we shall see in the next section. Alkemade presented some theoretical arguments predicting that such a detector should have satisfactory properties.

Northrop, in 1953, and later Hilsum and Harding (*13*), in 1961, described an infrared image converter based on a property of semi-conductors: namely, that there is an absorption edge with a fairly abrupt change from transparency to opacity over a short range of wavelengths. The region of low absorption is at wavelengths longer than the absorption edge. The absorption edge is steep and it is temperature sensitive, shifting to longer wavelengths as the temperature is raised. Selenium is such a semiconductor, having its absorption edge at a very convenient wavelength in the visible spectrum, close to the yellow emission lines of a sodium vapor lamp. The transmission of the sodium light by a thin film of selenium depends on the temperature of the film. Thus, if the infrared radiation from a warm body is imaged on a selenium film that has been coated with a partially absorbing metallic layer, the image can be observed visually or photographically by viewing a sodium lamp through it. Hilsum and Harding suggested that the contrast could be improved if the background radiation was reduced by incorporating the selenium film in an interference filter.

Some unpublished experiments were performed by the author with F. Waska to determine whether the absorption-edge image converter could be used as a spectroscopic infrared detector. The layer of selenium was deposited on a collodion film, followed by an infrared-absorbing layer of antimony. A film was inserted in each beam of a

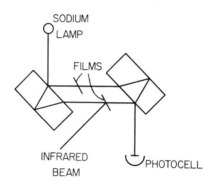

Fig. 11.8   Experimental selenium-film detector.

Jamin interferometer (Fig. 11.8) that was adjusted for destructive interference. When one film was irradiated with infrared radiation, its transmittance to sodium light was reduced. The destructive interference was then not complete, and the light passed by the interferometer was detected with a photocell. The results of the early experiments were not discouraging, but certainly did not produce a superior detector.

IV. PHOTOCONDUCTIVE DETECTORS

Photoconductors, like bolometers, change their electrical resistance when radiation falls on them. However, the mechanism is completely different. Photons incident on a photoconductor excite electrons from a nonconducting condition into a conducting state. Photoconductors are therefore quantum detectors, not thermal detectors. Consequently, photoconductors are more sensitive and respond faster than thermal detectors, but they have a well-defined wavelength limit, beyond which they will not respond.

Photoconductors are made of semiconducting materials. There are three general classes of photoconductors: intrinsic, extrinsic, and free electron. We describe the mechanism operating for each of these in the following sections.

A. Intrinsic Photoconductors

In free atoms or molecules, electrons are confined to fairly well-defined orbits and energy states. In solids, however, the interaction of the atomic forces of neighbors has the effect of smearing and over-lapping the energy levels of some electrons into bands. In nonconductors, the electrons are still closely associated with individual molecules or atoms. In metallic conductors, some electrons are held jointly by the assembly of atoms and are free to migrate throughout the substance. These are conduction electrons, and under the influence of an electric field they comprise the current conducted through the metal.

Semiconductors are an intermediate case. Normally, most of the electrons are more-or-less localized in individual atoms or molecules, in the so-called valence band. (See Fig. 11.9.) There are enough places

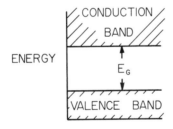

Fig. 11.9   Energy bands in a semiconductor.

available in the valence band to hold all of the electrons. If an electron is somehow given sufficient energy, for example, from an incident photon, it can be promoted into an excited energy level in the conduction band. It leaves behind its empty site — a positive hole — and eventually it falls back into an empty site in the valence band. While it is in the conduction band, the electron is free to meander through the material and serve as a carrier of electric current. The conductivity depends on the number of electrons in the conduction band and the mobility of these electrons. If the electrons have been promoted to a conducting state by absorbed photons, the phenomenon is photo-conductivity. As the effect is a property of the pure material itself, it is termed intrinsic photoconductivity. There are no energy states available to the electron in the region between the valence and conduction bands. The width of this region is the intrinsic energy gap E_G of the semiconductor. Clearly, the electrons must be given at least sufficient energy to cross this gap if they are to be promoted into the conduction band. Therefore, a photon must have a frequency of at least

$$\nu = E_G/h$$

to effect the conductivity of a photoconductor. The cutoff does not occur abruptly, of course, but it is a fairly steep function of frequency. It is also temperature sensitive, moving to higher frequencies with decreasing temperature. This is because of changes in the width of the forbidden band gap with temperature.

The first of the modern photoconducting near-infrared detectors was a vacuum-deposited film of thallium sulfide. It is limited to wavelengths shorter than $1.1\,\mu$. This was followed by detectors made from evaporated or chemically deposited layers of lead sulfide, lead selenide, and lead telluride. PbS and PbSe are useful to about 3 and $4.5\,\mu$, respectively, at room temperature, and somewhat shorter wavelengths when cooled. PbTe is used only at liquid-nitrogen temperature or lower, because at room temperature the rate of spontaneous promotion of electrons into the conduction band is too high. PbTe is useful to about $5.2\,\mu$. More recently single crystals of indium arsenide and indium antimonide have become available. Detectors are made by grinding or etching plates of these materials to a film a few microns thick. Indium arsenide and indium antimonide can be used to about 3.5 and $7.6\,\mu$, respectively, at room temperature. Very recently intrinsic photoconductors have been made of compounds of mercury, cadmium, and tellurium.

The detailed theory of photoconductivity is very complicated, but the responsivity of a photoconductor can be discussed in simple terms. A monochromatic radiation beam carrying power W falls on a detector of area A. The probability that a photon will promote an electron into the conduction band is the quantum efficiency q. The number of conduction electrons formed per unit area is therefore

$$n = qW/Ah\nu$$

The change in conductivity of the detector is proportional to n

$$\Delta\sigma = \beta n$$

and the change in resistance of the detector is dependent on $\Delta\sigma$ and also on the ratio of distance between electrodes l and the width of the detector b.

$$\Delta R = -\frac{l\Delta\sigma}{b} = -\frac{\beta q W}{h\nu b^2}$$

A current is passed through the detector and the change in resistance is measured as a voltage change

$$|\Delta v| = Ki|\Delta R|$$
$$= Ki\beta q W / h\nu b^2$$

where the constant K depends on the circuit in which the detector is used. The responsivity of the detector is therefore

$$r = \frac{\Delta v}{W} = \frac{Ki\beta q}{h\nu b^2}$$

It would seem that the responsivity increases with decreasing optical frequency until the limiting frequency is reached, that it is most favorable with a long thin detector configuration having the electrode on the ends ($l \gg b$), and that it increases with current. However, raising the current also increases the temperature, which reduces β. Furthermore, the noise is found to increase as l/b^2, so there is no real advantage in the long thin configuration.

The performance of a well-constructed photoconducting detector is limited by fluctuations in incident radiation and current noise. With values of current which give optimum performance, current noise is considerably more important than Johnson noise, and the latter can therefore be disregarded. The contribution to the minimum detectable power from current noise W_{mi} can be discussed by showing that the mean-square voltage fluctuation in a strip carrying a current is

$$\overline{\Delta v_i^2} = \frac{Ci^2l}{b^3}$$

where C is a constant and l and b are the dimensions of the strip [Smith, et al. (1)]. This produces a fluctuation in the measured voltage in terms of the circuit parameter K,

$$\overline{v_i^2} = K^2\overline{\Delta v_i^2}$$

W_{mi} can now be written, using the responsivity,

$$W^2_{mi} = \frac{\overline{v_i^2}}{r^2} = \frac{CAh^2\nu^2}{\beta^2 q^2}$$

W_{mi} is proportional to the square root of the area A of the detector. It is independent of the current itself and the circuit parameter.

The contribution to minimum-detectable power from radiation fluctuations is determined by invoking the result that the root-mean-square fluctuation in a stream of photons is equal to the number of photons in the stream. The fluctuation in the number of conduction electrons produced in the detector depends on the quantum efficiency, which depends on optical frequency, and also on the spectral frequency distribution of the incoming radiation. The low-frequency cutoff of a photoconductor renders it insensitive to a large portion of the thermal radiation from its surroundings at room temperature. Thus, the limiting NEP imposed on a photoconducting detector by radiation fluctuation is significantly less than in thermal detectors. It can be reduced further by cooling the detector and surrounding it with cooled shields.

The radiation noise power is also proportional to the square root of the detector area, so it is appropriate to define detectivity in terms of the D^* parameter, though D^* is now a strong function of wavelength. The detectivities of some commonly used intrinsic photoconductors are plotted in Fig. 11.10.

The response time of photoconductors is determined by the time an

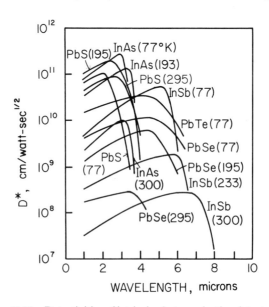

Fig. 11.10 Detectivities of intrinsic photoconducting detectors.

electron remains in the conduction band before it falls back into the valence band. This varies considerably from one photoconducting material to another, and it also depends on the temperature, since it is longer at low temperatures. Some typical response times of intrinsic photoconduction are given in Table 11.1. Extrinsic photoconductive

TABLE 11.1
Response Times of Intrinsic Photoconductive Detectors

Detector	Temperature (°K)	Typical response time (μsec)
PbS	300	300–3000
	195	500–3500
	78	1000–5000
PbSe	300	1–10
	195	10–100
	78	10–150
PbTe	78	10–30
InSb	295	0.05–0.2
	195	1–10
	195	1–10
	78	1–10
Tl$_2$S	295	500

detectors are much faster than thermal detectors, with response times measured in microseconds rather than milliseconds.

The photoelectromagnetic (PEM) effect has been used as a variation of photoconductive detectors. Radiation is incident on a slab of semiconducting material (Fig. 11.11). It is absorbed and electron–hole pairs are formed. These occur mostly near the surface, so a concentration gradient is produced in a direction normal to the surface. The electrons and holes diffuse into the interior of the slab. If a magnetic

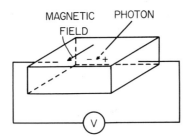

Fig. 11.11 Photoelectromagnetic effect.

field is applied in a direction parallel to the surface, the trajectories of the diffusing charges are deflected, electrons in one direction and positive holes in the other. Thus, a potential is induced across the length of the slab. PEM detectors of indium antimonide can be used at room temperature to wavelengths of nearly 8 μ. The detectivity at 6.2 μ is 3×10^8 and the time constant is 0.2 μ sec.

B. Extrinsic Photoconductors

The limit to the spectral range of sensitivity of an intrinsic photo-conductor is the width of the forbidden gap, which is determined by the nature of the pure material. Practical intrinsic photoconductors are presently limited to the wavelength region shorter than about 10 μ.

A pure semiconductor can be doped with a small concentration of impurity atoms. These atoms may have energy states in the forbidden energy gap of the host material. This is illustrated in Fig. 11.12. An

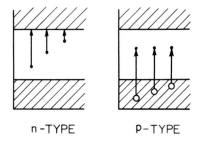

n -TYPE   P-TYPE

Fig. 11.12   Energy bands and impurity levels in extrinsic photoconducting detectors.

electron contributed by an impurity atom can be promoted into the conduction band by a photon of lower energy than is required to cross the energy gap of the host. This is an *n*-type photoconductor. On the other hand, an electron can be promoted from the valence band of the host material and accepted by the impurity atom. The hole remaining behind in the host lattice is the carrier of positive current. This is a *p*-type photoconductor. A photoconductor prepared with impurity atoms is called an impurity or extrinsic semiconductor.

The host material of an extrinsic photoconductor must be highly purified initially. So far, only germanium has been used. Germanium has been doped with gold, mercury, copper, zinc, boron, gallium, antimony, and cadmium. The concentration of dopant is fairly critical for optimum performance. These detectors, because they respond to relatively long wavelength radiation, must be cooled to reduce noise.

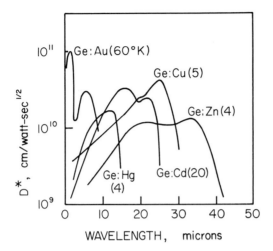

Fig. 11.13   Detectivities of extrinsic photoconducting detectors.

Germanium doped with gold requires liquid-nitrogen cooling (77°K), mercury and cadmium require liquid hydrogen (35°K), and copper, zinc, boron, gallium, and antimony need liquid helium (4°K). Some of these (Ge:Sb, Ge:B, and Ge:Ga) are useful to about $120\,\mu$. The detectivities of some extrinsic photoconductors are plotted in Fig. 11.13.

Note that a detector with a long-wavelength cutoff is not necessarily the best choice for shorter wavelengths. There is no one best detector for the entire infrared spectrum.

The time constant of extrinsic photoconductors is potentially about a microsecond. However, the resistance of these photoconductors is very high, and the capacitance of the input circuit is often sufficiently high to lengthen the actual time constant somewhat.

An infrared photodetector can be fabricated by forming an interface between a p- and an n-type semiconductor called a p–n junction. Incoming photons are absorbed, causing an excess of electrons on one side of the junction and an excess of positive holes on the other side. A net current flows across the junction to neutralize these excesses. The p–n junction thus functions as a photovoltaic detector.

C. Free-Electron Photoconductors

The impurity states in n-type indium antimonide are unusual in that they fall in the conduction band. This material therefore behaves in a sense as a metallic conductor, and the mechanism of photoconductivity is different from the intrinsic or extrinsic photoconductor. Energy of the incident photon stream is absorbed by the free electrons of the conduction band, their mobility is thereby increased, and the conductivity of the material increases. The free-electron photoconductor has been called a hot electron bolometer (14). The detector is used at a very low temperature (2°K) that is attained by pumping on the liquid-helium coolant. In addition, a magnetic field of about 7 kG is sometimes applied. The purpose is to raise the sensitivity so that the noise from the detector will at least approach the amplifier noise.

Because free-electron absorption is frequency-dependent, being proportional to λ^2, the free-electron detector is specifically a far-infrared detector limited to wavelengths longer than about $200\,\mu$. The wavelength of maximum sensitivity is about $1000\,\mu$. With a magnetic field applied, the NEP of the detector is about 5×10^{-12} W at $1000\,\mu$. The time constant is $0.2\,\mu$sec.

In a very strong magnetic field the free electrons follow circular orbits, much as the electrons in a cyclotron. The frequency of rotation in these orbits, or cyclotron resonance frequency, is given by

$$\omega_c = Be/m^*$$

where B is the magnetic field and m^* is the effective mass of the electron. An electron traveling in such an orbit can preferentially absorb radiation of frequency ω_c and the absorption is no longer limited by the λ^2 relationship. Thus, a cyclotron resonance detector can be a very sensitive *tunable* detector (15). However, the required fields are high, amounting to 10 kG at $200\,\mu$, 14 kG at $100\,\mu$, and 76 kG at $26\,\mu$. Cyclotron resonance detectors are cooled to 4°K. They offer a noise-equivalent power of 10^{-11} W at $150\,\mu$ and 0.5×10^{-10} W at $26\,\mu$, and they have a response time of less than a microsecond.

The last detector we describe in this chapter is the Josephson junction, announced by Grimes et al. (16) in 1966. A Josephson junction consists of a pair of superconducting materials separated by a very thin junction, perhaps an oxide layer. A current flows through the junction even when no potential is applied. The junction is therefore called a tunneling junction. The imposition of a potential causes a further current, according to the characteristic sketched in Fig. 11.14. When far-infrared radiation is allowed to fall on the junction, the characteristic shifts as shown. Thus, if the junction is biased in the region where it is sensitive to absorbed radiation, incident modulated radiation can be detected by measuring the induced alternating current.

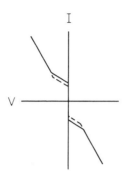

Fig. 11.14 Current–voltage characteristics of a Josephson junction. The characteristic shifts to the dashed line when radiation is detected.

A noise-equivalent power of 10^{-14} W has been attained in the far infrared.

REFERENCES

1. R. A. Smith. T. E. Jones, and R. P. Chasmar. *The Detection and Measurement of Infrared Radiation*, Oxford Univ. Press, London and New York, 1957; 2nd ed., 1968.

2. E. H. Putley. *J. Sci. Inst.*, **43**, 857 (1966).

3. H. Levinstein. "Infrared detectors." *Applied Optics and Optical Engineering*, Vol. II, R. Kingslake, ed., Academic Press, New York, 1965.

4. M. Czerny. W. Kofink, and W. Lippert. *Ann. Physik*, **8**, 65 (1950).

5. H. Murmann. *Z. Physik*, **54**, 741 (1929).

6. W. Woltersdorff *Z. Physik*, **91**, 230 (1934).

7. R. A. Hanel. *J. Opt. Soc. Am.*, **51**, 220 (1961).

8. J. Cooper. *Rev. Sci. Inst.*, **33**, 92 (1962).

9. M. J. E. Golay. *Rev. Sci. Inst.*, **18**, 347, 357 (1947).

10. M. Czerny. *Z. Physik*, **53**, 53 (1929).

11. G. W. McDaniel and D. Z. Robinson. *Appl Opt.*, **1**, 311 (1962).

12. C. T. J. Alkemade, *Physica*, **20**, 433 (1954).

13. C. Hilsum and W. R. Harding, *Infrared Phys.*, **1**, 67 (1961).

14. E. H. Putley. *Proc. Inst. Elec. Electron. Engr.*, **54**, 1096 (1966).

15. M. A. C. S. Brown and M. F. Kimmitt. *Infrared Phys.*, **5**, 93 (1965).

16. C. C. Grimes. P. L. Richards, and S. Shapiro. *Phys. Rev. Letters*, **17**, 431 (1966).

CHAPTER 12

Electronic Systems of Infrared Spectrophotometers

I. INTRODUCTION

We are now ready to study how the various electronic "black boxes" of an infrared spectrophotometer work together. We shall trace the path of a signal from the point at which it is transformed into an electrical signal by the detector to the point at which it is traced into an ink line by the recorder.

II. PREAMPLIFIER AND AMPLIFIER

The feeble signal from the detector must be amplified considerably before it can be used to drive a recorder or a servomotor. A thermo-couple, for example, might have a noise-equivalent power of 10^{-10} W and a responsivity of several volts per watt. Thus, a signal as small as 10^{-10} V must be treated by the amplifier. Depending on the sensitivity of the output device, this might call for an amplification factor as high as 10^{11}.

Because of the weakness of the signal and the tendency of lead wires to pick up noise, it is the usual practice to place a preamplifier of one or several stages very close to the detector. The signal can then be amplified to a safer level before it is transmitted to the main amplifier.

The preamplifier is very carefully designed and constructed to minimize internally generated noise and pickup. The tubes (or transistors in a few newer designs), and especially the very first input tube, are of a type selected for low noise characteristics. The tube may even be selected from a large sample of nominally identical tubes for minimum noise. The locations of the components and wires in the preamplifier are arranged to minimize pickup. The lead from the detector to the preamplifier is, of course, shielded, and the entire preamplifier might be mounted in a grounded mumetal enclosure to exclude stray electric and magnetic fields. The lead from the preamplifier to the main amplifier is also shielded. The preamplifier chassis might be mounted on vibration damping pads to reduce the possibility of microphonic signals from the vibrations set up by choppers or rotating beam splitters.

Coupling of the detector to the preamplifier often presents special problems. The best radiation thermocouples have low electrical resistance — typically a few ohms — and the Johnson noise corresponding to this resistance is the main source of noise from the detector. The best input tubes also generate some noise, and the tube noise at moderate frequencies is typically equivalent to the Johnson noise from a resistor of tens to hundreds of kilohms. The noise of the tube increases as $1/f$ at lower frequencies. This is the so-called flicker noise. On the other hand, the response of thermal detectors decreases with increasing frequency because of the slow response time. Thus, a compromise in chopping frequency is necessary. For optimum performance, the detector, rather than the tube, should be the limiting noise source. This is accomplished by coupling the detector to the preamplifier through a transformer as in Fig. 12.1. The voltage generated by the detector, when it is exposed to chopped radiation, varies periodically from v to $v + \Delta v$. The current in the detector is really pulsating dc, rather than ac. The grounded center tap of the transformer provides a reference point, so the output voltage signal varies periodically from

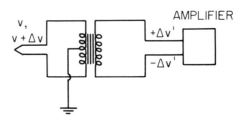

Fig. 12.1   Transformer coupling of a detector.

$-\Delta v'$ to $\Delta v'$, and the current is true ac. Alternatively, a center-tapped output transformer or resistance divider following the amplifier can be used to perform the same function. We have mentioned in Section 10.IV that an impedance Z_i in the primary circuit of a transformer is transformed to an equivalent impedance

$$Z_0 = Z_i n^2$$

in the secondary circuit, where n is the turns ratio of the transformer. Considered from the viewpoint of noise generation, we know that the Johnson noise voltage generated by the detector resistance R_D is

$$\overline{v_{J,D}^2} = 4kTR_D\Delta f$$

The transformer increases the noise to

$$(\overline{v_{J,D}^2})' = n^2\overline{v_{J,D}^2}$$

The tube, on the other hand, generates a noise equivalent to a resistance R_T

$$\overline{v_T^2} = 4kTR_T\Delta f$$

Thus, if we wish the noise from the detector to dominate and set the limit to the noise-equivalent power, we require

$$(\overline{v_{J,D}^2})' > \overline{v_T^2}$$

or

$$n^2 \cdot 4kTR_D\,\Delta f > 4kTR_T\,\Delta f$$

Hence, a transformer having a turns ratio in excess of

$$n = (R_T/R_D)^{1/2}$$

is needed. Furthermore, the impedance of the transformer itself must be high enough to avoid loading the source. That is, we wish to maximize voltage transfer. The transformer also gives us the benefit of amplification equal to n, and moreover, by use of suitable circuitry, a certain amount of tuning to the frequency of the signal can be introduced, which helps to filter out undesired noise of other frequencies. The use of a transformer as a coupling device is, of course, restricted to ac systems, and this is the second important reason for preferring chopped radiation systems over dc systems. The first reason, as we have mentioned in the previous chapter, is the freedom from drift due to ambient temperature changes. Transformer coupling can be used in a system with unchopped radiation by converting the dc output from the detector to ac. This is the basis of the so-called breaker-type

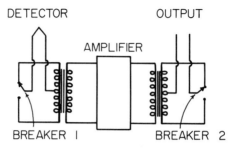

Fig. 12.2 Breaker amplifier.

amplifiers used, for example, in the early Perkin–Elmer 12B spectrophotometer (Fig. 12.2). The first breaker chops the input signal, making it ac; the second breaker is synchronized with the first and reconverts the amplified ac signal to dc.

Detectors having resistances somewhat in excess of the noise-equivalent resistance of the preamplifier tube can be coupled directly to the tube. Detectors of very high resistance, however, may require the insertion of a cathode follower, or some similar circuit, to avoid unfavorable loading by the amplifier.

The main amplifier can be located some distance from the pre-amplifier, provided it is connected with shielded cable. It may be in a separate cabinet or built into the structure of the spectrophotometer. In the latter case, the heat dissipated by the tubes can have a beneficial effect by maintaining the temperature of the monochromator above ambient and by protecting the optics from atmospheric humidity.

The requirements to be met by the main amplifier are fairly strict. It must not introduce any noise to corrupt the signal. It must have a very high gain, as we have indicated above, but it must be stable. This usually means that a great deal of negative feedback is used. Finally, if the spectrophotometer is to be usable as a single-beam instrument, the amplifier must be linear in its response. Fortunately, the frequency range is not very broad because the chopping rate of infrared instruments is low.

III. DEMODULATOR AND FILTER

The ac electrical signal leaving the amplifier usually must be converted to dc before it can be recorded or used to drive a servo. This is done by a rectifier or demodulator. Consider the primitive rectifier

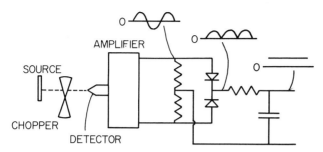

Fig. 12.3 Incoherent demodulator.

shown in Fig. 12.3. An ac signal is delivered by the amplifier. The diodes alternately pass each half-wave of the alternating current, converting it to a full-wave rectified signal. Finally, an *RC* filter smooths out the ripple, leaving a dc signal. This is an *incoherent* demodulator because it is not sensitive to the frequency or the phase of the incoming ac signal. It is not used much in practical spectro-photometers because it indiscriminately rectifies noise as well as signal and because double-beam spectrophotometers make use of the phase of the input signal.

The system depicted in Fig. 12.4 is a *coherent* or *synchronous* demodulator. It uses switches in place of the diodes of Fig. 12.3. The switches are synchronized with the chopper in such a way that one switch is open and the other closed during alternate half-cycles of the ac signal. They function in the same way as diodes in producing a full-wave-rectified signal and a positive dc signal from an ac signal (Fig. 12.4A). However, if the phase of the incoming signal is shifted by 180° with respect to the switches (as in Fig. 12.4B), a negative full-wave-rectified signal and a negative dc signal are produced. It is clear from Fig. 12.4C that a signal shifted in phase by 90° produces no dc signal at all. Thus the coherent rectifier tends to discriminate against signals not of the proper phase. Moreover, it also tends to reject signals not of the proper frequency. Figure 12.4D illustrates this for a signal of half the proper frequency. Because noise is a mixture of signals of random phase and frequency, the synchronous demodulator acts as a filter. It rectifies and transmits the desired signal, but rejects some of the noise.

Let us put this discussion on a mathematical basis. Consider for the moment that the action of the synchronous rectifier is to multiply an incoming signal by the simple demodulation function

$$r(t) = \sin \omega t \qquad (12.1)$$

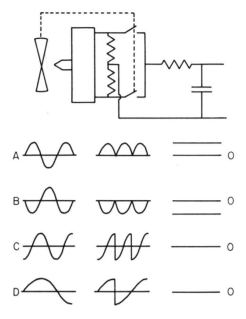

Fig. 12.4 Synchronous demodulator.

which serves to change the sign of the signal every half-cycle of frequency ω. If the incoming signal is

$$s(t) = s_0 \sin \omega t$$

it is modified by the demodulator to

$$s'(t) = s(t)r(t) = \tfrac{1}{2}(s_0)(1 - \cos 2\omega t)$$

The signal $s'(t)$ consists of a dc component of magnitude equal to half the amplitude of the original signal and an ac component of twice the frequency of the original signal. The ac component is discarded by the filter following the demodulator.

Now, consider a signal, which might be a sample of noise, of frequency ω' and phase ϕ,

$$n(t) = n_0 \sin (\omega' t + \phi)$$

The modified noise signal is

$$\begin{aligned}
n'(t) = n(t)r(t) = \tfrac{1}{2}(n_0)\{ &\cos \phi [\cos (\omega' - \omega)t \\
&- \cos (\omega' + \omega)t] - \sin \phi [\sin (\omega' - \omega)t \\
&- \sin (\omega' + \omega)t]\}
\end{aligned} \qquad (12.2)$$

Phase discrimination by the synchronous demodulator is seen in the case $\omega = \omega'$ for which Eq. (12.2) reduces to

$$n'(t) = \tfrac{1}{2}(n_0)[\cos \phi - \cos (2\omega t + \phi)] \tag{12.3}$$

The dc portion of $n'(t)$ is reduced by multiplication with the cosine of the phase. The amplitude of the double-frequency ac portion is not affected, but it is removed by the filter.

In the case $\omega \neq \omega'$ we might as well set $\phi = 0$. Equation (12.2) is then

$$n'(t) = \tfrac{1}{2}(n_0)[\cos (\omega' - \omega)t - \cos (\omega' + \omega)t] \tag{12.4}$$

and $n'(t)$ consists of two ac components of frequencies equal to the sum and difference of ω and ω'. If $\omega' \gg \omega$, these ac components are removed by the filter.

The description of the demodulator action by Eq. (12.1) was a convenient approximation. The actual demodulation function is more likely to be a periodic succession of positive and negative trapezoids. Equation (12.1) could be replaced by the Fourier series representation of the demodulation function:

$$r(t) = \sum_{n=1}^{\infty} a_n \sin n\omega t$$

But this merely adds terms to our equations which contain frequencies that are multiples of ω, and the filter removes them. We are therefore justified in retaining only the first term of the Fourier series. Similarly, a steady optical signal is converted to a periodic trapezoid by the chopper, but in this case too the multiple frequencies are removed by the filter.

However, if the optical signal is not steady but a slowly varying function of time $s_0(t)$ generated by a scanned spectrum, the modulated signal is

$$s(t) = s_0(t) \sin \omega t \tag{12.5}$$

The frequency content or spectrum of $s_0(t)$ is represented by its Fourier transform $S_0(\omega)$. As $s_0(t)$ is a slowly varying function in time, it contains frequencies in a fairly narrow band about zero (i.e., dc).

This is illustrated in Fig. 12.5A. The modulated signal, Eq. (12.5), appears as in Fig. 12.5B. We showed in deriving Eq. (12.4) that each frequency of $S_0(\omega')$ is converted to $\omega' - \omega$ and $\omega' + \omega$ by multiplication with $\sin \omega t$. That is, the Fourier transform of $s(t) \sin \omega t$ is just the Fourier transform of $s_0(t)$ shifted by $+\omega$ and $-\omega$. This process of generating sum and difference frequencies is performed again when $s(t)$ is subjected to treatment by the synchronous demodulator, and the

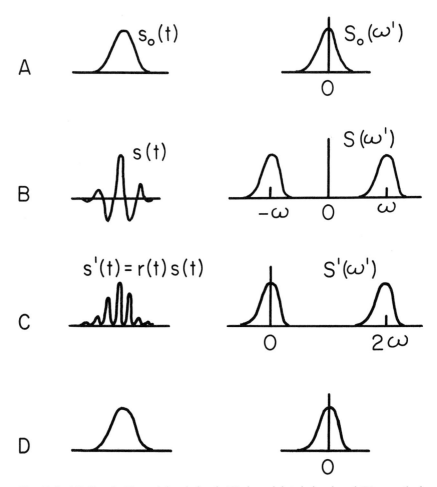

Fig. 12.5 (A) Signal, (B) modulated signal, (C) demodulated signal, and (D) smoothed signal in time and frequency domains.

effect is to shift the transform of $s(t)$ back to $\omega = 0$, and also to $+2\omega$ and -2ω, as in Fig. 12.5C. Finally, the filter smooths $s'(t)$, as in Fig. 12.5D, which is equivalent to eliminating the portion of the Fourier transform centered at $+2\omega$ and -2ω. All of this can be put on a more rigorous mathematical form, but this is postponed until Chapter 13.

We can now readily see that the filtering of noise signals of low frequency can be done only at a sacrifice, because the same low frequencies of $s(\omega')$ will be removed simultaneously. To avoid distorting the signal, its frequency content must then be restricted to even lower frequencies. This means that the use of a greater amount of filtering requires a slower spectrum-scanning rate. This too is discussed in greater detail in the next chapter. Finally, it is clear from Fig. 12.5C that if the signal contains a frequency greater than the chopping frequency ω, the portion of $s'(\omega')$ centered about $\omega' = 0$ will overlap the portion centered about 2ω. This will distort the recorded spectrum.

We have not yet discussed the physical construction of the synchronous demodulator. A direct approach is to operate mechanical switches by means of a cam coupled to the chopper. The function of switches and cams are sometimes combined in the form of slip rings or commutators. Magnetic reed switches operated by rotating magnets have been used, as have photodiodes that are turned off and on by a beam of light interrupted by a rotating sector disk. The diode bridge or ring demodulator (Fig. 12.6) is synchronized by means of a reference signal from a generator that is coupled to the chopper. The generator switches the diodes on and off, and thus switches the signal through alternate paths in the bridge. A similar device uses solid-state switches such as field-effect transistors. Provision for synchronization of demodulators by mechanically phasing some rotating member is always incorporated in a spectrophotometer and should be checked by the spectroscopist occasionally, especially if a loss of sensitivity or complete breakdown is experienced.

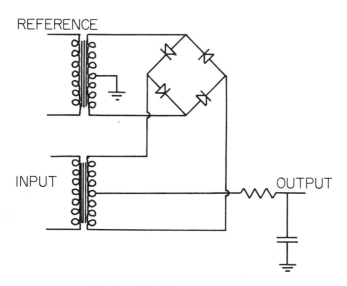

Fig. 12.6 Ring diode demodulator.

The electrical network immediately following the demodulator has two purposes. It is, first of all, a low-pass filter which passes the information of low frequency contained in the signal while it rejects higher frequencies. Thus, it discriminates against noise in favor of the desired signal. It also completes the job started by the demodulator by rejecting the double-frequency ripple superposed on the dc signal. The first function is generally called noise filtering. The filter is represented schematically by a simple *RC* circuit in Fig. 12.3 and 12.4. Often the noise filter is just this simple, but more sophisticated forms are beginning to be used. Noise filtering is necessarily accompanied by some distortion of the desired spectrum. The discussion of this subject is taken up in Chapter 13.

The second function of the filter, suppression of the ripple of double the modulation frequency, is performed by a narrow band-rejection filter. The most commonly used form is the twin-T or parallel-T filter shown in Fig. 12.7. The circuit rejects a narrow band of frequencies centered at

$$\omega_0 = 1/2RC \text{ rad/sec}$$

We now have a demodulated and smoothed signal. If our spectrophotometer is a single-beam instrument, we are ready to record this signal as a function of wavelength. This is done in modern instruments with a recording potentiometer. The principle of the recorder has been described in Section 10.VI. If our spectrophotometer is a double-beam instrument, the system is a bit more complicated.

IV. DOUBLE-BEAM SYSTEMS

There are three general classes of double-beam systems: optical null, ratio computing, and memory. Let us discuss these separately.

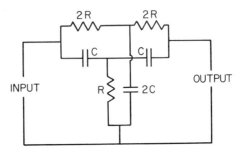

Fig. 12.7   Twin-T filter network.

A. Optical Null Systems

The chopper of a single-beam spectrophotometer periodically interrupts the radiation from the source, creating an ac signal from the dc radiation beam. Stated another way, the radiation from the source (modified by the sample) is periodically compared with radiation from the chopper blade. The difference in these two sources of radiant power is the amplitude of the ac signal. Except in unusual cases, which we discuss in a later chapter, the radiation from the source is greater than the radiation from the chopper, or equal to it at wavelengths where the sample is totally absorbing. The synchronous demodulator converts the ac signal to a dc signal that ranges from zero to positive values.

In a double-beam spectrophotometer, the chopper is replaced with a beam switcher that alternately samples the radiation from the sample and reference beams. The difference in these two levels of radiation power is the amplitude of the ac signal. In contrast with the single-beam case, the sample beam can deliver more or less power than the reference beam. If the sample beam contains more power than the reference beam, the ac signal is converted to a positive dc signal by the synchronous demodulator. This is illustrated in Fig. 12.8A. If, on the other hand, the reference beam delivers more power, the situation is as in Fig. 12.8B. The ac signal is shifted 180° in phase and the demodulated signal is negative. This distinction is important to the operation of an optical null system.

The demodulated and smoothed signal is now remodulated by an electrical chopper (Fig. 12.9). It is usually convenient to drive the chopper magnetically from the main power line. The modulated signal, also shown in Fig. 12.8, differs in phase by 180°, depending on whether the dc signal is positive or negative. This modulated signal is amplified, the wave shape is refined if necessary, and the resulting voltage is applied to one winding of an ac control motor. The other winding — that is, the reference winding — is excited from the same source used to drive the modulator. When the signal is zero, the motor does not turn;

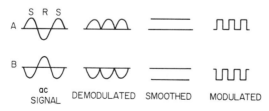

Fig. 12.8   Signal treatment in a double-beam optical null spectrophotometer.

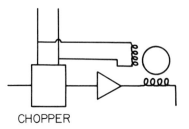

CHOPPER

Fig. 12.9   Driving a control motor with a chopped signal.

when it is not zero and in phase with the voltage applied to the reference winding, the motor turns in one direction; when it is 180° out of phase with the reference winding, the motor turns in the opposite direction. This is, of course, just the way in which the recording potentiometer operates as described earlier. In the optical null spectrophotometer, the motor controls the position of an optical attenuator in the reference beam. If the radiation power in the reference beam exceeds the power in the sample beam, the attenuator is driven farther into the beam until the two powers are equal and the motor stops. If the power in the sample beam is greater, the attenuator is withdrawn. The system is constantly seeking a condition of equal power in the two beams: hence, the name optical null.

A pen can be mechanically positioned by the same mechanism that controls the attenuator. The line traced by the pen when wavelength is scanned is the transmission spectrum of the sample with respect to the transmission of the reference material. Absorbers common to both beams, such as atmospheric CO_2 and H_2O and solvents intentionally placed in the reference beam, do not cause an ac signal and therefore do not contribute to the spectrum. Their absorption is canceled.

In more sophisticated, and versatile, spectrophotometers the pen is not coupled mechanically to the attenuator; it is driven through an additional servosystem. A potentiometer is attached to the attenuator in such a way that the position of the attenuator controls the position of the wiper of the potentiometer. A fixed potential is impressed on the potentiometer. Thus, the potential at the wiper is proportional to the position of the wiper, and therefore to the transmittance of the sample. This potential is introduced into a servosystem much like the one just described for the attenuator. It is modulated, amplified, and applied to one winding of a control motor, which positions the recorder pen. By varying the potential applied to the potentiometer any portion of the transmittance scale can be expanded to full scale. If the potentiometer is wound to generate a logarithm function instead of a linear function, the spectrophotometer can be made to record absorbance.

It is necessary to maintain the optical power in the reference beam at a relatively constant level in order than a small displacement from equilibrium will generate a sufficiently large restoring force to drive the servo, but not so large as to effect its stability. This is done by varying the slit width with wavelength. Means of accomplishing this are discussed a little later.

Recall from an earlier chapter that the dispersion of a monochromator can be doubled by double-passing the radiation. As we pointed out single- and double-passed radiations leave simultaneously through the exit slit, but a chopper placed inside the monochromator serves to discriminate between them. This technique was first applied to a single-beam instrument. Extension to double-beam instruments is more difficult because the beam is already modulated by the beam switcher, but it is not impossible. Let us discuss some examples.

In the system suggested by Rochester of Grubb–Parsons and described by Orr (l), the radiation at the intermediate slit image is chopped at twice the beam-switching frequency f (see Fig. 12.10). The radiation reaching the detector consists of double-passed radiation modulated as in Fig. 12.10B, and superposed single-passed radiation as in Fig. 12.10A. The signal B can be considered to be a superposition of two signals, Figs. 12.10C and D, of frequencies $2f$ and f. The amplitude of C is

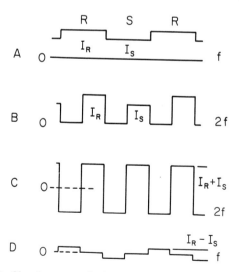

Fig. 12.10 Signal treatment in the double-passed system of Rochester.

$\frac{1}{2}(I_R + I_S)$, the average of the amplitudes of the sample and reference signals. The amplitude of D is $\frac{1}{2}(I_R - I_S)$. The signal D is just the error signal desired to drive the optical attenuator in the reference beam. Notice that D is 45° out of phase with the single-passed radiation signal A. Hence, a synchronous demodulator of frequency f and phased with D will detect the error signal due to double-passed radiation and ignore the signal due to single-passed radiation. If the phasing of the demodulator is not exact, some of the single-passed radiation will be detected, causing serious error in the spectrum.

A similar system was proposed by Daly (2) of Unicam. He uses a chopping rate of half the beam-switching frequency (see Fig. 12.11). The signal at the detector is the superposition of a signal of frequency f with amplitude $I_R - I_S$, a signal of frequency $f/2$ of amplitude $I_R + I_S$, and another signal of frequency $f/2$, 90° out of phase and of amplitude $I_R - I_S$. A demodulator of frequency $f/2$ synchronized with the last of these extracts the difference signal $I_R - I_S$. If the phasing of the demodulator is not exact, a portion of the sum signal of frequency $f/2$ will be detected. Both the Daly and the Rochester systems require two chopping frequencies, complicating the compromise imposed by slow detector response and flicker noise.

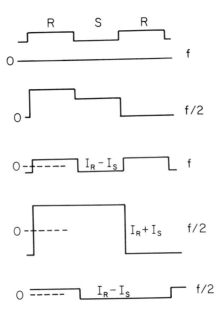

Fig. 12.11 Signal treatment in the double-passed system of Daly.

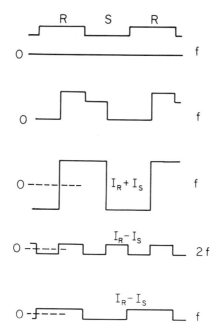

Fig. 12.12 Signal treatment in the double-passed system of Coates and Scott.

A scheme devised by Coates and Scott of Perkin–Elmer and discussed by Golay (3) is shown in Fig. 12.12. The double-passed radiation is chopped with the same frequency as the beam-switching frequency but 90° out of phase. The signal at the detector is a superposition of two signals of frequency f with amplitudes equal to the sum and the difference of the sample and reference beams, and also a difference signal of frequency $2f$. The demodulator detects only the latter. Small phase errors merely reduce the amplitude of the difference signal in this system, without picking up other signals. However, imperfections in the beam-switching mirror, for example, could produce a signal of frequency $2f$, which would be detected.

Double-passed, double-beam systems are roughly $\sqrt{2}$ as wasteful of energy as single-passed systems because of the necessary use of beam choppers in addition to beam switchers. On the other hand, this is more than recovered by the increase in spectral dispersion. Doubled dispersion permits the slits to be opened twice as wide for equivalent resolution and this increases energy at the detector by a factor of 2.

A basic disadvantage common to all optical null systems is that the linearity of such systems depends on the precision of the reference beam attenuator. The ratio computing systems, discussed next, were developed to overcome this problem.

B. Ratio Computing Systems

A variety of ratio computing or ratio recording systems have been proposed and many of them have been constructed and tested. We shall discuss a selection of them in this section. Ratio computing systems can be classified as frequency discriminating systems, phase discriminating systems, or memory systems. The first two are called *time-sharing* systems because sample and reference beams are separated in space (usually) and periodically compared at a rapid rate. The optical null systems are therefore also time sharing. Memory systems are called *space sharing* because the sample and the reference are placed in the same beam at different times. A common characteristic that distinguishes ratio computing systems from optical null systems is the elimination of a reference beam attenuator. The ratio of sample beam power to reference beam power is computed directly from electrical signals.

Typical of the frequency discriminating systems is the simple scheme shown in Fig. 12.13. Sample and reference beams are exposed alternately at frequency f with a shuttered or blank section in between to establish a zero level. The signal at the detector is thus as depicted in Fig. 12.13A. There are numerous ways of computing I/I_0. The signal can be considered to be the sum of the sample and reference signals in Figs. 12.13B and C. A synchronous demodulator switches on and off, thereby extracting the reference signal I_0. A second demodulator extracts I. The demodulators can be simple synchronous switches that send the appropriate signals into separate channels, as in Fig. 12.14. The ratio

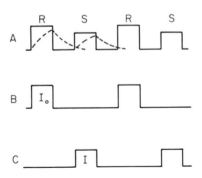

Fig. 12.13   Signal treatment in an elementary ratio computing system.

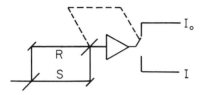

Fig. 12.14   Synchronous demodulator switch.

I and I_0 can be computed by standard means, for example, by impressing I_0 across the slide wire of a recording potentiometer and applying I to the input. The servosystem of the recorder will drive the wiper to the point on the slide wire where a signal equal to I appears. The position of the wiper, and therefore the position of the recorder pen, indicates the fractional part of I_0 represented by I, or $T = I/I_0$.

This simple system is adequate when rapidly responding detectors are used, but there is an inherent error for slow detectors. Thermal detectors, as we have seen, have long time constants and the signals sketched in Fig. 12.13 as square waves are really exponentially rising and decaying signals, shown as dotted lines. The sample beam demodulator, in addition to measuring the sample signal at a given instant, is also measuring a portion gI_0 of the still-decaying reference signal, a portion g^2I of the previous sample signal, and so on. The magnitude of g depends on the decay rate relative to the beam-switching frequency. We say that there is cross talk between sample and reference signals. The reference demodulator is acting similarly. The ratio computed by the system is

$$T' = T\frac{1-g}{1+Tg}$$

We consider only terms through g. Clearly T' is an erroneous representation of T. When T vanishes, T' is apparently equal to g. By offsetting the zero and adjusting the 100% level, T' can be made correct at $T = 0$ and 1. The apparent transmittance then becomes

$$T' = \frac{I + gI_0 + g^2I + \cdots}{I_0 + gI + g^2I_0 + \cdots} \approx \frac{I + I_0 g}{I_0 + Ig} = \frac{T+g}{1+Tg}$$

The more refined system of Hornig *et al.* (*4*) is sketched in Fig. 12.15. A modified system, proposed by Golay (*3*), uses narrower blank sections. The analysis and the succeeding analyses follow those given by Golay. The signal can be considered equivalent to a signal of frequency f (Fig. 12.15B) added to a signal of frequency $2f$ (Fig. 12.15C).

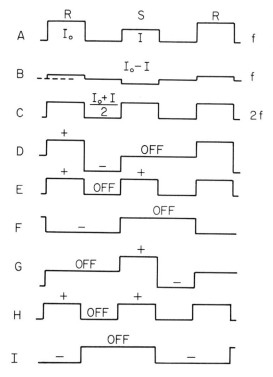

Fig. 12.15   Signal treatment in the ratio computing system of Hornig *et al.*

The $2f$ component has an amplitude of $\frac{1}{2}(I_0 + I)$. The f component has an amplitude $I_0 - I$, but the effective signal amplitude is actually $(I_0 - I)/\sqrt{2}$. This is because only the first harmonics of the signals are used. The f component is zero half of the time, and so its first harmonic is reduced by $1/\sqrt{2}$. This can be shown by expanding the signals in Fourier series. A demodulator with the switching pattern shown in Fig. 12.15D is synchronized with the signal. This pattern is equivalent to a synchronous demodulator of frequency $2f$ plus another one of frequency f shifted in phase $-45°$ from the signal of Fig. 12.15B. These are shown in Figs. 12.15E and F. This demodulator therefore extracts the $2f$ component of Fig. 12.15C plus the f component of Fig. 12.15B, with the amplitude of the latter reduced by a factor $(-1/\sqrt{2})$ because of the phase shift [cf. Eq. (12.3)]. The net signal is thus

$$\frac{I_0 + I}{2} - \left(\frac{1}{-\sqrt{2}}\right)\left(\frac{I_0 - I}{\sqrt{2}}\right) = I_0$$

A second demodulator with the switching pattern of Fig. 12.15G is equivalent to the sum of the patterns of Fig. 12.15H and I. Pattern H is the same as E and I is the same as F, but shifted $+45°$ relative to B. The net signal from this demodulator is

$$\frac{I_0 + I}{2} - \left(\frac{1}{\sqrt{2}}\right)\left(\frac{I_0 - I}{\sqrt{2}}\right) = I$$

Signals representing I and I_0 are thus available separately and the ratio can be calculated as before.

The switching mechanism used to produce these switching patterns is shown in Fig. 12.16. The amplifier provides two outputs that are equal in magnitude but of opposite sign. The switches S_1 and S_2 operate at the beam-switching frequency. When the sample beam is exposed, the two amplifier outputs are switched to the two poles of S_3 which samples them at twice the beam-switching frequency. When the reference beam is exposed, the outputs are switched to S_4. The system as we have described it so far suffers from the same cross-talk difficulty we described above. However, the addition of a resistor R permits a portion h of the difference signals to be fed back, so the apparent sample signal is

$$I' = I + gI_0 + h(I'_0 - I')$$

and the apparent reference signal is

$$I'_0 = I_0 + gI + h(I' - I'_0)$$

These equations are solved to find

$$I' = \frac{I(1 + h + gh) + I_0(g + h + gh)}{1 + 2h}$$

and

$$I'_0 = \frac{I(g + h + gh) + I_0(1 + h + gh)}{1 + 2h}$$

The calculated transmittance, I'/I'_0 equals the correct transmittance I/I_0 if

$$g + h + gh = 0$$

Therefore, if the resistor R is adjusted to make

$$h = \frac{-g}{1 + g}$$

the system will give correct transmittance values in spite of cross talk.

The radiation signal of Figs. 12.13A and 12.15A is equivalent to a beam-switching operation at frequency f superposed on a chopping frequency $2f$. Other chopping frequencies can be used. The chopping

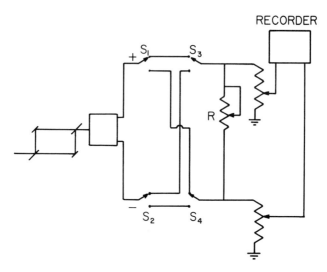

Fig. 12.16   Switches in system of Hornig *et al.*

frequency is generally called the carrier frequency. It can be quite a bit higher than the beam-switching frequency only if a fast-responding detector is used. Notice also that the bandwidth of the amplifier must be widened to accommodate the carrier frequency, and this increases the noise at the demodulator.

A somewhat different system can be constructed in which the sample and reference beams are chopped at different frequencies, with no superposed beam-switching frequency. This is accomplished in the Cary Model 90, for example, by using a stationary beam splitter and beam recombiner, a $13\frac{1}{3}$-cps sample beam chopper and a $26\frac{2}{3}$-cps reference beam chopper. The chopping frequencies can have any ratio, provided the ratio is not an odd number. Synchronous demodulators extract sample beam and reference beam signals separately.

The Halford–Savitzky (5) system used in Perkin–Elmer Models 13 and 301 spectrophotometers is a phase system. This method also requires a stationary beam splitter. In the Perkin–Elmer instruments, the top half of the slit is illuminated with radiation from the sample beam and the bottom half with radiation from the reference beam. Sample and reference beams are chopped at the same frequency f, but with a phase difference of 90°. Two synchronous demodulators of frequency f, phased to match the sample and reference beam signals, extract I and I_0. The ratio is calculated with a servorecorder in the usual way.

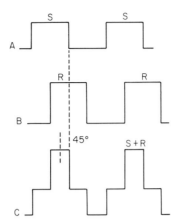

Fig. 12.17 Signal treatment in the ratio computing system of Halford and Savitzky.

The ratio calculation can be done more directly if an approximation is acceptable. The sample and reference beams, chopped 90° out of phase, are depicted in Figs. 12.17A and B, respectively. The combined signal, shown in Fig. 12.17C, for equal sample and reference beams, is shifted in phase 45° relative to the reference beam. The amount of phase shift depends on the relative amplitude of sample and reference. In fact, if the fundamental chopping frequency is f, the combined signal is

$$S = I_0 \sin 2\pi ft + I \cos 2\pi ft$$

from which the amplitude is

$$S = (I^2 + I_0^2)^{1/2}$$

and the phase is

$$\phi = \tan^{-1} I/I_0 \approx I/I_0$$

The phase angle ϕ can be measured directly with a standard electronic instrument. Thus the approximate transmittance can be read directly. Actually the approximation is not bad, as $\phi = \tan \phi$ within 1% for angles less than 45°. By restricting the permitted range of ϕ, the accuracy can be improved. Golay (6) devised a system in which the sample and reference beams used $\frac{2}{5}$ and $\frac{3}{5}$ of the slit height, respectively, thus limiting ϕ to $\tan^{-1} \frac{2}{3}$.

By eliminating the reference beam attenuator and its mechanical imperfections, the ratio computing systems would seem to provide spectrophotometry of higher potential accuracy. There are still difficulties associated with the ratio computers, however. Thermal detectors are slow in their response, so there can be "cross talk" between reference and sample channels in some systems, although there are ways of reducing the importance of this, as we have described. Moreover, because of the necessity for either superposed chopping or stationary beam splitting, the ratio computers are more wasteful of energy and, therefore, are usually somewhat lower in signal-to-noise ratio than optical null instruments. The relative noise properties of various spectrometric systems have been discussed in a more mathematical way by Golay (3) and the interested reader should refer to his paper. It should be kept in mind that Golay assumes perfect chopping and demodulation in his analysis.

C. Memory Systems

The memory principle was proposed by Avery (7) and used in the Beckman IR-3. The spectrum with the beam empty or containing a reference material is recorded and stored in a memory such as magnetic tape. This spectrum is referred to by the instrument while a sample is being run, and the ratio is computed and recorded. The device is therefore a space-sharing ratio-computing system. It apparently does not suffer the energy loss inherent in time-sharing ratio computers, but it is slower because of the necessity of making two runs. This is really equivalent to the energy loss. Moreover, memory systems cannot compensate for short-term variations in source brightness, etc.

V. Slit Programming Systems and Adaptive Systems

The radiant power supplied by the source of an infrared spectrophotometer varies enormously as a function of wavelength. It is necessary to compensate for this in order neither to exceed the allowable input signal nor to fall below a practical signal level. In single-beam spectroscopy this means keeping the recorder pen on scale. In double-beam work, it means keeping the servo loop gain at the desired level.

Various unsatisfactory methods of controlling the power level at the entrance slit can be considered, such as slit-height masks or aperture diaphragms, but these waste energy without providing a useful return. In other words, they degrade the performance of the instrument at all wavelengths to the level of performance at the worst wavelength. Control of the slit width is a satisfactory method because resolution is improved at wavelengths where power can be discarded. Resolution is optimum at every wavelength for a given fixed radiation power input.

In the early days of point-by-point measurements with single-beam instruments and galvanometers, the spectroscopist could adjust the slit width at each wavelength with the sample removed from the beam, measure the galvanometer deflection, insert the sample, again measure the deflection, and calculate the ratio. When recording spectrometers with unchopped radiation came into use, the operator recorded a short spectral section of background signal with a slit width giving a pen deflection varying between perhaps 30 and 100% of full-scale pen travel. He next inserted the sample and repeated the same spectral section. The slit width was then adjusted and the operation repeated for the next spectral section. This section-by-section method was necessary anyway because the dc instrument drifted badly, and the zero level had to be reestablished periodically.

Chopped beam instruments no longer required frequent zero checks, and the periodic stop to reset slit width was a nuisance, so methods of automatically controlling the slit width were invented. Initially, there were mechanical cams coupled to the wavelength drive motor with a rider actuating the slit-adjusting mechanism. In some cases, the cam was a flat metal plate, contoured to program the slit width for constant signal level. The Perkin–Elmer instruments used a helical cam with a string wrapped about it to rotate the slit-adjusting shaft. Fixed cams continue to be used in low-cost spectrophotometers, double-beam as well as single-beam. However, they suffer from inflexibility. It is inconvenient to vary the program for different resolutions and changes in source brightness and the condition of optical components cannot be compensated.

In some of the more sophisticated instruments, the mechanical cam has been supplanted by the electrical cam-controlled servosystem. A typical system is shown in Fig. 12.18. The programmer contains a number of adjusting potentiometers, $R_1, R_2, \ldots, R_5$, arranged in parallel and with a voltage applied across them. The wiper of each is at a fraction of the applied potential determined by its position. R_1 is an interpolating potentiometer. It is a multitapped potentiometer, having a series of taps or contacts along its length with each tap connected to the wiper of one of the adjusting potentiometers. The wiper of R_1 is driven by the wavelength-driving mechanism of the spectrometer. When the instrument is at wavelength λ_1, the wiper is at the potential determined by the wiper of R_1, at λ_2 it is at the potential picked off R_2, and so on. At intermediate positions of the wiper of R_1, the potential is the interpolated value between the potentials at adjacent taps. In order for the interpolation to be linear and also to avoid undue interdepen-

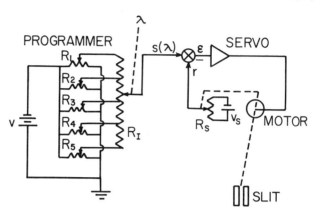

Fig. 12.18   Electrical cam method of programming slit width.

dence of the settings of the potentiometers, it is desirable for the resistance of the individual segments of R_1 to be somewhat higher than the resistance of the adjusting potentiometers. Thus with proper setting of the adjusting potentiometers any curve can be approximated by a series of line segments. The curve in this case is the slit width required to maintain constant energy, but other uses can be found for this device as well.

The function $s(\lambda)$ generated in this way is applied to the input of a servomechanism where it is compared with a reference signal r. The difference, $\epsilon = s - r$, after suitable amplification, drives a motor that positions the wiper of a potentiometer R_s. The wiper picks off a portion of a voltage v_s, and this is the reference voltage r. The position of the wiper of R_s is therefore proportional to $s(\lambda)$. The slit mechanism is driven by the same motor and therefore the slit width is also proportional to $s(\lambda)$. The adjusting potentiometers are set by moving the wiper of R_1 from tap point to tap point (i.e., by setting the instrument at the corresponding wavelengths) and adjusting for the required pen deflection with the spectrophotometer running single beam. If there is some interdependence in potentiometer settings, one or more additional adjustments may be necessary. Usually it is sufficient to maintain the energy constant within a factor of 2.

Tapped potentiometer programmers of this type are flexible enough to permit compensation for slow source or detector changes in time, or for the difference between sources or detectors when they are replaced. However, the interval between taps is too coarse to permit adjustment for rapid changes in background energy, as when an absorbing material is present in the reference beam.

It would be advantageous to have a slit-adjusting mechanism that would automatically adjust itself and do it with sufficient sensitivity to compensate for background absorption. This is an example of an adaptive servosystem. If a signal proportional to the reference beam energy is available, such a system can be constructed as sketched in Fig. 12.19. A fixed but adjustable reference voltage v_r is introduced into the servosystem and subtracted from the signal v_0 to form an error signal $\epsilon = v_r - v_0$. The error signal is amplified and made to drive the motor that opens and closes the slits. The reference beam energy I_0 varies with the slit width and the signal v_0, which is proportional to the reference beam energy, is subtracted from v_r, thus completing the loop. The slits are thereby automatically adjusted for a constant reference beam energy, which can be selected by adjusting the magnitude of v_r.

The necessary signal representing the reference beam energy is directly available in all of the amplitude-type ratio computing systems and in the phase systems of the Halford–Savitzky type described in the preceding section. It is also available in memory-type systems. Hence, ratio computing systems normally use self-adjusting slit programs. Optical null systems do not have a signal proportional to reference beam energy as an essential feature of their operation, and so they usually have cam-operated slit programs. However, if the beam is chopped at twice the beam-switching frequency $(2f)$, a signal occurs with frequency $2f$ and amplitude equal to the sum of the reference and sample beams. This has been discussed in connection with Rochester's double-passed system and is illustrated in Fig. 12.10. This signal can be extracted with a synchronous demodulator of frequency $2f$. As the reference beam and sample beam energies are equal in an optical null system, the signal can be used to maintain the desired constant energy. Moreover, the precision requirements of a ratio computing system need not be met because we merely ask that the energy be constant within a factor of perhaps 2. Consequently, the $2f$ chop-

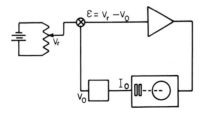

Fig. 12.19 Adaptive slit-control servo.

ping does not have to interrupt the beam 50% of the time; a very brief interruption of the beam every half-cycle of the beam-switching operation is sufficient. If the detector signal fails to decay completely during this interruption it is of no consequence, as a signal approximately proportional to the reference beam energy is all that is required. Therefore, the price in energy lost by chopping is not at all important. In fact, the normal beam-switching mirrors usually provide the beam interruption anyhow and no additional choppers are required.

These systems for adjusting slit width are examples of adaptive servosystems. They adapt the instrument to changes in conditions, either by anticipating the change and preprogramming the variable parameter (slit width), or by sensing changes and automatically changing the parameter as required. There are other examples of adaptive systems used in infrared spectrophotometers. Let us discuss a couple of them.

If a spectrum is scanned rapidly, the finite response time of the spectrophotometer systems prohibits it from faithfully reproducing the shape of the spectrum. Bands will be shifted, weakened, and distorted. This is discussed in detail in Chapter 13, but for now it should be evident that this can be ameliorated either by scanning more slowly or by reducing the response time of the instrument.

The distortion depends on the relationship between scanning rate and the observed bandwidths of the spectrum, and the observed bandwidth depends on the spectral bandwidth of the monochromator and the natural half-widths of the bands. For most efficient use of the monochromator, the scanning rate should be varied to cover a constant number of spectral bandwidths per unit time interval. In some instruments the scanning rate is programmed to accomplish this.

Speed suppression systems slow the scanning rate when the instrument falls behind the spectrum and return it to its normal scanning rate after it has caught up. Regions where there are no bands are scanned rapidly. This is obviously a much more economical use of instrument time than scanning the complete spectrum at a reduced rate. The speed suppression system used on some early instruments consisted of a tachometer coupled to the pen drive mechanism which fed back a voltage to the wavelength drive motor proportional to the pen speed. Thus, when the pen was traveling fast, either up scale or down scale, the scanning rate was decreased. Unfortunately, at the band peaks where the pen was not moving, the scanning rate increased and the tips of bands acquired a squared-off appearance. Newer optical null instruments extract an error signal proportional to the difference between the sample and reference beam energies and use it as a measure of how well the instrument is following the spectrum. If the error signal is zero, the spectrum is being followed exactly. If it is large, obviously the error in the recorded spectrum is large at that point. One method of making use of the error signal is illustrated in Fig. 12.20. The error signal is extracted following the synchronous demodulator, before it enters the comb servo amplifier. It is filtered, modulated, and amplified to a level high enough to light a small lamp. The brightness of the lamp depends on the magnitude of the error signal. The lamp illuminates a photodiode whose resistance is high in the dark, but falls when the photodiode is exposed to light. The photodiode shunts the current supplied to the wavelength motor and thereby controls its speed, slowing it down when the error signal is large. The time constant of the system is adjusted to maintain some speed suppression when the error signal decreases at the peaks of sharp bands.

The same system can be used to adapt the response time of an optical null spectrophotometer to changes in error signal. In this case the time constant is reduced if the instrument falls behind the spectrum, permitting it to catch up rapidly. Of course, the noise level rises at the same time, but this occurs usually on the steep slopes of absorption

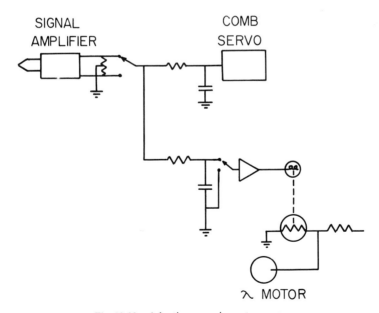

Fig. 12.20   Adaptive scanning-rate servo.

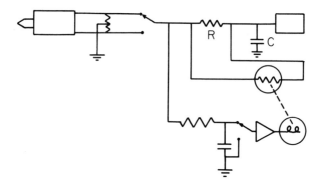

Fig. 12.21   Adaptive time-constant-control servo.

bands where the noise is not evident. A typical system is sketched in Fig. 12.21. In this case, the photodiode is shunted across the resistor of the RC filter. The resistance of the combination is reduced when the error signal is large, and the time constant, which is the product of the resistance and the capacitance, is decreased.

REFERENCES

1. S. F. D. Orr, *J. Opt. Soc. Am.*, **43**, 709 (1953).

2. E. F. Daly, *Nature*, **171**, 560 (1953).

3. M. J. E. Golay, *J. Opt. Soc. Am.*, **46**, 422 (1956).

4. D. F. Hornig, G. E. Hyde, and W. A. Adcock, *J. Opt. Soc. Am.*, **40**, 497 (1950).

5. A. Savitzky and R. S. Halford, *Rev. Sci. Inst.*, **21**, 203 (1950).

6. M. J. E. Golay, *J. Opt. Soc. Am.*, **45**, 430 (1955).

7. W. H. Avery, *J. Opt. Soc. Am.*, **31**, 633 (1941).

CHAPTER 13

The Electromechanical Transfer Functions
of Infrared Spectrophotometers

I. INTRODUCTION

In Chapter 7 we discussed the spectral modulation transfer function and the slit function of a monochromator, and the role played by these functions in distorting the shape of recorded spectra. In Chapter 10 we considered the transfer functions and impulse response functions of electrical networks. We saw that a simple RC circuit introduces a phase delay and an amplitude reduction into a sinusoidal input signal; that is, it distorts the signal. In this chapter we discuss the transfer function of a complete spectrophotometer due the the electrical and mechanical components. We then investigate the distortion of recorded spectra produced by the nature of the electromechanical transfer function. The reader may wish to read the reviews of Rautian (1) and Seshadri and Jones (2).

II. THE MATHEMATICAL FORM OF THE ELECTROMECHANICAL TRANSFER FUNCTION

Let us first consider a hypothetical single-beam spectrophotometer, assuming for simplicity that all signals, optical and electrical, are dc.

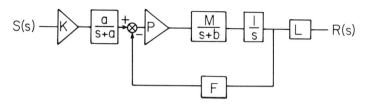

Fig. 13.1 Block diagram of a single-beam spectrophotometer.

That is, we disregard the action of the radiation chopper and demodulator. Ratio-computing double-beam instruments can be treated in the same way as single-beam instruments, and do not require a separate discussion. The block diagram for the system transfer function is shown in Fig. 13.1 in terms of Laplace transforms.

The input signal $s(t)$ is an optical signal with transform $S(s)$. The input is converted to an electrical signal by the detector and amplifier. It is filtered and then introduced into the recorder. If an error signal $\epsilon = v_S - v_R$ exists, it is amplified by the pen servo amplifier and impressed on the windings of the servomotor. The motor responds by turning at a rate $d\theta/dt$. The constants in the transfer function of the motor, $M/(s + b)$, can be written in terms of the properties of the motor and associated gears, and so on (see Clark (3), for example). The angular position of the motor shaft, θ, is the integral of $d\theta/dt$, and so the corresponding contribution to the transfer function is $1/s$. A potentiometer with a fixed impressed voltage is mounted on the motor shaft. A voltage v_R which is proportional to θ is picked off the slide wire and subtracted from v_S to generate the error signal, and the loop is complete. The pen position is proportional to θ, and provides the output $r(t)$. The system transfer function can be written in terms of the feedforward and feed back parts of the loop by inspection. Using the equation of Section 10.VII,

$$H(s) = F(s)/G(s) \tag{13.1}$$

we find

$$H(s) = \frac{R(s)}{S(s)} = \frac{LKa/(s + a) - PM/s(s + b)}{1 + PMF/s(s + b)}$$

$$= \frac{KLPMa}{[s(s + b) + PMF](s + a)} \tag{13.2}$$

The theory of control systems treats the behavior of poles and zeros of the transfer function in the complex s plane as the system parameters are varied. In the transfer function of Eq. (13.2) the pole of the noise

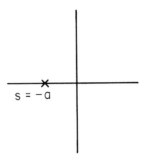

Fig. 13.2   Noise filter pole at $s = -a$ in the s plane.

filter at $s = -a$ contributes an exponentially decaying impulse response function with time constant $1/a$. This is unaffected by changes within the recorder loop (Fig. 13.2). As the loop gain PMF of the pen servo is increased, the two poles of the recorder loop follow the paths shown in Fig. 13.3. They can be (1) separate and real, giving an overdamped response or a simple exponential response if one of them is dominant; (2) coincident, giving a critically damped response; or (3) complex, giving an underdamped response. The noise filter contribution is expected to be dominant in infrared instruments, in the sense that its time constant is significantly longer than the response time of the rest of the system.

The optical null double-beam spectrophotometer is necessarily more complicated. The detector and signal amplifier are inside a servo loop, as shown in the block diagram of a hypothetical instrument in Fig. 13.4. The input optical signal from the sample beam is compared with the reference beam signal by the detector and its transformer. The differ-

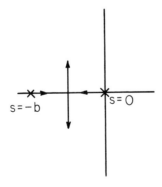

Fig. 13.3   Locus of recorder loop poles with loop-gain changes.

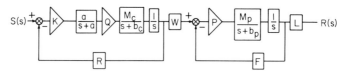

Fig. 13.4 Block diagram of a double-beam optical null spectrophotometer.

ence ϵ is amplified, filtered, and further amplified by the comb servo amplifier. It drives the comb servomotor, whose shaft rotational position determines the position of the reference beam attenuator comb. The attenuator transmits a portion R of the reference beam energy to the detector. The shaft position of the comb servomotor is converted into an electrical signal with a potentiometer, and this signal is treated by the pen servo just as in the single-beam configuration.

The contribution to the transfer function by the pen servo loop is the same as in the single-beam case. The transfer function of the comb servo loop can be written by inspection of the block diagram using Eq. 13.1. The overall transfer function is

$$H(s) = \frac{R(s)}{S(s)} = \frac{KaQM_cW}{s(s+a)(s+b_c) + KaQM_cR}$$

$$\cdot \frac{PM_p}{s(s+b_b) + PM_pF} \cdot L \qquad (13.3)$$

The behavior of the poles of the pen servo loop is just as in the single-beam case (Fig. 13.5), but the noise filter is now part of the comb servo loop and its contribution changes as the loop parameters are varied. The behavior of the three poles of the comb loop is shown in Fig. 13.6. As the loop gain $KaQM_cR$ is increased, one of the poles moves uni-

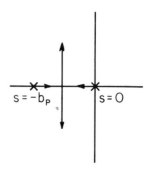

Fig. 13.5 Locus of pen servo poles.

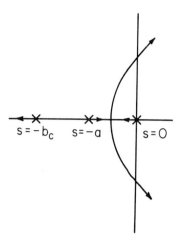

Fig. 13.6 Locus of comb servo poles.

formly to the left and contributes nothing of significance to the response. As the other two poles travel the paths shown, the system changes from overdamped to critically damped to underdamped. Unlike the single-beam case, these poles can cross into the right half-plane, making the system unstable.

It is important to observe a fundamental distinction between single- and double-beam spectrophotometers. The response of the single-beam instrument is governed by the time constant of the noise filter, which is independent of the other parameters of the system. The response of an optical null double-beam instrument varies with the loop gain and is thus dependent on parameters such as amplifier gain K and reference beam energy R.

Often, elements and even complete feedback loops are added to control systems in order to modify the response or stability of the system. Because of the friction and inertia of the mechanical parts of a servosystem, it is desirable to have a high loop gain, so a small error signal can produce a large driving signal to start the parts moving. Once underway, however, it might be desirable to ease off on the driving force to avoid overshoot when the error signal is reduced to zero. This can be accomplished with a rate feedback loop. A signal proportional to the speed of the servomotor is fed back to reduce the magnitude of the error signal. The fed-back signal is zero before the motor begins to move and increases as the motor speeds up. A tachometer generator coupled to the motor shaft can be used to generate

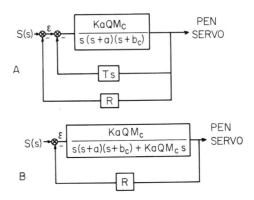

Fig. 13.7   Equivalent block diagrams of rate-controlled pen servo.

the signal. The block diagram of the comb servo part of the system is given in Fig. 13.7A. The contribution in the individual blocks of Fig. 13.4 have been combined in one block. The signal proportional to the rate of travel of the motor is Ts, because the Laplace transform of the differential operator is s. The inner loop is simplified with the aid of Eq. (13.1), reducing the block diagram to the form shown in Fig. 13.7B. The overall transfer function, including the pen servo, is

$$H(s) = \frac{KaQM_cW}{s(s+a)(s+b_c) + KaQM_cs + KaQM_cR}$$
$$\cdot \frac{PM_pL}{s(s+b_b) + PM_pF}$$

An alternate system, used in the Beckman IR-4 series of instruments, is the rate feedback damping circuit of Braymer (4). The system is shown schematically in Fig. 13.8A. Not only does the comb potentiometer relay input information to the pen servo, as in Fig. 13.4, but it also feeds back the signal in a small loop within the large reference beam comb loop. This system is presented in equivalent form in Fig. 13.8B by redefining ground potential and the signs of the potential across the comb slide wire. The system is reduced further in Fig. 13.8C by replacing the comb servo amplifier and motor with its transfer function, and the optical input with a fictitious generator. The transfer function of the circuit containing R_1, R_2, and C can be written with reference to Fig. 13.9 by writing the sum of the currents at the node

$$\frac{E_\epsilon(t) - K[E_s(t) - \theta(t)E_2]}{R_1} + \frac{E_\epsilon(t)}{R_2} + \frac{d}{dt}[E_\epsilon(t) + \theta(t)E_2]C = 0$$

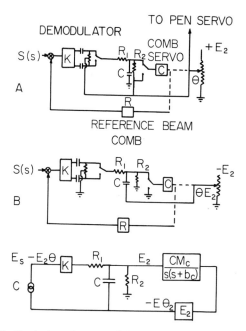

Fig. 13.8 Equivalent diagrams of the rate feedback system of Braymer.

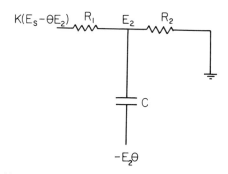

Fig. 13.9 Equivalent circuit from rate feedback system.

This equation can be written in terms of the Laplace transforms

$$\frac{E_\epsilon(s) - K[E_s(s) - \theta(s)E_2]}{R_1} + \frac{E_\epsilon(s)}{R_2} + Cs[E_\epsilon(s) + \theta(s)E_2] = 0$$

or

$$\frac{KE_s}{R_1} - \theta\left[Cs + \frac{K}{R_1}\right]E_2 = \left[\frac{R_1 + R_2}{R_1 R_2} + Cs\right]E_\epsilon$$

Now we substitute the comb servo transfer function

$$\theta(s) = \frac{QM_c}{s(s+b_c)}E_\epsilon(s)$$

and write

$$\frac{\theta(s)}{E_s(s)} = \frac{K/R_1}{s[(s+b_c)/QM_c][Cs+(R_1+R_2)/R_1R_2]+E_2(Cs+K/R_1)}$$

The overall transfer function of the instrument, including the pen servo, is thus

$$H(s) = \frac{K/R_1}{s[(s+b_c)/QM_c][Cs+(R_1+R_2)/R_1R_2]+E_2(Cs+K/R_1)}$$

$$\cdot \left[\frac{PM_pL}{s(s+b_b)+PM_pF} \right] \tag{13.4}$$

The behavior of the poles of Eq. (13.4) in the s plane is sketched in Fig. 13.10. The pole that remains on the negative real axis is usually dominant and provides an exponential response, as we see in the next section.

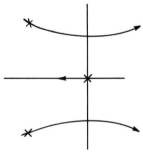

Fig. 13.10 Locus of poles in rate feedback system.

III. Measurement of the Transfer Function

The transfer function, as suggested by the nature of the Fourier or the Laplace transform, can be thought of as the complex frequency response of a system. The Fourier transform of a sinusoidal input function of frequency ω_0 is a pair of Dirac delta functions

$$\mathscr{F}\{\sin \omega_0 t\} = \delta(\omega-\omega_0)+\delta(\omega+\omega_0) = \begin{cases} 1 & |\omega| = |\omega_0| \\ 0 & |\omega| \neq |\omega_0| \end{cases}$$

We shall consider only positive frequencies.

With such an input, the output of a linear system is

$$R(\omega) = H(\omega_0) = |H(\omega_0)| e^{j\omega_0 t}$$

Thus, the output is also sinusoidal, with amplitude and phase equal to that of the transfer function at frequency ω_0. Conversely, if the amplitude and phase of the output of a linear system are measured as functions of the frequency of a sinusoidal input, the transfer function is thereby determined. We now consider a method of doing this.

Because the detector is part of the feedback loop of a double-beam spectrophotometer, the sinusoidal input must be in the form of a sinusoidally varying optical signal. More realistically, because negative optical signals cannot be generated, an input of the form

$$S(t) = \sin^2 \tfrac{1}{2}(\omega_0 t) = \tfrac{1}{2}(1 - \cos \omega_0 t)$$

is used. A sine-squared varying infrared signal can be generated with a properly shaped aperture in a rotating sector disk. However, it is not inconvenient to work with visible light in most infrared instruments if the grating or Littrow mirror can be properly positioned. This permits the use of a rotating polarizer to produce the desired signal.

A device used by the author (5) in the Beckman research laboratories to generate sine-squared modulated visible light has a rotating polarizer and fixed analyzer. The polarizer is a sheet of Polaroid mounted between glass disks, 10 cm in diameter. A square of sheet Polaroid serves as the analyzer. In order to obtain the wide range of frequencies required, namely 0.003–5.5 cps, the polarizer is mounted on the output shaft of a gear box with a selection of nine gear clusters. The input shaft is turned by an ac control motor. A dc tachometer is coupled to the motor, and its output is measured with a voltmeter to determine the modulation frequency. The device is placed in the sample compartment of a spectrophotometer. The amplitude of the output of the instrument is measured from the recorder trace.

In order to measure the phase shift between input and output signals, a miniature light bulb is installed in front of the rotating Polaroid. Its light is detected by a cadmium sulfide detector behind a Polaroid analyzer. The signal from the cadmium sulfide detector is placed on one axis of an oscilloscope, and a signal taken from the recorder potentiometer of the spectrophotometer is placed on the other axis. The phase is measured from the Lissajous figure. For the most part the amplitude of the transfer function provides all the information that is needed. Should phase information be required, it can also be obtained by calculation from the amplitude information, as the two are not independent in so-called minimum-phase linear systems. They are, in fact, Hilbert transforms of one another.

A particularly meaningful presentation of the transfer function is in the form of the Bode plot in which the amplitude in decibels (20 times the logarithm of the amplitude) is plotted against log ω. This has been discussed in Section 10.VII. The amplitude plot for a simple exponential system approaches two lines asymptotically, a straight line of zero slope at lower frequencies and a straight line of -20 dB/decade at higher frequencies. The lines intersect at a frequency ω_c rad/sec, which is related to the decay time τ through the relationship

$$\tau = 1/\omega_c$$

At ω_c the amplitude is -3 dB. Other poles add further changes in slope of -20 dB/decade at their corresponding frequencies. The phase change in a simple exponential system varies from $0°$ at low frequencies to $-90°$ at high frequencies, passing through $-45°$ at ω_c. Additional poles contribute their own phase shifts, which are additive.

A. Transfer Function of a Single-Beam Spectrophotometer

The experimentally determined amplitude of the electromechanical transfer function of a Beckman IR-9 operated as a single-beam instrument is plotted in Fig. 13.11 for the four nominal response periods, 1, 2, 8, and 32 sec, of the noise filter. The actual periods (recall that a

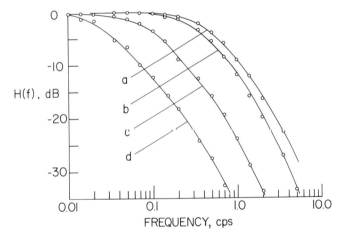

Fig. 13.11 Electromechanical transfer function of a single-beam spectrophotometer with nominal noise filter response periods of (a) 1, (b) 2, (c) 8, and (d) 32 sec.

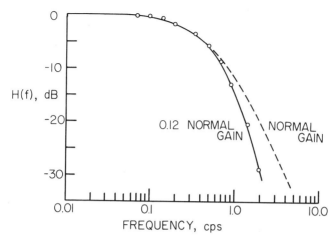

Fig. 13.12 Effect on transfer function of reducing pen servo gain.

period is four time constants) measured from the -3-dB points are 1.4, 1.9, 7.1, and 25,6 sec. Evidence of a second pole is seen in the neighborhood of 1 cps, where the slope increases. This is one of the two poles of Fig. 13.2 associated with the pen servosystem. When the gain of the pen servo amplifier (P) is reduced, this pole moves closer to the origin, corresponding to a longer time constant. Figure 13.12 is the experimental transfer function of the IR-9 with the noise filter set for a 2-sec response period, but with the pen servo gain reduced by a factor of 0.12 from its normal value. This makes the pen servo pole nearly coincident with the noise filter pole, and a loss of higher frequency information is suffered.

B. Transfer Function of a Double-Beam Spectrophotometer

The transfer function for double-beam operation of a Beckman IR-9 is shown in Fig. 13.13. An aperture was inserted in the reference beam to compensate for the losses imposed by the polarizer. The measured response periods, from the -3-dB points, are 1.7, 2.2, 7.0, and 27.9 sec. At the faster response times, the pole on the real axis of Fig. 13.6 is moved to the left, and the two complex poles respond by moving to the right. They appear in the measured transfer function with 1- and 2-sec response periods as typical resonancelike features near 2 cps. The effect of reducing the comb servo amplifier gain (Q) to 0.1 of its normal value is seen in Fig. 13.14. In contrast with the single-beam case (Fig. 13.12), the response time of the dominant pole

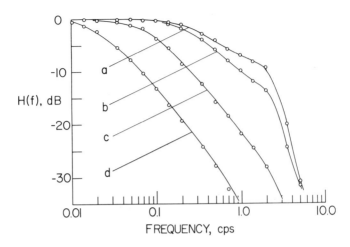

Fig. 13.13 Electromechanical transfer function of a double-beam spectrophotometer with nominal noise filter response periods of (a) 1, (b) 2, (c) 8, and (d) 32 sec.

now also changes because of its involvement in the servo loop. In this example the response period increases from 2.2 to 2.9 sec.

The effect of modifying the loop gain by changing the reference beam energy is shown in Fig. 13.15 for a nominal 2-sec response period. A convenient measure of the reference beam energy in Beckman spectrophotometers is the ratio of single-beam to double-beam pen deflection for the same gain, slid width, etc. With a single-beam to

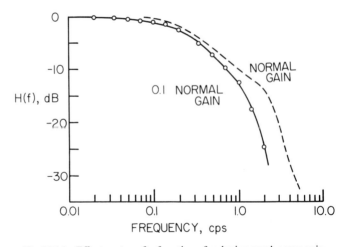

Fig. 13.14 Effect on transfer function of reducing comb servo gain.

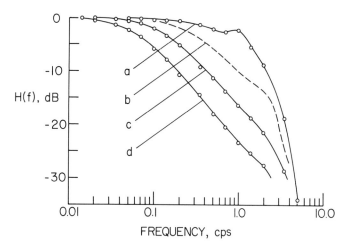

Fig. 13.15   Effect on transfer function of reduced reference beam energy. Single-beam/ double-beam ratios are (a) 2, (b) 1, (c) $\frac{1}{2}$, and (d) $\frac{1}{4}$.

double-beam energy ratio of 1/2, the response period increases to 4.9 sec, an increase by a factor of 2.2. With a single-beam to double-beam ratio of 1/4 the response period is 10.2 sec, an increase by a factor of 4.6. Thus, to a fair approximation, the response period increases in the inverse ratio of the single-beam to double-beam energy ratio. The fourth curve of Fig. 13.15 is for a single-beam to double-beam ratio of 2. Here the effect of the complex poles (1 cps) is so pronounced as to render meaningless a definition of time constant in terms of the −3-dB point.

IV. NOISE FILTERS

The dominant noise in a properly designed and maintained infrared spectrophotometer originates in the detector. Whether it is Johnson or shot noise, it has a uniform spectral distribution, meaning that the noise power in any unit increment of frequency is constant. The noise in an infrared instrument is filtered in two ways. First by cross correlation in the synchronous rectifier, which retains only the noise in a band around the demodulation frequency. This was discussed in Section 12.III. The second type of noise filtering is usually done by a passive RC network that passes low frequencies and stops high frequencies. Its action necessarily introduces distortions into the recorded information.

We have already given the Laplace transform transfer function of the simple single-exponential decay filter:

$$e^{-at} \leftrightarrow \frac{1}{s+a} \tag{13.5}$$

where a is the reciprocal of the time constant. In terms of the Fourier transform, the transfer function is

$$e^{-at} \leftrightarrow \frac{1}{j\omega+a} \tag{13.6}$$

Transfer functions of some more complicated filters are given as we proceed. Transfer functions of standard RC networks may be found in numerous textbooks on control systems or electrical engineering, along with procedures for designing networks to realize desired transfer functions.

The noise power transmitted by a linear system is, by definition, the mean square of the noise amplitude. If "white" noise (i.e., frequency independent) of spectral power density k^2/π per unit interval of frequency is introduced at the input of the filter described by Eq. (13.6), then the transmitted noise power is

$$\langle N^2 \rangle = \frac{2a^2k^2}{\pi} \int_0^\infty \frac{d\omega}{\omega^2+a^2} = k^2a \tag{13.7}$$

The filter has been "normalized" to make $H(O) = 1$. The noise in the output is also obviously proportional to the gain G of the amplifier. Therefore, the root-mean-square noise is given by

$$N_{rms} \propto G\sqrt{a} = \frac{G}{\sqrt{\tau}} \tag{13.8}$$

and is inversely proportional to the square root of the time constant. This relationship is easily verified experimentally by calculating the rms noise from recorded noise records.

Equation (13.8) represents the noise in electrical units (i.e., volts). If a spectrophotometer is operated double beam with a single-beam to double-beam energy ratio

$$M = SB/DB$$

different from unity, then the noise in terms of units of pen displacement is expanded in the inverse ratio. Moreover, the response time is now different from the nominal or single-beam response time τ_0. In fact, we showed experimentally (Fig. 13.15) that for filters of the type we are considering here

$$\tau \approx \tau_0/M \qquad (13.9)$$

so the noise can be written

$$N_{\text{rms}}^{(\text{pen})} \propto \frac{G}{M(\tau_0/M)^{1/2}} = \frac{G}{(M\tau_0)^{1/2}} \qquad (13.10)$$

This equation is of course applicable as well to instruments for which the single-beam to double-beam ratio is not readily measured. In such cases M could be interpreted as pen displacement per volt of signal, for example. The ratio M, as we said, is nothing more than a convenient measure of the reference beam energy or the closed-loop gain. It follows that M can be written

$$M \propto GW^2T \qquad (13.11)$$

where W is the monochromator slit width and T is the transmittance of the reference beam; G is equivalent to K in Eq. (13.3) and W^2T is equivalent to R. Now Eq. (13.10) can be rewritten with Eq. (13.11) to read

$$N_{\text{rms}}^{(\text{pen})} \propto \frac{\sqrt{G}}{W(T\tau_0)^{1/2}} \qquad (13.12)$$

We now turn to the question of distortion of the output of a linear system, and in particular, distortion of spectral line shapes by the noise filter. This distortion is termed *tracking error*. The problem has been treated by a number of workers. Brodersen (6) considered a spectral line with a Gaussian contour distorted by filtering with a simple exponential (single time constant) network and also with a critically damped network. Schubert (7) discussed a Gauss profile distorted by a simple exponential filter and he considered the optimum choice of parameters. Frei and Gunthard (8) treated both Gauss- and Lorentz-shaped lines scanned with a rectangular slit function and further distorted with a simple exponential filter. Schreiber (9) also discussed Gauss- and Lorentz-shaped lines distorted by a simple exponential filter, but he included the effect of a triangle slit function. Stewart (5) investigated both theoretically and experimentally the distortion of a Gauss-shaped line by a number of more complicated filter networks. The discussion of the remainder of this section is based on that study.

The Gauss function is selected for mathematical convenience and because many spectral lines resemble, at least superficially, a Gaussian curve. Moreover, there is some evidence that the slit functions of typical monochromators have Gaussian contours for intermediate slit widths.

A. Simple Exponential Filter or Lag Filter

The simple exponential filter can be constructed of a single resistor and capacitor (see Section 10.II). The transfer function and impulse response are given by Eqs. (13.5) and (13.6). A Gaussian spectral band has a contour

$$s(t) = \exp\left(-c^2 t^2 / 2\sigma^2\right) \tag{13.13}$$

in a time representation, where we suppose the line is scanned at a rate of c cm^{-1}/sec and where the peak is arbitrarily placed at $t = 0$. The line has a peak height of unity and a variance σ^2 (wavenumbers squared) in terms of which its half-width (width at half-maximum intensity) is

$$\Delta\nu_{1/2} = 2.35\sigma \text{ wavenumbers}$$

The treatment is given for an emission line, but a trivial modification extends it to absorption lines.

The convolution of the impulse response and the spectral line can be evaluated by direct integration, using a standard form found in tables of integrals,

$$r(t) = \int_{-\infty}^{t} a \exp\left[-a(t-t')\right] \exp\left(-\frac{c^2 t'^2}{2\sigma^2}\right) dt'$$

$$= \sqrt{\frac{\pi}{2}} \frac{\sigma a}{c} \exp\left(-\frac{c^2 t^2}{2\sigma^2}\right) \exp\left[\frac{c^2}{2\sigma^2}\left(\frac{\sigma^2 a}{c^2} - t\right)^2\right]$$

$$\cdot \left[\operatorname{erfc} \frac{c}{\sqrt{2}\sigma}\left(\frac{\sigma^2 a}{c^2} - t\right)\right] \tag{13.14}$$

The factor in brackets is the complementary error function

$$\operatorname{erfc} x = \frac{2}{\sqrt{\pi}} \int_{x}^{\infty} e^{-\xi^2} d\xi = 1 - \operatorname{erf} x$$

and erf x is the error function. In order to display Eq. (13.14) graphically let us make some substitutions. We define

$$K = \frac{\sigma a}{\sqrt{2}c} = \frac{\Delta\nu_{1/2} a}{3.32 c} \tag{13.15}$$

Hence, $1/K$ is 3.32 times the scanning rate expressed as half-bandwidths per time constant. We also substitute

$$x = \frac{\nu}{\sqrt{2}\sigma} = \frac{ct}{\sqrt{2}\sigma} \tag{13.16}$$

With this substitution, Eq. (13.4) becomes

$$r(x) = \sqrt{\pi}Ke^{-x^2} \exp\left[(K-x)^2\right] \mathrm{erfc}(K-x) \qquad (13.17)$$

This equation is plotted in Fig. (13.16) for selected values of K.

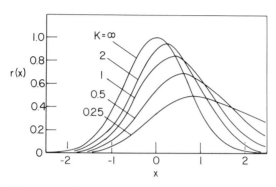

Fig. 13.16 Distortion of a Gaussian band by a lag filter.

The distorted spectral line $r(t)$ suffers in three respects: its peak is shifted; its peak intensity is reduced; and its symmetric shape is lost. The first two effects are not independent, for if $s(t)$ is any input (which does not even have to be symmetric) introduced into a simple exponential filter, we can differentiate the output

$$r(t) = a \int_{-\infty}^{t} \exp\left[-a(t-t')\right]s(t')\,dt'$$

to find

$$\frac{dr(t)}{dt} = -a^2 \int_{-\infty}^{t} \exp\left[-a(t-t')\right]s(t')\,dt' + as(t)$$

$$= a(-r(t) + s(t))$$

The peak of the output curve occurs where $dr(t)/dt = 0$, and therefore

$$r(t) = s(t) \qquad (13.18)$$

Equation (13.18) states that the peak of the distorted output curve falls on the contour of the input curve. This can be readily seen in the curves of $r(t)$, for a Gaussian $s(t)$, of Fig. 13.16. It is also easily demonstrated experimentally.

Now let us find the centroid $\bar{t}$ of the output curve $r(t)$ (i.e., the "center of gravity"). We take advantage of a theorem on Laplace transforms which states that for a normalized function $r(t)$

$$\Delta\bar{t} = -\lim_{s\to 0}\frac{\partial R(s)}{\partial s} \qquad (13.19)$$

where $R(s)$ is the Laplace transform of $r(t)$.

From Eq. (13.5) for the transfer function, the transform of the output is

$$R(s) = \frac{aS(s)}{s+a}$$

The time lag of the centroid is therefore

$$\Delta\bar{t} = \frac{1}{a}[S(0) + jaS'(0)] \qquad (13.20)$$

Now $S(0) = 1$, because we have specified that $s(t)$ is normalized and $S'(0) = 0$ for any $s(t)$ having its centroid at $t = 0$. This will be true for any symmetric function. Hence, Eq. (13.20) becomes

$$\Delta\bar{t} = 1/a = \tau \qquad (13.21)$$

This important result states that the centroid of a symmetric function, when filtered by a simple exponential filter, is shifted by an amount equal to the time constant of the filter. It is sometimes stated that the peak of $r(t)$ is shifted by τ, but this is only approximately true. It is a good approximation when the shift is small. To the extent that this approximation holds, we can write for the peak shift of a Gaussian line

$$\Delta\bar{t} \approx 1/a = \tau \quad \text{or} \quad \Delta\nu = c\tau$$

and for the peak intensity, from Eq. (13.13),

$$I \approx \exp(-c^2/2\sigma^2 a^2) = \exp(-c^2\tau^2/2\sigma^2) \qquad (13.22)$$

An experimental example of tracking error in a spectrophotometer using a simple lag noise filter is shown in Fig. 13.17. Two bands in the absorption spectrum of indene were scanned at various rates and with various time constants to provide data at a succession of values of K. The agreement with peak heights calculated for Gaussian-shaped bands is good. Calculated peak height errors for the lag filter and other filters discussed in the following sections are plotted in Fig. 13.18.

B. Overdamped Filter or Lag-Lag Filter

The impulse response of a lag-lag filter is

$$h(t) = \frac{ab}{a-b}[e^{-at} - e^{-bt}]$$

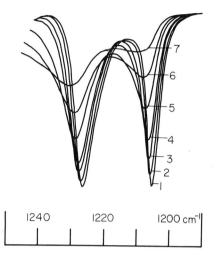

Fig. 13.17 Absorption bands of indene scanned with a lag filter, with the following values of time constant (seconds) and scanning rate (cm^{-1}/min), respectively: (1) $\frac{1}{4}$, 8; (2) $\frac{1}{4}$, 80; (3) $\frac{1}{4}$, 200; (4) 2, 80; (5) 2, 200; (6) 8, 80; (7) 8, 200.

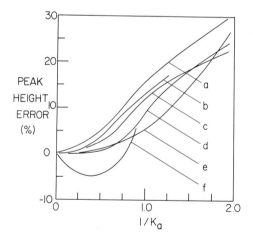

Fig. 13.18 Peak height errors for various filters: (a) lag filter; (b) lead-lag-lag filter with $K_a/K_d = 1.5$; (c) critically damped filter; (d) integrator; (e) second-order Butterworth filter; (f) lead-lag-lag filter with $K_a/K_d = 2.0$.

and its transfer function is

$$H(s) = \frac{ab}{b-a}\left[\frac{1}{s+a} - \frac{1}{s+b}\right] = \frac{ab}{(s+a)(s+b)}$$

The output for a Gaussian input can be written immediately from the results of Section 13.IV. A

$$r(x) = \sqrt{\pi}\left(\frac{K_a K_b}{K_b - K_a}\right)e^{-x^2}$$

$$\cdot \{\exp\left[(K_a - x)^2\right] \operatorname{erfc}(K_a - x) - \exp\left[(K_b - x)^2\right] \operatorname{erfc}(K_b - x)\}$$

$$(13.23)$$

where x is defined by Eq. (13.16) and K_a and K_b are defined as in Eq. (13.15).

The frequency response of a lag-lag filter at high frequencies decreases at twice the rate of a simple lag filter, that is, -40 dB/decade compared with -20 dB/decade. Thus the lag-lag filter better eliminates high-frequency noise, and it is natural to ask whether it can provide improved signal-to-noise ratio with no sacrifice in band distortion, or reduced distortion with no increase in noise. Let us consider the noise power passed by a lag-lag filter. From the Fourier transform

$$H(\omega) = \frac{ab}{(j\omega + a)(j\omega + b)}$$

we find

$$|H(\omega)|^2 = \frac{(ab)^2}{\omega^4 + \omega^2(a^2 + b^2) + (ab)^2}$$

So the output noise power, for white-noise input of density k^2/π, is

$$\langle N^2 \rangle = \frac{2k^2}{\pi} \int_0^\infty \frac{(ab)^2 \, d\omega}{\omega^4 + \omega^2(a^2 + b^2) + (ab)^2}$$

$$= \frac{abk^2}{a + b} \qquad (13.24)$$

Comparision of Eqs. (13.24) and (13.7) shows that a lag-lag filter with time constants $1/a$ and $1/b$ transmits the same noise power as a lag filter $1/a'$ if

$$a' = \frac{ab}{a + b}$$

We call these filters *noise-equivalent*.

The lag-lag filter is mathematically equivalent to two lag filters in series, although it cannot be physically constructed this way unless the two sections are isolated by a buffer amplifier. Therefore, following the discussion leading to Eq. (13.20), the shift of the centroid of $r(t)$ is the sum of the time constants of the separate lag sections, and to a good approximation, as in Eq. (13.21),

$$\Delta \bar{t} \approx \frac{1}{a} + \frac{1}{b} = \tau_a + \tau_b$$

But

$$\frac{1}{a} + \frac{1}{b} = \frac{1}{a'}$$

where $1/a'$ is the time constant of the noise-equivalent simple lag circuit. It follows that noise-equivalent lag-lag and simple lag circuits produce the same centroid shift and roughly the same peak shift. Moreover, using Eq. (13.22), the peak intensity is

$$I \approx \exp\left[\frac{-c^2}{2\sigma^2}\left(\frac{1}{a^2} + \frac{1}{b^2}\right)\right] = \exp\left[\frac{-c^2}{2\sigma^2}\left(\frac{b^2 + a^2}{a^2 b^2}\right)\right]$$

$$= \exp\left[\frac{-c^2}{2\sigma^2}\left(\frac{1}{a'}\right)^2\right]$$

which is the same as the peak intensity of the noise-equivalent lag filter. Thus there seems to be no advantage in using a lag-lag filter in trying to decrease small tracking errors. This result may be more sensible when we consider that although the lag-lag filter transmits amplitude information at higher frequencies than the noise-equivalent lag filter, the phase shift is greater at high frequencies, approaching $180°$ for the lag-lag filter and only $90°$ for the lag filter.

The extension of this discussion to filters consisting of more than two lag filters in series is trivial, but the approximation involved in setting the centroid shift equal to the peak shift is worse.

C. Critically Damped Filter

A critically damped filter is a lag-lag filter having two coincident poles, or time constants. The transfer function is thus

$$H(s) = \frac{a^2}{(s+a)^2}$$

and by inversion of the Laplace transform, the impulse response is

$$h(t) = a^2 t e^{-at}$$

The output for a Gaussian input can be found from Eq. (13.23) by finding $\lim_{b \to a} r(t)$. The result is

$$r(x) = 2\sqrt{\pi} K^2 e^{-x^2}\left\{\frac{1}{\sqrt{\pi}} - (K-x)\exp\left[(K-x)^2\right]\operatorname{erfc}(K-x)\right\}$$

$$= 2K[Ke^{-x^2} - (K-x)r_{\text{lag}}(x)] \tag{13.25}$$

The second form of Eq. (13.25) is written in terms of a Gaussian band $r_{\text{lag}}(x)$ distorted by a simple exponential lag filter. Equation (13.25) is plotted for several K values in Fig. 13.19.

By extension of the argument used in the previous section it is easy to show that a critically damped filter produces the same centroid shift as a noise-equivalent lag filter with $a' = \frac{1}{2}a$. To the approximation that the centroid shift equals the peak shift, it also produces the same intensity loss. A comparison of Figs. 13.19 and 13.16 verifies this, although the critically damped filter is seen to be slightly superior to the noise-equivalent lag filter.

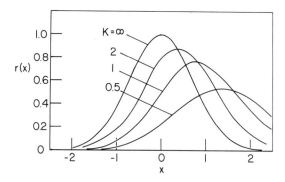

Fig. 13.19 Distortion of a Gaussian band by a critically damped filter.

A simple passive critically damped RC filter circuit cannot be made because the poles cannot be made to coincide. A critically damped active filter can be constructed as illustrated in Fig. 13.20 using an operational amplifier. An experimental comparison of a lag filter and a noise-equivalent critically damped filter shows the latter to be somewhat superior in that its lower peak height error permits a scanning

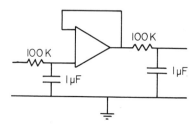

Fig. 13.20 Active critically damped filter.

speed increase of perhaps 25%. This is in agreement with the theoretical prediction of Fig. 13.19. Brodersen (6) has also demonstrated this property of the critically damped filter.

D. Underdamped Filter

An underdamped filter has two complex poles, which are complex conjugates of each other. The transfer function can be written variously

$$H(s) = \frac{a^2 + b^2}{(s + a + jb)(s + a - jb)} = \frac{a^2 + b^2}{s^2 + 2as + (a^2 + b^2)}$$

$$= \frac{a^2 + b^2}{(s + a)^2 + b^2}$$

and the impulse response is

$$h(t) = \frac{a^2 + b^2}{b} \sin bt \, e^{-at}$$

The parameter ζ, given by

$$\zeta = \frac{a}{(a^2 + b^2)^{1/2}} = \frac{K_a}{(K_a^2 + K_b^2)^{1/2}}$$

is the damping ratio. The K's are once more defined as in Eq. (13.15). When $\zeta = 1$, the system is critically damped; $\zeta < 1$ corresponds to an underdamped system; and $\zeta > 1$ corresponds to an overdamped system. The damping ratio for a system with moderate overshoot generally is about 0.7. A system having $\zeta = 0.7$ overshoots by about 4% when a dc signal is suddenly imposed. If ζ is reduced to 0.5, the overshoot is about 17%. The system thus responds to a suddenly applied constant signal by overshooting and undergoing damped oscillations as it approaches its final state.

An underdamped filter with $\zeta = 1/\sqrt{2} = 0.707$ is a Butterworth filter of second order. The order refers to the number of poles. Butterworth filters have their poles evenly distributed on a semicircle with its center at the origin of the s plane. Butterworth filters have optimally flat transfer functions. That is, they have a nearly uniform response in the passband and a sharp cutoff with attenuation increasing at a rate of 20 dB/decade/pole.

The white noise power transmitted by an underdamped filter is

$$\langle N^2 \rangle = k^2 a / 2\zeta^2$$

for an input noise of spectral power density k^2/π.

The output from a Gaussian input can be found by substituting $K_a + jK_b$ for K_a and $K_a - jK_b$ for K_b in Eq. (13.23), or by direct time integration of the convolution with $h(t)$. The result of the substitution is

$$r(x) = \frac{\sqrt{\pi}(K_a^2 + K_b^2)}{-2jK_b} e^{-x^2} \exp[-K_b^2(K_a - x)^2]$$

$$\cdot \{\exp[2jK_b(K_a - x)] \operatorname{erfc}(K_a - x + jK_b)$$

$$- \exp[-2jK_b(K_a - x)] \operatorname{erfc}(K_a - x - jK_b)\} \qquad (13.26)$$

From the properties of error functions we can write (10)

$$\operatorname{erfc}(K_a - x + jK_b) = 1 - \operatorname{erf}(K_a - x + jK_b)$$

$$= 1 - \operatorname{erf}^*(K_a - x - jK_b) = \operatorname{erfc}^*(K_a - x - jK_b)$$

With this result we introduce the real and imaginary parts of $\operatorname{erfc}(K_a - x - jK_b)$ into Eq. (13.26), write $\exp[-2jK_b(K_a - x)]$ and $\exp[2jK_b(K_a - x)]$ in terms of sines and cosines, and thus find

$$r(x) = \frac{\sqrt{\pi}(K_a^2 + K_b^2)}{K_b} \exp(-K_b^2) e^{-x^2} \exp(K_a - x)^2$$

$$- [R(x) \sin 2K_b(K_a - x) + I(x) \cos 2K_b(K_a - x)] \qquad (13.27)$$

where $R(x)$ and $I(x)$ are, respectively, the real and imaginary parts of

$$\operatorname{erfc}(K_a - x) - jK_b$$

Tables (10) are available for the real and imaginary parts of the function

$$w(z) = e^{-z^2} \operatorname{erfc}(-jz) \qquad z = x + jy$$

In order to find values for R erfc and I erfc, we note that

$$w(x + jy) = \exp[-(x^2 - y^2 + 2jxy)] \operatorname{erfc}(y - jx)$$

or

$$\operatorname{erfc}(y - jx) = \exp[(x^2 - y^2) + 2jxy] w(x + jy)$$

$$= [\exp(x^2 - y^2)] (\cos 2xy + j \sin 2xy) w(x + jy)$$

It follows that

$$R \operatorname{erfc}(y - jx) = [\exp(x^2 - y^2) [\cos 2xy \, Rw \, (x + jy) - \sin 2xy \, Iw(x + jy)]$$

and

$$I \operatorname{erfc}(y - jx) = [\exp(x^2 - y^2) [\sin 2xy \, Rw(x + jy) + \cos 2xy \, Iw(x + jy)]$$

Equation (13.27) is plotted in Fig. 13.21.

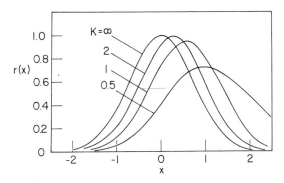

Fig. 13.21   Distortion of a Gaussian band by a second-order Butterworth filter.

We can investigate the centroid shift for any symmetric input function by using the theorem of Eq. (13.19). We find

$$\Delta \bar{t} = \frac{2a}{a^2 + b^2} = \frac{2\zeta^2}{a}$$

The design of a filter using an operational amplifier is comparatively easy. The overall transfer function of the system is

$$H(s) = Z_f(s)/Z_i(s)$$

where $Z_f(s)$ is the transfer function of the network in the feedback loop and $Z_i(s)$ is the transfer function of the input network.

A design for an underdamped filter is shown in Fig. 13.22. The

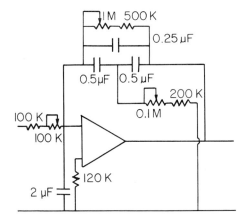

Fig. 13.22   Active underdamped filter of the second-order Butterworth type.

network in the feedback loop contributes a term

$$Z_f(s) = \frac{K(1+as)}{s^2 + 2\zeta\omega_n s + \omega_n^2}$$

The input network introduces a pole at $1/R_0 C_0$ to cancel the zero at $1/a$ in $Z_f(s)$, so the overall transfer function is

$$H(s) = \frac{K}{s^2 + 2\zeta\omega_n s + \omega_n^2}$$

Figure 13.23 is a comparison of experimental peak height errors suffered by two bands in the indene spectrum when scanned with a second-order Butterworth filter at various rates against the errors observed with a noise-equivalent lag filter. The Butterworth filter is clearly superior, in agreement with the theoretical prediction.

E. Lead-Lag-Lag Filter

Up to now we have considered only the influence of the poles of the transfer function. The introduction of a zero into the transfer function

$$H(s) = \frac{ab}{d} \frac{s+d}{(s+a)(s+b)} \qquad d < a, b \qquad (13.28)$$

produces a phase lead, in contrast with the phase lags of the poles.

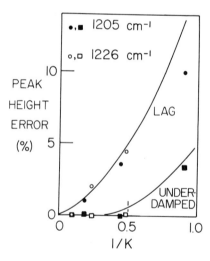

Fig. 13.23 Comparison of the peak height errors observed in the spectrum of indene scanned with a lag filter and with a second-order Butterworth filter.

The zero also produces an increase in the amplitude response of $+20$ dB/decade in contrast with the attenuation of the lag filter. The transform of the derivative of a function is s times the transform of the function itself. Thus a zero in the transfer function causes a fraction of the first derivative of the function to be added to the function. The impulse response corresponding to Eq. (13.28) is thus

$$h(t) = \frac{ab}{b-a}\left\{(e^{-at}-e^{-bt})+\frac{1}{d}(-ae^{-at}+be^{-bt})\right\}$$

$$= \frac{ab}{b-a}\left[\left(1-\frac{a}{d}\right)e^{-at}-\left(1-\frac{b}{d}\right)e^{-bt}\right]$$

The output for a Gaussian input can be written from the output $r_0(t)$ for a lag-lag filter, Eq. (13.23) and its derivative,

$$r(t) = r_0(t)+\frac{1}{d}\frac{dr_0(t)}{dt}$$

or

$$r(x) = \sqrt{\pi}\frac{K_aK_b}{K_b-K_a}e^{-x^2}\left\{\left(1-\frac{K_a}{K_d}\right)\exp[(K_a-x)^2]\operatorname{erfc}(K_a-x)\right.$$
$$\left.-\left(1-\frac{K_b}{K_d}\right)\exp[(K_b-x)^2]\operatorname{erfc}(K_b-x)\right\}$$

The action of the lead-lag-lag filter is illustrated in Fig. 13.24. The addition of the derivative r_0' adds signal to the left of the peak of r_0 and subtracts signal to the right of the peak. The result is a better representation of the original signal $s(t)$. This technique has been described as peak sharpening or resolution enhancement (11, 12) in the recent literature. It is in reality the intentional introduction of a distortion in an attempt to compensate for another distortion.

If the two poles of $H(s)$ are coincident, the impulse response and the output for a Gaussian input can be found by calculating the limit as

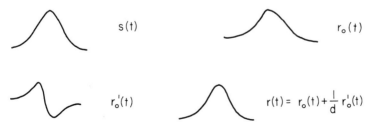

Fig. 13.24 Action of a lead-lag filter.

$b \to a$. The resulting equations are

$$h(t) = \frac{a^2}{d} e^{-at}[1 + (d-a)t] \tag{13.29}$$

and

$$r(x) = \sqrt{\pi} K_a^2 e^{-x^2} \left\{ \frac{2}{\sqrt{\pi}} \left(1 - \frac{K_a}{K_d}\right) \right.$$

$$+ \exp[(K_a - x)^2]\, \text{erfc}(K_a - x)\left[\frac{1}{K_d} - 2(K_a - x)\left(1 - \frac{K_a}{K_d}\right)\right]\right\}$$

$$= \left(1 - \frac{K_a}{K_d}\right) r_{\text{crit. damp}}(x) + \frac{K_a}{K_d} r_{\text{lag}}(x)$$

Thus, a Gaussian band distorted by a lead-lag-lag filter with equal poles can be described in terms of a Gaussian distorted by a critically damped filter and by a simple exponential lag filter. Clearly, from the form of Eq. (13.29), this result is valid in general, for a band of any shape.

Gaussian bands distorted by lead-lag-lag filters having $K_a/K_d = 2$ and 1.5 are plotted in Fig. 13.25. For $K_a/D_d = 1.5$ there is some overshoot at intermediate K values, but the peak height error remains small

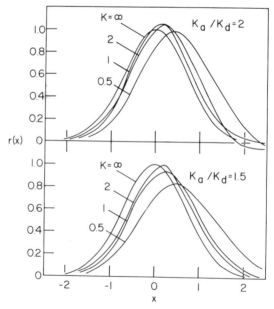

Fig. 13.25 Distortion of a Gaussian band by lead-lag-lag filters with $K_a/K_d = 1.5$ and 2.0.

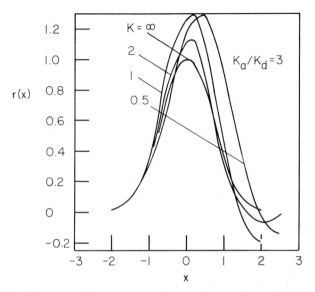

Fig. 13.26   Distortion of a Gaussian band by a lead-lag-lag filter with $K_a/K_d = 3.0$.

even at small K values. Moreover, the peak shift remains small. Figure 13.26 shows a similar set of curves for a lead-lag-lag filter having $K_a/K_d = 3$. In this case the overshoot is excessive.

Now let us investigate the expected distortion of any symmetric input function filtered by a lead-lag-lag filter. The centroid displacement is

$$\Delta \bar{t} = \frac{1}{abd}(-ab + ad + bd)$$

and in case $a = b$

$$\Delta \bar{t} = \frac{2}{a} - \frac{1}{d} = 2\tau_a - \tau_d$$

In a lead-lag-lag circuit $d < a$, or $\tau_d > \tau_a$. Therefore, the peak shift (centroid shift) is always less than the shift produced by a noise-equivalent simple lag filter. In fact, $\Delta \bar{t}$ can be made zero by choosing $d = \frac{1}{2}a$, as in the example of Fig. 13.25. The *peak* displacement evident in the figure is due to the distinction between centroid and peak.

The white noise power transmitted by a lead-lag-lag filter, for an input noise of spectral power density k^2/π, is

$$\langle N^2 \rangle = \frac{k^2 ab(ab + d^2)}{(a+b)d^2}$$

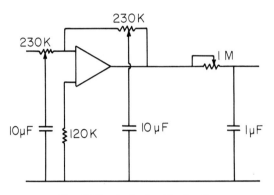

Fig. 13.27   Active lead-lag-lag filter.

which reduces to

$$\langle N^2 \rangle = \frac{k^2 a}{2d^2}(a^2 + d^2)$$

when $a = b$.

A typical lead-lag-lag network is shown in Fig. 13.27. Two filters of this type with different values of the ratio K_a/K_d were designed and used in scanning two bands in the spectrum of cyclobexane. Peak height errors, together with errors from a noise-equivalent lag filter, are plotted in Fig. 13.28. The negative peak height errors incurred with the lead-lag-lag filters represent overshoot. It is evident that for a moderate amount of overshoot, a lead-lag-lag filter permits faster scanning than a noise-equivalent lag filter.

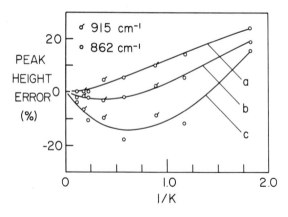

Fig. 13.28.   Peak height errors observed in the cyclohexane spectrum with: (a) lag filter; (b) and (c) lead-lag-lag filters.

F. Adaptive Filters

We shall not attempt to review the theoretical treatment of adaptive or time-varying systems. An adaptive system in a spectrophotometer adjusts itself to prevent the error signal from exceeding a preset level. It can do this either by slowing the scanning speed (speed suppression) or by reducing the time constant (period suppression) when the error signal increases. In the latter function the noise filter of the instrument might be an adaptive lag filter consisting of a single-stage RC filter in which the resistor is a photodiode.

G. Sampled Data Filters and Computers

All of the noise filters discussed so far have in common the property of functioning in real time. They perform some combination of weighted integration and differentiation of past information in order to estimate the present value of the input. There is necessarily a lag between the input and the estimated output. This alone is not particularly serious because it is easy to correct for small peak shifts. However, the information is further distorted by frequency-dependent attenuation and phase shift.

In order to avoid this distortion, one would like to use a filter that introduces no phase shift and that attenuates higher frequencies in such a manner as to minimize distortion and maximize the signal-to-noise ratio. Filters of this type can be defined mathematically but are difficult to realize as electrical circuits. However, the mathematical prescription for such a filter can be utilized in the form of a computer program.

The output information from a spectrum can be sampled at discrete times and memorized by a computer. The computer then smooths the data and forms the final output spectrum. There are two general approaches, termed by Whitney (13) smoothing in the large, or spectrum filtering, and smoothing in the small. One form of spectrum filtering entails the computations of the Fourier transform of the spectrum. The transform is multiplied by a filtering function, which might be simply a uniform function up to a cutoff frequency beyond which it has zero value. This technique was demonstrated by King et al. (14) long ago. The inverse transform is then computed and is presented as the smoothed spectrum. This process can be carried out with a general-purpose digital computer.

This type of noise filtering is natural in Fourier spectroscopy with

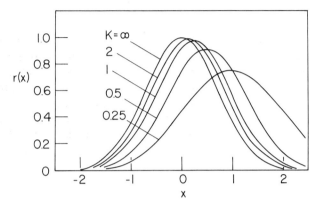

Fig. 13.29 Distortion of a Gaussian band by a running-mean integrating filter.

interferometers. The effect of noise in Fourier spectroscopy has been discussed by Schubert (*15*) and by Williams and Chang (*16*).

Smoothing in the small is done by averaging sampled values from neighboring points in the spectrum. The average can be a simple running mean or a mean weighted by some polynomial or other function. Figure 13.29 and the curve labeled "integrator" (d) in Fig. 13.18 are for a running mean. Savitzky and Golay (*17*) have described an example of this technique, in which the spectral data are sampled and digitized for future treatment by digital computer. They use weighting functions appropriate for fitting data to polynomials of specified order, with least-mean-squared error.

A third method of smoothing makes use of a so-called computer of average transients. A spectrum is scanned repeatedly and sampled output data are stored in the memory of a digital computer. Values from each subsequent scan are added to the values already stored. After M scans, the amplitudes of the true input spectrum have accumulated to M times the values from a single scan, while the noise has increased by only $\sqrt{M}$ because of its random character. Thus, after 100 scans the signal-to-noise ratio is increased by 10 times. This technique has been applied to spectrophotometry by Cuthrell and Schroeder (*18*) and by Bluhm *et al.* (*19*).

H. Optimum Noise Filters

There are three approaches to the problem of optimum noise filtering. The most elementary approach is to assume a filter type and optimize its parameters on the basis of certain criteria. A somewhat more difficult problem is the derivation of an optimum transfer function and the synthesis of actual networks or devices to realize that function. Finally, and most difficult of all, is the determination and realization of general optimum systems that might be nonlinear, time-varying, or adaptive.

Petrash (20) and Schubert (21) have presented parameter-optimization solutions. Their starting assumptions and optimization criteria differ, but among their conclusions is agreement that a spectrum should be scanned at a rate of about 0.2 spectral slit width per time constant when using a lag filter.

The problem of deriving an optimum transfer function evidently has not been solved for a spectrophotometer. We can require that the difference between the recorded spectrum and the actual spectrum must be minimized in the sense of least-mean-squared differences. This will lead to the Wiener–Hopf integral equation for the optimum transfer function in much the same way that our treatment of Section 7.III.A led to an integral equation for an optimum spectral modulation transfer function. The solution of the Wiener–Hopf equation is complicated. It is discussed, for example, by Lee (22) and by Wainstein and Zubakov (23), among others.

V. Avoiding Distortion in a Recorded Spectrum

One way to deal with distortion in recorded spectra is to prevent it from occurring in the first place by proper choice of instrument parameters. The usual procedure is to select the monochromator slit width required to resolve the spectrum and then set the amplifier gain, according to some procedure, appropriate to the slit width. For example, in some instruments this is done by adjusting the gain for full-scale single-beam pen travel. The response period is then chosen to maintain the noise below a tolerable level. Finally, a sharp line in the spectrum is recorded several times at successively faster scanning speeds, and the fastest scanning speed consistent with permissible tracking error is chosen. This procedure can be very time consuming, especially at the slower scanning speeds, so a set of standard conditions is usually determined and used for most routine spectra. Systematic methods for adjusting spectrophotometers have been discussed by Potts and Smith (24).

Having determined satisfactory values for the parameters under one set of conditions, it is desirable to have a mathematical relationship from which values appropriate to other conditions can be predicted.

Let us derive such a relationship between gain, speed, period, and resolution. Equations of this sort have been discussed by Daly and Sutherland (25), by Zbinden and Baldinger (26), and by Nikitin (27).

We assume that the instrument response can be approximated by a simple lag filter. We have seen, in the foregoing discussion, that the peak intensity of a recorded spectral line depends on a parameter that is identified with the scanning rate in half-widths of the line per time constant. The half-width is, of course, the observed bandwidth (OBW) after broadening by the monochromator, not the half-bandwidth (HBW) of the spectral line itself. That is, we say that for a given accuracy of presentation we can scan no faster than a specified number of OBW's per time constant. In symbols

$$c \propto \frac{OBW}{\tau} \tag{13.30}$$

We have also demonstrated (Fig. 13.15) that the time constant of an instrument operating double beam is approximately inversely proportional to the single-beam to double-beam energy ratio M, expressed in Eq. (13.9). With Eq. (13.11) substituted, τ is found in terms of gain G, slit width W, and reference beam transmittance T,

$$\tau \propto \frac{\tau_0}{GW^2T}$$

Now if a very narrow line is being scanned, the observed bandwidth is proportional to the slit width. On the other hand, if the line is very broad, the OBW is nearly independent of slit width. Hence, we have two limiting cases for Eq. (13.30),

$$c \propto \frac{W^3GT}{\tau_0} \qquad HBW \ll W \tag{13.31}$$

$$c \propto \frac{W^2GT}{\tau_0} \qquad HBW \gg W \tag{13.32}$$

These are the equations that give the dependence of the permitted scanning speed on gain, slit width, time constant, and reference beam transmittance.

We also derived an expression for the root-mean-square noise in a recorded spectrum expressed in Eq. (13.12). This equation can be combined with Eq. (13.31) or (13.32) to obtain

$$\frac{W^5N^2T^2}{c} = const \qquad HBW \ll W \tag{13.33a}$$

$$\frac{W^4 N^2 T^2}{c} = \text{const} \qquad \text{HBW} \gg W \qquad (13.33\text{b})$$

Equation (13.33a) tell us, for example, that in order to double the resolution and maintain the same tracking error and noise we must scan 32 times slower. They also tell us that if we wish to expand the transmittance scale by a factor of 10 and maintain the same noise and tracking error, we must either scan 100 times more slowly or sacrifice resolution by a factor of 2.5.

Clearly, the correct choice of parameters can very well entail either excessive scanning time, or high noise level, or a sacrifice of resolution.

VI. CORRECTION OF DISTORTED SPECTRA

Sometimes spectral distortion is unavoidable. If a good computational method of correcting distortion is available, it might even be desirable to record a distorted curve with a minimum use of spectrophotometer time, and to correct it later. An alternative approach is to record a spectrum with a minimum of noise-filter distortion but with a high noise level. The filtering is performed later.

Let us indicate briefly how the process of deconvolution might be applied to recover distorted information. The recorded spectrum after distortion of an input signal by the system is given by

$$r(t) = \int_{-\infty}^{\infty} h(t - t') s(t') \, dt'$$

The record is further treated by convolution with another function $k(t)$. This can be done, for example, on a digital computer. The new output is $v(t)$

$$v(t) = \int_{-\infty}^{\infty} k(t - t') r(t') \, dt' \qquad (13.34)$$

or

$$v(t) = \int_{-\infty}^{\infty} \int_{-\infty}^{\infty} h(t' - t'') k(t - t') s(t'') \, dt'' \, dt' \qquad (13.35)$$

The Laplace transform of Eq. (13.35) is

$$V(s) = H(s) K(s) S(s)$$

If we wish $v(t)$ to be identical with the original undistorted input, then we must choose $k(t)$ by setting

$$K(s) = \frac{1}{H(s)}$$

If our spectrophotometer can be approximated by a simple lag filter then

$$K(s) = \frac{1}{a/(s+a)} = \frac{s+a}{a}$$

and, from Eq. (13.34)

$$S(s) = V(s) = \frac{s+a}{a} R(s)$$

or

$$s(t) = v(t) = r(t) + \frac{1}{a} \frac{dr(t)}{dt} \tag{13.36}$$

Equation (13.36) is a formula for the operation we must perform on the recorded spectrum to recover the true form of the input. The operation $H(s)K(s)$ is equivalent to no filtering at all, and hence the recovered spectrum would be badly corrupted by the noise derived from the second term in the equation. We can avoid this by replacing Eq. (13.36) with

$$v(t) = r(t) + \frac{1}{a} \left(\frac{dr(t)}{dt} \right)_{\text{smoothed}}$$

where the derivative has been smoothed in some sense.

The distortion of a recorded signal occurs in the first place because the noise filter is necessarily an asymmetric device, unable to make use of future information. The smoothing of dr/dt, however, can be done in a symmetric fashion in order to suppress noise but not asymmetrically distort the information.

The procedure just outlined is complex, and although potentially useful, has not been developed to the point of routine use. We ask, however, if we cannot take advantage of some of the characteristics of distorted spectra discussed in earlier sections to at least approximately correct them.

We have seen that the centroid of a symmetric peak scanned with a simple lag filter is shifted in time by the time constant of the filter, independently of the width of the band. The peak is shifted by very nearly the same amount for moderate distortion. The peak frequency is thus shifted by

$$\Delta \bar{\nu} \approx c\tau$$

Thus, the spectroscopist, in setting his instrument, should offset the recorder paper by the product of the scanning speed and the time constant in order to correct for the frequency shift. This is not practical when an adaptive system is used to vary the scanning speed or the time constant.

Unlike the centroid shift, the peak height of a distorted band depends on its shape and width. If it is known that the band is Gaussian, or if we are willing to assume this as an approximation, and if the width is known, then for $t = 1/a$ and therefore $x = 1/2K$, Eq. (13.17) for the distorted band becomes

$$r(0') = \sqrt{\pi}K \exp(K^2 - 1)\,\text{erfc}\!\left(K - \frac{1}{2K}\right) \qquad (13.37)$$

and the corrected peak height may be calculated by dividing the observed peak height by $r(0')$. Bear in mind that the peak height of an absorption band, as we are using the term, is measured from the base to the peak, not from the zero line to the peak.

Equation (13.37) is plotted in Fig. 13.30. The computation at higher values of K is carried out by making use of the asymptotic expansion

$$\text{erfc}\,z \approx \frac{1}{\sqrt{\pi}}\frac{e^{-z^2}}{z}\sum_{k=0}^{\infty}\frac{(-1)^k(2k)!}{k!(2z)^{2k}}$$

Only three terms are needed for $z \geq 2$. Also plotted on this figure are experimental values of the two indene bands of Fig. 13.17. Evidently, peak heights can be corrected in this way for K down to about 1.25, that is, scanning speeds up to about one-fourth of an observed bandwidth per time constant or one observed bandwidth per period. At this speed the peak height error is about 13%. The error is about 3% at a scanning speed of 0.3 observed bandwidth per period. The deviation of the experimental values from the calculated curve in Fig. 13.30 at

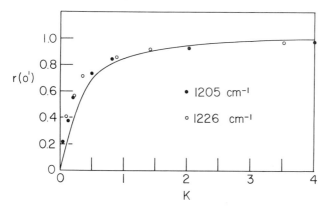

Fig. 13.30 Calculated [Eq. (13.37)] and measured values (indene) of band intensity at $t = 1/a$.

low K values is expected because of the departure of the peak position from the centroid of the distorted band.

We might also attempt to correct peak intensity information by measuring the intensity of the band at the frequency where the peak would appear if scanned very slowly, that is, at $t = 0$. Equation (13.17) then becomes

$$r_0 = \sqrt{\pi} K e^{K^2} \operatorname{erfc} K \qquad (13.38)$$

This is expected to be less satisfactory because of the greater uncertainty in measuring intensity values on the side of a recorded band rather than at the peak.

REFERENCES

1. S. G. Rautian, *Sov. Phys. Usp.*, **66**, 245 (1958).
2. K. S. Seshadri and R. N. Jones, *Spectrochim. Acta*, **19**, 1013 (1963).
3. R. N. Clark, *Introduction to Automatic Control Systems*, Wiley, New York, 1962.
4. N. B. Braymer, U. S. Pat. 2,978,625, Apr. 4, 1961.
5. J. E. Stewart, *Infrared Phys.*, **7**, 77 (1967).
6. S. Brodersen, *J. Opt. Soc. Am.*, **43**, 1216 (1953).
7. M. Schubert, *Exptl. Tech. Physik.*, **6**, 203 (1958).
8. K. Frei and Hs. H. Gunthard, *J. Opt. Soc. Am.*, **51**, 83 (1961).
9. G. Schreiber, *Z. Angew. Physik*, **18**, 221 (1964).
10. M. Abramowitz and I. A. Stegun, "Handbook of mathematical functions," *NBS Appl. Math. Ser.* 5, 1964.
11. L. C. Allen, H. M. Gladney, and S. H. Glarum, *J. Chem. Phys.*, **40**, 3135 (1964).
12. J. W. Ashley, Jr. and C. N. Reilley, *Anal. Chem.*, **37**, 626 (1965).
13. C. A. Whitney, *Astrophys. J.*, **137**, 327 (1963).
14. G. W. King, E. H. Blanton, and J. Frawley, *J. Opt. Soc. Am.*, **44**, 397 (1954).
15. M. Schubert, *Optik*, **18**, 147 (1961).
16. R. A. Williams and W. S. C. Chang, *J. Opt. Soc. Am.*, **56**, 167 (1966).
17. A. Savitzky and M. J. E. Golay, *Anal. Chem.*, **36**, 1627 (1964).
18. R. E. Cuthrell and C. F. Schroeder, *Rev. Sci. Inst.*, **36**, 1249 (1965).
19. A. L. Bluhm, J. A. Sousa, and J. Weinstein, *Appl. Spectry.*, **18**, 188 (1964).
20. G. G. Petrash, *Opt. Spectry.*, **6**, 516 (1959).
21. M. Schubert, *Exptl. Tech. Physik*, **6**, 203 (1958).
22. Y. W. Lee, *Statistical Theory of Communication*, Wiley, New York, 1960.
23. L. A. Wainstein and V. D. Zubakov, *Extraction of Signals from Noise*, Prentice-Hall, Englewood Cliffs, N.J., 1962.
24. W. J. Potts and A. L. Smith, *Appl. Opt.*, **6**, 257 (1967).
25. E. F. Daly and G. B. B. M. Sutherland, *Proc. Phys. Soc.*, **A62**, 205 (1949).
26. R. Zbinden and E. Baldinger, *Helv. Phys. Acta*, **26**, 111 (1953).
27. V. A. Nikitin, *Opt. i. Spektroskopiya.*, **4**, 523 (1958).

Photometric Accuracy of Infrared
Spectrophotometers

I. Introduction

The very name spectrophotometer implies an instrument to be used for the quantitative measurement of power in a spectrum of radiant energy. The difficult measurement of absolute power—radiance or irradiance—is discussed in Chapter 17. In absorption spectrophotometry the measurement is considerably simplified to one of relative transmittance, and the question of photometric accuracy becomes one of photometric linearity. This is because the spectroscopist is free to establish a level of unity transmittance and a level of zero transmittance and make his measurements relative to those levels. If the instrument responds linearly to changes in radiant power, then the measurements of transmittance are "accurate," provided certain other conditions are met.

In this chapter we first discuss the origins of photometric errors, and then describe methods of determining photometric accuracy.

II. Origins of Photometric Errors

Let us first survey the sources of inaccuracies, and then discuss some of them in greater detail.

The zero transmittance level of a spectrophotometer is usually established by inserting an opaque shutter in the beam and adjusting the recorder to indicate zero. However, a zero level established in this way may not be the same as the zero associated with a very strong absorption band of a sample. The shutter might emit radiation at a different rate than the sample, or it might reflect radiation into the monochromator. Moreover, an opaque shutter stops all radiation from the source, but if the monochromator permits stray radiation to reach the detector, the zero level will not be reached even at wavelengths of very strong absorption. These effects are discussed in Sections III and IV.

The 100% transmittance level is established by removing absorbers from the beam and perhaps inserting a blank window equivalent to an empty cell or a cell filled with solvent. It is usually necessary to determine this level at each wavelength at which measurements are to be made. Once more, emission and stray radiation can affect the setting. Nonequivalence of the reference cell or blank certainly influences quantitative measurements. If the sample is a solid, particularly a thick solid or a solid of high refractive index, it affects the focus of the beam and reflects radiation. A liquid of high refractive index in a cell can do likewise. Multiple internal reflections can produce erroneous readings. These sample effects are discussed in Section V.

Amplifier nonlinearity in a single-beam spectrophotometer affects photometric accuracy in an obvious manner. The spectroscopist should bear in mind that the linearity of an amplifier very likely depends on the gain or signal level. Saturation can occur in the first stage when very strong signals are encountered, as with wide monochromator slits and correspondingly low amplifier gain. Amplifier linearity also affects the response of ratio-computing systems, and so does the crosstalk effect discussed in connection with the Hornig system in Section 12.IV.B. Amplifier linearity does not directly affect the accuracy of optical null spectrophotometers, but the properties of reference beam attenuators do, and often are even more serious. This is discussed in Section VI.

We have discussed in Chapter 7 how the monochromator distorts the shape and peak intensity of recorded spectral lines. This has a direct and important effect on photometric accuracy, and should always be considered in choosing slit widths. Recall that the effective bandwidth of the monochromator should be no greater than about two-tenths of the half-bandwidth of the bands of the sample if peak height errors are to be kept under 2%. Atmospheric absorption can distort the slit function in a way that varies with wavelength, and this can introduce strange artifacts in a spectrum. The effect on photometric errors has been discussed by Goodwin and Johnson (*1*).

An equally important source of photometric error is the dynamic or tracking error associated with the finite scanning rate. This was discussed in Chapter 13. Remember that a scanning rate of 0.2 observed bandwidth per response period (0.05 OBW per time constant) produces a peak height error of 1% in an instrument with a simple lag noise filter.

Wavelength calibration errors can affect photometric accuracy if measurements are to be made at a specific wavelength setting. It is usually best to scan over a peak and make the measurement at the observed maximum. Measurements made on the side of a steep slope are particularly difficult to control. The effect of tracking error in such measurements is obvious.

Finally, photometric accuracy is affected by the limit to precision imposed by noise. Often, repeat traces over the same peak serve to improve the precision of measurement.

III. STRAY RADIATION

Let us first see how stray radiation affects photometric measurements with an optical null double-beam instrument. Consider a beam of radiation delivering power G from the spectrometer source in the frequency interval passed by the monochromator. The sample beam transmits a portion of this radiation given by $T_S T_R G$, where T_S is the transmittance of the sample and T_R is the transmittance of an absorber which is common to both beams, for example, a solvent or an atmospheric component. An additional amount of radiation U is detected because of stray radiation in the monochromator. Thus the radiation associated with the sample beam is

$$J_S = T_S T_R G + U = (T_S T_R + u)G \qquad (14.1)$$

where u expresses U as a fraction of G.

The reference beam transmits an amount of radiation from the source given by $T_R T_C G$, where T_C is the transmission of the reference beam attenuator or comb. The stray radiation associated with the reference beam is also attenuated by the comb, so we have for the radiation in the reference beam

$$J_R = T_R T_C G + U' T_C = (T_R + u')T_C G \qquad (14.2)$$

U', the stray radiation associated with the reference beam, is not necessarily equal to U because the sample absorbs some of the radiation which would otherwise appear as stray radiation.

The instrument adjusts the position of its reference beam attenuator to make $J_S = J_R$, from which the transmittance of the comb is found, and therefore the transmittance indicated by the spectrophotometer,

$$T_C = \frac{T_S T_R G + U}{T_R G + U'} = \frac{T_S T_R + u}{T_R + u'} \qquad (14.3)$$

Ideally, T_C and T_S are equal, but this occurs only when u and u' vanish.

Note first of all that T_C is a *linear* function of the transmittance of the sample, so there is no curvature in the relationship between T_S and T_C. When an absorption band of the sample is so strong that T_S approaches zero, the comb transmission is

$$T_C(0) = \frac{u}{T_R + u'} \qquad (14.4)$$

If u' is not too large and the reference beam has no strong absorbers ($T_R = 1$), we have

$$T_C(0) \approx u$$

This gives us a convenient, though approximate, way of measuring the proportion of stray radiation present. That is, we insert a sample having a very strong absorption band and observe the apparent transmittance by which the instrument fails to agree with a zero level established with an opaque shutter. The presence of T_R in the denominator of Eq. (14.4) makes the apparent zero dependent on the transmission of a common absorber.

In the absence of an absorption band of the sample, when T_S is unity the comb transmission is

$$T_C(1) = \frac{T_R + u}{T_R + u'}$$

If the contributions of stray radiation to the two beams are equal, we find $T_C(1)$ equal to unity, and all of the error is introduced by a displacement of the zero end of the transmittance scale. When u and u' differ, $T_C(1)$ can be either greater or less than unity; $T_C(1) < 1$ when $u < u'$, which is usually the case. The effect of stray radiation on photometric accuracy is plotted in Fig. 14.1.

In general, in the presence of stray radiation the position of the reference beam attenuator depends on the transmittance of a common

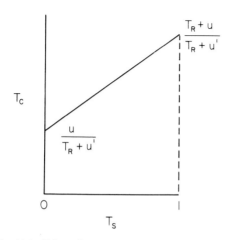

Fig. 14.1   Effect of stray radiation on photometric accuracy.

absorber, T_R, and particularly so in the case in which u and u' differ. This is often why it is difficult to obtain perfect cancellation of atmospheric absorption bands when some samples are studied, even though cancellation is obtained with the sample removed.

Single-beam instruments and ratio-computing instruments of course do not have reference beam attenuators. Nevertheless, exactly the same photometric errors occur in the presence of stray radiation. The radiation associated with the sample is given by Eq. (14.1) and the radiation associated with the reference is, in place of Eq. (14.2),

$$J_R = (T_R + u')G$$

Either the spectroscopist or the spectrometer must compute the ratio to find the apparent transmittance of the sample,

$$T_{S'} = \frac{J_S}{J_R} = \frac{T_S T_R + u}{T_R + u'}$$

and this is the same as Eq. (14.3).

The special problem of measuring stray radiation in the presence of extraneous radiation is discussed below in Section IV.C.

IV. EXTRANEOUS RADIATION

The effect of extraneous radiation is similar to stray radiation. Stray radiation, as we know, occurs when radiation takes an unplanned path in the monochromator following scatter, reflection from a baffle, etc. By extraneous radiation we mean thermal radiation *other* than that from the source. All of the elements of the spectrophotometer emit and absorb thermal radiation as they try to reach thermal equilibrium with their surroundings. Some of this radiation reaches the detector and produces a signal resulting in photometric error. In this section we show how the magnitude of this error can be estimated. This problem has been discussed by Stewart (2).

TABLE 14.1

Values of r, c, b, m, or f^a

$\nu(cm^{-1})$	Temperature		
	298°K	300°K	310°K
50	0.168	0.170	0.176
100	0.152	0.153	0.159
200	0.121	0.122	0.129
400	0.073	0.075	0.080
500	0.056	0.057	0.062
700	0.031	0.032	0.036
1000	0.012	0.012	0.014
1500	0.002	0.002	0.003

^aFor various temperatures, relative to a blackbody source at 1600°K.

Recall that the thermal power radiated by an object is the product of the emittance of the object times the power that would be emitted by a blackbody at the same temperature. For convenience in the analysis to follow, the radiant power that this hypothetical blackbody would emit in a prescribed wavelength interval and into a prescribed solid angle is written as a fraction of the power radiated by the spectrometer source, G. The emittance of a body is given by Kirchhoff's law in terms of its reflectance and transmittance. For example, a sample having transmittance T_s and reflectance R_s has emittance $(1 - T_s - R_s)$. A blackbody at the temperature of the sample emits radiant power at the rate sG. The radiant power emitted by the sample is thus written $(1 - I_s - R_s)sG$. The dependence of the source emission G on the wavelength and temperature is given by the Planck radiation law and the spectral emittance of the source. The fraction s depends on the wavelength and the temperatures of both the source and the sample. Values of s, assuming a black-body source at temperatures of 1000, 1300, and

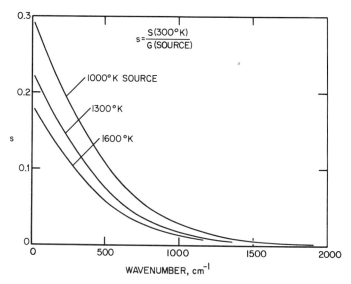

Fig. 14.2 Relative emission from a blackbody at 300°K compared with blackbody sources at 1000, 1300, and 1600°K.

1600°K and a sample at 300°K, are plotted in Fig. 14.2. Evidently problems of extraneous radiation are expected to become much more serious in the far infrared where s can become a significant fraction indeed. Minami *et al.* (3) have devised a special electronic system to cope with the problem in a far-infrared spectrophotometer.

In the following discussion we use the symbol s in representing the radiation from the sample, in the manner just described. Similarly, r is used for radiation from a reference cell or sample, c for the reference beam attenuator or comb, b for a shutter used to establish the zero level, m for the monochromator, and f for the radiation chopper or modulator. Values of these parameters are also read from Fig. 14.2 and, in addition, values for temperatures of 298, 300, and 310°K and a range of wavenumbers are given in Table 14.1. The corresponding upper case letters, used as subscripts on R or T, represent the reflectance or transmittance of the designated optical element. Thus T_S is the transmittance of the sample, R_B is the reflectance of the shutter, and so on.

A. Single-Beam Spectrophotometers

We now discuss a number of spectrometer systems. Consider first the single-beam system shown schematically in Fig. 14.3. The instru-

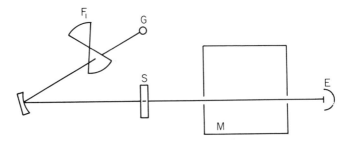

Fig. 14.3   Single-beam spectrophotometer.

ment contains a chopper placed immediately following the source at F_1 or, in the case of double-passed monochromators (Fig. 14.4), between the entrance and exit slits at F_2. In the first case, with the chopper at F_1, extraneous radiation from sample cells or shutters, or radiation emitted by the monochromator toward the detector, cannot be modulated and therefore is not detected. The chopper itself both radiates and reflects radiation originating elsewhere. We shall assume that the source of such reflected radiation is the monochromator itself or a wall or other element at the same temperature as the mono-chromator. The radiation signal at the detector when the sample cell is in place is thus

$$J_S = GT_aT_S - (1 - R_f)fGT_aT_S - mGR_fT_aT'_aT_S^2 \qquad (14.5)$$

where T_a is the transmittance of the atmospheric path between the source and the detector, T'_a is the transmittance from monochromator to chopper, and the other symbols have already been defined. Equation (14.5) states that the signal at the detector originates in: (1) the radia-tion from the source, attenuated by sample and atmospheric absorption; (2) radiation emitted by the chopper which is also attenuated and which appears with a minus sign because it is out of phase with the radiation from the source and thus reduces the signal; (3) radiation emitted by

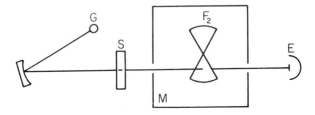

Fig. 14.4   Single-beam, double-passed spectrophotometer.

the monochromator or walls of the instrument, reflected by the chopper, and thus attenuated by two passes through the sample and a different atmospheric path T_a'. We assume that the chopper is so oriented that radiation emitted by the sample is not reflected back to the monochromator and detector.

A similar equation can be written for a reference material inserted in the instrument in place of the sample,

$$J_R = GR_aT_R - (1 - R_f)fGT_aT_R - mGR_fT_aT_a'T_R^2$$

T_R can, of course, be unity if the 100% level is established with an empty beam.

If the zero level is established by inserting an opaque shutter between the chopper and the detector, no modulated radiation can reach the detector and $J_0 = 0$.

The apparent transmittance of the sample is found from

$$T_{app} = \frac{J_S}{J_R} = \frac{T_S[1 - (1 - R_f)f - mR_fT_a'T_S]}{T_R[1 - (1 - R_f)f - mR_fT_a'T_R]} \tag{14.6}$$

Equation (14.6) reduces to the proper value

$$T_{app} = \frac{T_S}{T_R}$$

if $R_f = 0$.

Evidently, the chopper of a single-beam, single-passed infrared spectrophotometer should be as nonreflecting as possible. Fortunately, this is consistent with the use of crystal choppers in far-infrared spectrophotometers for control of stray radiation and grating-order overlap. Nonvanishing values of R_f do not introduce error in T_{app} when T_S is either zero or equal to T_R, but can affect photometric measurements of intermediate values. As an extreme example, let us take a highly reflecting chopper with $R_f = 1$, and also for simplicity let $T_a' = T_R = 1$. In the far infrared values of $m = 0.2$ are not improbable, as seen in Fig. 14.2. In this case, according to Eq. (14.6), a sample of actual transmittance $T_S = 0.50$ can have an apparent transmittance $T_{app} = 0.56$. This is a significant error indeed.

A double-passed, single-beam spectrophotometer has its chopper inside the monochromator, as in Fig. 14.4. We designate by T_a the atmospheric transmittance from source to sample cell T_a' from sample to detector, T_a'' from sample to whatever element of the monochromator is considered to be emitting (e.g., the slit jaws), and T_a''' from chopper to detector. The signals seen by the detector when sample, reference, or a shutter are placed in the beam are, respectively,

$$J_S = GT_aT_a'T_S + E_ST_a' + E_MT_a'T_a''R_S - E_MT_a'''$$
$$J_R = GT_aT_a'T_R + E_RT_a' + E_MT_a'T_a''R_R - E_MT_a'''$$

and

$$J_0 = E_BT_a' + E_MT_a'T_a''R_B - E_MT_a'''$$

The measured transmittance of the sample is

$$T_{app} = \frac{J_S - J_0}{J_R - J_0}$$

Let us assume that the chopper and the emitting element of the monochromator are nonreflecting and at the same temperature, that the sample, reference, and shutter are all at the same temperature, and that neither the reference nor the shutter are reflecting. With these conditions, the apparent transmittance of the sample is

$$T_{app} = \frac{T_S}{T_R} + \frac{R_s(mT_a'' - s)}{T_R(T_a - s)} \tag{14.7}$$

The error in transmittance vanishes if the sample does not reflect ($R_s = 0$), or in the absence of atmospheric absorption, if the monochromator and shutter are at the same temperature ($T_a'' = 1$, $m = s$). Let us suppose that a highly reflecting material (e.g., germanium) having $R_s = 0.64$ and $T_s = 0.36$ is measured at 400 cm^{-1}. If there is no atmospheric absorption and if the monochromator temperature is 310°K and the sample temperature is 300°K, Eq. (14.7) tells us that T_{app} is 36.4%, or 0.4% too high.

B. Double-Beam Spectrophotometers

A typical double-beam optical null spectrophotometer with photometer optics arranged in a box configuration is sketched in Fig. 14.5. The transmissions of atmospheric absorbers in various sections of the optical path are defined in the figure. The fraction of the reference beam attenuator inserted into the beam is f_c. We assume there is no trimming attenuator in the sample beam. We can write for the ac signal seen through the sample beam contributions from sample transmission, sample emission, and reflected monochromator emission,

$$J_S = GT_aT_a'T_S + E_ST_a' + E_MT_a'^2R_S$$

Similarly, for the reference beam

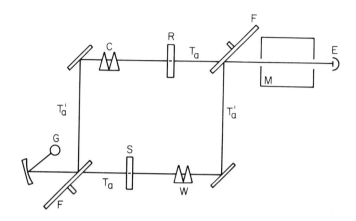

Fig. 14.5 Double-beam spectrophotometer.

$$J_R = GT'_a T_a T_R f_c + E_c T_R T_a + E_R T_a + E_M T_a^2 R_R$$

which includes the transmittance of the attenuator and also its emission. The servomechanism of the instrument positions the attenuator to make $J_S = J_R$, and therefore at balance

$$f_c = [T_S T'_a(T_a - s) + R_S T'_a(mT'_a - s) + T_R T_a(r - c) + R_R T_a(r - mT_a) + T'_a s - T_a r]/T_R T_a(T'_a - c)$$

We now discuss this equation from the standpoints of the zero, transmittance, and compensation errors.

At a wavelength where the sample is opaque, but with a finite reflectance, the reference neither reflecting nor absorbing, and the optical path free from atmospheric absorption, the reference beam comb is positioned at

$$f_c^0 = \frac{R_S(m - s) - (c - s)}{1 - c} \tag{14.8}$$

The departure of this value of f_c^0 from zero is a zero error, and it originates in the temperature differences between the monochromator and the sample and between the reference beam comb and the sample. It is usual for the comb to be at the same temperature as the monochromator, $m = c$, and hotter than the sample, $m > s$. In this case Eq. (14.8) becomes

$$f_0 = \frac{(m - s)(R_S - 1)}{1 - c} \leqslant 0$$

That is, the comb tends to be driven below zero, although a mechanical stop often prevents this. When stray radiation is present, a falsely optimistic measure of its magnitude is obtained. For example, if the attenuator is 10°C hotter than a nonreflecting sample ($R_s = 0$), the error amounts to 0.8% at 50 cm^{-1}, 0.4% at 700 cm^{-1}, and 0.2% at 1000 cm^{-1}.

Now let us consider transmittance errors. Once more we assume a spectral region free of atmospheric absorption, a nonreflecting and nonabsorbing reference material, and a monochromator and reference beam attenuator at the same temperature ($m = c$). The position of the attenuator when a sample is inserted is

$$f_c = T_S\left(\frac{1-s}{1-m}\right) - \left(\frac{m-s}{1-m}\right)(1-R_S)$$

Evidently, the apparent transmittance of the sample depends on its reflectance and temperature and on the temperature of the monochromator.

Finally, we discuss compensation errors. If the sample and reference beams contain identical partial absorbers, the balance position of the comb should not depend on their transmittance. Failure to meet this condition results in compensation error. Let us first consider a spectral region free of atmospheric absorption and a sample which is slightly less transparent than a reference, but which has the same reflectance, so

$$T_S = T_S^0 - \delta T_S \qquad T_R = T_S^0 \qquad R_S = R_R$$

In this case, the reference beam comb balances at

$$f_c' = 1 - \frac{T_S^0 + R_S - 1}{T_S^0}\left(\frac{s-r}{1-c}\right) - \frac{\delta T_S}{T_S^0}\left(\frac{1-s}{1-c}\right) \qquad (14.9)$$

whereas the correct transmittance is

$$f_c = 1 - \frac{\delta T_S}{T_S}$$

The compensation error in Eq. (14.9) depends largely on the temperature difference between sample and reference.

Now consider a region of spectrum where atmospheric absorbers attenuate significantly. Suppose that the sample and reference are identical and at the same temperature. The balance position of the comb is

$$f_c = 1 - \left(\frac{1-T_S}{T_S}\right)\left[\frac{s(T_a-T_a')}{T_a(T_a'-c)}\right] + R_S\left[\frac{s(T_a-T_a')-m(T_a^2-T_a'^2)}{T_S T_a(T_a'-c)}\right]$$

Only in the case of a symmetric photometer design having $T_a = T_a'$ does compensation error for atmospheric absorbers vanish for all values of T_S and R_S.

There are other double-beam optical null spectrophotometers having photometer configurations differing from that of Fig. 14.5. The photometer system sketched in Fig. 14.6, for example, is found in the Perkin–Elmer Model 21 series. The power in the sample beam is

$$J_S = GT_a T_a' T_S + E_S T_a' + E_M T_a'^2 R_S$$

and in the reference beam is

$$J_R = GT_a T' T_R f_c + E_c T_a + E_R T_a' f_c + E_m T_a'^2 f_c R_R$$

The reference beam comb is balanced at

$$f_c = \frac{T_S(T_a-s) + R_S(mT_a'-s) + (s-c)}{T_R(T_a-r) - R_R(r-mT_a') - (c-r)}$$

The system of Fig. 14.7 is used in the Beckman IR-5 series. The sample beam power is

$$J_S = GT_a T_a' T_S + E_S T_a' + E_M T_a'^2 R_S$$

and the reference beam power is

$$J_R = GT_a T_a' T_R f_c + E_c T_R T_a' + E_R T_a' + E_M T_a'^2 R_R$$

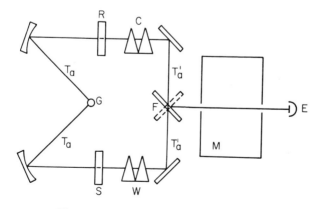

Fig. 14.6 Double-beam spectrophotometer.

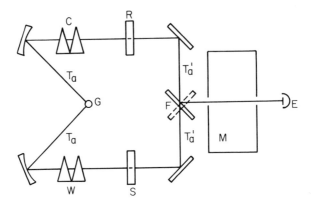

Fig. 14.7   Double-beam spectrophotometer.

The reference beam comb balances at

$$f_c = \{T_S[T_a-s] + R_S[mT'_a-s] + T_R(r-c) \\ + R_R(r-mT'_a) + (s-r)\}/T_R(T_a-c)$$

It is recognized that radiation from intentionally heated samples (or radiation to intentionally cooled samples) can lead to serious photometric errors in double-beam optical null instruments, especially in the far infrared. The whole problem can be avoided by substituting a non-chopping beam-recombining mirror in place of the rotating mirror usually used in optical null instruments. A fabricated strip mirror of the type described in Chapter 4 can be used. An elaborate fabricated mirror is used in the Cary Model 90 ratio recording spectrophotometer. The extraneous radiation is further simplified in ratio recording instruments with the elimination of the reference beam attenuator. A simpler two-element mirror can be used to split either the pupil or the slit into two parts—one used for the sample and one for the reference beam. The Halford–Savitzky ratio-recording system and the Beckman IR-102 and Microspec use this approach. The Beckman IR-11 far-infrared instrument provides a means of stopping the beam-combining mirror (which is at the entrance pupil) in the half-way position in order to accomplish this.

C. Measurement of Stray Radiation

Because reference beam attenuators are unreliable at low transmittance values, stray radiation should be measured with the spectrophotometer operating in the single-beam mode if possible. Consider the configuration of Fig. 14.5 with an opaque shutter of reflectivity R_B in the reference beam. It is assumed that there is no atmospheric absorption and that the shutter and sample are at the same temperature. The ac signal at the detector is

$$J = J_S - J_R = GT_S + E_S + E_M R_S - E_B - E_M R_B$$

or

$$\left(\frac{J}{G}\right)_{\text{sample}} = T_S(1-s) + (R_S - R_B)(m-s)$$

If there is no sample in the sample beam, the signal is

$$\left(\frac{J}{G}\right)_{\text{open}} = (1-s) - R_B(m-s)$$

and if a shutter of reflectance R'_B is placed in the sample beam, the signal is

$$\left(\frac{J}{G}\right)_{\text{shutter}} = (R'_B - R_B)(m-s)$$

The apparent transmittance of the sample is therefore

$$T_{\text{app}} = \frac{(J/G)_{\text{sample}} - (J/G)_{\text{shutter}}}{(J/G)_{\text{open}} - (J/G)_{\text{shutter}}}$$

$$= \frac{T_S(1-s) + (R_S - R_B)(m-s) - (R'_B - R_B)(m-s)}{(1-s) - R_B(m-s) - (R'_B - R_B)(m-s)} \quad (14.10)$$

When $R_B = R'_B = 0$ this relationship reduces to

$$T_{\text{app}} = T_S + R_S\left(\frac{m-s}{1-s}\right)$$

Stray radiation in a monochromator is determined from the apparent transmission of an absorption band that is known to be effectively opaque at the wavelength of interest and transmitting at other wavelengths. In the presence of an amount of stray radiation $U = uG$, Eq. (14.10) becomes for $T_S = 0$ and $R_B = R'_B = 0$

$$T^{(1)}_{\text{app}} = \frac{R_S(m-s) + u}{1-s+u}$$

If the sample does not reflect and S is not too large, T_{app} is very nearly equal to the stray radiation. However, R_S can be large, especially in the vicinity of strong absorption bands. $T^{(1)}_{\text{app}}$ will then indicate an erroneously high value for u. For example, a sample that is opaque at $700\ cm^{-1}$ but reflects 25% will indicate a stray light level about 0.1% too high. If the sample beam is now blocked by placing a shutter between the sample and the source, a different apparent transmittance is observed,

$$T_{app}^{(2)} = \frac{R_S(m-s)}{1-s+u}$$

The difference between these values of T_{app} is a much better approximation to u than $T_{app}^{(1)}$ alone, because the effect of sample reflection is subtracted out. Thus

$$\Delta = T_{app}^{(1)} - T_{app}^{(2)} = \frac{u}{1-s+u} \approx u$$

V. SAMPLE AND SAMPLE CELL EFFECTS

The effect of finite resolution of a monochromator on apparent transmittance was discussed in Chapter 7. We have discussed in the preceding section the effects on the signal when the sample or sample cell emits or reflects extraneous radiation into the monochromator. We should also consider the effects of internal reflection of the primary spectrometer beam within the cell. With reference to Fig. 14.8, we can write for the radiant power emerging from a cell filled with liquid

$$\begin{aligned}
J(T) &= J_0(1-R)^2T + J_0(1-R)^2T^3R^2 + J_0(1-R)^2T^5R^4 + \cdots \\
&= J_0(1-R)^2T\,[1 + T^2R^2 + T^4R^4 + \cdots] \\
&= \frac{J_0(1-R)^2T}{1-T^2R^2}
\end{aligned} \tag{14.11}$$

We have assumed that the reflection R occurs only at the interface between the cell window and air, that all rays escaping the cell in the direction of the detector are detected, and that the rays from succes-

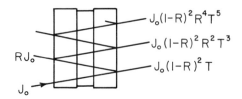

Fig. 14.8 Internal reflection in a sample cell for liquids.

TABLE 14.2
Apparent Transmittance of a Liquid
Sample in a Cell with Reflecting
Windows

T	T_{app}		
	$R = 0.05$	$R = 0.15$	$R = 0.40$
0	0	0	0
0.25	0.249	0.245	0.212
0.50	0.499	0.492	0.438
0.75	0.749	0.743	0.692
1.00	1.000	1.000	1.000

sive passes through the cell are in phase and add without interference. If a transparent liquid material is substituted for the sample, the power is

$$J(1) = \frac{J_0(1-R)^2}{1-R^2} = J_0\left(\frac{1-R}{1+R}\right)$$

Hence, the apparent transmittance of the sample is found to be

$$T_{app} = \frac{J(T)}{J(1)} = T\left(\frac{1-R^2}{1-T^2R^2}\right) \tag{14.12}$$

Values of T_{app} are tabulated in Table 14.2 for cell windows of low, medium, and high reflectivity. The error in apparent transmittance becomes significant for moderate reflectance.

If the relationship between optical path length and wavelength is such that the alternate contributions to $J(T)$ in Eq. (14.11) are out of phase, then the equation for transmitted power is

$$J'(T) = \frac{J_0(1-R)^2T}{1+T^2R^2}$$

Thus, the spectrum is modulated by an interference fringe system of wavelength spacing given by

$$\Delta\lambda = 2bn$$

where b is the thickness and n the refractive index of the sample. The amplitude modulation of the fringes is given by

$$\frac{J(T)-J'(T)}{J(T)+J'(T)} = T^2R^2$$

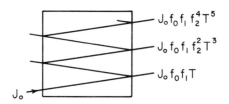

Fig. 14.9   Internal reflection at the sample–window interface in a gas cell.

Now let us consider the effect of reflection from the interface between the sample and the cell window. This is very important in the case of a gas sample (refer to Fig. 14.9). A fraction f_0 of the radiation approaching the first window is reflected back toward the source. A fraction f_1 of the radiation approaching the second window is reflected back toward the source from either the first or second surface of the window. A fraction f_2 of the radiation approaching the first window from the right is reflected back through the sample. Proceeding just as before, we find for the radiation transmitted out of the cell

$$J(T) = J_0[f_0 f_1 T + f_0 f_1 f_2^2 T^3 + f_0 f_1 f_2^4 T^5 + \cdots]$$
$$= J_0 f_0 f_1 T / (1 - T^2 f_2^2)$$

The apparent transmittance of the sample is

$$T_{\text{app}} = \frac{J(T)}{J(1)} = T\left(\frac{1 - f_2^2}{1 - T^2 f_2^2}\right) \qquad (14.13)$$

which is much like Eq. (14.12). An expression for f_2 can easily be derived by considering Fig. 14.10. The reflectance at the sample–window interface is R_1 and the reflectance at the window–air interface is R_2. We have

$$f_2 = R_1 + R_2(1 - R_1)^2 + R_2^2 R_1(1 - R_1)^2 + R_2^3 R_1^2(1 - R_1)^2 + \cdots$$
$$= R_1 + (1 - R_1)^2 R_2(1 + R_2 R_1 + R_2^2 R_1^2 + \cdots)$$

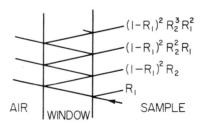

Fig. 14.10   Internal reflection within a window.

TABLE 14.3

Apparent Transmittance of a Gaseous Sample in a Cell with Reflecting Windows

T	T_{app}		
	$R = 0.05$	$R = 0.15$	$R = 0.40$
0	0	0	0
0.25	0.248	0.234	0.172
0.50	0.496	0.474	0.367
0.75	0.747	0.727	0.619
1.00	1.000	1.000	1.000

$$= R_1 + \frac{R_2(1-R_1)^2}{1-R_2 R_1} = \frac{R_1 - 2R_1 R_2 + R_2}{1 - R_1 R_2}$$

When $R_1 = R_2$, as in the case of a gas sample, Eq. (14.13) becomes

$$T_{app} = T\left[\frac{1 + 2R - 3R^2}{1 + 2R + R^2(1 - 4T^2)}\right]$$

Values of T_{app} are tabulated in Table 14.3, for comparison with the entries of Table 14.2. The error is clearly larger in the case of the gas cell.

In the event that the sample to be studied is a solid slab, the apparent transmission can be measured only relative to the empty radiation beam,

$$T_{app} = J(T)/J_0$$

and from Eq. (14.11) this can be written

$$T_{app} = T\left[\frac{(1-R)^2}{1-T^2 R^2}\right] \tag{14.14}$$

Values of T_{app} are given in Table 14.4. The errors are quite large in this case. If the reflectivity is not too high, the denominator of Eq. (14.14) can be set equal to unity; T_{app} can then be corrected easily if the reflectivity is known or is calculated from the refractive index. If the reflectivity is not small, T can be calculated from the somewhat more complicated expression

$$T = \frac{[(1-R)^4 + 4R^2 T_{app}^2]^{1/2} - (1-R)^2}{2R^2 T_{app}}$$

TABLE 14.4
Apparent Transmittance of a
Reflecting Solid Sample

T	T_{app}		
	$R = 0.05$	$R = 0.15$	$R = 0.40$
0	0	0	0
0.25	0.226	0.181	0.091
0.5	0.451	0.363	0.188
0.75	0.678	0.549	0.297
1.00	0.905	0.739	0.429

Now let us turn to a different kind of problem associated with the cell. Let us suppose that the cell (or a solid sample) is fabricated with an average thickness b_0 but with a wedge amounting to Δb over the width W (see Fig. 14.11). The thickness measured at a distance x from the center is therefore $b_0 + (\Delta bx/W)$. A sample of absorption coefficient a is placed in the cell. The radiant power transmitted by the cell, ignoring reflection is

$$J = \frac{J_0}{W} \int_{-W/2}^{+W/2} \exp\left[-a\left(b_0 + \frac{\Delta bx}{W}\right)\right] dx$$

$$= J_0 e^{-ab_0} \left[\frac{\exp\left(a\Delta b/2\right) - \exp\left(a\Delta b/2\right)}{a\Delta b}\right]$$

So the measured transmittance in terms of the nominal transmittance for a thickness b_0 is

$$T = 2T_0 \left[\frac{\sinh\left(a\Delta b/2\right)}{a\Delta b}\right]$$

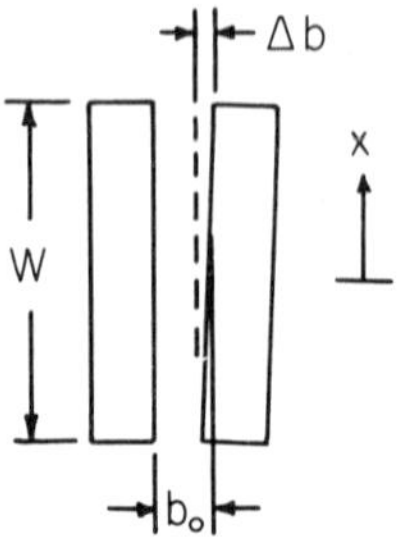

Fig. 14.11 Wedged sample cell.

$$\approx T_0\left[1-\frac{1}{24}(a\Delta b)^2\right] \qquad \text{for } a\Delta b \text{ small}$$

The error in T is

$$\Delta T = T_0 - T = \frac{T_0}{24}(a\Delta b)^2$$

Thus, if $T_0 = 0.37$, which means that $ab_0 = 1$, and if $\Delta b/b_0 = 0.1$, then $\Delta T \approx 0.37 \times 0.0004$, which is an insignificant error. Even in the extreme case of $\Delta b/b_0 = 1$, the error is only about 4% of T_0. Some experiments with wedged samples have been reported by Olsen et al. (4).

These equations apply equally well to a cell whose thickness varies randomly but uniformly in the range $b_0 - \frac{1}{2}(\Delta b) \leqslant b \leqslant b_0 + \frac{1}{2}(\Delta B)$. We conclude that cells constructed with the precision usually obtained ($\Delta b/b_0 \leqslant 0.05$) do not contribute significantly to photometric nonlinearity.

We have discussed the cell effects as though all rays in the spectrometer beam were parallel. In fact, as we have seen, nonaxial rays must be present even in so-called collimated beams. Thus, the cell presents a distribution of thickness to the radiation beam. Consider a cell placed in a converging beam of radiation as in Fig. 14.12. The aperture with sides X and Y represents a focusing element of focal length f. The length of a typical ray emanating from a point (x, y) of the aperture is

$$l = (x^2 + y^2 + f^2)^{1/2}$$

and the path length through a cell of nominal length b_0 is

$$b(x, y) = \frac{b_0}{f}l = \frac{b_0}{f}(x^2 + y^2 + f^2)^{1/2} \qquad (14.15)$$

We disregard refraction in the sample which would reduce b to a value somewhat closer to b_0; that is, the present discussion is confined to

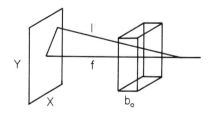

Fig. 14.12 Converging beam through a cell.

gaseous samples. If the absorption coefficient of the sample is a, the radiation power transmitted by the sample is

$$J = \frac{J_0}{XY} \int_{-Y/2}^{Y/2} \int_{-X/2}^{X/2} \exp\left[-ab(x,y)\,dx\,dy\right]$$

For spectrophotometers of moderate aperture, X and Y are sufficiently smaller than f, so that Eq. (14.15) for $b(x,y)$ can be written as a binominal approximation,

$$b(x,y) \approx \frac{b_0}{f}\left[f + \frac{x^2+y^2}{2f}\right]$$

With this approximation, the transmitted power is

$$J = \frac{4J_0}{XY} e^{-ab_0} \int_0^{X/2} \exp\left(-ab_0 x^2/f^2\right) dx \int_0^{Y/2} \exp\left(-ab_0 y^2/f^2\right) dy$$

$$= \frac{J_0}{ab_0} e^{-ab_0} \frac{\pi f^2}{XY} \operatorname{erf} \frac{(ab_0)^{1/2}X}{2f} \operatorname{erf} \frac{(ab_0)^{1/2}}{2f} Y$$

The ratios f/X and f/Y are the arc tangents of the convergence angle in the x and y planes. The error in transmittance in a converging beam for a sample with $ab_0 = 1$ is plotted in Fig. 14.13 as a function of convergence angle.

Finally, we consider the effect of the presence of a cell or solid

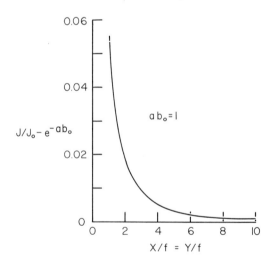

Fig. 14.13 Transmittance error in a converging beam.

sample on the focus of the spectrometer. We showed in Section 3.V.A that insertion of a plate of thickness t and refractive index n into a converging beam shifts the point of convergence by an amount $t[1 - (1/n)]$. Usually the source of a spectrophotometer is imaged on the slit, or, conversely, a slit image is placed on the source. The source is a narrow cylinder and a severe displacement of the image can cause the slit to be incompletely filled. Consequently, a low transmittance reading can be obtained for a solid sample, particularly if it is quite thick or has a high refractive index. Of course, if the instrument is incorrectly adjusted, insertion of the plate might improve the focus and a high reading will be obtained. The condition of the instrument can be checked by inserting a block of transparent material of known reflectivity and verifying that Eq. (14.14) holds.

VI. The Reference Beam Attenuator

The reference beam attenuator of optical null spectrophotometers is the most serious contribution to photometric inaccuracy. When the attenuator is in the form of a multitooth comb, the departure from straightness of the edges of the teeth limits the photometric linearity. In a typical instrument the comb may be 25 mm in overall height and have 5 teeth. In order to attain a photometric accuracy of 0.1%, the tolerance on the combined width of the teeth at any given position is 0.025 mm. The tolerance on each tooth, allowing for randomness in the error, is $(0.025/5)\sqrt{5} \approx 0.01$ mm. It is bad enough that the comb must be fabricated so accurately, but it must also be carefully guarded against damage in use which might nick or warp the teeth. Even a dust particle on the edge of a tooth can produce a kink in the photometric response when a narrow slit is imaged on the comb.

The presence of a slit image on the comb causes another problem. The slit is also usually imaged on the detector and the response of many infrared detectors is quite nonuniform over the surface of the target. For example, a wire-type thermocouple may have its target supported by four thin wires that also serve as heat conductors between the target and its surroundings. Consequently, a plot of the response of the detector measured along its length appears typically as in Fig. 14.14. Plots of this kind can be made by passing a narrow slit along the length of a magnified image of the detector and illuminating the detector through the slit. As a consequence of this variation in sensitivity, a perfectly linear comb placed at an image of the detector can produce a

RESPONSE

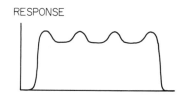

Fig. 14.14   Nonuniform response of a detector.

nonlinear response. Nonuniformity of illumination of the slit produces a similar effect.

In an attempt to avoid the problems associated with the discrete nature of the comb-type attenuator, Tarbet and Daly (5) use a rotating star-shaped disk (see Fig. 9.20). The rapidly rotating disk presents the detector with an effectively uniformly attenuated radiation beam, provided the speed of rotation is large with respect to the time constant of the detector. The attenuation is varied by moving the axis of the disk with respect to the radiation beam as shown.

The author once fabricated a reference beam attenuator in the form of a single wedge opening (a single-tooth comb). The edges of the wedge were made of thin, flexible metal that could be deformed by a row of 12 adjustable cams. The object was to adjust the attenuator for linear response and thus compensate for all of the errors due to nonuniformity of detector response or beam brightness. The adjustment was tedious and the results not particularly encouraging. It is also possible to calibrate the transmitting potentiometer which provides the pen servo with information about the position of the comb. This can be done by using a multitapped potentiometer with a network of adjustable resistors, similar to the arrangement used for programming the slit width. The difficulty with this adjustable system is that the nonlinearity of the attenuator is likely to be wavelength (or slit width) dependent.

There is an inherent source of nonlinearity common to most reference beam attenuators with the exception of the venetian-blind and rotating-sector types. This has been discussed by Kudo (6). It occurs because of the finite width of the radiation beam. If the attenuator is ideally linear and the detector is perfectly uniform, the rate of change of attenuation with comb position is uniform up to the point where the apex of the openings arrives at the edge of the beam (b in Fig. 14.15). At this point the rate of change of attenuation is only half as great as when the apex travels across the width of the beam. This leads to a relationship between attenuation and comb position as shown in Fig.

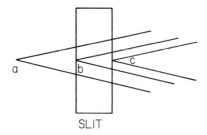

Fig. 14.15 Opening of a comb at its narrow end relative to slit image.

14.16. A typical comb might be 15 cm long, so the change in slope occurs at 4% transmittance when a slit 6 mm wide is used, and at 0.1% T when a narrower 0.15-mm slit is used. Incidentally, this explains why it is seldom desirable to place the attenuator at an image of the pupil of the monochromator, which is, in general, much wider than the slit. Sometimes the ends of the openings of the comb are staggered. This makes the change in slope less abrupt, but spreads it over a wider range. In any case, the fabrication of attenuators is particularly difficult at the narrow openings. Combs are rarely reliable below about 10% T, and consequently transmission measurements at the low end of the scale are to be viewed with suspicion. As a corollary, electrically expanded scale presentation of the lower end of the transmittance range is usually to be avoided because the low transmission end of the attenuator comb is used. It is generally better to expand the spectrum of dense samples by placing an optical attenuator in the reference beam.

All of these difficulties associated with the reference beam attenua-

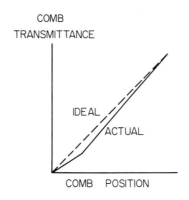

Fig. 14.16 Photometric error introduced by a comb.

tors of optical null instruments limit their photometric accuracy to perhaps 2% of full scale for the transmission range from 10 or 15 to 100% T. This is adequate for many purposes. The precision of measurements is much better in modern instruments and sometimes it is desirable to calibrate the transmission response of a spectrophotometer. The possibility of doing this is discussed in the next section.

VII. MEASUREMENT OF PHOTOMETRIC ACCURACY

A. Beer's Law Test

Before turning to some acceptable methods for the determination of photometric accuracy, let us dispose of a method that is widely used, but has little merit. The method in question is the so-called Beer's law test. The procedure is to measure the transmittance of two individual samples and then to place both the samples in the beam simultaneously, in tandem. The argument is that in a linear instrument the transmittance of the tandem samples is the product of the transmittances of the separate samples. Equivalently, the absorbance of the tandem samples is the sum of the absorbances of the separate samples. There is nothing wrong with this statement, unless we are concerned with strange effects such as reflections between samples. It does not follow, however, that in a nonlinear instrument the transmittance of the tandem samples is not the product of the separate transmittances. Consider an instrument which indicates a transmittance T_a' for a sample whose actual transmittance is T_a, and for which the following relationship holds:

$$T_a' = T_a{}^p \qquad p > 0 \qquad (14.16)$$

Similarly, sample b has an apparent transmittance

$$T_b' = T_b{}^p$$

The combined actual transmittance of the samples in tandem is

$$T = T_a T_b$$

and the apparent transmittance is

$$T' = (T_a T_b)^p = T_a' T_b'$$

According to Beer's law test, this proves that the instrument is linear. But the instrument is not linear, so the Beer's law test is not a valid one. Moreover, relationships of the type given by Eq. (14.16) are not unusual as spectrometer responses. A family of these curves is shown in Fig. 14.17.

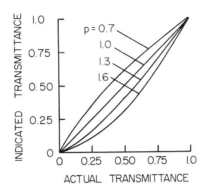

Fig. 14.17   Response curves for which Beer's law is obeyed.

We must conclude that the Beer's law test is useless as a means of establishing the photometric linearity of a spectrophotometer. Nevertheless, it does serve a useful purpose. An instrument that passes the Beer's law test can be used safely to perform quantitative analyses, provided the absorptivities of the materials in question are determined from measurements made on that instrument and not transferred from some other spectrometer.

B. Standard Filters

It would be useful to have available a filter or set of filters of known transmittance for establishing the linearity of response of a spectrophotometer. The desired properties of standard filters are fairly restrictive. These properties are the following:

(1) The transmittance must vary slowly with wavelength or not at all in order that the influence of the spectral bandwidth of the monochromator is not significant.

(2) A wide range of transmittance values must be available to establish the response of the instrument over its entire transmittance scale.

(3) The reflectivity of the filters must be low to avoid problems associated with extraneous radiation. This requirement eliminates the use of thin metallic films on transparent substrates.

(4) The filters must not be very thick, lest they defocus or deviate the spectrometer beam.

(5) The material of the filters must be durable, homogeneous, capable of being worked to good optical quality, and not highly sensitive to environmental conditions such as temperature or humidity.

Wire or etched grids are not satisfactory for use as standard filters. Their transmittance depends on their position and orientation in the spectrometer beam in some instruments.

Possibly the best compromise proposed so far is the calcium fluoride plate. It has been used by the author with W. Kaye in the research laboratories of Beckman Instruments (7, 8). Calcium fluoride has an absorption edge in the mid infrared. It has a reasonably shallow slope, so sensitivity to resolution is not great, but careful wavelength calibration is required. A wavenumber error of 0.5 cm^{-1} can introduce a photometric error of 0.2% T. The absorption lines of the 10-μ ammonia band provide a convenient set of wavelength calibration points falling along the absorption edge. Other wavelength standards can be used. Indene has been found to be useful provided a means is used for interpolating between the known bands. The refractive index of calcium fluoride is known, so the contribution of reflectivity to attenuation can be calculated and added to the contribution from absorption of a crystal of measured thickness. The absorptivity of CaF_2 is somewhat temperature sensitive (7, 9), necessitating careful temperature control. The transmittance increases by as much as 0.2% of full scale per 1°C temperature rise.

The major problem with standard filters is, of course, that they must somehow be standardized. This remains one of the problems of infrared spectroscopy which is not completely solved. Some progress has been made in the visible region of the spectrum in establishing the transmittance of filters by making use of the inverse-square law. A source and detector are mounted on an optical bench. The light flux falling on the detector varies inversely with the square of its distance from the source. Attenuation by a filter is compensated by repositioning the source, and the transmittance of the filter is calculated from the distances. This method has not been used in the infrared.

The transmittance of two or more polarizers can be calculated from the angle between their optic axes. Bennett (10) has used a polarizing attenuator to establish the photometric response of spectrometers in the visible region of the spectrum. This technique also has not been used in the infrared.

C. Apertures

An aperture containing several openings can be placed in the radiation beam. The openings are uncovered separately and in combinations, and the extent to which the spectrometer response is additive is taken as an indication of its linearity. This method has been used by Ward (*11*) and by Kramer and Spott (*12*). The method of apertures seems at first glance to be the same as the Beer's law test, but there is a fundamental difference. The radiant power reaching the detector by way of one of the openings is independent of the effect of other openings. In the Beer's law test, on the other hand, the radiation passing through one filter is further attenuated by the other filters. The aperture method is a valid test of spectrophotometric linearity provided certain potential difficulties are avoided. The aperture should be illuminated incoherently, of course, in order that radiant power passed by the separate openings is additive. This is usually the case, but the use of laser illumination is not advisable. The aperture should not be placed at a position conjugate with the detector, because the response of the detector can very well depend on how it is irradiated. For example, if the target of a thermal detector is supported at the ends, the flow of heat from the center depends on whether the ends are warmed by incident radiation. Usually the slit of the monochromator is imaged on the detector, so the aperture should be placed at an image of the pupil of the monochromator. Finally, the usual precautions regarding reflection and emission are to be taken. It is best to place the aperture at a position following the radiation chopper. This complicates its use with double-beam instruments.

D. Rotating Sectors

A sector disk having slots cut radially through it can serve as a photometric standard. The disk is placed in the beam and rotated rapidly about its axis. It interrupts the beam a fraction of the time equal to the ratio of closed to open areas. The action of the rotating sector is similar to the aperture of Section VII.C, except that the radiation is partitioned in time rather than in space.

The sector can be constructed so as to be continuously variable in transmittance from 0.5 to 0. This is scaled to cover the 1.00–0 range by initial setting of the gain or slit width of a single-beam spectrophotometer or by inserting an attenuator in the reference beam of a double-beam instrument. Such a sector was described in Section 9.VII. Alternatively, a set of fixed sector disks of different transmittances can be used.

The sector transmits a beam of radiation that is switched from on to off to on at fixed intervals. The spectrometer system is asked to integrate this pulsating beam and present a dc signal proportional to the integrated intensity within the linearity of the system. This means that the chopping rate of the sector must be fast relative to the response time of the system. This is an extension of Talbot's law to electrical detectors. If the system response can be characterized by a single dominant time constant τ, its impulse response is

$$h(t) = e^{-t/\tau}$$

and its transfer function, the Fourier transform of $h(t)$ normalized to unity at $\omega = 0$, is

$$H(\omega) = \frac{1/\tau}{j\omega + 1/\tau}$$

The ratio of the amplitude of output to input, for a sinusoidal input of frequency ω, is given by

$$|H(\omega)| = \left[\frac{(1/\tau)^2}{\omega^2 + (1/\tau)^2}\right]^{1/2}$$

When $\omega \gg 1/\tau$ this equation becomes

$$|H(\omega)| = \frac{1}{\omega\tau}$$

If we desire the ripple in the output to be less than 0.1% and if the time constant of the system is 0.5 sec, then the chopping rate must be at least 2000 rad/sec. A sector with ten openings can provide this rate if it is rotated at about 1800 rpm. This speed is easy to achieve.

We ask now, can the sector method be easily used to measure linearity of the detector? A typical thermal detector, such as a thermocouple, has a time constant of perhaps 10 msec. Thus, if a ten-bladed sector is rotated at 1800 rpm, a 50% ripple (roughly) is present in the output from the detector. A measure of detector linearity cannot be obtained under this condition. Indeed, if the detector has a very rapid response, such as those of photomultipliers and photoconductors, the output will be nearly fully modulated, with no integration provided by the detector at all. In this case, the output from the detector is switched almost completely on and off with the passing of each blade of the sector. Even if the detector is extremely nonlinear, this will affect only the amplitude of the pulses. The integrating circuitry will provide an output proportional to the on–off ratio. Thus, the rotating sector is not suitable for measuring the photometric linearity of moderately fast detectors, but, on the other hand, nonlinearity of the detector does not affect the linearity of an optical null spectrophotometer. Thus the rotating-sector method is easily applied to measuring nonlinearities in optical null systems associated with comb nonlinearities. It has been used by Stewart (7), Kramer and Spott (12), Fischer and Nillius (13), Shimozawa and Wilson (14), Jones (15), and Levin and King (16).

E. Electrical Methods

There are several electrical methods of determining photometric linearity. Some instruments are so designed that they can operate easily in either a single- or a double-beam mode. The instrument is adjusted initially to have full-scale recorder pen deflection both single and double beam. An attenuator is inserted in the sample beam and the single- and double-beam pen deflections are compared. This is repeated for a range of attenuations. The attenuator might be the ordinary sample beam attenuator that is provided for beam balance adjustment. It is probably better to avoid the discrete nature of the sample beam attenuator comb and use an absorbing sample instead. A CaF_2 window is a useful attenuator whose transmittance is changed by changing wavelength. Remember that the single- and double-beam 100% levels must be reestablished with each setting. This procedure is a convenient method for determining or calibrating the linearity of the comb, if the single-beam response is known or assumed to be linear by some other method.

A related procedure is to arrange the spectrometer to operate single beam through what is normally the reference beam. This can usually be done by merely installing a reversing switch at the synchronous demodulator. The normal input to the comb-positioning motor is interrupted and the comb is driven at a uniform rate into the beam. This can be done by attaching a small synchronous motor to the comb drive mechanism. In this way the transmittance of the comb may be directly recorded as a function of its position, again assuming linearity of the single-beam response of the instrument. It should be recognized that another limitation exists in this method because of possible eccentricities in the gears of the comb mechanism, which after all needn't be very free from eccentricity to perform their normal function.

The linearity of the spectrophotometer system can be measured independently of the properties of the detector by injecting electrical signals of varied amplitude and appropriate frequency and waveform into the preamplifier just following the detector (12). This can be applied to double-beam systems of both the optical null and the ratio-computing type by injecting the electrical signal in place of the sample signal. The synthetic sample signal can be applied at various places, depending on the specific nature of the instrument.

REFERENCES

1. A. R. Goodwin and F. A. Johnson, *J. Sci. Inst.*, **42**, 418 (1965).

2. J. E. Stewart, *Appl. Opt.*, **2**, 1141 (1963).

3. S. Minami, H. Yoshinaga, and K. Matsunaga, *Appl. Opt.*, **4**, 1137 (1965).

4. A. L. Olsen, K. B. La Baw, and L. W. Nichols, *J. Opt. Soc. Am.*, **54**, 813 (1964).

5. C. S. C. Tarbet and E. F. Daly, *J. Opt. Soc. Am.* **49**, 603 (1959).

6. K. Kudo, *Sci. of Light*, **4**, 85 (1955).

7. J. E. Stewart, *Appl. Opt.*, **1**, 75 (1962).

8. J. E. Stewart, *Appl. Opt.*, **2**, 1303 (1963).

9. W. Kaiser, W. G. Spitzer, R. H. Kaiser, and L. E. Howarth, *Phys. Rev.*, **127**, 1950 (1962).

10. H. E. Bennett, *Appl. Opt.*, **5**, 1265 (1966).

11. W. Ward, unpublished work at Cary Instruments.

12. L. Kramer and S. Spott, *Exptl. Tech. Physik*, **11**, 319 (1963).

13. H. Fischer and O. Nillius, *Exptl. Tech. Physik.* **9**, 68 (1961).

14. J. T. Shimozawa and M. K. Wilson, *Spectrochim. Acta*, **22**, 1591 (1966).

15. R. N. Jones, *J. Japan. Chem.*, **21**, 609 (1967).

16. I. W. Levin and W. T. King, *J. Chem. Phys.*, **37**, 1375 (1962).

CHAPTER 15

Experimental Methods of
Infrared Transmission Spectroscopy

I. INTRODUCTION

In this and the remaining chapters we discuss various experimental methods used in infrared spectroscopy. No attempt will be made to present a complete bibliography because the literature is simply too vast. In 1960 the author published a review (*1*) of sample handling methods which contains a bibliography of over three hundred entries. The reader is referred to that list for literature references and titles of papers that describe methods mentioned only briefly here.

The majority of infrared studies are in the area of transmission spectroscopy. The term transmission spectroscopy is preferred to absorption spectroscopy because only the radiation transmitted by the sample is measured. The radiation removed from the beam by the sample is partly reflected and partly scattered. It is only after the reflected and scattered parts are accounted for that the absorption by the sample can be stated.

In this chapter we discuss some of the sample handling methods and devices used in infrared transmission spectroscopy. We also discuss some of the techniques of quantitative measurement of absorbance.

II. Methods for Liquid Samples

Liquid samples may be either pure materials or solutions of liquids, solids, or even gases in liquid solvents. The choice of solvent is limited by chemical properties and solubilities and, of great importance, by infrared transparency. In general, single solvents that are transparent over very wide spectral ranges are not available, but complementary pairs can often be found. Carbon tetrachloride and carbon disulfide are the most common pair. Carbon tetrachloride has a very strong region of absorption between 730 and 820 cm^{-1}, but elsewhere it is reasonably transparent. Carbon disulfide, on the other hand, is useful in the 770-cm^{-1} region, but absorbs strongly at about 1500 and also near 2250 cm^{-1}. Sometimes some chemists' tricks are invoked to improve solubilities. For example, acids can be made soluble in carbon tetrachloride and carbon disulfide by the addition of 0.5 to 2% triethylamine.

Even chemically inert solvents can significantly affect the spectrum of a solute. To cite some examples, the absorption band near 3000 cm^{-1} in the spectra of alcohols is associated with the stretching vibration of the O—H group. In pure liquid the band is broad and strong, but in dilute solutions the hydrogen bonds between adjacent alcohol molecules are broken and the band is replaced by a weaker and narrower one at a somewhat higher frequency. The carbonyl stretching band of molecules containing C=O groups is usually broader and more intense in a chloroform solution than in carbon tetrachloride due to an interaction with chloroform molecules. Some molecules, such as 1,2-dichloroethane, have two or more stable rotational isomers because of the occurrence of hindered rotation about a saturated bond. The spectra of the isomers usually differ in many respects. If one isomeric form is more polar than another, their relative concentrations are affected by the nature and amount of the solvent.

Water is a poor infrared solvent because it absorbs very strongly over most of the infrared and because the solubility of many alkali halide window materials is high in water. However, it is an important medium in inorganic and biological chemistry. By using thin cells of appropriate material it is sometimes possible to obtain infrared spectra of aqueous solutions. The rate of attack on alkali halide windows can be reduced by presaturating the solution with the window material.

Liquid samples for infrared transmission studies must be confined in a definite layer between transparent windows. The assembly of windows, spacer, body, and so on, is called a cell. The choice of window material is dictated by consideration of optical transparency, strength, chemical inertness and solubility, size of material available, and, of course, cost. The selection, properties, and preparation of window materials were discussed in Section 5.I.

A typical liquid sample has a thickness between 0.01 and 0.10 mm. Some very weak or very dilute absorbers require a thicker cell, and occasionally the strongest bands in a spectrum can be resolved only in a thinner cell or in a more dilute solution.

The simplest cells are the demountable variety. They consist of two windows clamped together with a metal or plastic spacer to define their separation. A typical demountable cell is illustrated in Fig. 15.1. A window and the spacer are placed on the plate P_1 and a drop of liquid is placed on the window inside the spacer. Care is required to prevent the liquid from running between the spacer and the window. With experience, the proper amount of liquid can be judged. The second window is placed on top of the spacer, the other plate P_2 is added, and the whole assembly is clamped together with screws. Such cells can be used only with nonvolatile samples. If the sample is very viscous and a qualitative spectrum is sufficient, the spacer is sometimes omitted. A variety of spacerless cell has a flat depression etched in one window. This can be done by brief exposure to a water–alcohol solution, for example, with the outer part of the window protected by a layer of grease, or by placing a saturated piece of blotting paper in contact with the window. In cross section, this cell appears as in Fig. 15.2. Sometimes a moat is etched or cut into the surface of a window with a sharp blade or an ultrasonic cutter, as in Fig. 15.3, to provide a volume for

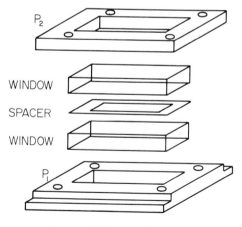

Fig. 15.1 Demountable cell.

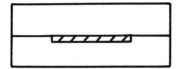

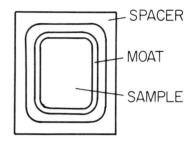

Fig. 15.2 Cross section of a demountable cell with an etched area.

Fig. 15.3 Demountable cell with a moat.

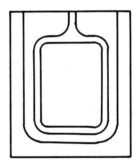

Fig. 15.4 Demountable cell with an extended moat.

surplus liquid. This prevents liquid from being trapped between the spacer and the window. A variation on this cell design, shown in Fig. 15.4, has the moat extended to the edge of the window. The cell can be assembled with a U-shaped spacer and the sample then introduced through the opening at the top. Even with these precautions, however, the thickness of a demountable cell is not highly reproducible.

Spacers for demountable cells are usually thin lead or copper foil, or plastic film such as polyethylene, Mylar, or Teflon. A loop of fine wire can be used as a spacer, and often gives better thickness reproducibility.

When close control of sample thickness is required or when volatile samples are studied, sealed cells must be used. The traditional construction of these cells is with amalgamated spacers. The surfaces of a lead spacer are wetted with mercury, the spacer is placed on one of the windows, and the cell is assembled. The soft amalgam conforms to the surface of the window, and as it hardens it forms a tight seal that will not permit volatile liquids to leak away. With well-polished, dust-free windows and care in assembling the cell, vacuum- and pressure-tight seals can be obtained. Very thin lead foil is difficult to amalgamate because the mercury tends to migrate completely through and destroy the gasket. Copper foil can be used for thin spacers. A copper spacer is first immersed in a solution of mercuric chloride in glacial acetic acid until it is thoroughly blackened. The black substance is wiped off with a tissue, leaving a shiny surface that is easily wetted with mercury.

One of the windows of the sealed cell is drilled before assembly (see Fig. 15.5) and a second amalgamated gasket is used to seal this window to a metal plate. The metal plate has drilled ports to match the holes in the windows, through which the sample is introduced. Usually the holes terminate in larger tapered holes. A hypodermic syringe is seated in the tapered hole for convenient insertion of the sample, and after the cell is filled the holes are plugged with Teflon stoppers or by other means. Some spectroscopists prefer to leave the syringe in place in the cell. It provides a reservoir of sample and helps keep the cell filled if minor leaks are present or if evaporation is taking place. When toxic or reactive samples are investigated, the cell body is made with fittings for attaching tubing, perhaps through needle valves. The sample handling procedure can be carried out on a manifold and material transferred in an inert atmosphere or under vacuum if necessary. Smith and Creitz (2) developed an efficient method for transferring a solution into a cell with air pressure in a way that minimizes changes in concentration due to differential evaporation.

The cleaning of sealed cells presents some difficulties. Volatile samples can be removed from the cell by suction into a syringe. Viscous

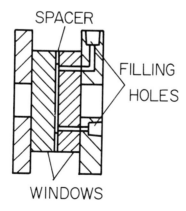

Fig. 15.5 Cross section of a sealed cell for liquid samples.

materials are often reluctant to leave the cell, particularly if the thickness is small, but they can be sucked out with a vacuum pump. A suitable trap should be provided to avoid contaminating the pump oil. After the sample is removed the cell must be flushed with a solvent until all traces of the sample are gone. Sometimes the sample or solvent migrates between the spacer and the cell window, to reappear later, and it is essential to remove all of it before the next sample is inserted. Probably every spectroscopist has discovered a new absorption band at about $770 \, cm^{-1}$ which later turns out to be carbon tetrachloride from an earlier cleaning. It is good practice to flush volatile samples and solvents from the cell with a stream of *dried* air or nitrogen. The windows are cooled by the evaporating liquid and unless precautions are taken water vapor from the air condenses on them and attacks their surfaces, eventually ruining them.

Unusual circumstances sometimes indicate the need for special cell materials. Soft metals, such as indium, and some polymers, can be used as gaskets and spacers. Sealing is accomplished by application of pressure, with no mercury required. Sometimes waxes, cements, or resins are used as sealing agents. Low-melting solder can be used to seal cells, but the alkali halide windows are easily cracked and so this technique is only rarely used. Soft and deformable window materials, such as silver chloride, can be tightly sealed to spacers and cell bodies by the application of sufficient pressure. Because silver chloride is relatively inexpensive, cheap cells can be made with plastic bodies. These can be discarded after use. Similar cells with high-density polyethylene windows are used for far-infrared spectroscopy of liquids.

By proper design, cell windows, seals, and filling fittings can be made to withstand several atmospheres of pressure. Gases that are liquefied at relatively low pressure can thus be studied in the liquid phase at room temperature. Thick windows of alkali halides, sapphire, and even diamond have been used at pressures up to 200,000 atm. It pays to be a bit careful with such cells. A cell filled with liquefied butene once failed at the National Bureau of Standards; the butene was ignited by the spectrometer source, and a potentially dangerous fire resulted. On another occasion a cell of liquefied dimethylamine leaked, giving the laboratory a lingering fishy odor.

Occasionally liquid samples are encountered which are exceptionally transparent over a fairly broad spectral region. This permits the use of sample cells of long path and thus provides very high sensitivity to minor constituents. Parts per million of some impurities have been determined in such samples as Freon, titanium tetrachloride, and liquid chlorine using cells up to 5 cm long. Glass tubes with windows cemented on the ends are used for cells. A particularly useful cell body is made from an Erlenmeyer flask with the bottom cut off.

One of the attractions of double-beam instruments is the ability to perform differential measurements. In its most simple form, variations in background due to atmospheric absorption, source variations, and so on, are removed, leaving a flat I_0 for the recorded spectrum. Next, the effects of reflections from the windows of the sample cell can be removed by placing a plain window of the same material in the reference beam. Absorption by the cell window is canceled by using a reference window whose thickness equals the combined thickness of the sample cell windows. An empty cell should not be used as a reference for two reasons. In the first place, the four window-to-air surfaces would overcompensate for reflection from the two window-to-air surfaces of the sample cell. The reflectitivity of the window-to-sample surfaces is presumably quite low. In the second place, interference in radiation multiply reflected back and forth in the narrow space between the windows of an empty cell produces fringes in the spectrum. In studying liquid solutions, the absorption bands contributed by the solvent can be canceled by placing a cell containing pure solvent in the reference beam. The path length of solvent in the reference beam must be the same as the equivalent path length in the sample beam. Therefore, the two cells are not of the same length for perfect cancellation, but their relative lengths are determined by the concentration of the solution. If two cells of known thickness are available, one can choose a concentration of solution to place in the thicker cell such that the amount of solvent in the path is the same as the solvent in the thinner cell in the reference beam. It is much more convenient to have a cell of adjustable thickness for reference use. A number of such cells have been devised with threaded barrels permitting path lengths to be varied from nearly zero to several millimeters. A simpler, but perhaps less versatile, arrangement uses a wedged spacer between rather wide windows. The cell is mounted in a carriage such that it can slide transverse to the beam direction. Thus, the thickness of the part of the cell in the beam can be selected. The range of thicknesses available with a wedge cell is limited to a factor of perhaps 2 or 3. The actual transmittance of the material in the cell is an average value over the width of the beam, W (see Fig. 15.6),

$$T = \int_{-w/2}^{w/2} e^{-ab(x)} \, dx$$

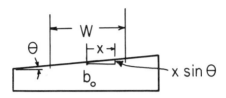

Fig. 15.6   Wedged cell.

where the cell thickness is given in terms of the wedge angle θ by

$$b = b_0 + x \sin \theta$$

The integral is easily evaluated to give

$$T = e^{ab_0} + \frac{2 \sinh (aw/2 \sin \theta)}{a \sin \theta}$$

Thus, perfect compensation is not obtainable, but for many applications the wedge cell is good enough.

In Sections 14.III and 14.IV we discussed the effects of stray radiation and extraneous emission on solvent compensation in double-beam instruments. To these sources of difficulty we should add the possibility of temperature differences in the two beams causing a difference in density of the sample and reference materials. Furthermore, the absorptivity and widths of absorption bands often depend on the nature of the solvent and concentration as we mentioned above. All of these factors sometimes make it very difficult to obtain complete solvent compensation, especially when an expanded abscissa is used to enlarge small differences. The spectroscopist must be beware of instrumental artifacts that might be mistaken for real absorption peaks. Occasionally, interactions and reactions between solvent and solute are studied with a double reference cell, sketched in Fig. 15.7. The proper thickness of solute and solvent are kept separated in the two cell sections, although both absorb the incident radiation. The thickness of the sample cell is equal to the combined thickness of the reference cell. The relative concentrations of solute and solvent in the sample cell is the same as the relative thickness of the two sections of reference cell. Differences in absorptivities or bandwidths or the occurrence of small amounts of reaction products show up in the differential spectrum.

A typical cell for liquids has an area of perhaps 3 cm². With a cell thickness of 0.1 mm a minimum sample volume of 0.03 ml is required to fill it, but dead space in the filling ports can easily increase this to several milliliters. If the cell is moved to a position of minimum beam

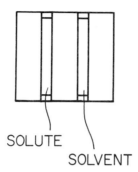

SOLUTE

SOLVENT

Fig. 15.7   Double reference cell.

size, usually a slit image, the area can be reduced somewhat. More-over, the volume associated with the filling ports can be minimized by using capillary tubing, as illustrated in Fig. 15.8. The tip of the filling tube is touched to the surface of the sample and the liquid is pulled into the cell by capillary action. With these so-called microcells, sample size can be reduced, typically, to 0.01 ml, or even less if the cell is made smaller than the beam size, with the attendant loss in energy this implies. The use of grooved windows, as described above and in Fig. 15.4, is advantageous in demountable microcells. The sample is picked up and transferred to the cell with a pipette having the end drawn to a capillary. A little pressure from the fingertip at the open top of the tube usually causes the liquid to move between the cell windows when the fine end of the tube is touched to them.

Further reduction in sample size requires the use of some sort of beam condenser to reduce the area of the radiation beam. If the sample

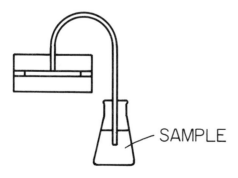

SAMPLE

Fig. 15.8   Microcell filled through capillary tube.

is to be placed between the source and the monochromator, as it almost always is in a double-beam infrared spectrophotometer, the reduction in the size of the beam is limited to a factor of 5 or 6. Greater reduction raises the power density of radiation in the sample to a level that causes samples to evaporate, boil, or even ignite. Low-power beam condensers can be refracting types, with lenses of KBr, for example, or off-axis reflecting types. Typical units are pictured in Figs. 15.9 and 15.10. In designing such systems it must be kept in mind that ideally the ultimate optical properties of the sample beam must not be altered, lest the reference beam no longer compensates. This cannot be accomplished in practice because the optical systems do introduce aberrations. Particularly, the refracting system has chromatic aberrations that change with wavelength preventing a flat I_0. As far as first-order paraxial optics are concerned, the positions and sizes of the slit image and the image of the aperture stop (i.e., grating or prism) in the vicinity of the source should not be shifted when the beam condenser is inserted in the sample beam.

Let us consider the design of the simple two-lens system of Fig. 15.9A. The first lens of focal length f_1 is placed a distance p_1 from the spectrometer slit (or an image of the slit). A reduced slit image is formed q_1 from the lens, according to the equation

$$q_1 = \frac{p_1 f_1}{p_1 - f_1}$$

The sample is placed in coincidence with this reduced image. The magnification of the image, which determines the sample size, is

$$m_1 = q_1/p_1$$

from which we deduce the required focal length

$$f_1 = \frac{p_1 m_1}{1 + m_1}$$

A second lens of focal length f_2 and distance p_2 from the reduced image reimages the slit at its original position,

$$q_2 = \frac{p_2 f_2}{p_2 - f_2} = -(p_1 + q_1 + p_2)$$

The geometry of Fig. 15.9A is constructed to satisfy the requirement that the slit must be reimaged at its original size, though inverted. That is,

$$m_2 = q_2/p_2 = -m_1$$

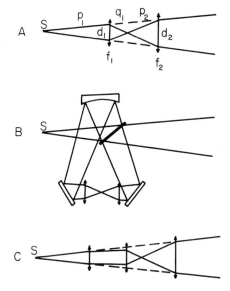

Fig. 15.9 Beam condensers: **(A)** two-lens system; **(B)** mirror system; **(C)** three-lens system.

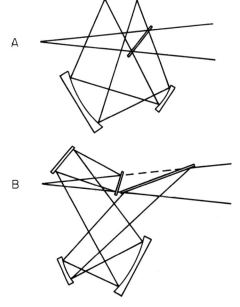

Fig. 15.10 Reflecting beam condensers.

This is equivalent to demanding that the rays leave the second lens with the same slope as they approached the first lens, and then intersect at the original slit. From the figure we can write

$$d_2 = \frac{p_2}{q_1}d_1 = \frac{p_1+q_1+p_2}{p_1}d_1$$

or

$$p_2 = \frac{q_1(q_1+p_1)}{p_1-q_1} = \frac{f_1p_1{}^2}{(p_1-f_1)(p_1-2f_1)} = \frac{p_1m_1(1+m_1)}{1-m_1}$$

which gives us the required distance from the reduced slit image to the second lens. The focal length of the second lens is found from the equation

$$f_2 = \frac{p_2q_2}{p_2+q_2}$$

with appropriate expressions for p_2 and q_2 substituted. The final expression for f_2 is

$$f_2 = \frac{f_1p_1{}^2}{(p_1-2f_1)^2} = \frac{p_1m_1(1+m_1)}{(1-m_1)^2}$$

Let us say, by way of illustration, that p_1 is 20 cm and that a fivefold reduction in the slit image is required. The necessary focal lengths and spacings are

$$f_1 = 3.3 \text{ cm} \qquad f_2 = 7.5 \text{ cm}$$
$$q_1 = 4 \text{ cm} \qquad p_2 = 6 \text{ cm}$$

If the slit is 2.5 cm long and the radiation beam leaving the mono-chromator is diverging at a $f/10$ rate, the first lens must be at least as large as

$$d_1 = 2.5 + \frac{20}{10} = 4.5 \text{ cm}$$

and the second lens must be

$$d_2 = 6.75 \text{ cm}$$

The necessary radii of curvature of these lenses are rather short. The curvature of a symmetric double convex lens, made of material of refractive index 1.5, from Eq. (3.33),

$$R_1 = f$$

The first lens is especially highly curved. However, higher refractive indices are available for use in the infrared, giving a more manageable shape. High-quality imaging is not to be expected; this device is a beam condenser, not a microscope.

Usually, beam condensers must be mounted in the normal sample area of the spectrophotometer and, therefore, the first lens is quite close to a slit image. In this case the focal length and consequently radii of curvature of the lens surfaces are found to be impossibly short for a reasonably long slit and demagnification. One solution to this problem is shown in Fig. 15.9B. The beam is brought out of its normal area with plane mirrors in order to increase the distance from the slit image. One of the mirrors is very thin and both sides are used. A similar arrangement using all reflecting optics is shown in Fig. 15.10A. This is a Perkin–Elmer design. The two focusing mirrors are ellipsoids because they are used highly off-axis. A sixfold slit image reduction is achieved. The device shown in Fig. 15.10B uses reflecting optics at a smaller off-axis angle and gives a fourfold image reduction. It is a design used by Beckman.

An alternate solution is shown in Fig. 15.9C. This is the Beckman three-lens system of White et al. (3). The initial object (slit image at S) is imaged virtually by the first lens at a convenient distance for the second lens to produce a reduced image. The requirements on the size and focal length of the second lens are now attainable with reasonable refractive indices. The final lens returns the rays to their original state. This beam condenser is designed for a fivefold reduction in image size.

When the beam size must be reduced by factors of more than 5 or 6 to accommodate very small samples, a Cassegrain condensing system offers some advantages. It does not introduce chromatic aberrations and it is used on-axis, permitting very short image distances. The optics of the Cassegrain system were discussed earlier in Section 3.III.C. The Perkin–Elmer system provides a sample size of 0.65×0.22 mm. It is used in dispersed radiation following the exit slit because a concentrated image of the full radiation from the source would heat most samples excessively.

Cells for microsampling of liquids are usually miniature forms of standard-sized cells. Occasionally, capillary tubing of silver chloride or polyethylene is used for very small samples. Membrane filters also can be used as sample cells. The liquid to be studied is trapped in the tiny holes of the porous filter. Filters made of various materials are available, but cellulose nitrate–acetate seems to be especially useful except in the 840- and 1070-cm^{-1} regions where it absorbs strongly. These filters are particularly valuable in trapping and observing effluents from gas chromatographs.

So far we have discussed microsampling in terms of reducing the required sample area by means of a beam condenser. We should also mention the possibility of reducing the thickness of the sample layer. Of course, this reduces the intensity of the absorption band. Sensitivity can be resorted, to some extent, by expanding the transmittance scale of the spectrometer. A useful technique due to Francis and Ellison (4) for the study of very thin layers is illustrated in Fig. 15.11. A film of sample is deposited on one of a pair of plane mirrors. The radiation is reflected back and forth many times between the two mirrors before it escapes and enters the monochromator. It is attenuated with each pass through the film and therefore the effective thickness of the film is the actual thickness multiplied by the number of passes.

It is often desirable to raise the temperature of a sample in order to reduce its viscosity, or melt it if it is normally solid, or to study some property as a function of temperature. Moderate increases in temperature of ordinary cells can be achieved by using hot air blowers or heat lamps. Larger and more controllable increases can be attained electrically. A typical heated cell design is sketched in Fig. 15.12. This cell is heated with constantan wire and can be used to temperatures up to about 120°C. Similar cells are useful to perhaps 200°C, although allowance must be made for the effects of emission from the hot sample (see Section 14.IV). It is easy to modify a standard liquid cell for use at elevated temperatures by substituting for one of the metal clamping plates a plate of transite having an annular groove milled in its surface. A heating element is coiled in the groove and packed with wet asbestos that hardens to hold the element in place and provide electrical insulation. The ends of the element protrude from the plate and are connected to a source of electrical current.

An unusual technique has been devised by Greenberg and Hallgren

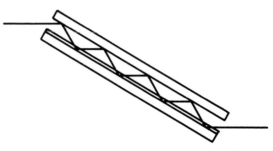

Fig. 15.11   Multipass method for thin liquid films.

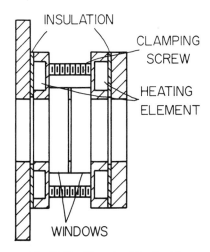

Fig. 15.12   Heated cell for liquids.

(5) for studying melted samples at very high temperature. A platinum or stainless steel wire grid serves as both heater and sample support. The melted sample is self-supported on the fine openings between wires of the grid. The grid is usually mounted inside an evacuable tube with ordinary windows for transmitting infrared radiation. A typical assembly is sketched in Fig. 15.13.

A discussion of cells used at temperatures below room temperature is deferred to a later section on solid samples.

III. Methods for Gaseous Samples

A simple cell for the study of gas samples is a cylinder of glass (or sometimes metal) closed at both ends with suitable windows and fitted with valves or stopcocks. A typical general-purpose gas cell has a path length of 10 cm. The cell windows can be sealed to the body with amalgamated lead or copper gaskets, wax, grease, cement, or soft metal or plastic gaskets. Sometimes the cell body is given a tapered shape and a rectangular cross section in order to conform to the shape of the beam (6). This is done in order to minimize the volume of gas required. Quite small gas cells, with short path lengths, can be mounted in beam condensers. It is usually desirable to evacuate the cell and then admit gas or vapor to the desired pressure. Sometimes, in making quantitative measurements on light molecules, the sample gas is pressured in order to broaden the rotational structure in the spectrum. But when it

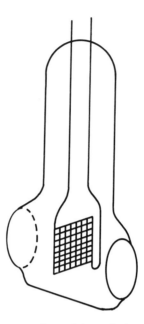

Fig. 15.13 Apparatus for studying melted samples on a grid.

is desired to resolve the rotational structure, a minimum of gas pressure is used. In the far infrared, polyethylene windows are usually used. These will maintain a vacuum if they are sufficiently thick. If very thin plastic windows are used for some reason, such as minimizing window absorption, they can be reinforced with a wire grid.

Very dilute or weakly absorbing gas samples require the use of long path lengths. If we attempt to produce a path of length l by collimating the radiation and passing it through a long pipe, we find that, in order to fill the monochromator with radiation, a mirror of diameter at least

$$D = l/F + h$$

is required, where F is the f/number of the monochromator and h is the height of the slit (see Fig. 15.14). For example, a 10-m path and an $f/10$ monochromator require a mirror of more than 1-m diameter. A light pipe can be used if the detector is placed at the end of the pipe, as in Fig. 15.15. This reduces considerably the volume of sample required, and this approach is frequently taken in the construction of nondispersive infrared analyzers. It is not well suited to spectrometers, however, particularly the double-beam type, and still requires a fairly large sample volume.

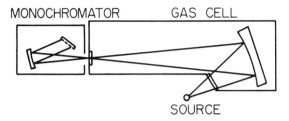

Fig. 15.14   Long-path gas cell.

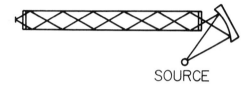

Fig. 15.15   Light-pipe gas cell.

The efficiency of a long-path gas cell can be improved significantly by constructing it as a multipassed cell, thus causing the same volume of sample to be used over and over. Figure 15.16 illustrates one form of multipass cell, based on a principle of J. U. White (7). The radiation enters the cell and is deflected to fill one of a pair of spherical mirrors at the end of the cell. This mirror forms an image of the monochromator

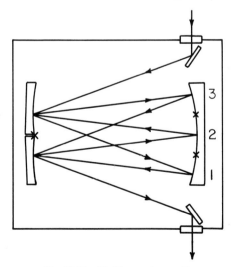

Fig. 15.16   Multipassed gas cell.

slit at the surface of the large spherical mirror at the other end of the cell. The radiation then diverges and fills the second of the pair of mirrors, which returns it to a different area of the large mirror, and so on for a number of passes through the cell. Eventually, the slit image falls off the edge of the large mirror and the beam is intercepted by the small corner mirror and led out of the cell. The number of passes is determined by the adjustment of one of the mirrors, which can be done from outside the cell. This particular cell can be adjusted for 0.1–10.0 m total path length and can be pressurized to 10 atm for an equivalent atmospheric pressure path length of 100 m. The optical efficiency at maximum path length is around 30 or 40%. Cells of this type have been constructed with path lengths of thousands of meters. White *et al.* (8) have designed a miniaturized multipass gas cell to be used with a beam condenser.

The sampling efficiency of gas cells can be defined as path length per unit volume. The efficiencies of a number of typical cells are given in Table 15.1.

TABLE 15.1

Efficiencies of Representative Gas Cells

Cell	Path length (cm)	Volume (ml)	Efficiency (cm/ml)
Standard	10	200	0.05
Multipass (7)	100	950	0.105
Multipass (7)	1000	4000	0.25
Multipass Microcell (8)	60	20	3.0
Minimum volume, single pass (6)	7.7	2.5	3.08
Minimum volume, beam condenser (6)	0.9	0.045	20

Gaseous samples have been studied by condensing them onto a cold window, by compressing them until they liquefy, by dissolving them in liquid solvents, or by adsorbing them on active surfaces. Gases can be studied at elevated temperature by heating a normal cell with a heat lamp or with a heated air stream, or by attaching heating elements to the cell. Once more, care is required to avoid emitted radiation from the hot sample. By properly locating the radiation chopper, absorption spectra of flames can be observed.

IV. METHODS FOR SOLID SAMPLES

The preparation of solids for infrared study is usually more difficult than preparation of liquids or gases because solids do not always readily conform to the shape required by the spectroscopist. Sometimes the problem is avoided by dissolving the solid in a suitable solvent or melting it in a heated cell. But the spectra of substances in the solid state usually differ from the spectra of the same material in the liquid state.

Occasionally, a solid sample is obtained directly in the form of a self-supporting film of size and thickness appropriate for infrared study. The film can be mounted in the radiation beam in a liquid cell frame, taped to the entrance port of the spectrometer, or held in place with a small magnet. A solid film mounted in a gas cell can be exposed to reactive gases and the spectrum of the reaction products can be studied as they form. The spectra of free films frequently show interference fringes due to internal reflections at the smooth surfaces. These can be reduced in amplitude or eliminated by roughening the surface with an abrasive, by coating the surfaces with a liquid or greasy material, or by submerging the film in a liquid-filled cell.

Solids can be formed into thin films in various ways. A solution in a volatile solvent can be poured onto a smooth surface, such as mercury in a dish, plastic, glass, metal foil, or polished metal. After the solvent has evaporated, perhaps speeded up with the application of heat, the remaining film can be stripped off. In the case of films formed on thin plastic sheets, the superposed spectra of film and plastic can be obtained if the latter is not objectionable. Film thickness can be controlled, more or less, by the concentration of the solution, by spreading the solution between spacer strips before evaporation of the solvent, or by spreading the solution with a doctor blade or some other notched tool. Films also can be cast on or between infrared transparent windows.

It is often difficult to remove the last trace of solvent from films cast from solution. To avoid this problem films can sometimes be formed from solid samples by rolling, by pressing between metal plates, foil, or plastic sheets with heat if necessary, by slicing thin sections with a microtome, or by melting the solid and allowing it to solidify on or between surfaces or windows. Plastic materials can sometimes be formed into thin layers by squeezing them between infrared transmitting windows. Somewhat harder materials, including some crystalline substances, can be formed by pressure between windows if the sample area is small relative to the window size. This usually requires a beam-condensing device, and considerable care is required to avoid cracking the windows.

The preparation of crystalline samples for infrared study is an art. If large crystals are available from some source, perhaps grown from the melt, they can be cleaved, ground, and polished following the usual optical practices for soft materials. When very thin sections are required, as is usually the case, it is difficult to grind the specimen without fracturing it. A thick crystal can sometimes be cemented to a plate, ground to the required thickness, and then floated off the plate in a dish of solvent that dissolves the cement but not the crystal. The crystal is then transferred to a mount. If the crystal is initially cemented to an infrared transmitting window, it need not be removed provided the cement layer is very thin or has no significant absorption bands.

Polycrystalline films can be deposited on cooled surfaces by vacuum sublimation. The sample can be condensed on an infrared transparent window or on a mirror. If conditions are ideal, thin *single* crystals can be grown by vacuum sublimation. Zwerdling and Halford (9) have grown specimens of benzene in this way in the apparatus illustrated in Fig. 15.17. Liquid samples can be mounted between windows and solidified by reducing the temperature. Some samples will form in an amorphous or crystalline state depending on how they are treated. The sample can be cooled rapidly or slowly, or it can be

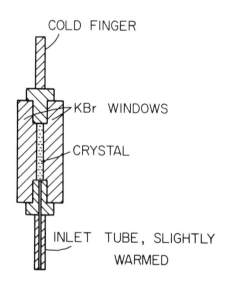

Fig. 15.17   Apparatus for growing crystals between windows.

cooled well below the freezing temperature and then warmed somewhat, or it can be cycled repeatedly through transition temperatures.

Many cells for the study of cooled samples have been built by in-
dividuals, and a number are available commercially. They take a variety
of forms, from simple boxes whose contents are cooled by a stream of
cold nitrogen, to elaborate cells cooled with liquid helium and capable
of reaching temperatures only one or two degrees above absolute zero.
A typical cell for liquid-nitrogen temperatures (10), developed at the
National Bureau of Standards, is sketched in Fig. 15.18. The sample
is mounted or condensed on a cold fringer cooled by liquid nitrogen
contained in a reservoir. This cell was constructed of glass. Metal is
also used in making some cells in the interests of greater durability,
but metal is a more efficient conductor of heat. The cell is evacuated
to prevent the outer windows from being cooled by conduction and
thus collecting water vapor. This cell has an unusual feature in the
silver chloride beam-condensing lenses, which permit the study of quite
small samples. It was intended for study of minor constituents extracted
from polluted air.

Cells for liquid-helium temperature are much more sophisticated,
and extreme precautions are taken against heat leaks. They usually

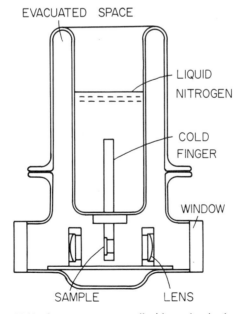

Fig. 15.18 Low-temperature cell with condensing lenses.

are of a double Dewar vessel design, with an inner vessel containing liquid helium and an outer vessel for liquid nitrogen. A typical design is sketched in Fig. 15.19. Fairly recently efficient open-cycle refrigeration units, based on the Joule–Thompson expansion of hydrogen, have become available. Small cryostats have been constructed with these which can reach temperatures approaching 15°K.

As a variation of these straight-through transmission cells, the sample is sometimes condensed on a cooled mirror surface. The reflected beam, observed at some convenient angle, is attenuated by two passes through the sample.

A little care is required in the selection of greases and waxes used in the construction and sealing of low-temperature cells. Constituents of low molecular weight sometimes sublime onto the cold surfaces and appear as strange bands in the infrared spectrum.

Powdered samples are very commonly encountered in infrared absorption spectroscopy and a number of techniques have been devised for handling them. Solids can be ground into fine powder with the usual laboratory mortar and pestle. A hard substance such as mullite is recommended to avoid contaminating the sample with material abraded from the mortar. Special grinders have been devised for difficult samples. Some make use of a diamond-surfaced spatula. Fine particles can be isolated from coarser ones when necessary by means of sedimentation in liquid, or in an air stream, by electrostatic

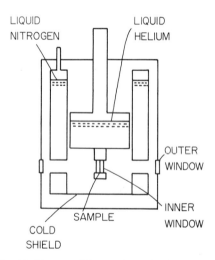

Fig. 15.19 Double Dewar cell for studies at liquid-helium temperature.

devices, or with graded sieves. Soft and rubbery samples can often be prepared by grinding with a powdered material that serves as an abrasive, preferably one that is transparent in the region to be studied, e.g., KBr. Sometimes a stubborn material can be ground by first cooling it in liquid nitrogen.

A layer of powder can be spread on a window or suspended in a volatile solvent that evaporates and leaves a layer of particles on the window. The spectrum is then observed by direct transmission. However, unless the particles are exceedingly small, energy losses due to scatter may be intolerable. Scatter losses can be reduced by suspending the particles in a medium of approximately the same refractive index. Recall that scattering is proportional to $[(n_1 - n_2)/(n_1 + n_2)]^2$. A commonly chosen medium is mineral oil or "nujol," and the technique has become known as the mineral oil or nujol mull technique. Other materials — liquids, greases, or waxes — are useful provided they are sufficiently transparent and have the proper refractive index. Often mineral oil is used for the complete spectrum except for the regions of strong C—H absorption near 3000, 1470, and 1390 cm^{-1}. These regions can be filled in with a perfluoro- or perchloro- oil. Mineral oil mulls are very useful for obtaining spectra of powdered samples in the far infrared. Mineral oil has no strong absorption bands in the far infrared, and moreover the particle size of the powder can be larger than in mid-infrared studies because of the inverse-fourth-power relationship between scattering and the ratio of particle diameter to wavelength.

In preparing a mull, the liquid is added sparingly to the powdered sample, and mixed thoroughly with it. Further grinding with the pestle will generally accomplish the desired objective of coating each particle separately with a thin layer of liquid. The resulting slurry or paste is transferred to a window, spread evenly across it, and covered with another window with enough pressure to ensure contact between window and liquid and to expel air bubbles. If too little liquid is used, the particles will not be coated and separated, and scatter will occur. If too much is used, the absorption bands of the liquid will be too prominent. Only practice will enable the spectroscopist to make acceptable mulls *most* of the time. The use of a cell window with an annular groove is helpful if a spacer is used in the demountable cell. This prevents sample material from being trapped under the spacer. Some workers grind the powder between a pair of salt windows that have previously been roughened with emery paper.

As we have seen, the refractive index of a sample is wavelength dependent and changes rapidly and sometimes by a substantial amount in the vicinity of an absorption band. It often happens that the refractive index of the powdered sample is not well matched to that of the liquid material, and consequently the level of scattered radiation is high. But near an absorption band of the sample, the indices can accidently become well matched (see Fig. 15.20) and the scatter abruptly vanish over a narrow wavelength interval. This phenomenon is called a Christiansen filter or window. As with all scattering effects it can be minimized by ensuring that the particles are ground as small as possible. On rare occasions, the Christiansen effect is deliberately induced to provide a narrow bandpass filter.

The chief disadvantage of the mull technique lies in the intrusion of the absorption bands of the mulling agent in the spectrum of the sample. A variation of the mull technique, the pellet technique, was developed to circumvent this difficulty. It was invented independently by several groups of workers: Schiedt and Reinwein (*11*), Stimson and O'Donnell (*12*), and Cook *et al.* (*13*). The powdered solid sample is mixed with fifty or a hundred times as much alkali halide powder, usually potassium bromide. The mixture is placed in a die and compressed to form a transparent disk or pellet. The pellet is mounted directly in the spectrometer. Alkali halides other than KBr can be used to obtain a better refractive index match in controlling scatter.

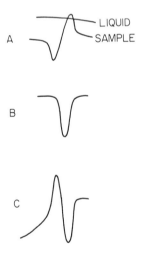

Fig. 15.20 Christiansen filter effect: (**A**) refractive indices; (**B**) transmittance of sample; (**C**) observed transmittance of suspension.

A number of dies have been devised and many have been described in the literature. Indeed, the announcement of the pellet technique seemed to fire the imagination of spectroscopists and a large effort was expended by many workers in perfecting and improving the method. A die can be a very simple arrangement, as in Fig. 15.21. It is suitable for making pellets from perhaps 1 cm in diameter down to micropellets of several millimeters which must be placed in a beam condenser for observation. Very small pellets can be prepared in an even simpler arrangement in which the powder is confined in an opening cut in a piece of blotting paper or in several layers of aluminium foil and pressed between polished metal plates. Pellets made in small dies tend to be cloudy, either immediately or after removal from the die, because of trapped air. Larger pellets must have the air removed by evacuation before pressure is applied. This complicates the construction of the die. Furthermore, while small pellets can be squeezed in an ordinary vise, large pellets require the use of much higher pressures and so hydraulic presses must be used. Even with air expelled before pressing, pellets usually become cloudy eventually. However, they remain mechanically stable and usually can be raised to high temperature or cooled to low temperature without fracturing.

The sample and the alkali halide powder can be mixed by grinding them together in a mortar, ball mill, or miniature ball mill of the Wig-L-Bug type. They can also be mixed by making a slurry of alkali halide in a solution containing the sample and permitting the solvent to evaporate, or by freeze-drying a solution of alkali halide and sample in water or some other solvent. It is necessary to start with dried alkali halide powder when mechanical mixing is used. It can be dried by freeze-drying or by heating in a vacuum oven. Inevitably a small absorption band is observed at $3450\,cm^{-1}$ in the spectra of pellets, with a weaker band at $1650\,cm^{-1}$. These are due to water that adsorbs on the surface of the particles. It is so tenacious that at various times

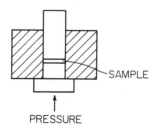

Fig. 15.21   Simple pellet die.

it has been proposed that the bands might indicate an actual reaction product between alkali halide and water, e.g., KOH. This can be disproved, however, by precipitating KBr from solution in D_2O. When the powder is subjected to standard drying procedures, all traces of OD absorption bands can be eliminated, although the weak O—H bands occur. Thus, the OD bands are not due to KOD, or some such product, and the OH bands are therefore not likely to be due to KOH.

Another misconception about alkali halide pellets has to do with the mechanism by which the pellet is formed. It is sometimes stated that the particles partially melt under the high local pressures occurring at particle boundaries. It is easy to melt a small quantity of KBr in a Pyrex test tube. A crystal of KBr added to the liquid in the test tube remains at the surface until it is entirely melted. Thus the density of solid KBr is less than that of liquid KBr at the melting point and, therefore, application of pressure *raises* the melting point.

At first it was hoped that the alkali halide pellet technique would replace the mineral oil mull method of preparing solid samples. It soon became clear, however, that the pellet method has some severe disadvantages of its own. Some spectroscopists refuse to use the pellet method altogether. Probably the greatest source of trouble is the tendency of many substances to react with alkali halides under pressure, or even during mixing, either immediately or after a long period of time. This is by no means limited to inorganic compounds. Organics, including some carbohydrates, amino acids, and even the unlikely compound thiourea, are found to react under certain conditions. Thiourea (*14*), for example, reacts with KBr, KI, and CsBr under moderate pressure, but not with NaCl, NaBr, NaI, and KCl. Thus, it is sometimes possible to avoid reaction by a proper choice of alkali halide. Some compounds merely transform to a different crystal habit under pressure, but this nevertheless can affect the appearance of the spectrum. There is some evidence that in the case of thiourea the alkali halide lattice is disrupted under pressure, permitting ions to intrude into the lattice of thiourea molecules, perturbing them and modifying their vibrational frequencies. The occurrence of reactions in alkali halide pellets can often be detected by comparing spectra with curves obtained from mineral oil mulls. But the spectroscopist must be sure that differences observed are really caused by reactions. Sometimes crystallites in mineral oil mulls can be preferentially oriented by sliding windows together, thus affecting the relative intensities or absorption bands. Sometimes too, samples are somewhat soluble in mineral oil, and solution spectra can be quite different from solid-phase spectra.

Molecules in crystal lattices are arranged in fixed positions and orientations. The directions of the oscillating electric dipoles associated with the molecular vibrations are therefore also fixed in direction. Only the component of radiation having its electric vector lying in the same direction as the dipole is absorbed. Hence it is possible to learn something of the direction of the dipole by studying a crystal with polarized incident radiation. When the crystal is oriented to maximize a given absorption band, the dipole moment of the band is most nearly parallel to the plane of (electric) polarization of the radiation. It is not always possible to obtain single crystals of sufficient size for infrared study, even though the availability of beam condensers has reduced the required size considerably. Often polycrystalline layers are prepared by condensation on a window in such a way that a preferred axis of the crystallite tends to line up in more or less the same direction. Sometimes crystallites can be made to do this by condensing them in a thermal gradient across the face of the window, or by starting the precipitation by application of a cold spot or a seed crystal at one point in the window. Sometimes the preferred growth of the crystallite is strongly influenced by the nature of the window, and different windows can give different results. There is even some evidence that certain windows can more effectively control the growth of crystallite if they are first rubbed in the preferred direction.

Polymers are not crystalline, in the usual sense, but they frequently contain long chains of repeated molecular units which can be more or less aligned in some way. The chains are sometimes aligned by rolling or stretching polymer films. Spectra obtained with polarized radiation parallel and perpendicular to the stretch direction can be quite different, and the differences offer useful clues to the orientation of certain functional groups attached to the polymer chain. Fibers are special cases of oriented samples. They can be studied with polarized radiation if they are first aligned in a common direction; this is a tedious process, but sometimes useful information results. The fibers can be studied dry if scatter permits, or they can be immersed in a liquid medium, or they can be carefully placed between layers of alkali halide powder and pressed into pellets. Samples of unoriented fibers can be prepared for study with unpolarized radiation by chopping them into short lengths with a sharp blade or in a mill and then mounting them in a pellet or mull.

The axes of molecules in solid films are often randomly arranged in the plane of the window on which they are deposited, but oriented to some extent in a direction normal to the window. For example, substances that crystallize in the form of flat platelets often tend to form with the platelets parallel to the window surface. This kind of orientation cannot be studied with polarized radiation, but it is often illuminating to compare the spectrum of such a sample with a spectrum of the same material in a condition of random orientation, as in a mull or pellet. Vibrations that tend to be perpendicular to the film surface absorb weakly in the spectrum of the film and more intensely in the random sample. Similar information can sometimes be obtained from films, including self-supporting or polymer films, by tilting them in the beam.

V. Quantitative Methods

Quantitative analysis by means of infrared absorption spectroscopy is based on the use of Beer's law

$$A = \log\frac{1}{T} = abc$$

which relates the transmittance T or absorbance A of a sample to its absorptivity a, thickness b, and the concentration c of the absorbing component. Before the concentration can be determined from a measurement of the transmittance, it is necessary to know the absorptivity and the thickness of the sample. If the sample contains several components, Beer's law is expanded to take the form of a set of simultaneous equations

$$A(\lambda_m) = b \sum_{n=1}^{k} a_n(\lambda_m)c_n, \quad m = 1, 2, \ldots, k \qquad (15.1)$$

with the summation extended over the k components. The absorptivities of each component are determined separately at each of k wavelengths λ_m, and the k components are determined by simultaneously solving the k equation of this type, one for each wavelength. The number of equations can be reduced by one if the condition

$$\sum_{n=1}^{k} c_n = 1$$

is used. In fact, analysis of a simple two-component mixture is usually made at only one wavelength.

The thickness of a liquid cell can be measured by weighing it empty and then filled with some liquid (provided proper account can be taken of dead volume), by comparing band intensities of a sample in a cell of known thickness with intensities in the unknown cell, or by using a micrometer if the thickness of the windows before assembly is known. Difficulties with the first two methods are obvious, and the last method is subject to error in thin cells because the windows may be etched with use, thus giving a longer path length. Fortunately, cell thicknesses can easily be measured interferometrically on an ordinary infrared spectrophotometer, if the inner window surfaces are reasonably flat and parallel and the spacer is not too thick. An empty liquid cell acts as a Fabry–Perot etalon, and the spectrum recorded with an empty cell in the beam is modulated with interference fringes due to constructive and destructive interferences of radiation multiply reflected from the inner window surfaces. Maxima occur when twice the cell thickness equals some multiple of the wavelength

$$2b = k\lambda_{max} \quad k = \text{integer}$$

with wavelength and cell thickness given in the same units. This equation can be written in terms of frequencies of the maxima,

$$k = 2b\nu_{max}$$

with ν_{max} expressed in (centimeter)$^{-1}$ and b in centimeters. In general, the value of k associated with each fringe is not known initially, so b cannot be determined from the position of one fringe. However, if the frequencies of two well-separated fringes are measured, ν_{max} and ν'_{max}, the difference between these frequencies gives

$$k - k' = 2b(\nu_{max} - \nu'_{max}) \quad \text{or} \quad \Delta k = 2b\,\Delta\nu$$

The left-hand side of this equation is merely one greater than the number of fringes lying between ν_{max} and ν'_{max}. The cell thickness is

$$b = \frac{1}{2}\frac{\Delta k}{\Delta\nu}$$

The use of this equation is illustrated in Fig. 15.22A. A more precise method uses a plot of ν_{max} versus fringe number, counted from some arbitrary starting place (Fig. 15.22B). The slope of the plotted line is $\Delta k/\Delta\nu$, and it can be measured more precisely from a large number of plotted points.

Thicker liquid cells are more difficult to measure. This is partly because the fringes are more closely spaced, and thus harder to resolve. Also, it is more difficult to maintain a uniform path length in a thicker cell because of the greater tendency to form a wedge-shaped sample space.

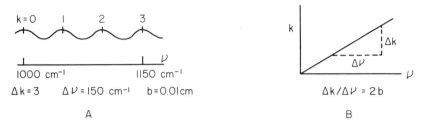

Fig. 15.22   Determination of cell thickness from interference fringes.

Gas cells are usually so thick that there is no possiblity of using interference fringes to measure the path length. But it is much easier to measure a 10-cm length with sufficient accuracy by using an ordinary scale.

The thickness of a solid film or crystal can be measured with a micrometer or calipers.

Once the thickness of a cell or sample is established, the absorptivity of a liquid substance can be determined from a measurement of its transmittance. If the material is studied in solution, then the concentration must also be known. In making up standard solutions some precautions should be observed. In the first place, the concentration must be expressed in a form equivalent to the number density of molecules of the sample per unit volume of solution. That is, the appropriate measure of concentration for absorption spectroscopy is molar concentration. This is true for liquids dissolved in a liquid solvent and also for solids and gases dissolved in a liquid solvent. If a component of a liquid solution is volatile, care must be taken to ensure that the concentration does not change after the solution is prepared.

Molar concentration is also appropriate for gas mixtures that can be described in terms of partial pressures. In studying gas mixtures it is important to control the temperature of the gases or to correct for the temperature, because of its effect on the pressure. It is also important to ensure that complete mixing of the gaseous components has occurred. Sometimes gas mixing is a very slow process, particularly in a vessel such as a gas cell. Some workers have incorporated a motor-driven stirring arrangement in their cells, and others have introduced glass beads that aid in gas mixing when the cell is shaken. In this case it is probably better to mix gaseous samples in an auxiliary vessel or cylinder that can be shaken vigorously without the danger of damaging expensive windows.

Molar concentration is appropriate for solid solutions, but these are not often encountered. However, when solid particles are suspended in a liquid or solid medium, the concentration in the medium is not significant. The important quantity is some measure of the number of molecules of the sample in the path. There is a complication in the quantitative measurement of absorption by particles, which is discussed a little later.

Now let us turn to the question of just how the transmittance at an absorption band is measured. This is not as straightforward as it might seem. The definition of transmittance, $T = I/I_0$, tells us to determine the ratio of transmitted radiation to incident radiation. Hence, if we adjust the spectrometer for full-scale recorder pen deflection with the sample removed, we have made $I_0 = 1$, and the pen position with the sample inserted gives us T directly. This is the so-called cell-in–cell-out method. But we want to cancel the attenuation due to reflection and absorption by the cell, so when we are studying gases I_0 should really be set with the empty cell in the beam. This is not proper when we are studying liquids, however, because an empty cell reflects radiation at four surfaces, whereas a cell with a liquid in it reflects significantly at only two surfaces. Hence for liquid samples, I_0 is established with a single blank window that is ideally of a thickness equal to the combined thicknesses of the windows of the sample cell. With solution samples, I_0 is established with the cell filled with solvent. If the concentration is high and if the solvent also absorbs to any extent at the wavelength of interest, this procedure is not strictly correct, because there is more solvent in the cell when I_0 is established than there is when I is measured. In double-beam spectrophotometry these methods are obviously equivalent to recording the spectrum with the sample cell in one beam and a blank window or a reference cell filled with solvent in the other. If a cell of variable thickness is used as the reference cell, its thickness can often be adjusted to compensate correctly for solvent absorption.

With solid samples and where partially overlapping bands are encountered, it is sometimes difficult to establish I_0. A useful alternative is the base-line method illustrated in Fig. 15.23. A straight line is constructed across the base of the band and I_0 is measured to this base line. The base line might be drawn between two points on the curve at defined wavelengths, or it might be constructed tangent to the curve on either side of the peak. Usually a little experimentation is necessary with solutions of a series of concentrations in order to determine the best approach.

In fact, whatever method of measurement is used, it is wise to study

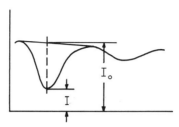

Fig. 15.23 Base line for measuring transmittance.

a set of synthetic solutions. These synthetic solutions serve two purposes. In the first place, they permit a certain amount of experimentation before choosing a technique. Equally important, the data from measurement on known solutions provide a working curve from which concentrations can be read from measured values of absorbance. This is usually to be preferred over calculating the concentration using a previously determined absorptivity and Beer's law. The reason for this is that quite often the working curve is indeed a curve rather than a straight line. In this case, we say that there are deviations from Beer's law.

The choice of cell thickness and of concentration of the sample in the solvent is usually at the disposal of the spectroscopist. Let us see if there is a guide to help him make an optimum choice. For ease of manipulation, we use Beer's law in the exponential form of Eq. (4.19). The sensitivity of transmittance to changes in concentration is found by differentiating Beer's law,

$$\frac{\partial T}{\partial c} = -Kbe^{-Kbc} = -KbT \qquad (15.2)$$

This equation can be interpreted as giving the error induced in a measurement of concentration by an error made in measuring the transmittance

$$\Delta c = \frac{\Delta T}{-TKb} = \frac{\Delta T}{T[(\ln T)/c]}$$

or

$$\frac{\Delta c}{c} = \frac{\Delta T}{T \ln T} \qquad (15.3)$$

This equation is plotted in Fig. 15.24, making the assumption that ΔT is a constant error or uncertainty, independent of the actual magnitude of the transmittance. This is certainly a reasonable assumption for

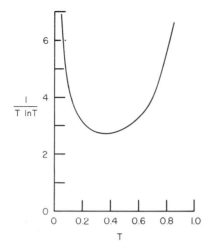

Fig. 15.24 Sensitivity of concentration determinations as a function of transmittance.

spectrometers using thermal detectors. The transmission for minimum error, set by adjusting the cell thickness, is found by differentiating Eq. (15.2) and setting the result equal to zero,

$$\frac{\partial^2 T}{\partial b \, \partial c} = -Ke^{-Kbc} + K^2 bce^{-Kbc} = 0$$

The solution of this equation is

$$Kbc = 1$$

and, therefore, the optimum transmittance is

$$T = 1/e \approx 36.8\%$$

However, as may be seen from Fig. 15.24, the exact value of T is not too important. Any choice of transmittance between 20 and 60% permits nearly optimum determinations.

The situation is somewhat different for detectors with shot noise, whose noise is proportional to the square root of the transmittance (noise $= k\sqrt{T}$). In this case, in place of Eq. (15.3) we have

$$\frac{\Delta c}{c} = \frac{k\sqrt{T}}{T \ln T} = \frac{k}{\sqrt{T} \ln T} = \frac{-k}{Kbce^{Kbc/2}}$$

The condition for minimum error in c is found from

$$\frac{\partial \, (\Delta c/c)}{\partial b} = \frac{ke^{-Kbc/2} - \frac{1}{2}Kbce^{-Kbc/2}}{Kb^2 ce^{-Kbc}} = 0$$

This gives us

$$\tfrac{1}{2}Kbc = 1$$

Therefore, the optimum transmittance in the case of shot noise is

$$T = 1/e^2 \approx 13.5\%$$

Again, the exact choice of T is not very critical. Purely shot-noise-limited detectors are not usually encountered in infrared spectroscopy.

Now let us turn to a discussion of several practical methods of quantitative analysis. We have mentioned the direct application of Beer's law in the general form of Eq. (15.1) for the simultaneous determination of several components of a solution. The selection of wavelengths for measuring transmittance is very important for the success of the method. With experience the spectroscopist learns to make this selection. In general, it is desirable, when possible, to select wavelengths at which each component absorbs without significant interference from other components. When overlapping bands cannot be avoided, it is often desirable to use a differential method in which the absorption by one component is canceled by means of a reference cell filled with just that component. The interested reader should study Robinson's paper (15) and the review by Hughes (16).

The transmittances at two wavelengths of a two-component mixture are given by

$$\frac{I(\lambda_1)}{I_0(\lambda_1)} = \exp\left[-K_1(\lambda_1)bc_1 - K_2(\lambda_1)bc_2\right]$$

and

$$\frac{I(\lambda_2)}{I_0(\lambda_2)} = \exp\left[-K_1(\lambda_2)bc_1 - K_2(\lambda_2)bc_2\right]$$

The ratio of these transmittances can be written

$$\frac{I(\lambda_1)}{I(\lambda_2)} = \frac{I_0(\lambda_1)}{I_0(\lambda_2)} \exp\left(-\{[K_1(\lambda_1) - K_1(\lambda_2)]c_1 + [K_2(\lambda_1) - K_2(\lambda_2)]c_2\}b\right)$$

or

$$\ln\left[\frac{I(\lambda_1)}{I(\lambda_2)}\right] = \ln\left[\frac{I_1(\lambda_1)}{I_0(\lambda_2)}\right] - \{[K_1(\lambda_1) - K_1(\lambda_2)]c_1 + [K_2(\lambda_1) - K_2(\lambda_2)]c_2\}b$$

If the substitution $c_2 = 1 - c_1$ is made, we have a linear relationship between c_1 and $\ln[I(\lambda_1)/I(\lambda_2)]$ (or $\log[I(\lambda_1)/I(\lambda_2)]$). The values of I_0 at λ_1 and λ_2 affect the intercept but not the slope. Thus, if for some reason it is inconvenient to measure or establish I_0, the transmittance ratio method is useful. As always, it is best to work from a working curve derived from known mixtures than to measure the absorptivities and hope that Beer's law is valid.

There are occasions in spectrophotometric analysis when there is uncertainty about the total amount of sample in the cell or about the cell thickness. This might happen, for example, when dealing with demountable cells, pellets, or mulls, or in cases where microquantities of solutions or gases must be transferred.

We can write the Beer's law expressions in the form of Eq. (4.20),

$$A(\lambda_1) = [a_1(\lambda_1)c_1 + a_2(\lambda_1)c_2]bC$$

and

$$A(\lambda_2) = [a_1(\lambda_2)c_1 + a_2(\lambda_2)c_2]bC$$

which include a factor C to account for the concentration of the sample mixture in the solvent or solid matrix, the total gas pressure, or deviations from the nominal cell thickness b. In the absorbance ratio method, we find the ratio of these equations,

$$\frac{A(\lambda_1)}{A(\lambda_2)} = \frac{a_1(\lambda_1)c_1 + a_2(\lambda_1)c_2}{a_1(\lambda_2)c_1 + a_2(\lambda_2)c_2}$$

The product bC conveniently cancels out, so there is no need to control these factors closely. If the wavelengths are chosen with care to minimize overlap, such that

$$a_1(\lambda_2) \approx 0 \qquad a_2(\lambda_1) \approx 0$$

then this equation becomes nearly linear in the concentration ratios

$$\frac{A(\lambda_1)}{A(\lambda_2)} \approx \frac{a_1(\lambda_1)}{a_2(\lambda_2)} \cdot \frac{c_1}{c_2}$$

It is sometimes useful to deliberately introduce a known amount of some substance into a sample before it is mulled, pelleted, or otherwise prepared for study. The additional material serves as an internal standard and quantitative measurements are made relative to it. The standard is chosen to provide a useful absorption band in a convenient spectral region, but to have no interference with absorption bands of the sample.

A variation of the absorbance ratio method is used in the study of certain temperature-dependent phenomena. Application to other equilibrium processes is evident. Let us consider a molecule, such as 1,2-dichloroethane, which can assume two isomeric forms by rotation of the end group about the carbon-carbon bond. There is a free-energy difference, $\Delta E = E_2 - E_1$, between the two forms. The relative population of the two forms is given by the Boltzmann relationships

$$c_1 = q_1 \left[\frac{\exp(-E_1/kT)}{\exp(-E_1/kT) + \exp(-E_2/kT)} \right]$$

and

$$c_2 = q_2 \left[\frac{\exp(-E_2/kT)}{\exp(-E_1/kT) + \exp(-E_2/kT)} \right]$$

where the q's are the degeneracies of the isomeric species. In the case of 1, 2-dichloroethane, q_1 for the *trans* form is 1 and q_2 for the *gauche* form is 2. We try to select absorption bands unique to each species, for example, by studying the temperature dependence of their intensities. The absorbances of these bands, assuming no overlap with bands of the opposite species, are written

$$A_1 = a_1 b c_1 = \frac{a_1 b q_1 \exp(-E_1/kT)}{\exp(-E_1/kT) + \exp(-E_2/kT)}$$

$$A_2 = a_2 b c_2 = \frac{a_2 b q_2 \exp(-E_2/kT)}{\exp(-E_1/kT) + \exp(-E_2/kT)}$$

The ratio of these is

$$\frac{A_1}{A_2} = \frac{a_1 q_1}{a_2 q_2} e^{\Delta E/kT}$$

and the natural logarithm of this ratio is

$$\ln\frac{A_1}{A_2} = \ln\frac{a_1 q_1}{a_2 q_2} + \frac{\Delta E}{kT}$$

The absorptivities are not known so we need a method of determining ΔE without them. This is done by measuring $\ln(A_1/A_2)$ at several temperatures and plotting against $1/kT$. The slope,

$$\frac{\partial(\ln(A_1/A_2))}{\partial(1/kt)} = \Delta E$$

gives us the free-energy difference.

So far we have confined transmittance measurements to the peaks of absorption bands, and this is usually the way in which routine quantitative analyses are performed. Sometimes, however, the integrated absorbance is more useful,

$$\int_{\text{band}} A \, dv = bc \int_{\text{band}} a \, dv$$

This might be the case, for example, when a measure of the total aliphatic C—H content of a sample is to be determined from the intensity of the 2900-cm^{-1} C—H band. The shape of this band varies somewhat from compound to compound because of slight differences in frequency of the contributing group, but, with certain limitation, the integrated absorbance is proportional to the number of C—H groups present. Moreover, the integrated absorbance is much less sensitive than the peak value to changes in monochromator spectral bandwidth.

The integrated band intensity is also useful in determining the dipole moment derivative of the normal vibrational mode responsible for the band. In this case, the quantity

$$\int_{band} a(\nu) \, d\ln\nu = \int_{band} \frac{a(\nu)}{\nu} \, d\nu$$

is sometimes more meaningful.

The measurement of integrated absorbance can be accomplished by numerical integration or planimeter integration of the recorded spectral record, remembering that it must either be recorded or re-plotted in the form $\ln(1/T)$ versus ν. An expedient occasionally used is to transfer the recorded band to a sheet of paper whose weight per unit area is known, cut out the band, and weigh it. Devices for recording the integrated signal directly are common. These devices are of two general types. A mechanical integrator based on the ball-and-disk principle can be attached to the recorder, or an operational amplifier with an integrating feedback capacitor can integrate the electrical output of the spectrometer. Neither device interferes with the normal recording of absorption spectra.

We should at least mention the idea of recording the logarithm of absorbance, that is,

$$\log\left(\log\frac{1}{T}\right) = \log abc = \log a + \log bc = \log a + \text{const}$$

This approach provides a record whose *shape* depends only on the absorptivity of the sample. But the curve is displaced by a constant amount that is determined by the path length and concentration. The method breaks down when $a = 0$.

In Chapter 14 numerous sources of photometric error were discussed, including the effects of stray radiation, extraneous emission, reflection from cells and samples, and error introduced by the spectrophotometer. We conclude this chapter by treating some additional errors associated with the sample.

We have already discussed the effect of the finite band of radiation passed by the monochromator on the peak height of a recorded band. Liquid samples rarely have bands as narrow as 3 or 4 cm^{-1} and usually are wider than 10 cm^{-1}. Remember the rule of thumb that the effective bandwidth of the monochromator should be one-fifth of the half-bandwidth or less. Even the smaller modern instruments that use diffraction gratings can provide resolution sufficient to reproduce accurately the intensities of these bands. This was not true of prism instruments, and accounts for a great deal of the variation of band intensities measured on the same sample with different prism instruments — variation often exceeding 25%. Infrared spectroscopists could never measure an absorptivity on one instrument and use the result to perform a quantitative analysis with another instrument. With modern grating instruments this limitation due to spectral bandwidth no longer exists for liquid samples, but even so it is good practice not to try to transfer intensity data between instruments.

Crystalline solid samples at very low temperature sometimes have very sharp bands that exceed the ability of small grating instruments to record them faithfully. Gas samples, at reduced pressure especially, have very sharp lines in their rotational final structure, lines as narrow as a few tenths of a wavenumber. Even the finest research spectrometers are hard put to reproduce these lines. For a qualitative and oversimplified illustration of how this affects the measured intensity of a band, refer to Fig. 15.25. The actual spectrum is represented by a sequence of infinitely narrow lines. The slit function of the monochromator is supposed to be triangular and to span several rotational lines, so only an envelope, with perhaps a hint of rotational structure due to spurious resolution, is recorded. The slit is set on the wavelength of maximum absorption. As the amount of sample in the

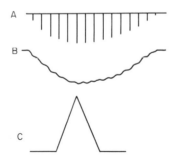

Fig. 15.25   Gas phase spectrum. A, actual spectrum; **B** recorded spectrum; **C**, triangular slit function.

cell is increased, absorption at the frequencies of the lines spanned by the slit increases until virtually complete absorption takes place at these frequencies. Further increase in the amount of sample can produce no additional absorption, even though the apparent absorption is not complete because of the radiation passed by the sample *between* the rotational lines, Beer's law certainly does not seem to be obeyed in this case.

Fortunately for those who wish to measure the absorption intensity quantitatively, the lines are broadened greatly at higher pressure. By intentionally pressure broadening a gas phase spectrum, the rotational lines can be made to merge with each other, and the envelope of the band is readily measured. This is the basis of the Wilson–Wells method (*17*).

The spectroscopist with an instrument of modest resolution, who observes only the envelope of a gas sample with no sign of rotational fine structure, should not assume that the sample is pressure broadened and therefore that "obedience to Beer's law" can be assumed. He should always determine a calibration curve with known samples.

As a final topic in this chapter let us consider the absorption of radiation by particles suspended in a medium, such as in a mull or pellet. For simplicity we assume that the index of refraction of the particles varies only slightly across an absorption band and that it is closely matched by the index of the suspension medium so scatter can be disregarded. We further assume the particles to be spheres, all with the same radius r (see Fig. 15.26). A parallel beam of radiation illuminates the sphere with irradiance I_0. The radiation leaving the cylindrical element is (*18*)

$$dE = L_0 \hat{n} \cdot \hat{j} d\sigma_1 \exp(y_2 - y_1) a$$

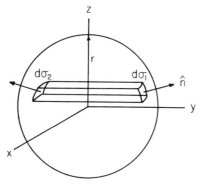

Fig. 15.26   Geometry of a sample particle.

where $\hat{n}$ is a unit vector normal (outward) to $d\sigma_1$, $\hat{j}$ is the unit vector in the y direction, and $\hat{n} \cdot \hat{j}$ is their scalar product. The projected cross-sectional area of this particle on the plane of the wave front is

$$A = \int_{s_1} \hat{n} \cdot \hat{j} \, d\sigma_1$$

and the mean transmittance of a single particle is

$$\left(\frac{I}{I_0}\right)_p = \frac{1}{AI_0} \int_{s_1} dE$$

$$= \frac{1 - (2ar + 1) \exp(-2ar)}{2a^2 r^2}$$

Now let us suppose that a mass m_p of sample of density ρ_p is embedded in a mass m_k of a medium having a density of ρ_k. The volumes occupied by sample and medium are $v_p = m_p/\rho_p$ and $v_k = m_k/\rho_k$, respectively. The number of particles in the disk is

$$n_p = \frac{v_p}{\frac{4}{3}\pi r^3} = \frac{3m_p}{4\pi r^3 \rho_p}$$

We can imagine the particles to be arranged in a pseudolattice with an average lattice constant

$$\lambda = \left[\frac{v_k + v_p}{n_p}\right]^{1/3}$$

The ratio of particle cross-sectional area to the cross-sectional area of a unit lattice cell is

$$\frac{\sigma_p}{\sigma_c} = \frac{\pi r^2}{\lambda^2} = \left[\frac{3\sqrt{\pi}m_p\rho_k}{4(m_k\rho_p + m_p\rho_k)}\right]^{2/3}$$

The transmittance of the cell is

$$\left(\frac{I}{I_0}\right)_l = \left(1 - \frac{\sigma_p}{\sigma_c}\right) + \frac{\sigma_p}{\sigma_c}\frac{[1 - (2ar + 1)\,e^{-2ar}]}{2a^2 r^2}$$

If the cross-sectional area of the pellet or mull is S, then the number of layers of these lattice units is

$$N = \frac{\sigma_c n_p}{S} = \frac{1}{rS}\left[\frac{3m_p}{4\pi\rho_p}\left(\frac{m_k}{\rho_k} + \frac{m_p}{\rho_p}\right)^2\right]^{1/3}$$

The transmittance of the whole array is

$$\frac{I}{I_0} = \left[\left(\frac{I}{I_0}\right)_l\right]^N$$

Figure 15.27 is a plot of the inverse transmittance of a typical sample as a function of absorptivity for very fine, medium, and coarse particle size. It demonstrates that absorption bands are apparently weaker when the particles are large and that a linear Beer's law relationship does not occur for large particles.

An analysis of the transmittance of an absorbing medium containing transparent particles leads to similar conclusions.

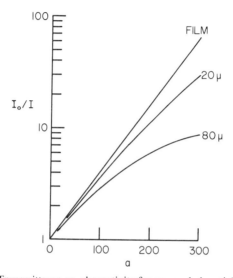

Fig. 15.27 Transmittance vs. absorptivity for suspended particles and a film.

REFERENCES

1. J. E. Stewart, "Sample preparation," *Encyclopedia of Spectroscopy*, G. L. Clark, ed., Reinhold, New York, 1960, pp. 546–569.
2. F. A. Smith and E. C. Creitz, *Anal. Chem.*, **21**, 1474 (1949).
3. J. U. White, S. Weiner, N. L. Alpert, and W. M. Ward, *Anal. Chem.*, **30**, 1694 (1958).
4. S. A. Francis and A. H. Ellison, *J. Opt. Soc. Am.*, **49**, 131 (1959).
5. J. Greenberg and L. J. Hallgren, *Rev. Sci. Inst.*, **31**, 444 (1960); *J. Chem. Phys.*, **33**, 900 (1960).
6. K. E. Stine, D. E. McCarthy, and H. J. Sloane, *Developments in Applied Spectroscopy*, Vol. 4, Plenum Press, New York, 1964, p. 121.
7. J. U. White, *J. Opt. Soc. Am.*, **32**, 285 (1942); J. U. White and N. L. Alpert, *ibid*, **48**, 460 (1958).
8. J. U. White, N. L. Alpert, W. M. Ward, and W. S. Gallaway, *Anal. Chem.*, **31**, 1267 (1959).
9. S. Zwerdling and R. S. Halford, *J. Chem. Phys.*, **23**, 2215 (1955).

10. J. E. Stewart, *Anal. Chem.*, **30**, 2073 (1958).

11. U. Schiedt and H. Reinwein, *Z. Naturforsch.*, **7b**, 270 (1952); U. Schiedt, *ibid*, **8b**, 66 (1953).

12. M. M. Stimson and M. J. O'Donnell, *J. Am. Chem. Soc.*, **74**, 1805 (1952).

13. E. S. Cook, C. Kreke, E. B. Barnes, and W. Motzel, *Nature*, **174**, 1144 (1954).

14. J. E. Stewart, *J. Chem. Phys.*, **26**, 248 (1957).

15. D. Z. Robinson, *Anal. Chem.*, **23**, 273 (1951); **24**, 619 (1952).

16. H. K. Hughes, *Appl. Opt.*, **2**, 937 (1963).

17. E. B. Wilson and A. J. Wells, *J. Chem. Phys.*, **14**, 578 (1946).

18. J. E. Stewart, *J. Res. NBS*, **54**, 41 (1955).

CHAPTER 16

Experimental Methods of Infrared Reflection Spectroscopy

I. Introduction

When a beam of radiation encounters a surface between two optically distinct media, a portion of it enters the second medium where it may be partly or completely absorbed, and a portion is reflected from the surface. Most routine infrared spectroscopy is done with the transmitted radiation, but a great deal of information about a sample can sometimes be obtained from a study of the reflected radiation. Activity in infrared reflectance spectroscopy has always been high among physicists interested in the optical properties of solids. In recent years chemists have begun to use reflectance measurements at an increasing rate.

In this chapter we treat three general types of reflectance phenomena, namely, specular, diffuse, and some aspects of internal reflectance. Reviews of the theory and methods of reflectance spectroscopy may be found in the books of Wendlandt and Hecht (1) and Harrick (2).

II. Specular Reflectance Measurements

The simplest experimental arrangement for measuring specular reflectance is illustrated in Fig. 16.1A. A pair of plane mirrors, M_1 and

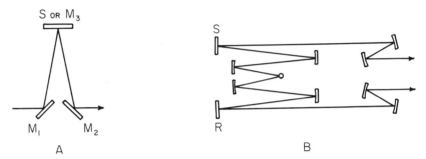

Fig. 16.1 Arrangements for measuring specular reflectance: (**A**) single beam; (**B**) double beam.

M_2, is inserted in the radiation beam. They deflect the beam to a reflecting sample S and return it to its original path. The effect on the total path length and focus of the photometer optics is usually negligible. The spectrum is recorded with the sample in position and also with a reference mirror substituted for the sample. The ratio of the two spectra is the reflectance of the sample relative to the reference. The reflectance of a good aluminized mirror is greater than 97.5% throughout the infrared region, so the use of such a mirror as a reference gives reflectance spectra which are reliable to 2% or so. Certainly this is good enough for qualitative reflectance spectra. Simple attachments of this type are available for insertion in a cell holder in the sample beam of a spectrophotometer. In principle an identical unit with reference mirror might be installed in the reference beam of a double-beam instrument. Barnes and Schatz (3) have modified the photometer optics of a Perkin–Elmer Model 21, as shown in Fig. 16.1B, to do double-beam reflectance measurements. With suitable beam-condensing elements, a fairly small sample can be used. A typical microreflectance unit using lenses, suitable for samples of 1×5 mm in area, is illustrated in Fig. 16.2. Attachments have also been constructed using concave mirrors.

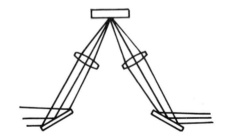

Fig. 16.2 Specular reflectance of a small sample.

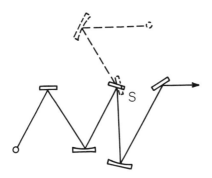

Fig. 16.3 Specular reflectance at a variable angle of incidence.

These simple devices are limited to one angle of incidence and reflectance, typically 20 or 30°. The reflectance is dependent on the angle, however, and it is often desirable to study reflectance as a function of angle. An arrangement for doing this with a single-beam spectrophotometer is given in Fig. 16.3. The source and source mirrors are rotated about an axis lying in the surface of the sample. Double-beam spectrometers present a little problem because of the difficulty of rotating the source. One solution is shown in Fig. 16.4A. Measurements at normal incidence can be made only by using a beam splitter, as in Fig. 16.4B.

Reflectance is also a function of the polarization of the incident beam except near normal incidence, so it is often necessary to insert a polarizer in the optical train. As usual, it is best to place the polarizer immediately before or following the sample to avoid the possibility of ellipticity produced by mirrors.

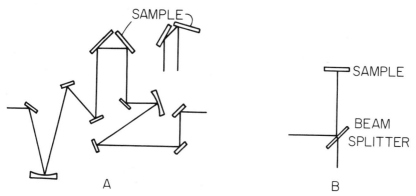

Fig. 16.4 (**A**) Apparatus for measuring specular reflectance with variable angle of incidence without deviating the beam. (**B**) Reflectance at normal incidence.

The sample whose specular reflectance is being measured must be well polished so that it will not scatter or diffusely reflect a significant portion of the incident radiation. There is some evidence that surface treatment effects the structure of the reflecting sample, and therefore optical constants determined from reflectance measurements may not be representative of the bulk material. The flatness of the surface is important only to the extent that curvature can affect the image of the source on the slit and reduce somewhat the radiant power entering the monochromator. Certainly, optical quality flatness need not be demanded. Samples which transmit to some extent can give erroneous reflectances because of reflection from the second surfaces, as illustrated in Fig. 16.5A. If the sample is sufficiently thick and the angle of incidence large, the beam reflected internally is displaced far enough from the primary beam so that it cannot enter the monochromator.

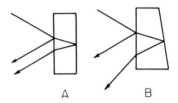

Fig. 16.5   Reflectance from transmitting samples.

The sample can be intentionally wedged to provide an angular separation between the two beams, as in Fig. 16.5B. If the sample is thin, the second surface can be roughened. The radiation reflected from it is diffuse, and its intensity drops rapidly (as the inverse square) along the optical path. Application of an absorbing paint helps to attenuate this troublesome radiation. Keep in mind, however, that a surface which strongly scatters visible radiation can be a pretty good specular reflector in the longer-wavelength infrared, and paint that is black to the eye might not be to the thermocouple.

   The measurement of *absolute* reflectance, as distinct from the methods discussed so far for reflectance relative to some standard, presents some difficulties. Variations on a method developed by Strong (*4*) are usually used (see Fig. 16.6). The form shown in Fig. 16.6B is used in the Cary spectrophotometer. With the reference mirror placed at M_3 the source is imaged on the monochromator slit and a reference spectrum is recorded. The sample is then inserted at S and mirror M_3 is repositioned at M_3', usually by rotating it about an axis. The total path

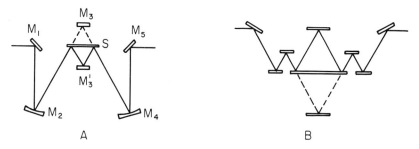

Fig. 16.6 Methods for measuring absolute specular reflectance.

length and the image of the source on the slit are just as they were before the sample was introduced, but the radiation beam has been attenuated by *two* reflections from the sample. The ratio of the sample record to the reference record is the square of the absolute specular reflectance. This has the effect of increasing the accuracy of measurements made on good reflectors, but it makes measurements of low reflectance more difficult.

A serious problem associated with Strong's method arises because of the sensitivity of the position of the image on the slit to alignment of the sample. In fact, the forms shown in Fig. 16.6 are usually used with an integrating sphere to collect the radiation, rather than directing it to a monochromator. This limits the method to the visible and near-infrared regions. Bennett and Koehler (5) have constructed an elaborate optical system to overcome this difficulty (see Fig. 16.7). The radiation leaves the exit slit of the monochromator and, with the sample in place, follows

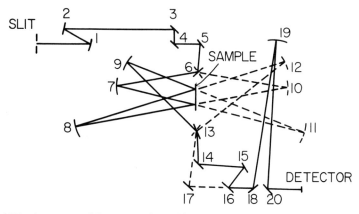

Fig. 16.7 Apparatus of Bennett and Koehler for measuring absolute specular reflectance.

the sequence of reflections: M_1, M_2, M_3, M_4, M_5, M_6, M_7, sample, M_8, sample, M_9, M_{13}, M_{14}, M_{15}, M_{16}, M_{18}, M_{19}, M_{20}, detector. An image of the exit pupil of the monochromator is formed at mirror M_2 and twice on the sample. With the sample removed the sequence following M_7 is M_{11}, M_9, M_{13}, M_{14}, and so on. The path length is unchanged and all mirrors are the same except that M_{11} replaces M_8. However, M_6 can be rotated to deflect the beam to M_{10} instead of to M_7. Mirror M_{12} takes the place of M_9. Mirror M_{13} is also rotated and M_{17} takes the place of M_{14}. Once more all mirrors except M_8 and M_{11} are used in the same way whether the sample is in place or not, but now the roles of M_8 and M_{11} are reversed; that is, M_{11} is used when the sample is in place and M_8 when it is not. The product of measurements made in the two configurations is the fourth power of the absolute reflectance of the sample. Once the relative reflectances of M_8 and M_{11} are established, it is no longer essential to use the second configuration, except for occasional checks because the reflectance of mirrors changes over a period of time. An important advantage of this system and the reason for going to such a complicated arrangement is that a tilt in sample orientation does not affect the image of the slit on the detector. This is easily seen by referring to Fig. 16.8, which shows the sample and mirror M_8 (or M_{11}).

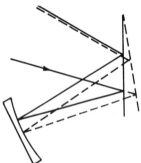

Fig. 16.8   Tilt compensation in apparatus of Bennett and Koehler.

For the most part reflectance measurements are made on solids. Occasionally liquids are studied, which requires either that the sample is placed in a horizontal container so its surface is observable from above or that the sample is placed in a cell and the reflectance relative to the window material is measured.

A nice application (6) of reflectance measurement is the determination from interference fringes of the thickness of thin films deposited or grown on a solid surface. This is important, for example, in the study of epitaxial layers on semiconductor materials. The situation is slightly more complicated than the analysis of the transmittance spectrum of a Fabry–Perot etalon or an empty cell (see Fig. 16.9). Consider ray 1

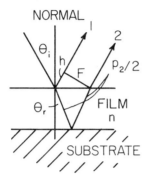

Fig. 16.9 Reflection by a film on a substrate.

which is reflected directly from the surface and ray 2 which is reflected from the interface between the film and the substrate. On the plane wave front F, ray 1 has traveled a distance

$$p_1 = 2t \tan \theta_r \sin \theta_i$$

while ray 2 has traveled a distance

$$p_2 = \frac{2t}{\cos \theta_r}$$

But ray 2 has been propagated in a region of refractive index n. Moreover, a phase shift $\Delta\phi_r$ is incurred by each ray upon reflection. The relative phase difference of the two rays on the front F is therefore

$$\Delta\phi = \frac{2\pi(p_2 - p_1)}{\lambda} + \Delta\phi_r$$
$$= \frac{4\pi t(n - \sin \theta_r \sin \theta_i)}{\lambda \cos \theta_r} + \Delta\phi_r \qquad (16.1)$$

In order for constructive interference to occur on the front the phase difference must obey

$$\Delta\phi = 2k\pi \qquad k = 1, 2, 3, \dots$$

Also, from Snell's law,

$$\sin \theta_r = \frac{\sin \theta_i}{n}$$

Equation (16.1) can thus be written

$$2\pi k = \frac{4\pi t}{\lambda}(n_2 - \sin^2 \theta_i)^{1/2} + \Delta\phi_r$$
$$= 4\pi t\nu(n^2 - \sin^2 \theta_i)^{1/2} + \Delta\phi_r$$

The thickness can be found from the number of fringes k in the frequency interval $\nu_2 - \nu_1$,

$$t = \frac{\Delta k}{2(\nu_2 - \nu_1)(n^2 - \sin^2 \theta_i)^{1/2}} \qquad (16.2)$$

if we are safe in assuming that n and $\Delta\phi_r$ are constant over the range $\nu_2 - \nu_1$. When the radiation is incident normally, Eq. (16.2) reduces to

$$t = \frac{\Delta k}{2(\nu_2 - \nu_1)n} \qquad (16.3)$$

and in the case of small θ_i and large n, Eq. (16.3) is a good approximation.

III. DIFFUSE REFLECTANCE MEASUREMENTS

Not all objects reflect specularly. Depending on surface condition, more or less of the reflected beam is reflected at angles other than the negative of the incident angle. This radiation is said to be diffusely reflected. A diffusely reflecting surface is said to reflect a parallel beam of incident radiation uniformly in all directions. That is, the apparent irradiance of the surface (radiant power per unit of *projected* area) is independent of the viewing angle. This statement is the basis of the *Lambert cosine law*. Lambert's law is found to hold experimentally in some but not all cases. It is an empirical law and not a fundamental consequence of electromagnetic theory.

Numerous methods have been devised to measure diffuse reflectance. Let us first discuss the integrating sphere. The idea of the integrating sphere is illustrated in Fig. 16.10. A sphere is made with three openings: one serves as an entrance port for incident radiation; another, directly opposite the first, contains the sample; the third opening contains a detector. The inside surface of the sphere is coated with a substance which is as closely as possible a perfect diffuse reflector. Radiation reflected from the sample in all directions is intercepted by the sphere and rereflected. Eventually it either is detected, lost out the entrance port, reflected again by the sample, or absorbed.

DETECTOR

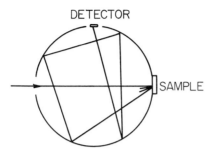

SAMPLE

Fig. 16.10   Integrating sphere for diffuse reflectance measurements.

The radiation diffusely reflected from any point on the sphere or sample is uniformly distributed over the surface of the sphere. This is readily seen from Fig. 16.11. The element of surface area dA_1 reflects

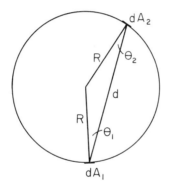

Fig. 16.11   Integrating sphere geometry.

an amount of radiant power J. From Lambert's cosine law, the radiant power leaving the surface in a direction θ_1 from its normal is $J \cos \theta_1 dA_1$. The power per unit area of wave front is reduced by the inverse square of the distance d to the surface element dA_2, and the projected area of this element presented to the beam is $dA_2 \cos \theta_2$. The intercepted radiant power is therefore

$$J(A_2) = \frac{J(A_1) \cos \theta_1 \cos \theta_2 \, dA_1 \, dA_2}{d^2}$$

So far we have not made use of the spherical nature of the surface of which dA_1 and dA_2 are parts. If we use this property, we can write

$$\cos \theta_1 = \cos \theta_2 = \frac{d}{2R}$$

and

$$J(A_2) = \frac{J(A_1)\, dA_1\, dA_2}{4R^2}$$

where R is the radius of the sphere. Thus the radiation received at dA_2 is independent of the position of dA_1 on the sphere. Let us continue with an elementary theory of the integrating sphere, following the treatment of Wendlandt and Hecht (1). The radiant power reflected from the sample is $J_0 r_s$, where r_s is the total reflectance of the sample and J_0 is the incident power. Of this an amount

$$J_1 = J_0 r_s \frac{b}{4\pi R^2}$$

is received directly by the detector through its port of area b. The remainder strikes the surface of the sphere somewhere other than in b and is reflected with an average reflectance r'_w, given by

$$r'_w = \frac{r_w(4\pi R^2 - b - p - c) + r_s p}{4\pi R^2} < 1 \tag{16.4}$$

We have used the symbol b to represent the area of the detector port, p the area of the sample port, and c the area of the entrance port. The radiant power reaching the detector after one reflection from the wall is

$$J_2 = \frac{J_0 r_s b}{4\pi R^2} r'_w$$

After two reflections

$$J_3 = \frac{J_0 r_s b}{4\pi R^2} r'^2_w$$

reaches the detector, and so on. The total power at the detector is therefore

$$J = \frac{J_0 r_s b}{4\pi R^2} (1 + r'_w + r'^2_w + \cdots)$$

$$= \frac{J_0 r_s b}{4\pi R^2 (1 - r'_w)} \tag{16.5}$$

The radiation available at the detector increases as $1/(1 - r'_w)$. In the visible and near-infrared regions sphere walls coated with magnesium oxide provide ample reflectivity. Coatings of flowers of sulfur permit working somewhat farther into the infrared (7). But unfortunately good coatings have not been devised for the longer wavelengths of the infrared, so coated integrating spheres are seldom used there. A new form of infrared integrating sphere has recently been described by Morris (8). He starts with a spherical glass flask which he alters by pressing a carbon tool into the heat-softened surface at closely spaced intervals. This forms a coarse lenticular inner surface of concave and convex elements which blend into each other. The inner surface is coated with aluminum. The aluminized lenticular surface has high reflectivity and angular dependence closely approximating an ideal diffuser. The lenticulated sphere is reported to be superior to a similar MgO-coated sphere at wavelengths longer than 1.5 μ, and it increases in efficiency at longer wavelengths.

There are other workable methods for measuring diffuse reflectance in the infrared. One method, shown in Fig. 16.12, uses a hemisphere

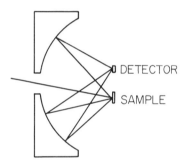

Fig. 16.12   Measurement of diffuse reflectance with Coblentz hemisphere.

coated with a good specular reflecting material such as aluminum. Such a device was used by Paschen and later by Coblentz (9). The radiation beam enters a port in the hemisphere and the slit of the monochromator is imaged on the sample, which is placed off the center of the sphere. Radiation from the sample is imaged by the hemisphere onto the detector on the other side of the center of the sphere.

This system is necessarily placed after the monochromator, but it is adaptable to double-beam operation in the following way: The top half of the slit image in the sample beam and the bottom half in the reference beam are masked. In the hemisphere, the top and bottom halves of the slit are formed sequentially as the beam-switching mirror rotates. The sample is placed to intercept the bottom half of the slit and a reference material to intercept the top half. Note that radiation emitted by the sample or reference is not chopped, so only the radiation reflected by them is detected.

The Coblentz hemisphere can be replaced with an ellipsoid or spheroid (*10*), with sample and detector placed at the foci, as in Fig. 16.13A or B. Two paraboloids can also be used, as in Fig. 16.13C. This

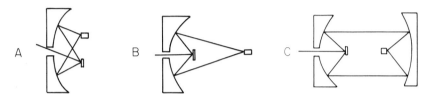

Fig. 16.13 Variations of Coblentz hemisphere.

has the advantage of placing the sample at a greater distance from the detector, which can be important if the sample must be heated. The effect of aberrations on the performances of hemispherical and ellipsoidal reflectometers has been discussed by Brandenberg (*11*). In general, it is very difficult to collect a large fraction of the reflected radiation, partly because of aberrations and partly because the angular field of view of most infrared thermal detectors is limited. White (*12*) has bypassed these problems by placing the source rather than the detector adjacent to the sample in the hemisphere. A chopper is placed between the source and the mirror in order to discriminate between radiation emitted by the sample and radiation reflected from it. As an alternate arrangement, the radiation leaving the hemisphere is also chopped. This permits measurement of the emitted radiation from the sample. The arrangements of Figs. 16.13A, B, and C can also be used with a source rather than the detector at one conjugate.

These techniques for measuring diffuse reflectance are relative methods, and they require a standard reflector. A method for measuring absolute diffuse reflectivity is due to Gier *et al.* (*13*). A hohlraum or blackbody is constructed as in Fig. 16.14. The cavity is designed to provide a good approximation to an ideal blackbody. The walls are heated electrically and can be raised uniformly to temperatures as high as 800°C. A sample is inserted through an opening in the wall and is cooled by a jet of cold water impinging on its back side. The interior of the blackbody is viewed through a water-cooled baffle. By tilting the blackbody, a measurement can be made of the radiation from the interior which is diffusely illuminating the sample and reflected out the opening. The ratio of this radiation to the radiation emitted from the

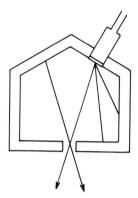

Fig. 16.14   Apparatus of Gier *et al.* for measuring absolute diffuse reflectance.

interior with the sample out of view gives the diffuse reflectance of the sample directly. The sample is viewed slightly off normal (5°), so radiation entering from outside the enclosure is not specularly reflected back to the monochromator and detector.

Samples can be prepared for diffuse reflectance measurement in various ways. Some materials, such as metal or plastic plates and the like, are automatically in the proper form. Paint or plastic samples can be coated on metal disks. Powder samples can be pressed into aluminum foil cups or suspended in sodium silicate or some other cement. Note, however, that the method of preparation can strongly effect the reflectivity of powders.

IV. Internal Reflectance Measurements

Infrared reflectance spectra, both specular and diffuse, of many inorganic samples show the kind of strong bands so gratifying to the spectroscopist. Reflectance spectra of organic compounds, on the other hand, are apt to be disappointing. This is because the variation of the optical parameters, index of refraction and absorption coefficient, is simply too small to provide large changes in reflectance in most instances.

Nevertheless, reflectance methods sometimes offer a certain amount of convenience, and some workers have used ordinary specular reflectance techniques to obtain *transmission* spectra by depositing a layer of a sample on a reflecting surface. The radiation reflected at the internal boundary between the sample and the metal is made to pass

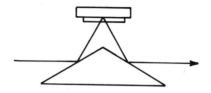

Fig. 16.15   Reflection technique for measuring a transmittance spectrum.

through the sample twice (see Fig. 16.15) and the transmission spectrum is recorded. This method has been extended to multiple passes, by methods such as the one illustrated in Fig. 16.16, for studying very thin films. However, scatter and reflection from the first surface of the sample layer also builds up with the number of passes, thus reducing the contrast of the absorption bands. Moreover, for near normal incidence there is a node in the electric vector of the electromagnetic field at the surface of a reflecting metallic mirror. Therefore, if the sample is much thinner than a half-wavelength, there will be only weak interaction between the radiation field and the sample. It has been suggested that the surface of the mirror should be roughened to increase the chance of radiation passing through a greater amount of sample. It is not clear that this is a generally useful approach because it also tends to decrease the thickness of some paths. Similarly, a thin dielectric layer evaporated on the metal between the mirror and the sample can increase absorption at those wavelengths at which it is a half-wavelength in optical thickness.

Usually, the incident radiation is not normal to the surface, and often a quite large angle of incidence is chosen. This increases the effective sample thickness somewhat, but this is not very important. Incidentally, it also spreads the beam out and requires a larger sample area, which is unfortunate in the study of microsamples. The important

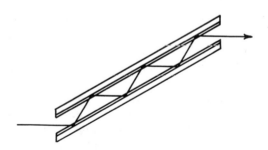

Fig. 16.16   Multireflection method.

advantage of a large angle of incidence is that, for radiation polarized with the electric vector parallel to the plane of incidence, the incident and reflected beams no longer cancel each other, and the node in the *E*-vector no longer occurs at the surface. This has been discussed by Greenler (*14*). A node still occurs for radiation polarized perpendicular to the plane of incidence.

It might appear that there is no advantage to be gained by depositing the sample on a dielectric rather than a metallic surface. The reflectivity between two dielectric surfaces is small under most conditions. However, there is one arrangement in which the reflectivity is very great indeed. That is when the radiation approaches the interface from the direction of the material of greater refractive index and the angle of incidence exceeds the critical angle for total internal reflection. In this situation an evanescent wave is propagated along the direction of the interface. Its extension into the optically rarer medium is damped exponentially and very rapidly, but nevertheless the wave is exposed to the medium effectively enough to be attenuated according to the absorption coefficient of that medium. Thus, if a sample of some material is intimately joined to the surface of an optically denser material, radiation that would otherwise be totally internally reflected at the surface is attenuated by the sample. This phenomenon is termed *attenuated total reflection* (ATR) and was first used by Fahrenfort (*15*) and by Harrick (*16*) in 1960 to obtain infrared absorption spectra of samples.

The optically dense medium can be in the form of a hemicylinder (Fig. 16.17A), a triangular prism (Fig. 16.17B), or a Dove prism (Fig. 16.17C). The prism or cylinder must be constructed of a material which is very transparent over a usefully wide region of the infrared, and it must have a refractive index which is higher than that of the sample. However, the extent of penetration into the sample, and therefore, the

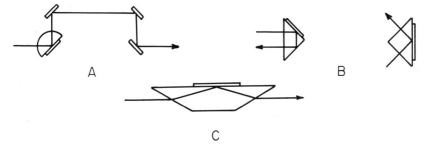

Fig. 16.17 Arrangements for measuring attenuated total reflectance.

band intensity, is greater if the indices are not much different. Therefore, some care in choosing the proper prism material for a given sample is required. KRS-5, silver chloride, germanium, silicon, and the Irtrans are often used, as well as CsBr and even NaCl for samples of lower refractive index. Several representative optical systems for the insertion of ATR units in a spectrophotometer are illustrated in Fig. 16.18. Some units permit the angle of incidence to be varied.

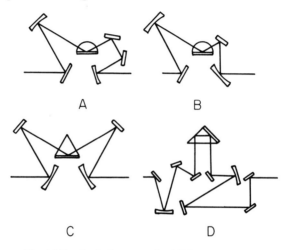

Fig. 16.18　Optical systems for ATR measurements.

Multipass ATR techniques permit observations on very thin films or weakly absorbing materials. Both sides of the plate are exposed to the sample. Some typical arrangements are shown in Fig. 16.19. The plates of Figs. 16.19B and C are used in-line and would not require any additional optics except that the shifting of the focus cannot be tolerated in some systems. The plate of Fig. 16.19A is used with an optical arrangement such as the double-beam system of the Wilks Scientific Corporation shown in Fig. 16.20. The plate of Fig. 16.19D can be inserted into a beaker containing a liquid sample. Harrick (*17*) has used this and other similar arrangements. The double plate of Fig. 16.19E, also due to Harrick, has a very useful property. If the plates are made of a material having a refractive index greater than 2, the geometry (*d*, *t*, and *θ*) can be chosen so that the focus of the beam is undisturbed. The number of reflections taking place at the prism–sample interface is (*18*)

$$N = \frac{d}{t} \csc \theta \cot \theta + 2(1 - \cot^2 \theta)$$

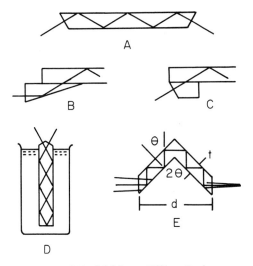

Fig. 16.19   Multipass ATR methods.

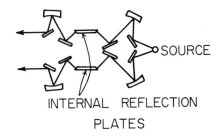

INTERNAL   REFLECTION

PLATES

Fig. 16.20   Wilks' double-beam ATR apparatus.

A sample studied by ATR can be in the form of a polished solid slab pressed against the back surface of the prism. If the surface is irregular, optical contact can be made by using a *thin* film of a liquid of low refractive index, but this is not necessary if partial contact is sufficient. Samples of fabric, thread, yarn, or fiber can also be pressed against the prism, as can samples of compacted powder. The sample can be in the form of a concentrated paste or suspension, or a viscous liquid spread on the prism. Less viscous or volatile liquids or solutions can be contained in a cell using the prism face as the window. Water solutions can be studied by ATR if the prism is made of water-insoluble material. This is useful in the analysis of inorganic or biological materials. Fine membrane filters are useful for collecting and supporting samples. The filter is pressed against the prism and the spectrum of the sample is obtained with the spectrum of the filter material superposed. Double-beam compensation techniques can be used to reduce the amount of interferences from the filter, or a metallic membrane filter can be used. Because the attenuation takes place in a very thin layer, multipassed ATR is a sensitive technique for the study of surface layers and adsorbed gases. The film can be formed on a substrate material which is then pressed against the crystal of the ATR unit.

If the sample thickness is greater than the penetration distance (usually several microns), variations in the thickness do not affect the intensity of the ATR bands. On the other hand, the intensity of bands associated with the components of a solution vary with concentration according to Beer's law, and successful quantitative analyses can be performed from ATR measurements.

The spectra observed with ATR techniques are similar to normal absorption spectra—sufficiently so that normal spectra can usually be used as reference curves for qualitative identification from ATR spectra, and vice versa. However, there are some differences. The shape of absorption bands is dominated by the absorption coefficient, but the shape of an ATR band is also influenced by the refractive index. This can introduce asymmetries in the band contour, just as in ordinary specular reflectance spectra. The spectroscopist should also be wary of differences which can occur because of orientation of the sample molecules when deposited on a crystal or substrate.

REFERENCES

1. W. W. Wendlandt and H. G. Hecht, *Reflectance Spectroscopy*, Wiley, New York, 1966.
2. N. J. Harrick, *Internal Reflection Spectroscopy*, Wiley, New York, 1967.
3. D. W. Barnes and P. N. Schatz, *J. Chem. Phys.*, 38, 2662 (1963).
4. J. Strong, *Procedures in Experimental Physics*, Prentice-Hall, Englewood Cliffs, N.J., 1944, p. 376.
5. H. E. Bennett and W. F. Koehler, *J. Opt. Soc. Am.*, 50, 1 (1960).
6. W. G. Spitzer and M. Tanenbaum, *J. Appl. Phys.*, 32, 744 (1961).
7. M. Kronstein, R. J. Kraushaar, and R. E. Deacle, *J. Opt. Soc. Am.*, 53, 458 (1963).
8. J. C. Morris, *Appl. Opt.*, 5, 1035 (1966).
9. W. W. Coblentz, *Bull. NBS*, 9, 283 (1913).
10. W. R. Blevin and W. J. Brown, *J. Sci. Inst.*, 42, 385 (1965).
11. W. M. Brandenberg, *J. Opt. Soc. Am.*, 54, 1235 (1964).
12. J. U. White, *J. Opt. Soc. Am.*, 54, 1332 (1965).
13. J. T. Gier, R. V. Dunkle, and J. T. Bevans, *J. Opt. Soc. Am.*, 44, 558 (1954).
14. R. Greenler, *J. Opt. Soc. Am.*, 55, 1571 (1965).

15. J. Fahrenfort, *Spectrochim. Acta*, **17**, 698 (1961).
16. N. J. Harrick, *J. Phys. Chem.*, **64**, 1110 (1960).
17. N. J. Harrick, *Appl. Opt.*, **4**, 1664 (1965).
18. N. J. Harrick, *Anal. Chem.*, **36**, 188 (1964).

CHAPTER 17

Experimental Methods of Infrared Emission
Spectroscopy

I. Introduction

Infrared emission studies were important in the early days of infrared spectroscopy. The theory of the spectral distribution of radiation from a blackbody is an essential cornerstone of modern quantum mechanics, and the experimental verification of the Planck law is based largely on infrared measurements. Infrared emission spectroscopy went through a long period of near dormancy until the period following World War II. At that time interest was aroused by the need for engineering data on thermal radiation from hot bodies, rocket exhausts, and so on.

More detailed accounts of infrared emission spectroscopy may be found in the writings of Penner (*1*), Tourin (*2*), Blau and Fischer (*3*), and Rutgers (*4*).

II. Standards of Emittance

If the radiation from an emitting sample is collected and imaged on the slit of a spectrometer, an emission spectrum can be recorded. This spectrum is not a true emittance spectrum of the sample for several reasons which are associated with the spectrometer itself. In the first place the response of the instrument, in volts of output per watt of radiation input, say, is in general not known. In any case it varies strongly with wavelength because of variations in detector response, efficiency of the mirrors and the grating or prism, and absorption by atmospheric H_2O and CO_2. Furthermore, emittance is defined in terms of radiant power per unit spectral interval (wavelength or wavenumber) and the spectral bandwidth of the monochromator depends on the dispersion of the instrument, the slit width, and also the broadening of the slit function due to diffraction and optical aberrations. All of these, except possibly the slit width, are wavelength dependent. Also, the emittance is defined in terms of power radiated into a unit solid angle. The angular acceptance of the monochromator might not be well defined, and unless a fixed defining aperture is used, it will vary with wavelength as the grating or Littrow mirror is rotated. Finally, the emittance is defined in terms of power radiated per unit of surface area of the sample. The sample is generally imaged on the slit of the monochromator, so the area studied varies with slit width. Not all of the radiation is successfully imaged on the detector because of aberrations in the detector optics.

If all of these factors were well defined, it would be possible to record an emission record and convert it to a spectral plot of radiant power emitted per unit of solid angle per unit of spectral increment for a unit area of specimen. The ratios of the values in this graph to the known values calculated from Planck's radiation law for a blackbody of the same temperature are, of course, the spectral emittances of the specimen.

It is very difficult to measure many of the required instrumental parameters. Emittance is therefore usually determined by recording an emission record for the sample and then recording a similar record for a blackbody at the same temperature as the sample, and under exactly the same conditions of optical geometry, slit width, and so on. The ratio is the emittance of the sample, with the instrumental parameters canceled out. Of course, the measurements might be carried out relative to a secondary standard, which has been compared with a blackbody.

In this discussion we use the term emittance to describe the thermal emission properties of an object. Emissivity is used in describing the properties of a material. A similar distinction is made between the terms reflectance and reflectivity.

Recall that a blackbody is opaque and has a reflectance of zero, and therefore an emittance of unity. A closed container into which radiation can somehow be introduced is a blackbody, because eventually the radiation will be absorbed even if the reflectivity of the wall material is only slightly less than unity. An approximate blackbody for laboratory use is naturally constructed as a closed container with a *small* opening for sampling the radiation within. Let us see how the size of the opening and the reflectance of the walls affect the reflectance of a laboratory blackbody. Assume that the device is constructed in the form of a sphere (Fig. 17.1) of total internal area B, with a single opening of area

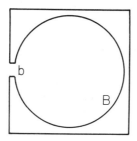

Fig. 17.1 Spherical blackbody.

b cut into the wall. The wall is assumed to be a diffuse reflector obeying Lambert's law and having a reflectance r_w. A radiation beam of power J_0 is introduced through the opening and we wish to calculate the power J returned through it. Evidently we can use the same expression derived for average reflectance of the integrating sphere [Eq. (16.4)] by letting the reflectance of the "sample" be r_w also and by imagining the detector port to be coincident with the entrance port. We have

$$r'_w = \frac{r_w(B-b)}{B}$$

Similarly, in place of Eq. (16.5), we find for the reflectance of the blackbody,

$$R = \frac{r_w b}{B(1-r'_w)} = \frac{r_w b}{B(1-r_w)+r_w b}$$

The emittance of the blackbody, $E = 1 - R$, can easily be written in terms of the emittance of the wall, $e_w = 1 - r_w$,

$$E = \frac{e_w}{e_w(1-b/B)+b/B} \qquad (17.1)$$

Equation (17.1) tells us that a very large sphere with a very small opening is a good approximation to an ideal blackbody even if the emissivity of the wall material is low. It also tells us that if the emissivity of the wall material is close to unity, the opening can be quite large, which should not come as a surprise. In practice, the size of the opening is determined by the size of the aperture of the rest of the optical system; the size of the sphere is limited by the power required to heat the wall and the difficulty of heating it uniformly if it is made very large; and the emissivity of the wall material is determined by the physical properties of available materials. For example, if we demand a laboratory blackbody of emittance 0.999, but the best wall material available has an emissivity of only 0.8, and the opening must be 25×6 mm, we find that the sphere must have a quite large radius of about 10 cm.

The equation for the emittance of a blackbody can be generalized to shapes other than spherical [this has been discussed by Gouffe (5), for example]. The generalized equation is

$$E = \frac{e_w[1 + (1 - e_w)(b/B - \Omega/\pi)]}{e_w(1 - b/B) + b/B} \tag{17.2}$$

where Ω is the solid angle subtended by the opening, as seen from the wall opposite it (Fig. 17.2). It is easy to see that in the case of the sphere

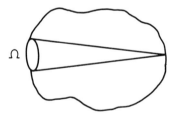

Fig. 17.2   Solid angle subtended by an opening in a blackbody.

$b/B = \Omega/\pi$, and Eq. (17.2) reduces to Eq. (17.1). The treatment has been further generalized by DeVos (6) to include departures from the Lambert law. More recent discussions are given by Nicodemus (7) and Fecteau (8).

Laboratory blackbody standards are usually formed in geometrically convenient shapes. As an alternate to the spherical cavity, a cylinder with a slot or circular opening in the wall or the end might be chosen, as in Fig. 17.3A or B. A wedge-shaped cavity, Fig. 17.3C, is an old form, commonly called the Mendenhall wedge. The conical cavity of Fig. 17.3D is a variation of the wedge. It is desirable to make the walls

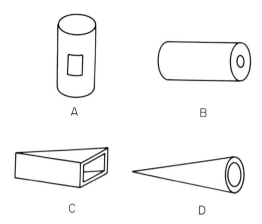

Fig. 17.3   Forms of laboratory blackbody standards.

of a material having a high emissivity, as we have seen. However, it is convenient to heat the walls electrically, and unoxidized metallic conductors do not normally have high emissivities. Carbon is a conductor of electricity and a good emitter and blackbodies have been made of hollow carbon cylinders, but carbon oxidizes in air at high temperatures. The source shown in Fig. 17.4 has a carbon body provided with a Vycor cylinder through which a stream of nitrogen flows to slow the oxidation. Even so, the carbon degenerates more or less

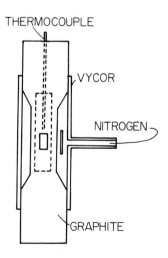

Fig. 17.4   Blackbody constructed from graphite rod.

quickly with use. The thermocouple inserted into the cavity is used to measure the temperature of the interior. Other materials are not so delicate. Rods of silicon carbide and other semiconductors can be used. Of course, it is not necessary to use a conducting material. Heaters can be applied to the outside of the walls or the blackbody can be placed in an oven. A sketch of a blackbody used at the National Bureau of Standards (9) is shown in Fig. 17.5. The core is made of Iconel metal.

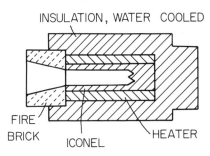

Fig. 17.5   Form of blackbody used at National Bureau of Standards.

Its interior was roughened by machining a fine screw thread on the surface. Then it was heavily oxidized, raising its emittance to around 0.9. The heating coil wrapped around the core can raise the temperature of the cavity to 1400°K. The exterior is water cooled. The calculated emissivity of the blackbody is 0.999 or greater.

Blackbody sources of modest size are available from several commercial suppliers. They offer temperatures ranging from less than room temperature to 3000°K. Operation at the highest temperature requires an atmosphere of inert gas. Extremely high temperatures (48,000°K) are possible in a device using an ablating wall material (10). It is intended for use in the visible and ultraviolet spectral region between 6000 and 1600 Å.

Blackbody reference sources to be used not far from room temperature have been fashioned from black paper rolled into a cone. Very-low-temperature blackbodies are sometimes required, as for example, in certain meteorological or astronomical studies of the radiation from the sky. These blackbodies might be cooled with liquid nitrogen.

Secondary standards of spectral emittance are very useful, not only because a blackbody source may be inconvenient to use, but because it is usually good practice to compare a sample of low emittance (for example) with a standard of low emittance. Silicon carbide and oxidized metal strips have often been used as secondary standards. Tungsten lamps are certified as standards of radiance (*11*) and irradiance (*12*) by the National Bureau of Standards for use from the near ultraviolet to the near infrared. However, the quartz envelopes used in constructing these lamps prevent them from being used at wavelengths longer than around 2.6 μ. Polished platinum metal has been proposed as a standard of low emittance.

III. Methods for Solid and Liquid Samples

It is sometimes more convenient to obtain a reflection spectrum than an emission spectrum of a specimen. The emittance of an opaque body is related to its reflectance by Kirchhoff's law. A relationship for a transmitting body is somewhat more complicated and requires a knowledge of transmittance as well as of reflectance in determining the emittance. The spectroscopist should be aware that the emittance is, in general, a function of temperature; so in order to apply Kirchhoff's law, the reflectance and the transmittance must be measured with the sample at the temperature for which the emittance is required. Precautions in the placing of the radiation chopper and in the nature of the beam combiner are required to ensure discrimination between the reflected or transmitted radiation and the thermal radiation emitted by the specimen.

The first step in the procedure for obtaining an emission spectrum from a specimen is to devise a way of collecting the emitted radiation and introducing it into a monochromator. The spectroscopist might simply replace the source of his spectrometer with the specimen or he might construct a transfer optics system. In either case, care should be taken to ensure that both the slit and the entrance pupil of the monochromator are filled with radiation. As usual, the best way to do this is with a visible light source illuminating the monochromator in the reverse direction, that is, from exit to entrance slit. A visible image of the pupil and the slit is observed and one or the other of these is made to fall on the specimen. It is easy to see whether the specimen is large enough to fill the aperture.

If the specimen has a discrete shape, such as a coiled wire or screen, it is difficult to illuminate an aperture uniformly, and a device for scrambling its image might be required. A roughened diffusing window or mirror or an integrating sphere can be used, but these elements have decided limitations in the infrared region, particularly at longer wavelengths.

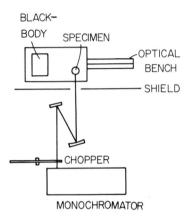

Fig. 17.6 Optical arrangement for comparing an emitting specimen with a blackbody standard.

Measurements requiring comparison with a reference source have been made in a variety of ways. In Fig. 17.6, a blackbody or other reference source is mounted on a carriage along with the specimen. Separate spectra are recorded of the reference source and the specimen by positioning first one and then the other before the monochromator. The radiation is modulated by a chopper, which is placed close to the monochromator and is therefore at the temperature of the slit. The conditions of amplifier gain, slit width, and so on, are the same for the two measurements. The ratio of the records is the emittance of the specimen if a blackbody reference is used and if both the reference and specimen are at the same temperature. Otherwise, corrections for the emittance and temperature of the reference must be made. The temperature of the blackbody can be held equal to that of the specimen by applying their respective thermocouple signals to the inputs of a feedback control circuit which controls the heating current to the blackbody. A shield, which may be water cooled, prevents the monochromator from being heated by radiation from the blackbody and specimen.

A "semi-double-beam" method (*13*) is illustrated in Fig. 17.7. A standard double-beam optical null photometer of the type shown in Fig. 17.7A is modified slightly, as shown in Fig. 17.7B. Spectra of the reference and the specimen are recorded in turn relative to the standard spectrometer source. The ratio of these spectra must be computed to obtain the emittance of the specimen, as before, but the computation is simplified because absorption due to atmospheric H_2O and CO_2 is canceled. Moreover, small variations in amplifier gain and slit width are

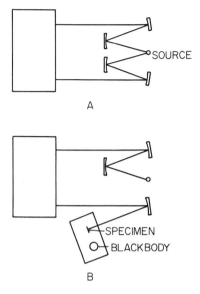

Fig. 17.7 Semi-double-beam method for emission measurements.

not significant. However, variations in the temperature of the source can introduce errors.

The computation of ratios in these methods can be performed by digital computer if the outputs are digitized and stored on punched tape, magnetic tape, or cards. Analog computers can also be devised for this purpose. In fact, the old Beckman IR-3 system would seem to be ideal for this purpose, and has indeed been used in this way. An alternative scheme, which has been successfully used for spectroradiometry in the visible region, makes use of the electrical function generator discussed in Section 12.V. The function generator attenuates the output and is so adjusted that the spectrum of a reference has the same shape as its emittance spectrum. Spectra of other specimens subsequently measured with the same function generator are automatically emittance spectra of the specimens.

True double-beam emittance spectra can be recorded by placing the specimen in one beam of a double-beam instrument and a reference or black body in the other, as in the simplified sketch of Fig. 17.8. Initial balancing of the two beams is done by placing identical specimens in each beam. An arrangement in use at the National Bureau of Standards (9) is shown in Fig. 17.9. It is based on the Halford–Savitzky ratio recording system. The comparison blackbody and either the

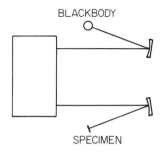

Fig. 17.8   Double-beam emittance measurement.

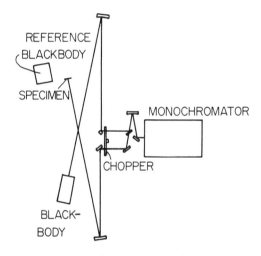

Fig. 17.9   Ratio computing system for emittance measurement.

specimen or the reference blackbody are imaged on the top and bottom halves, respectively, of the pupil of the monochromator.

All of these methods have been described for the case of normal emission, that is, emission in a small solid angle normal to the surface. They are easily modified to measure emission at angles other than those normal to the surface, but still in a small solid angle. The collection of radiation over a hemisphere requires that the specimen be placed in an integrating sphere (or hemisphere) or at the focus of a Coblentz sphere or ellipsoid. These methods are difficult in the infrared, where energy is always scarce.

The measurement of the emittance of a weak sample or a specimen at low temperature is an especially demanding process. The technique of spectral accumulation can be useful. The spectrum is scanned repeatedly, and values at selected wavelengths are digitized and stored in a memory core. The spectrum signals add directly with each scan, but the noise signals, being incoherent, add as the square root of the number of scans. Hence, after n scans the signal-to-noise ratio is improved in proportion to $\sqrt{n}$.

The use of an interferometer is especially attractive under these conditions because of the additional enhancement of signal-to-noise ratio by the Fellgett factor and the advantageous optical luminosity. Emission spectra of films of organic samples at only a few degrees above ambient have been observed by Low and Coleman (14) using a Block Associates interferometer with a coherent information added (Coadder, as Block calls it). A similar instrument is used in planetary and stellar spectroscopy (15).

The most direct way of heating a conducting solid sample for emission spectroscopy is by passing a current through it. Oxidation can be retarded by use of an inert gas atmosphere. The temperature of the surface can be measured by attaching a thermocouple junction. The wires should be small so as not to alter the local temperature materially by conducting heat away. Care must also be taken in mounting or supporting the specimen in order to avoid temperature gradients across the observed surface. Some workers, for example the group at the National Bureau of Standards (9), suspend the specimen in a cylindrical chamber or furnace whose walls are blackened and cooled to suppress reflection and emission. The sample is also mounted off the axis of the chamber to prevent radiation from reflecting from the walls back to the specimen. Baffles are installed inside the furnace to retard convection currents. Reflectors surround the ends of the specimen strip and reduce heat loss by radiation to compensate for the loss by conduction through the electrical contacts. In this way good temperature uniformity is maintained in the central part of the strip where observations are made. There is some evidence, at least in the case of tungsten, that electrical heating alters the surface structure and changes the emittance of the specimen.

The emissivity of certain materials used for coating can be found by painting them on metal strips and heating the strips electrically. A temperature-sensing thermocouple should be attached to the coating, not to the underlying metal, because steep temperature gradients can be supported by some insulating materials. If the coating is not thick enough to be opaque, the radiation from the metal surface will also be observed, modified by passage through the coating.

Highly transparent specimens are difficult to study because it is hard to heat them uniformly without observing radiation from the heater. Moderate temperatures can be obtained by mounting a specimen on a heated stage such as is used for transmission measurements of hot samples. A blast of heated air directed on the surface of the sample can also be used. For higher temperature, the hot air can be replaced with a flame, but it is difficult to avoid radiation from the flame and degradation of the surface. The specimen can be mounted on a highly reflecting and therefore poorly emitting metal surface and heated electrically. Samples of organic materials that are to be studied at fairly low temperatures of 100 or 200°C can be deposited on a metal plate or foil by evaporation from solution or from a melt.

The infrared thermal emission spectra of liquids can also be studied, although usually at fairly low temperatures, that is, rarely higher than several hundred degrees. A liquid sample can be confined to a heated cell normally used for transmission studies. One window of the cell might be replaced with a reflecting metal plate. Greases or waxes can be smeared or otherwise deposited on a plate or foil. The interferometric method, with information accumulation, has been applied to liquid samples, as, for example, in the identification of effluent fractions from a chromatograph.

This discussion has so far treated thermal radiation exclusively. Luminescence radiation is also studied in the infrared. Quantitative measurements can be made by comparison with the radiation from a blackbody of known temperature, but of course the concept of emissivity has no meaning. Luminescence can be excited by exposure to a beam of electrons (cathodoluminescence), a beam of photons (photoluminescence), or by passage of an electric current (electroluminescence). The mounting of samples is dependent upon the nature of the phenomena under study. Often the sample is cooled to the temperature of liquid nitrogen or helium in order to sharpen the transitions by freezing out thermal motion.

Lasers are special examples of luminescent samples. Solid-state lasers are usually either crystals or glasses doped with impurities and excited optically, or diode junctions which emit recombination radiation when an electric current passes through the junction. The study of laser radiation can seriously tax the resolving power of a monochromator because the transitions are often very sharp, producing narrow lines. Furthermore, because of the coherence of laser radiation, the slit function of the monochromator is not the same as it is with incoherent radiation. Laser radiation, unlike thermal radiation, is generally directional or evenly highly collimated. This might cause problems in filling the entrance pupil of a monochromator, although diffraction at a narrow entrance slit spreads the beam. If necessary, a diffuse reflector can be used to introduce the radiation.

IV. METHODS FOR GASEOUS SAMPLES

The methods used with gaseous samples are in some respects quite different from those appropriate for condensed samples. In the first place, reflectance measurements cannot be used in place of emittance measurements. Also, the narrow lines emitted by some gaseous samples, in contrast with the broad bands usually associated with thermal emission by solids, can place severe demands on the resolving power of a monochromator. Finally, the methods used to excite gaseous samples to emit are generally distinct from those used for solids and liquids.

The radiation emitted by a fairly cool gas can be observed without going to any particular trouble to heat it. The radiation levels are low and the interferometric methods are especially valuable in this case. Thus Chamberlain et al. (16) have recorded the radiation from ammonia gas at 39°C (radiation from a cloud can be detected and compared with empty sky), and Low and Clancy (17) have used their coadding interferometer to detect air pollution components in a plume arising from a smoke stack. In the latter work an 8-in. reflector telescope, used to collect the radiation, permits isolating very small areas for observation.

Goldring and Deeds (18) have considered how the reemission of absorbed radiation by gaseous molecules can affect the observed transmittance. The reemitted radiation is resonance–fluorescence radiation and it has been the same frequency as the absorbed radiation. At sufficiently low pressures, virtually all excited molecules can reemit before they lose their energy by collision. The radiation is emitted more or less isotropically, but a small portion is in the forward direction and is detected, thus directly increasing the measured transmittance. The reemitted radiation can be absorbed and reemitted again, of course, and it can encounter the cell walls and be either lost or scattered back into the cell. All of these possibilities make for a very complicated problem which was treated by using Monte-Carlo methods on a digital computer. The theory was applied to the absorption spectrum of HF. An interesting experiment suggests itself (Fig. 17.10): If an absorbing gas

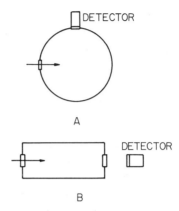

Fig. 17.10   Hypothetical experiment for discriminating against reemitted radiation.

at low pressure is contained in an efficient integrating sphere (Fig. 17.10A), the measured absorptance of the gas should be very small compared with the absorptance for the same path length and pressure in an ordinary cell (Fig. 17.10B).

Gases that can be raised to elevated temperatures without degradation can be heated in an oven. For example, the emission from the vapor of B_2O_3 at 1750°K has been observed by Dows and Porter (*19*). They placed (or produced) B_2O_3 in an open Alundum tube 10 cm long which was radiantly heated by Globar heaters. A small amount of radiation from the walls was observed, but a discrete line from $B_2^{11}O_3$ was observed at 2013 cm^{-1} with a shoulder at 2059 cm^{-1}; $B_2^{10}O_3$ emitted a line at 2114 cm^{-1}. The spectrum was scanned from 4000 to 700 cm^{-1}, but no other lines were observed.

Gases can be raised to very high temperatures for a brief instant in a shock tube. A typical optical arrangement for measuring absorption is shown in Fig. 17.11. The sample gas is placed in one chamber separated by a membrane of aluminium or Mylar from the pressuring gas in a second chamber. When a sufficiently high pressure is built up, the membrane is ruptured and a shock wave travels down the tube. The temperature of the gas abruptly rises to perhaps 2000°K as the wave front passes. The passage of the wave front takes only about a millisecond. If the emission spectrum is to be recorded, a very fast scanning spectrometer must be employed. Brief segments of the spectrum can be recorded in successive experiments. Fast scanning techniques are also applicable to other transient phenomena such as explosions and fast chemical reactions.

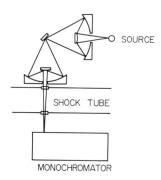

Fig. 17.11 Optical arrangement for observing absorption by gas in a shock tube.

The temperature of the constituents of a transient system can be determined from measurements at a single wavelength. The spectrometer in the experiment of Penzias *et al.* (*20*) simultaneously records the absorptance and the radiance at a fixed wavelength. Because of the location of the chopper, the absorptance is measured as an ac signal and the radiance as a dc signal. The signals are illustrated in Fig. 17.12. The radiance is determined by means of a separate calibra-

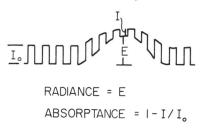

RADIANCE $= E$

ABSORPTANCE $= 1 - I/I_o$

Fig. 17.12 Simultaneous measurement of radiance and absorptance.

tion of the system with a blackbody standard. The ratio of radiance to absorptance, according to Kirchkoff's law for a nonreflecting specimen, is the blackbody radiance, and from this the temperature can be calculated.

That is,

$$\frac{\text{sample radiance}}{\text{blackbody radiance}} = \text{emittance} = \text{absorptance}$$

from which

$$\frac{\text{sample radiance}}{\text{absorptance}} = \text{blackbody radiance}$$

This method is especially suited to transient phenomena because absorptance and radiance are measured simultaneously. An alternate method, useful when the experimenter has more time, is sketched in Fig. 17.13, where E_3 is the radiance of the sample, E_1 the radiance of the source, and E_2 the combined radiation due to source emission,

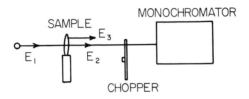

Fig. 17.13   Measurement of radiance and absorptance.

sample emission, and sample absorption. In terms of the sample absorption. In terms of the sample absorptance A

$$E_2 = E_3 + E_1(1-A)$$

from which

$$A = \frac{E_2 - E_3}{E_1} - 1$$

But E_3 is also given by Kirchhoff's law,

$$E_3 = AE_b = Ac_1\lambda^{-5}\left[\left(\exp\frac{c_2}{\lambda T}\right) - 1\right]^{-1}$$

where E_b is the radiance of a blackbody at the same temperature, as given by Planck's law. Thus, from experimental measurements of E_3 and A, the temperature can be calculated.

The emission spectrum of a flame can be recorded by using a burner or torch suitably constructed or modified to permit combustion of the gas under study. The collection of radiation can be made more efficient by placing a mirror behind the flame, with its center of curvature at the center of the flame, as in Fig. 17.14. This cannot be done with an opaque sample, of course.

A flat flame burner of the type used in the study of cool flames, as for example, in the study by Donovan and Agnew (21, 22) is shown in Fig. 17.15. The premixed components flow upward through honeycomb matrices and glass beads with a uniform velocity over a large area. A screen at the top helps stabilize the flame which exists between the windows. Either absorption or emission can be studied. An optical

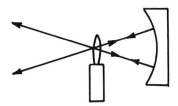

Fig. 17.14   Collection of radiation from a flame.

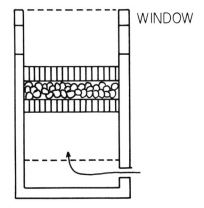

Fig. 17.15   Flat flame burner.

system is used to image the vertical slit of the monochromator on horizontal sections of the flame.

The emission from the flame of the exhaust of a rocket or jet engine, or farther downstream, from the hot gases of the exhaust, can be studied with an infrared spectrometer. As suggested earlier, fast reactions and explosions can be followed with a rapid-scan spectrometer.

Gases can be raised to very high temperatures and thereby made to emit radiation by means of radio-frequency dielectric heating, or electrodeless discharge as it is often called. The experimental arrangement of Cohen *et al.* (*23*) is sketched in Fig. 17.16. The external plate-like electrodes surround a quartz tube containing the gas. Windows are of sapphire or some similar material capable of withstanding high temperature. The electrodes are connected to a high-frequency genera-tor which is operating at perhaps 10 Mc/sec and generating a kilowatt or so of power. The rf field causes the molecules to undergo rapid trans-lational motion with frequent collisions and they are thereby raised to effective temperatures as high as 2000°K. They also tend to decompose,

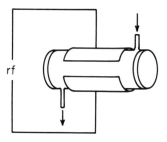

Fig. 17.16   Electrodeless discharge for excitation of gases.

so often a flowing system is used, with the gas sample being constantly replaced. An arrangement of this sort might also presumably be used for studying the absorption spectrum of excited molecules. Internal electrodes can be used to form a discharge too, of course, but corrosion of the electrode by reactive species in the gas can be troublesome.

REFERENCES

1. S. S. Penner, *Quantitative Molecular Spectroscopy and Gas Emissivities*, Addison-Wesley, Readings Mass., 1959.
2. R. H. Tourin, *Spectroscopic Gas Temperature Measurement*, Elsevier, Amsterdam, 1966.
3. H. H. Blau and H. Fischer, *Radiative Transfer from Solid Materials*, Macmillan, New York, 1962.
4. G. A. W. Rutgers, "Temperature radiation of solids," *Handbuch der Physik*, XXVI, Springer, Berlin, 1958, pp. 129–170.
5. A. Gouffe, *Rev. Opt.*, **24**, 1 (1945).
6. J. C. DeVos, *Physica*, **20**, 669 (1954).
7. F. E. Nicodemus, *Appl. Opt.*, **7**, 1359 (1968).
8. M. L. Fecteau, *Appl. Opt.*, **7**, 1363 (1968).
9. W. N. Harrison, J. C. Richmond, E. K. Plyler, R. Stair, and H. K. Skramstad, *WADC Tech. Rep.* 59-510, Aug. 1959 and Nov. 1960.
10. R. Goldstein and F. N. Mastrup, *J. Opt. Soc. Am.*, **56**, 765 (1966).
11. R. Stair, R. G. Johnston, and E. W. Halbach, *J. Res. NBS*, **64A**, 291 (1960).
12. R. Stair, W. E. Schneider, and J. K. Jackson, *Appl. Opt.*, **2**, 1151 (1963).
13. J. E. Stewart and J. C. Richmond, *J. Res. NBS*, **59**, 405 (1957).
14. M. J. D. Low and I. Coleman, *Spectrochim. Acta*, **22**, 369 (1966).
15. L. C. Block and A. S. Zachor, *Appl. Opt.*, **3**, 209 (1964).
16. J. E. Chamberlain, J. E. Gibbs, and H. A. Gebbie, *Nature*, **198**, 874 (1963).
17. M. J. D. Low and F. K. Clancy, *Environ. Sci. Tech.*, **1**, 73 (1967).
18. H. Goldring and W. E. Deeds, *Symposium on Molecular Structure and Spectroscopy*, Ohio State University, Columbus, Ohio, Sept. 6–10, 1966.

19. D. A. Dows and R. F. Porter, *J. Am. Chem. Soc.*, **78**, 5165 (1956).
20. G. J. Penzias, S. A. Dolin, and H. A. Kruegle, *Appl. Opt.*, **5**, 225 (1966).
21. R. E. Donovan and W. G. Agnew, *J. Chem. Phys.*, **23**, 1592 (1955).
22. R. E. Donovan, *J. Chem. Phys.* **27**, 324 (1957).
23. D. Cohen, R. Lowe, and J. Hampson, *J. Appl. Phys.*, **28**, 737 (1957).

CHAPTER 18

Advanced Methods of Infrared Spectroscopy

I. INTRODUCTION

In this final chapter we discuss a number of advanced methods of infrared spectroscopy. The topics selected are typical, not exhaustive, and, in keeping with the main theme of the book, the instrumental aspects are emphasized. It is hoped that the reader, in surveying these methods, will be able to apply them to his problems and to devise new problems which might be solved by using infrared spectroscopy.

II. THE EFFECT OF THE ENVIRONMENT OF THE SAMPLE

A great deal of research activity in infrared spectroscopy is devoted to studies of the effects of environment on the infrared spectrum of a material. Specifically, we refer to the influences of temperature, pressure, solvent, and concentration, and electric, magnetic, or electromagnetic fields.

A. Temperature

The temperature of a sample influences its infrared spectrum first of all by determining whether it is in the solid, liquid, or gas phase. The absorption bands of crystalline solids usually become quite sharp at low temperatures because excited vibrational states of molecules and lattices are less densely populated than at higher temperatures. This is also true of liquids, but the observed effects are not so striking because very low temperatures are usually not attainable. The vibration-rotation bands of gases are temperature sensitive because the population density of the rotational levels is temperature dependent and because line widths are strongly influenced by the rate of collision between molecules.

When distinct species can exist in a sample and the equilibrium among the species is a thermal one, then the change in the infrared spectrum with temperature can provide useful information about the nature of the components and their relative free energies. We have discussed this in connection with rotational isomers. Other systems which suggest themselves are those with hydrogen-bonded molecules, en-ol tautomers, and so on. The force constants of molecules which determine the vibrational frequencies of free molecules are not temperature dependent. However, the force constants of crystal lattices are temperature dependent because of the thermal expansion of the material and, consequently, the frequencies of the lattice vibrations are temperature dependent also. The frequencies of the Reststrahlen bands of the alkali halides, for example, shift to longer wavelengths with increasing temperature. The absorbance at wavelengths less than the Reststrahlen frequencies increases with increasing temperature.

The temperature dependence of the lattice vibrations of certain ferroelectric materials is especially striking. In $SrTiO_3$ (*1*), for example, a vibrational mode is observed at $100\ cm^{-1}$ at room temperature. The frequency varies with temperature according to the relationship

$$\nu^2 \propto T - T_0$$

It is observed at $40\ cm^{-1}$ at a temperature of $93°K$.

The absorption edges of semiconductors are temperature sensitive, and usually, but not always, shift to shorter wavelengths with reduced temperature. The magnitude of the shift is typically about $3\ cm^{-1}/°C$. At elevated temperatures the density of free carriers (electrons in the conduction band) increases and therefore so does absorption by free carriers. This is most noticeable in the far infrared because free-carrier absorption is proportional to the square of the wavelength.

One of the predictions of the modern theory of superconductors is that below the transition temperature T_c, where superconductivity occurs, a forbidden energy gap is created. The width of the gap is given by

$$E_g = 3.5\,kT_c$$

where k is Boltzmann's constant. The temperature T_c is never greater than about 18°K, and for most superconductors is much smaller; so transitions of electrons across the energy gap give rise to an absorption edge in the very far infrared at frequencies in the 10- to 30-cm^{-1} region. At frequencies below the absorption edge, the transmittance should rise, falling again to zero at frequencies approaching direct current where the reflectivity is unity. Such absorption edge behavior has been observed in the transmission spectra of very thin superconducting films of lead, tin, mercury, and indium by Tinkham (2, 3), Glover (2), and Ginsberg (3). Their measurements verified the formula, with a numerical constant of from 3 to 4. Richards and Tinkham (4) determined the energy gap of bulk superconductors by reflection measurements, extending the measurements to include vanadium, tantalum, and niobium. The experiments involve the detection of small differences in reflectivity between the normal and the superconducting states. In order to magnify the observed effect, an integrating conical cavity was constructed of the superconducting metal, as shown in Fig. 18.1. Radiation from the monochromator is transferred through a light pipe, condensed by a light cone, and led into the cavity. It is reflected numerous times inside the cavity before finally being detected by the bolometer. The multiple reflections from the cavity walls

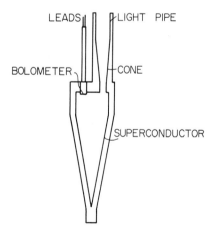

Fig. 18.1   Superconducting cavity.

enhance the effect of small changes in reflectivity. (See the discussion of integrating spheres, Section 16.III.) The cavity wall is made superconducting by immersing it in liquid helium. The material is made normal, that is, nonsuperconducting, without changing its temperature by applying a magnetic field.

B. Pressure

Increasing the pressure of a gas sample increases the frequency of collisions between molecules, and this causes the rotational lines to be broadened. This is often done intentionally in intensity studies and quantitative analysis to make the line width compatible with the resolving power of the spectrometer. At extremely high pressure the interactions between molecules may become vigorous enough to significantly disturb their structures. In this way some vibrations that are normally infrared inactive might be made to absorb infrared radiation. Absorption spectra of hydrogen, oxygen, nitrogen, and carbon dioxide gases at high pressure have been observed by Welsh et al. (5) and by Crawford et al. (6, 7). Oxholm and Williams (8), Asselt and Williams (9), and Smith et al. (10) have reported infrared absorption by liquid oxygen and nitrogen.

Solids have been studied at very high pressures. Wier et al. (11) and Whatley et al. (12) have used a cell with small diamond windows which transmit pressure as high as 50 kbars to the sample and permit the transmission of infrared radiation. The cell is sketched in Fig. 18.2. Diamond cells are now available from commercial suppliers. Other window materials have been used, including NaCl (13), MgO, quartz, and sapphire (14). Solids can undergo phase changes at high pressures, and the corresponding spectral changes supply useful clues to the nature of these changes. Organic liquids have been solidified at high pressures of up to 100 kbars by Brasch (15). Hydrogen-

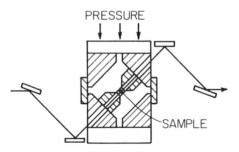

Fig. 18.2   Cell for studying samples under high pressure.

bonded crystals have been studied by Reynolds and Sternstein (*16*) and the effects of changes in the O—H and N—H bonds at high pressure (25 kbars) are observable near 3400 cm^{-1}. The effect of pressure on the O—H band of butanol solutions was observed by Drickamer and Fishman (*17*).

The absorption edges of semiconductors are expected to shift under pressure because of the change in lattice spacing. This has been observed experimentally in only a few cases (*18*).

C. Solvent and Concentration

The environment of molecules in solution is determined by the solvent and the concentration. Molecules which are hydrogen bonded into loosely bonded dimers and polymers when in the pure liquid state are separated into isolated molecules in dilute solutions in nonhydrogen-bonding solvents. Bands contributed by free molecules and H-bonded molecules can thus be distinguished.

Intermolecular forces need not be as well defined as hydrogen bonds to affect the infrared spectrum. Local electric fields due to permanent or induced electric dipoles also perturb vibrational frequencies and intensities. For example, the carbonyl stretching vibration near 1750 cm^{-1} in acetylated sugars is much wider and more intense in chloroform solution than in carbon tetrachloride.

The difference in electrical environment in different solvents and concentrations can affect equilibria among rotational isomers. The 1,2-dihalogen-ethanes have two isomeric forms, *trans* and *gauche*. The gauche form has a permanent dipole moment. In a polar solvent the gauche form is relatively more stable and therefore more populous than in a nonpolar solvent.

The optical properties and infrared spectra of semiconductors are dependent on the kind and concentration of impurity atoms. Usually the intrinsic energy gap is not sensitive, an exception being *n*-type indium antimonide. Excitation of electrons from impurity atoms to the conduction band are observable as an absorption continuum or, at very low temperature, as discrete lines.

D. Electric Fields

If the local internal electric fields in a sample can affect the infrared spectra so profoundly, one might expect that an externally applied field could produce similar effects. Long ago, in 1932, Condon (*19*) derived a quantum mechanical theory from which he concluded that a sufficiently high electric field would alter the selection rules in such a way that Raman-active vibrations would become infrared active. This effect would be of great assistance to infrared spectroscopists, but the field strength required was rather high (thousands of volts per centimeter) and no experiments were done for a long time. In 1950 Woodward (20) and Kastler (21) discussed the effect, again theoretically. Dows and Buckingham (22) have also given a theory.

In 1953 Crawford and Dagg (23) observed absorption by hydrogen molecules induced by an electric field. Their cell had electrode plates 85 cm long, separated by 1.4 mm. A potential of 14,000 V was applied to the electrodes. Hydrogen was introduced to a pressure of 2000 psi. The spectrum was observed with a grating spectrometer using a lead sulfide cell. Later work on hydrogen was reported by Crawford and MacDonald (24) and by Terhune and Peters (25).

More recently, in 1966, Anastassakis et al. (26) observed electric-field-induced infrared absorption in diamond. The Raman transition at 1336 cm^{-1} was made infrared active by an ac field of 1.2×10^5 V/cm. A grating spectrometer (Perkin–Elmer 12C) was used with a thermocouple and a coherent amplifier locked in to the frequency and phase of the applied field. An absorption of 2.5% was observed with a signal-to-noise ratio of 4 to 1. In addition a 4-cm^{-1} frequency shift was observed. Negative results were obtained with a sample of silicon at a field strength of 10^4 V/cm. Higher fields produced arcing.

Presumably, the equilibrium between a polar and a nonpolar rotational isomer can be perturbed by applying an electric field, and it might be possible to detect this from the infrared spectrum. The limit to the field strength that can be applied is determined by the dielectric strength (breakdown field) of the sample. An elegant way to detect small differences in absorption is by means of a phase-sensitive (lock-in) amplifier. In this instance, an ac electric field might be applied to the sample with the signal from the detector synchronously rectified in phase with the electric field. No radiation chopper is used. The normal absorption spectrum will not be observed, but the difference in absorption with and without the electric field might be detected. This method has been used to detect absorption by excited molecules and is discussed further below. It has evidently not been used to detect isomer equilibria in an electric field, as we propose here. It would be an interesting experiment.

Shifts in the frequencies of the O—H bands of alcohols have been induced with an electric field by Handler and Aspnes (27). They used an ac field of 52 kV/cm and synchronous detection. Related experiments immediately suggest themselves, such as the study of ferro-electric compounds, and so on.

The *Kerr effect* and the *Pockels effect* are well known and are frequently studied and used in the visible spectral region. If an electric field is applied to certain specimens in a direction normal to the direction of travel of the incident beam of radiation, as in Fig. 18.3,

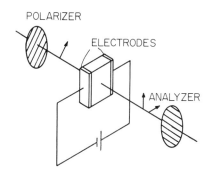

POLARIZER

ELECTRODES

ANALYZER

Fig. 18.3   Experimental arrangement for studying Kerr effect.

the field induces a birefringence in the sample. If the beam is plane polarized initially at an angle of 45° relative to the electric field, it will be found to be elliptically polarized when it leaves the sample. This is the *Kerr electrooptic effect.* The amount of ellipticity depends on the field strength, and of course on the nature of the specimen. An analyzer placed as shown, with its polarizing axis crossed with the axis of the polarizer, will extinguish the radiation when no field is applied to the specimen. When the field is applied, and the transmitted beam is made elliptically polarized, a portion of the radiation will be passed by the analyzer. The ellipticity can be measured by either preceding or follow-ing the sample with a quarter-wave device and placing an analyzer after the sample. The orientation of the quarter wave device and analyzer are adjusted until the radiation is extinguished. The eccen-tricity of the ellipse, representing the elliptically polarized radiation, and the rotation of the major axis of the ellipse can be determined from the angular setting of the quarter-wave plate and analyzer.

The *Kerr effect* is often used to make high-speed shutters and light modulators for the visible region. The liquid nitrobenzene is a frequent choice of material for visible Kerr shutters, and it appears to be the only liquid whose Kerr effect has been studied in the infrared [by Charney and Halford (*28*) in 1958].

An electric field applied to certain classes of uniaxial crystals causes them to become biaxial. A beam of plane-polarized radiation traveling along the optic axis of a uniaxial crystal does not become elliptically polarized, although the plane of polarization may rotate. With the application of an electric field parallel to the axis (i.e., in the direction of propagation of the radiation, as in Fig. 18.4), the crystal

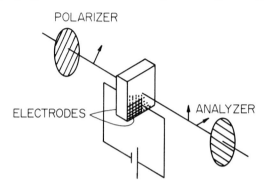

Fig. 18.4 Experimental arrangement for studying Pockels effect.

becomes biaxial and birefringent to the radiation. This is the *Pockels effect* or electric double refraction. The electric field can be applied by depositing semitransparent metal electrodes on the sample or by placing grid electrodes on the surfaces. The magnitude of the birefringence is proportional to the field strength and the path length through the sample. Note, however, that for a fixed potential applied to the surfaces of the sample, the field strength is inversely proportional to the path length. Therefore, the birefringence is proportional to the applied potential and independent of the path length. The effect is, of course, strongly dependent on the nature of the sample. The Pockels effect is also used in making shutters and modulators. The crystals ammonium dihydrogen phosphate and potassium dihydrogen phosphate have been used in the near infrared.

If an atom or molecule having degenerate energy levels is placed in an electric field, the degeneracy is removed in some cases, and the levels split. This can be observed as a splitting of emission or absorption lines and is called the *Stark effect*. The effect is observed frequently in the visible and ultraviolet and also in the microwave regions, but rarely in the infrared because of the relatively high resolution required to detect it. Infrared Stark effect has been observed by Sakurai and Shimoda (*31*) using, instead of a monochromator, a helium–neon laser emitting an extremely narrow line at $3.39\,\mu$. A very narrow region of spectrum (0.15 cm^{-1}) was explored by scanning the laser with a magnetic field. A dependence of absorption on the magnitude of an electric field applied to the sample is due to Stark effect.

An electric field applied to a semiconductor can affect the width of the forbidden energy gap. Thus a shift in the wavelength of the absorption edge can be produced with an electric field. This is the *Franz–Keldysh effect*.

E. Magnetic Fields

The application of a magnetic field to a specimen can alter its effect on an incident beam of infrared radiation. Reviews of magnetooptical effects have been given by Palik (*29, 30*) and Henvis (*30*). Usually rather substantial magnetic fields are required, thus requiring the use of large electromagnets. Fields greater than 100 kG are used, for example, in the National Magnet Laboratory in Cambridge, Massachusetts. Such magnets are large and massive. The spectroscopic instruments are usually installed where the magnet is located, rather than carrying the magnet to the spectrometer. Superconducting solenoid magnets can be used to generate fields up to about 80 kG and they are substantially less bulky than conventional electromagnets. Careful design of transfer optics is required to carry an infrared beam from the source through the sample in the magnet pole gap and into the monochromator.

A beam of plane-polarized radiation can be passed through a specimen, with a magnetic field applied along the direction of propagation, either by enclosing the specimen in a coil, as in Fig. 18.5A, or by placing it between the pole pieces of a magnet, as in Fig. 18.5B. The specimen becomes optically active in the magnetic field. The plane of polarization rotates in proportion to the field strength and path length. This is the *Faraday effect*, or magnetooptical activity. Its variation with wavelength is magnetooptical rotatory dispersion. The Faraday

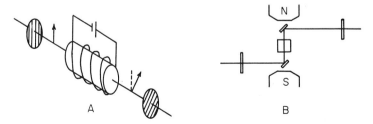

Fig. 18.5 Experimental arrangements for studying Faraday effect.

effect has been studied for some organic molecules, for example, by Hediger and Günthard and Wyss and Günthard (*32*). The Faraday effect is also exhibited by free carriers in semiconductors. With the magnetic field vector pointing in the same direction as the direction of propagation of the beam, electrons rotate the plane of polarization in a counterclockwise sense, looking toward the source, provided the frequency is greater than the cyclotron frequency (the cyclotron frequency is discussed below). The sense of rotation is reversed when the magnetic field is opposite to the direction of propagation. Positive carriers and electrons rotate the plane of polarization in opposite directions. The effective mass of free carriers can be determined from Faraday-effect measurement.

Plane-polarized radiation can be considered as a superposition of left- and right-circularly polarized radiation. If there is a difference in refractive index for the two circularly polarized components in a magnetic field, the resulting plane-polarized beam rotates its plane of polarization as it is propagated, and the Faraday effect is observed. This can also be observed as a difference in the frequency of interference fringes of a thin semiconductor film in a magnetic field observed with left- and right-circularly polarized radiation. If, in addition, there is a difference in absorption of the circularly polarized components, the resultant is elliptically polarized. This is *magnetocircular dichroism*. This has been measured in semiconductors, but evidently has not yet been detected in vibrating molecules or lattices.

Faraday effect can be measured by manual setting of a polarizer or analyzer. Special spectrometers have been devised for making these measurements automatically. The instrument of Yoshimoto and Mochida (*33*) is illustrated schematically in Fig. 18.6. Radiation from the source is dispersed with a monochromator. It is then separated into two beams with a rotating mirror, passed through polarizers *P* and *P'*, and recombined. The polarizers are oriented with their axes 90° apart, so the sample is illuminated alternately with radiation beams polarized normal to each other. An analyzer *A* follows the sample. When the sample does not rotate the plane of polarization, and the analyzer is oriented with its axis bisecting the axes of *P* and *P'*, switching of the beams does not produce a modulation of the signal at the detector. When the magnetic field is applied, the plane of polarization is rotated by the sample. The radiant power passed by the analyzer is increased during the time the beam is traversing one polarizer, and decreased when the other polarizer is used. An ac signal is thereby generated. The signal can be synchronously rectified and used to drive

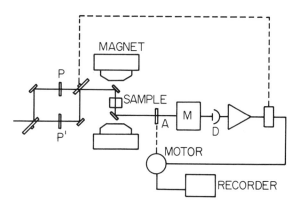

Fig. 18.6 Faraday effect spectrophotometer.

a servomotor which rotates the analyzer until the ac signal ceases. Yoshimoto and Mochido have demonstrated an angular sensitivity of 0.01°. A magnetorotation spectrum is recorded when the monochromator is scanned. A slit servo maintains the energy level of the instrument by adjusting the slit width to preprogrammed values.

A slightly different approach is used in an instrument described by Wyss and Günthard (*34*). Instead of switching the radiation beam between two differently oriented polarizers, a single polarizer is used, and it is oscillated at 1 cps with a small (8°) angular amplitude. A phase-sensitive amplifier treats the signal and drives a servo which positions the analyzer. The analyzer position is repeatable within 0.015° (root mean square).

An infrared spectropolarimeter has also been described by Pidgeon and Smith (*35*).

If a specimen is placed between the pole pieces of a magnet, as in Fig. 18.7, it becomes doubly refractive. Incident radiation is polarized

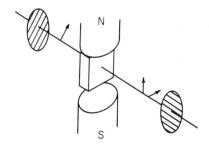

Fig. 18.7 Experimental arrangement for studying Voigt effect.

45° to the direction of the magnetic field. After passing through the sample, the radiation is elliptically polarized. This is the *Voigt effect*, or magnetic double refraction. It is similar to the Kerr electrooptic effect. It has evidently not been observed for molecular or lattice vibrations. The effect in semiconductors permits the determination of the effective mass and concentration of the free carriers.

In some liquids with permanent or induced magnetic moments, the application of a magnetic field causes them to be aligned with respect to the field. If the molecules are also optically anisotropic, the liquid becomes doubly refracting, as in the Voigt effect. This is called the *Cotton–Mouton effect*, and it does not seem to have been observed in the infrared.

Energy levels of atoms or molecules split in a magnetic field because the angular momentum vector can assume only discrete directions relative to the magnetic field, and the energy of the system depends on the plane of electron orbits relative to the field. This leads to a splitting of emission or absorption lines, or the *Zeeman effect*. It is the magnetic analog of the Stark effect. It can be observed in the emission spectra of atoms. The absorption edge of semiconductors can be shifted by applying a magnetic field. In indium antimonide, the shift amounts to an increase of about $1.8 \text{ cm}^{-1}/\text{kG}$, and fairly intense fields are required to observe it.

The valence and conduction bands of a semiconductor coalesce into so-called Landau levels in a magnetic field, giving rise to discrete absorption or reflection transitions. This is the *oscillatory magnetooptic effect*. These are observed very close to the absorption edge, and quite thin films are required to make the region accessible for transmission measurements. Strong fields are required to make the observation. This effect is also observed by studying the radiation reflected from a specimen in a magnetic field.

The *Kerr magnetooptic effect* is also observed in reflected radiation. A beam of plane-polarized radiation is incident upon a specimen whose surface is normal to the direction of an applied magnetic field. The reflected radiation is elliptically polarized. The plane of polarization of the incident beam is either in or normal to the plane of incidence to avoid confusion with the ellipticity that would otherwise be introduced in a polarized beam incident at a nonnormal angle even without the field.

Free carriers execute a circular motion in a magnetic field, much like the action of a cyclotron in forming a beam of electrons in a circular orbit. This is called *cyclotron resonance*. The frequency, in wavenumbers, is

$$\nu_c = \frac{eH}{m^*c}$$

where e is the charge of the free carrier and m^* its effective mass, c is the velocity of light, and H is the magnetic field strength. If the lifetime of the carriers is greater than a period of cyclotron rotation, the effect can be observed as an absorption or reflection maximum at frequency ν_c. Cyclotron resonance measurements are very important in establishing the effective mass of carriers. When the radiation is propagating parallel to the magnetic field, electrons absorb left-circularly polarized radiation, and holes absorb right-circularly polarized radiation, making it possible to discriminate between electron and hole absorption. When the magnetic field is normal to the direction of radiation propagation, only the component of the incident beam having its electric vector perpendicular to the magnetic field is absorbed.

Cyclotron resonance frequencies usually occur in the far infrared. A field of 60 kG is required to cause cyclotron resonance at about 250 cm^{-1} in InSb. Pulsed fields of 300 kG have been used at 1000 cm^{-1}. Cyclotron resonance can be observed by scanning the spectrum with a fixed magnetic field or by changing the field strength while observing a fixed wavelength. The availability of infrared lasers makes the latter approach especially desirable in studies of cyclotron resonance and other magnetooptical effects.

There is a characteristic plasma frequency associated with a plasma of free carriers. It is given by

$$\nu_p = \frac{e}{\sqrt{\pi c}} \left(\frac{n}{m^*}\right)^{1/2}$$

where e and m^* are, as before, the charge and the effective mass, respectively, and c is the velocity of light. The quantity n is the number of free carriers per unit volume. At frequencies above ν_p the medium is largely transparent, but at lower frequencies incident radiation is highly reflected. There is a minimum in the reflectivity at the plasma frequency. The plasma frequencies of metals lie in the ultraviolet. In semiconductors the plasma frequency is usually in the infrared, at frequencies considerably less than 1000 cm^{-1}. The minimum in the reflectivity in the neighborhood of ν_p shifts when a magnetic field is applied. This is called a *magnetoplasma effect*. With negatively charged free carriers and a magnetic field in the direction of radiation propagation, the plasma frequency for right-circularly polarized radiation shifts to lower frequencies, and for left-circularly polarized radiation it shifts to higher frequencies. The shifts are reversed for a reversal in field direction or in sign of the carriers. The magnitude of the shift is

$$\Delta \nu_p = \pm \nu_c + \frac{1}{8} \frac{\nu_c^2}{\nu_p}$$

where ν_c is the normal cyclotron resonance frequency. Observation of magnetic plasma shifts thus provide an alternate method of measuring the cyclotron resonance frequency.

F. Electromagnetic Fields

The effect of an incident electromagnetic field in producing excited species is discussed in a later section.

Interesting effects have been observed in the visual spectral region when an intense light beam from a laser falls on certain materials. Suppose the material is nonlinear in the sense that the dependence of its polarization on field strength contains higher-order terms,

$$p = \alpha E + \alpha' E^2 + \cdots$$

If the electric field of the incident radiation varies with frequency ν, the induced polarization will have components of frequency 2ν, 3ν, and so on. If the coefficients of the higher-order terms are large enough and if the incident beam is strong enough, harmonics of the incident frequency can be observed.

These observations have now been extended to the infrared. Patel (36), at the Bell Telephone Laboratories, applied the radiation from a CO_2 laser at 945 cm^{-1} to a sample of tellurium. The incident power from the laser was 0.17 W. Patel observed, in addition to the 945-cm^{-1} radiation transmitted by the sample, 10 μW of radiation at a wavelength of 1890 cm^{-1}. He was later able to observe harmonic generation in many crystals using a Q-switched CO_2 laser delivering 10kW in 300-μsec pulses. A copper-doped germanium detector was used because of its fast response time.

When the radiation from two lasers of different frequencies is introduced into a nonlinear substance, in addition to the harmonics of each of them, sum and difference combination frequencies occur. This has not yet been applied to the infrared, but the suggestion has been made that a tunable source of far-infrared radiation might be developed in this way. It has also been suggested (37–39) that the coherent molecular vibrations and plasmas generated by stimulated Raman scattering should be accompanied by discrete emission in the infrared.

Finally, we conjecture that with the availability of intense lasers for the infrared, such as the carbon dioxide, hydrogen, or cyanide lasers, many of the electrooptical phenomena discussed above might become observable under laser illumination.

III. INFRARED SPECTROSCOPY OF EXCITED STATES OF MOLECULES

After it has absorbed a quantity of energy, a molecule exists in an excited electronic, vibrational, or rotational state. It can return to the ground state by reemitting the energy in a single or in several steps. This is fluorescence radiation. It can also redistribute the energy into thermal energy by colliding with other molecules. It is interesting to measure the spectrum of the reemitted radiation and the time required for it to be emitted. The properties of an excited molecule are, in general, different from the properties of the same molecule in the ground state, and it is often desired to measure these. Molecules can also be in excited states by virtue of being raised to a high temperature or by undergoing a chemical change which leaves them temporarily in an excited and unstable state. In this section we describe some examples of infrared spectroscopy of excited states.

A. Infrared Fluorescence and Decay Times

Most vibrationally and rotationally excited molecules in the gas phase at moderate pressures lose their excess energy very rapidly by collision before a significant amount of fluorescence radiation is emitted. This appears as heat and raises the temperature of the gas. If N_0 molecules are suddenly excited at time t_0 the number of molecules remaining in the excited state after a time $t - t_0$ is

$$N = N_0 \exp\left[-\frac{(t-t_0)}{\tau}\right]$$

where τ is the decay time. This is quite analogous to the time constant of an electrical circuit, and can be measured with analogous techniques. That is, we can irradiate a gas with an impulse of radiation and measure the temperature as a function of time. Alternatively, we can irradiate the gas periodically and measure the periodically varying temperature. The phase shift varies as a function of frequency and can be used to determine the decay time.

The impulse response method might be carried out (see Fig. 18.8) by flowing a gas through a long tube at a known rate and illuminating it

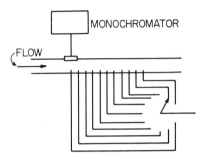

Fig. 18.8   Arrangement for studying luminescence decay in a flowing sample.

at a convenient point with a steady beam of radiation of the desired wavelength. The temperature of the gas is measured with sensitive detectors at known distances downstream from the point of illumination. This gives the information needed to compute τ. This method has apparently not been used in actual experiments.

The phase-shift method is exemplified by the apparatus of Fig. 18.9. A beam of radiation is chopped at a desired modulation frequency. The gas within the enclosure responds by displaying a sinusoidally varying temperature and pressure, which is monitored by observing the distortion of a delicate membrane either capacitively or optically. Instruments of this type are called spectrophones. They have been used, for example, by Jacox and Bauer (40) who studied relaxation processes in carbon dioxide.

In some molecules radiative decay dominates over collisional de-excitation even at elevated temperatures. In this case the intensity of fluorescence radiation is measured to determine the decay time.

In the case of carbon monoxide the delay is unusually long, amounting to about 33 msec. It has been observed in three different ways. We discuss these briefly because they may have general usefulness, although they are most easily applied to slowly decaying molecules.

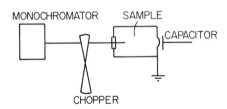

Fig. 18.9   The spectrophone.

Hooker and Millikan (*41*) excited CO molecules in a shock tube and observed the emitted radiation at the frequencies of the fundamental and the first overtone vibrations. The intensity was studied as a function of time after the shock passed. Later Millikan (*42*) excited a rapidly streaming CO sample by exposing it to an intense beam of infrared radiation. The reemitted radiation was observed at various positions downstream from the point of excitation. McCaa and Williams (*43*) also excited carbon monoxide with an infrared source (a carbon arc) but discriminated between exciting and fluorescence radiation with a double chopper arrangement, sketched in Fig. 18.10. The cell is surrounded by a rotating cylindrical chopper. Exciting radiation enters through the sector chopper at the end of the cylinder. Fluorescence radiation is observed through the slit opening in the side of the cylinder, out of phase with the exciting radiation.

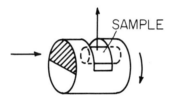

Fig. 18.10   Double chopper for separating exciting and fluorescence radiation.

The use of lasers in exciting vibrational fluorescence of molecules has been reviewed by Moore (*44*).

B. Absorption by Excited Molecules

A molecule in its ground state absorbs a photon and is raised to an excited state. If the excited molecule encounters another photon before it decays back to the ground state, it might be raised to an even higher excited state. The absorption spectrum of an excited molecule can be vastly different from the spectrum of a molecule in the ground state. Thus, the difference bands in a vibrational spectrum are found at frequencies

$$\nu_d = \nu_1 - \nu_2$$

where ν_1 is the usual frequency of a vibration and ν_2 is the vibrational frequency of the excited molecule. The intensity of the difference bands of a sample can be modified by varying the relative concentration of molecules in the excited state. This is usually done by changing the temperature of the sample. In principle, it could also be done by permitting the molecules to absorb radiation from an intense beam. This possibility was discussed earlier in Section II.F. Similar remarks apply to the rotational spectra of molecules and to the rotational fine structure in the vibration–rotation bands.

Hexter (45) investigated the vibrational spectra of several molecules in excited electronic states. These are molecules which display phosphorescence at low temperature. They are excited to an electronic singlet state by absorption of a quantum of ultraviolet light (see Fig. 18.11). A small portion of the molecules falls into an electronic triplet

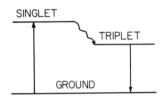

Fig. 18.11 Phosphorescence radiation from excited triplet level.

state through some radiationless transition involving exchange of energy with neighboring molecules. These molecules are unable to decay promptly to the singlet ground state because of selection rules. Eventually, after an average delay of milliseconds to seconds, they do fall to the ground state with the emission of a quantum of phosphorescence radiation. While in the excited triplet state they can absorb infrared radiation, which causes them to vibrate. The vibrational spectrum is quite different from the spectrum of molecules in the ground electronic state. Only a small fraction of the molecules is in the excited state at a given instance. It is necessary to discriminate between the spectrum of excited molecules and the spectrum of ground-state molecules. Hexter modulated the exciting ultraviolet radiation, but not the infrared beam. The number of excited molecules in the cell varied periodically, while the much higher number of ground-state molecules was essentially constant. Thus, absorption of the infrared beam by excited molecules was modulated, but not absorption by ground-state molecules. A phase-sensitive amplifier locked in to the exciting radiation was used to extract the signal. Hexter was able to observe and assign the spectra of acetone and naphthalene. The $C{=}O$ stretching vibration of acetone, for example, which is found at $1720\,cm^{-1}$ in the electronic ground state, shifts to $1300\,cm^{-1}$ in the first excited triplet state.

C. Excited Chemical Species and Free Radicals

Molecules can be promoted into excited, unstable states during a chemical reaction. Molecular fragments or free radicals can also be formed during chemical reactions. The vibrational spectra of these species can sometimes be observed during the reaction. For example, the emission or absorption spectra of molecules in flames, explosions, or shock waves are studied, sometimes with a fast scanning spectrometer.

Alternatively, unstable molecules and free radicals might be collected and stored for later, more leisurely observations. This is made possible by the matrix-isolation method (46, 47). The free radicals are swept from the reaction zone with a stream of inert gas. The mixture impinges on a very cold window, cooled with liquid helium in a Dewar cell. The mixture promptly freezes and the free radicals are trapped and immobilized in a matrix of frozen inert gas. It is an art to adjust the concentration, flow rate, and so on, in order to trap a sufficient number of radicals to permit observation and to prevent them from recombining or reverting to a stable form. Radicals can also be formed in the cell by photolysis of a sample of a low concentration of stable molecules trapped in an inert matrix.

Molecules adsorbed on a surface are held by forces which may be physical (i.e., electrical) or chemical. The study of the infrared absorption spectra of adsorbed molecules can offer a clue into the nature of these forces [see (33) and (34) of Chapter 1]. The substrate is usually a strongly absorbing material, making the study of a monomolecular adsorbed layer very difficult. In one of the first studies of this type Plisken and Eischens (48) used a bed of extremely fine particles of silica on which various gases were adsorbed. Much of the infrared spectrum was inaccessible because of the strong bands of silica, but in the clear regions it was possible to observe, for example, that the $C\!=\!C$ band of ethylene loses its double-band character. Evidently, a real chemical band is formed between the hydrocarbon and the substrate. This is particularly important in studies of chemisorption on catalysts.

IV. ELECTRON BEAMS AND PLASMAS

An ionized gas is called a plasma. In Section II of this chapter we described some experimentally observable effects due to the interaction of radiation with free electrons in solids. In particular, we mentioned cyclotron resonance, the plasma frequency, and various electrooptic and magnetooptic effects. Related phenomena are observed with free electrons in a gaseous plasma.

A. Plasmas

The plasma frequency for conduction electrons is in the ultraviolet in metals and in the infrared in semiconductors. Its location depends on the density of free charges. A plasma frequency also exists for gaseous plasmas. At frequencies lower than the plasma frequency, radiation is strongly reflected, just as in metals. If the electron density is less than about 3×10^{14} cm^{-3} the plasma frequency is found in the microwave region, and microwave techniques are commonly employed in plasma diagnostics. However, in denser plasmas with 10^{14}–10^{16} electrons per cubic centimeter, the plasma frequency is in the longer-wavelength infrared, and infrared measurements are appropriate. Brown et al. (49) have devised a far-infrared interferometer operating at 33 cm^{-1} to measure these plasmas. They determine the phase shift of the infrared radiation as it passes through the plasma. There is a plasma frequency for the ions of a gaseous plasma also, but it occurs at radio frequencies.

Cyclotron resonance is also observed when gaseous plasmas are subjected to magnetic fields. Once more, the effect can be observed in the infrared with sufficiently dense plasmas and intense fields.

Bremsstrahlung is the name applied to radiation associated with collisional acceleration of charged particles (the word translates into "braking radiation"). The intensity of bremsstrahlung at a given frequency and the value of the plasma frequency are related. Kimmett and Niblett (50) measured the far-infrared bremsstrahlung in the 40- to 140-cm^{-1} region from a theta-pinched deuterium plasma as a means of determining the electron density in the plasma. The pinch plasma is confined by a magnetic field. It is a pulsed device, and a fast indium antimonide detector is used to time-resolve the measurements.

The Russian workers Dushin et al. (51) observed bremsstrahlung in the 1050- to 5500-cm^{-1} region from an electrodeless induction discharge in hydrogen, using a gold-doped germanium detector at liquid-nitrogen temperature. They obtained the plasma density from their measurements.

Faraday rotation has been observed in a theta-pinched plasma by Craig et al. (52). Their experimental arrangement is sketched in Fig. 18.12. They use a He–Ne laser as a source of 2950-cm^{-1} radia-

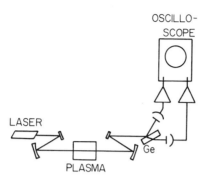

Fig. 18.12   Observation of Faraday rotation in a plasma.

tion. A germanium plate at the Brewster angle separates parallel and perpendicular components of the radiation transmitted by the plasma. These components are detected separately with two indium antimonide photovoltaic detectors, and the Faraday rotation is determined from the relative intensities. The plasma is pulsed in 3.5 μsec.

B. Electron Beams

A beam of electrons emits radiation if the electrons are accelerated or if they interact with matter in some fashion. If a beam passes very close to the surface of a dielectric, the dielectric provides a mechanism for removing energy from the beam in the form of electromagnetic radiation. This is Cerenkov radiation. It has been observed at a wavelength of 1.25 cm in the microwave region. It is also observed in the visible region when, for example, highly energetic charged particles enter the moderating water of an atomic reactor. Infrared Cerenkov radiation has evidently not yet been observed.

The possibility of producing radiation similar to Cerenkov radiation by passing an electron beam through, or past, a periodic structure has been discussed by several authors, especially Frank (53), Motz (54), and Salisbury (55). Smith and Purcell (56) succeeded in producing visible radiation by passing a very energetic beam of electrons past the surface of a diffraction grating. Ishiguro and Tako (57) reported on the observation of infrared Smith–Purcell radiation. More or less rigorous theories of Smith–Purcell radiation have been given by numerous authors. Let us briefly review an elementary theory based on Huygen's principle with reference to Fig. 18.13. An electron passes with velocity v across a grating of spacing b. A positive electrostatic

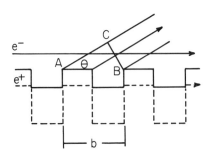

Fig. 18.13 Smith–Purcell radiation.

image of the electron is induced beneath the metal surface. As the electron passes the lands and grooves of the grating, the image moves in and out as shown, thereby producing a pulsating electric dipole. An oscillating dipole emits radiation. The time required for a wave front to propagate from A to C is $b \cos \theta / c$. The time for the electron to travel from A to B is b/v. Therefore, a plane wave is generated and propagates away from the grating whenever

$$\frac{b}{v} - \frac{b \cos \theta}{c} = \frac{k}{f} = \frac{kc}{\nu}$$

where $k = 1, 2, \ldots$ is the order number, analogous to the diffraction order, and $f = \nu/c$ is the frequency (in cycles per second) of the radiation.

The author has tried unsuccessfully to repeat the experiment of Ishiguro and Tako. Rigorous theory (58) tells us that the electron beam must pass extremely close to the surface of the grating. This imposes a serious experimental difficulty, partly because charge tends to collect on the grating and repel the beam, and partly because electrons striking the surface cause severe damage very quickly.

It is interesting to consider a grating coated with a thin film of conducting material having a high mobility and long mean free path, for example, a superconducting metal or indium antimonide at liquid-helium temperature. Will Smith–Purcell radiation be generated by a stream of conduction electrons passed through the film as they oscillate up and down the hills and dales of the grating?

V. MEASUREMENTS OF OPTICAL CONSTANTS IN THE INFRARED

The optical properties of isotropic materials are describable in terms of two parameters. These parameters might be the refractive index and the absorption coefficient or the real and imaginary dielectric constants. The amplitude and phase change of an electromagnetic wave that has traversed or been reflected by a substance is expressed in terms of these parameters. In this section we discuss some of the methods used in determining the infrared optical constants.

Let us start with homogeneous, isotropic transmitting samples. The absorption coefficient of a single plane-parallel solid sample or a liquid sample in a cell is most directly calculated from the Bouguer–Lambert law

$$I/I_s = e^{-Kb}$$

This is *not* correct because it assumes that attenuation of the incident radiation is only by absorption, neglecting reflection losses. However, if two or more thicknesses of sample are measured, the absorption coefficient can be determined from a more general equation which includes reflection

$$I/I_0 = (1-R)^2 e^{-Kb}$$

The absorption coefficient can be found from

$$-\frac{\partial \ln (I/I_0)}{\partial b} = K$$

This equation is not strictly correct either, because it neglects multiple internal reflections. It is a good approximation for weakly reflecting specimens.

When account must be taken of multiple reflections, but interference does not take place, the general equation

$$\frac{I}{I_0} = \frac{(1-R)^2 e^{-Kb}}{1-Re^{-Kb}}$$

is appropriate. The reflectivity must be determined somehow and introduced into this equation.

The obvious and direct way of measuring the refractive index of a transmitting material is by measuring the angles of incidence and refraction and making use of Snell's law. A solid sample can be ground into a prism of known apex angle and a liquid can be placed in a hollow fabricated prism. The hollow-prism method for liquids was used many years ago by Rubens (59) and later by Pfund (60) and Pittman (61). The prism is mounted on a spectroscope table. An infrared detector replaces the eyepiece and the entrance slit is illuminated with monochromatic radiation from a monochromator. For each wavelength, the angle of minimum deviation is searched out and the refractive index is calculated from the prism equation

$$\sin \theta'_1 = n \sin \alpha/2$$

Jaffe (62) has used a hollow prism followed by a Fabry–Perot etalon to generate a set of fringes for fiducial markers. The prism deviates the beam. By rotating the etalon, the beam is returned to its original state, and by counting fringes, the displacement of the beam position is determined. If the material is strongly absorbing, this method is not valid and not practical anyhow. Even when the absorption coefficient is small, the transmission of a prism at its base may be small, necessitating the use of a small prism angle or the limitation of measurements to the sharp end of the prism.

Another method of determining the refractive index is from the angle of total internal reflection. This method has been used successfully by Jaffe and his co-workers (63, 64). In the case of liquid samples the apparatus of Fig. 18.14 can be used. A thin layer of sample is

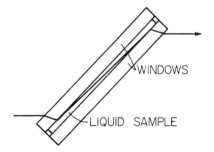

Fig. 18.14   Refractive index determination by total internal reflection.

placed between windows of refractive index which is known and which is less than the index of the sample. This method might also be used for solid slabs or evaporated films in optical contact with the windows, perhaps with a contacting liquid. The assembly is rotated until an angle is reached at which the beam can no longer escape because it is totally internally reflected. The refractive index of the specimen is calculated from the cutoff angle. In absorbing liquids the absorption coefficient must also be considered in calculating the cutoff angle. There is some uncertainty in setting the angle because the angle of incidence of the beam necessarily has a finite range of values. Furthermore, a fringe-like behavior of the cutoff edge is observed because of multiple reflections with interference. These are the *Herschel Fringes.*

One can also devise methods for measuring refractive indices based on Brewster's angle for dielectrics or the pseudo-Brewster angle at conducting surfaces.

Kagarise and Mayfield (65, 66) constructed a Fabry–Perot etalon with plates of germanium- or bismuth-coated calcium fluoride. Its spectrum is scanned and the positions of the Edser–Butler interference fringes are measured. The etalon is then filled with a liquid sample and the shift in fringe position is determined. From these the refractive index may be calculated. If necessary, successively more concentrated solutions of the sample can be used to follow the shifting fringes while maintaining identification of their order number. Account must be taken of the phase shift at the reflecting surfaces.

A similar method can be used for determining the refractive index of a plane-parallel plate of solid material. In this case the exact order of interference of the fringe is not known. The index can be determined from the fringe equations for adjacent fringes,

$$2nd = k\lambda_1 \quad \text{and} \quad 2nd = (k+1)\lambda_2$$

from which

$$n = \frac{1}{2d}\frac{\lambda_1\lambda_2}{\lambda_1 - \lambda_2}$$

Moss (67) points out that this measurement gives erroneous results if n depends linearly on wavelength.

There are two generally used methods for measuring the refractive index of a gas: by measuring the refraction in a hollow prism or the phase change in an interferometer. Jaffe (68) has reviewed some early as well as recent measurements. Legay (69) uses a Michelson interferometer with a gas cell in one arm and illuminated with a grating monochromator. Lithium fluoride lenses and a lithium fluoride-selenium beam splitter limit his measurements to the near infrared. Legay either introduces gas slowly while recording fringes at a fixed wavelength as a function of pressure or he fills the cell and scans wavelength to record a channeled spectrum. Jaffe et al. (70) use a high-resolution grating monochromator to illuminate a hollow prism containing the gas under study.

The optical constants of reflecting samples are determined by making use of Fresnel's equations. The refractive index of a nonabsorbing specimen in air can be determined from a measurement of the reflectivity at normal incidence

$$R = \left(\frac{n-1}{n+1}\right)^2$$

If the absorption coefficient is not zero, two measurements are sufficient to determine both n and k. There are several practical ways of doing this. The reflectivity of linearly polarized infrared radiation is measured at several angles of incidence (at least two, in principle) and n and k are evaluated from the appropriate Fresnel equations. A numerical or graphical solution is generally necessary. This method was used by Collins and Bock (71). Avery (72) used a similar method, except that he determined the ratio of reflectivities for perpendicular- and parallel-polarized incident radiation. These procedures are not very good for highly reflecting samples such as metals.

The techniques of attenuated total reflectance (73–75) are also used for determination of the optical constants of normally weakly reflecting substances. The ATR spectrum is obtained at two angles of incidence.

The optical constants can also be determined from the ratio of reflectivities of parallel- and perpendicular-polarized radiation and the phase difference at a single angle of incidence; that is,

$$R_P/R_S = \tan \psi \, e^{j\Delta}$$

This is the ellipsometry method which is very successful and powerful in the visible region. In ellipsometric measurements an incident beam is plane polarized at 45° to the plane of incidence. The parallel and perpendicular components of the reflected radiation are reflected to different extents and are shifted in phase by different amounts, thus resulting in elliptically polarized radiation. The ellipticity is determined by placing a quarter wave plate and an analyzer in the reflected beam and adjusting their azimuths until the radiation is extinguished. The quantities $\tan \psi$ and Δ are calculated from the ellipticity, and the optical constants are calculated from ψ and Δ. The quarter wave plate might be omitted and the ellipticity determined from measurements of reflected radiation as a function of angle of the analyzer. Beattie and Conn (76) devised a method of this type in which only analyzer settings of −45, 0, 45, and 90° are required. The −45° setting is used only as a symmetry check. This method is subject to errors that are introduced by partial polarization from other components in the optical system. Roberts (77) avoids this problem by using a third polarizing element to plane polarize the radiation from the source. The polarizer preceding the sample is set at −45, 0, 45, and 90°, and the analyzer is set at 0 and 90°. The values of $\tan \psi$ and Δ are determined from the intensities measured at the various combinations of settings.

Ellipsometric measurements are of great usefulness in the visible region in studies of the properties (including thickness) of thin films.

This application does not seem to have been made in the infrared region; ellipsometry is much more difficult in the infrared because of poor signal-to-noise ratio encountered in that region.

The amplitude of the reflectivity, R, and phase shift ϕ of the reflected beam are not independent. They are related by a Kramers–Kronig relationship, for normal incidence,

$$\phi(\nu) = \frac{\nu}{\pi} \int_0^\infty \frac{\ln R(\nu')}{\nu^2 - \nu'^2} \, d\nu'$$

This means that from a complete knowledge of the reflectivity at *all* frequencies, the phase can be computed. The quantities n and k can then be calculated from R and ϕ, using

$$k = \frac{2\sqrt{R}\sin\phi}{(1-\sqrt{R})^2 + 4\sqrt{R}\sin^2\phi/2}$$

and

$$n = \frac{1-R}{(1-\sqrt{R})^2 + 4\sqrt{R}\sin^2\phi/2}$$

This method is also called the Robinson–Price (78) method. It is very frequently used in the study of solid dielectrics, semiconductors, and metals. It can also be used for samples confined to a cell provided the effect of the window is taken into account.

Of course, reflectivity data are not available for all frequencies, as is demanded by the Kramers–Kronig integral. Usually some assumptions must be made about the contributions from very high frequencies and also from the very low frequencies when they are missing. Procedures are discussed by Bowlden and Wilmshurst (79), for example, for dealing with narrow reflection bands, and by Miloslavskii (80) for treating metals when only a limited experimental range is available.

Some other methods can be used for the determination of infrared refractive indices. Welber (81) has embedded powdered samples in pellets or alkali halides. The particle size is made large in order to produce strong scattering. At the wavelength where the refractive index of the sample matches that of the matrix material, scatter vanishes and a transmission maximum is observed. This is the Christiansen filter effect. From measurements of the wavelength of the maxima and the known refractive index of the matrix, the index of the sample at these discrete wavelengths is known.

The Fourier spectrometer or interferometer can be used for the determination of optical constants of transmitting and reflecting samples. We have mentioned the work of Legay in measuring the dispersion of gases with a Michelson interferometer. This method does not involve the inversion of a Fourier transform. Loewenstein (*82*) has analyzed the channeled spectrum of transparent solid plates with a Fourier spectrometer, also without the necessity of inverting a transform. If dispersion is very weak, the channeled spectrum is modulated with fringes of constant spacing

$$\Delta \nu = 1/2bn$$

The interferogram, therefore, consists of a central maximum at zero path difference and a sequence of sharp spikes at path differences of

$$x = 2kbn \qquad k = 1, 2, \ldots$$

In fact, the spikes are distorted into peaks whose shapes or signatures are determined by the dispersion.

Chamberlain *et al.* (*83*) and Bell and Russell (*84*) insert the sample into one beam of a Michelson interferometer. The sample can be a transmitting solid or a liquid or gas in a cell. It can also be a reflecting sample which replaces one of the Michelson mirrors. A Fourier analysis of the interferogram yields both amplitude and phase information and, therefore, permits the determination of both n and k.

The discussion of this section has so far been restricted to isotropic samples. The analysis of nonisotropic materials is somewhat more complicated. Polarized incident radiation is required for the study of dichroic samples. Optically active materials which exhibit rotatory dispersion are studied using techniques described earlier for the Faraday effect. Materials which are circularly dichroic can also be studied using techniques described earlier, but such studies have evidently not yet been fruitful.

VI. Infrared Spectroscopy of Remote Objects

Astronomical spectroscopy is the oldest example of the spectroscopic study of remote objects. A very good review of infrared astronomy by Murray and Westphal (*85*) has recently appeared in *Scientific American*. The two most serious problems encountered in astronomical spectroscopy in the infrared are weakness of the signal and interference by absorbing and scattering components of the earth's atmosphere. Astronomers go to great lengths to obtain an improvement of $\sqrt{2}$ or so in signal-to-noise ratio. Sinton (*86*), for example, uses two detectors and alternates the image of a star or planet on the top and bottom halves of the entrance slit of his monochromator.

Fourier spectroscopy with interferometers is attractive to astronomers because of the luminosity advantage. Indeed, the pioneer in this field, P. Fellgett, is an astrophysicist. J. and P. Connes (87) are presently recording, interferometrically, and analyzing some truly remarkable near-infrared planetary spectra.

The problem of atmospheric interference is partially ameliorated by locating observatories on mountaintops. However, even at an altitude of 4000 m a significant amount of absorbing atmosphere remains between the spectrometer and the sample. Some solar spectra from the Jungfraujoch (88) have been reproduced by Plass and Yates in the *Handbook of Military Infrared Technology.*

In order to get closer to their samples, and farther from the earth's atmosphere, spectroscopists have mounted spectrometers in airborne vehicles. Among the first efforts, crude by today's standards, was an infrared spectrometer mounted in a British mosquito bomber in 1952 (89). An airborne spectrometer described in 1958 (90) was much more effective. More recently jet aircraft have been fitted with a large variety of scientific instruments, including infrared spectrometers, for studies in connection with such events as solar eclipses.

Aircraft are limited in altitude. Balloons can penetrate to higher levels, and several experimenters have flown infrared spectrometers in balloons. In John Strong's (91) earlier efforts an operator went along with the spectrometer, but a human spectroscopist is really an extravagance in such operations. More recent experiments have been made with automated systems using gondola stabilizers and star or sun trackers for pointing the telescope. Both monochromators and interferometers have been used. Measurements have been made of solar spectra, planetary atmospheres, and also of the radiance of the earth. The February 1967 issue of *Applied Optics* features a number of articles on balloon-borne optics.

Even higher altitudes are attained by artificial satellites and space vehicles, and infrared experiments are performed on board some of these sophisticated systems.

Infrared spectroradiometers are used as land-based instruments for the study of other kinds of remote objects such as missiles and reentering satellites.

VII. NONINFRARED METHODS

In this last section let us mention a few noninfrared techniques for gathering information similar to that obtained from infrared spectroscopy.

Optical spectroscopy is done in regions other than the infrared. In the ultraviolet and visible regions the transitions are primarily electronic in nature. However, under suitable conditions, a finer structure can be observed corresponding to changes in vibrational levels occurring simultaneously with the electronic transitions. Often, the vibrational frequencies of electronically excited molecules can be obtained in this way from absorption or fluorescence spectra.

Raman spectroscopy is also done in the visible region and occasionally in the ultraviolet or near infrared. Monochromatic light scattered by a sample contains weak components of lower frequency. The frequency shifts correspond to vibrational and rotational levels of the molecules, which have become excited during the scattering process. Even weaker components are found at higher frequencies because some of the molecules were in an excited state to begin with and fall to the ground state. If a high-powered laser is used to excite the molecules, stimulated Raman spectra are observed. In this case the Raman-scattered lines are coherent.

Absorption spectroscopy can be done at very long wavelengths in the microwave region. Microwave spectra of gases are rotational spectra. Rotational constants of light polar molecules can be measured with greater accuracy in the microwave region than from rotation–vibration bands in the infrared.

Radio-frequency resonance measurements, such as nuclear magnetic resonance, provide information about the chemical bonds and structure of molecules, although vibrational frequencies are not obtained. Similarly, X-ray and neutron diffraction measurements are used in determining the structure of crystals, while mass spectroscopy offers some evidence about the structure of molecules.

A stream of slow thermal neutrons can be scattered by the atoms of a sample, particularly by the hydrogen atoms. Much as in the Raman effect, the scattering is partially inelastic and a determination of the energy-loss spectrum of the neutrons provides information about the vibrational frequencies of the molecules of the sample. It is a particularly useful tool for determining low-frequency vibrations of molecules and lattice phonons. Neutron-scattering spectroscopy has been applied to such diverse studies as low-frequency vibrations of polymers, hindered rotation in liquids, and rotational spectra of gases.

Interactions of beams of electrons with molecules have also been studied, and features associated with vibrational states are observed in electron scattering and absorption spectra. Photoelectron spectroscopy also yields vibrational structure. In photoelectron spectroscopy the molecules are illuminated with radiation capable of ionizing them. The energies of the emitted electrons are measured with a grid- or a deflection-type velocity analyzer. Definite peaks are observed at discrete energies and the fine details can be associated with vibrational state.

An interesting effect is observed in low-temperature tunneling junctions. These junctions consist of a layer of superconducting metal separated from a layer of nonsuperconducting metal by a thin metal-oxide film. Solid-state physicists observe the tunneling of electrons through the oxide layer when an electric field is applied across the junctions. The electrons may be inelastically scattered during their journey. At very low voltages, typically less than 50 mV, fine structure is observed in a plot of current vs. voltage (actually, d^2I/dv^2 is plotted). This is associated with the energy states of the superconducting metal. At higher potentials, from 0.05 to 0.5 V, corresponding to the region from 400 to 4000 cm^{-1} in the infrared, the electrons are scattered by the oxide layer and also by impurity molecules which may be included in the oxide layer either accidentally or intentionally. Jaklevic and Lambe (92), who discovered this effect, introduced water, propionic acid, acetic acid, and some deuterated compounds into their junctions. The curves resemble infrared vibrational spectra, and display recognizable O—H and C—H stretching and bending modes. Vibrations associated with functional groups of heavy atoms, such as the C=O group, are not observed. A theory of the effect is given by Scalapino and Marcus (93).

Some of the techniques discussed in this section are in widespread and routine use. Others are, at present, interesting phenomena. It is important for the infrared spectroscopist to recognize that there are many techniques for obtaining information about the structure and dynamics of matter. Because of differences in the physical effects used, various techniques are likely to supplement one another. The wise spectroscopist makes use of all the information he can obtain.

REFERENCES

1. A. S. Barker and M. Tinkham, *Phys. Rev.*, **125**, 1527 (1962).
2. R. E. Glover and M. Tinkham, *Phys. Rev.*, **104**, 844 (1956); **108**, 243 (1957).
3. D. M. Ginsberg and M. Tinkham, *Phys. Rev.*, **118**, 990 (1960).
4. P. L. Richards and M. Tinkham, *Phys. Rev.* **119**, 575 (1960).
5. H. L. Welsh, M. F. Crawford, and J. L. Locke, *Phys. Rev.*, **76**, 580 (1949).

6. M. F. Crawford, H. L. Welsh, and J. L. Locke, *Phys. Rev.*, **75**, 1607 (1949).
7. M. F. Crawford, H. L. Welsh, J. C. F. MacDonald, and J. L. Locke, *Phys. Rev.*, **80**, 469 (1950).
8. M. L. Oxholm and D. Williams, *Phys. Rev.*, **76**, 151 (1949).
9. R. V. Asselt and D. Williams, *Phys. Rev.* **79**, 1016 (1950).
10. A. L. Smith, W. E. Keller, and H. L. Johnston, *Phys. Rev.*, **79**, 728 (1950).
11. C. E. Wier, E. R. Lippincott, A. van Valkenburg, and E. N. Bunting, *J. Res. NBS*, **A63**, 55 (1959).
12. L. Whatley, E. Lippincott, A. van Valkenburg, and C. Weir, *Science*, **144**, 968 (1964).
13. H. G. Drickamer, R. A. Fitch, and T. E. Slykhouse, *J. Opt. Soc. Am.*, **47**, 1015 (1957).
14. W. F. Sherman, *J. Sci. Inst.*, **43**, 462 (1966).
15. J. W. Brasch, *Spectrochim. Acta*, **21**, 1183 (1965).
16. J. Reynolds and S. Sternstein, *J. Chem. Phys.* **41**, 47 (1964).
17. H. G. Drickamer and E. Fishman, *J. Chem. Phys.*, **24**, 548 (1956).
18. T. S. Moss, *Optical Properties of Semi-Conductors*, Butterworth, London, 1959.
19. E. U. Condon, *Phys. Rev.*, **41**, 759 (1932).
20. L. A. Woodward, *Nature*, **165**, 198 (1950).
21. A. Kastler, *Compt. Rend.*, **230**, 1596 (1950).
22. D. A. Dows and A. D. Buckingham, *J. Mol. Spectry.*, **12**, 189 (1964).
23. M. F. Crawford and I. R. Dagg, *Phys. Rev.*, **91**, 1569 (1953).
24. M. F. Crawford and R. E. MacDonald, *Can. J. Phys.*, **36**, 1022 (1958).
25. T. W. Terhune and C. W. Peters, *J. Mol. Spectry.*, **3**, 138 (1959).
26. E. Anastassakis, S. Iwasa, and E. Burstein, *Phys. Rev. Letters*, **17**, 1051 (1966).
27. P. Handler and D. E. Aspnes, *J. Chem. Phys.*, **47**, 473 (1967).
28. E. Charney and R. S. Halford, *J. Chem. Phys.*, **29**, 221 (1958).
29. E. D. Palik, *Appl. Opt.*, **2**, 527 (1963); **6**, 597 (1967).
30. E. D. Palik and B. W. Henvis, *Appl. Opt.*, **6**, 603 (1967).
31. K. Sakurai and K. Shimoda, *Japan. J. Appl. Phys.*, **5**, 938 (1966).
32. H. J. Hediger and H. H. Günthard, *Helv. Chim. Acta*, **37**, 1125 (1954); H. R. Wyss and H. H. Günthard, *ibid*, **49**, 660 (1966).
33. H. Yoshimoto and Y. Mochida, *Sc. of Light*, **14**, 73 (1965).
34. H. R. Wyss and H. H. Günthard, *Appl. Opt.*, **5**, 1736 (1966).
35. C. R. Pidgeon and S. D. Smith, *Infrared Phys.* **4**, 13 (1964).
36. C. K. N. Patel, *Phys. Rev. Letters*, **15**, 1027 (1965); **16**, 613 (1966).
37. E. Garmire, F. Pandarese, and C. H. Townes, *Phys. Rev. Letters*, **11**, 160 (1963).
38. R. Loudon, *Proc. Roy. Soc. (London)*, **A82**, 393 (1963).
39. F. De Martini, *J. Appl. Phys.* **37**, 4503 (1966).
40. M. E. Jacox and S. H. Bauer, *J. Phys. Chem.*, **61**, 833 (1957).
41. W. J. Hooker and R. C. Millikan, *J. Chem. Phys.*, **38**, 214 (1963).
42. R. C. Millikan, *J. Chem. Phys.*, **38**, 2855 (1963).
43. D. J. McCaa and D. Williams, *J. Opt. Soc. Am.*, **54**, 326 (1964).
44. C. B. Moore, in *Fluorescence: Theory, Instrumentation, and Practice*, G. G. Guilbault, ed., Dekker, New York, 1967, Chap. 3.
45. R. M. Hexter, *J. Opt. Soc. Am.*, **53**, 703 (1963).
46. E. Whittle, D. A. Dows, and G. C. Pimentel, *J. Chem. Phys.*, **22**, 1943 (1954).
47. E. D. Becker and G. C. Pimentel, *J. Chem. Phys.*, **25**, 224 (1956).
48. W. A. Plisken and R. P. Eischens, *J. Chem. Phys.*, **24**, 482 (1956).
49. S. C. Brown, G. Bekefi, and R. E. Whitney, *J. Opt. Soc. Am.*, **53**, 448 (1963).

50. M. Kimmett and V. Niblett, *Proc. Phys. Soc.*, **82**, 530 (1963).
51. L. A. Dushin, V. I. Kononenko, O. S. Pavlichenko, and I. K. Nikolskii, *Opt. Spectry.*, **19**, 378 (1965).
52. J. P. Craig, A. F. Gribble, and A. A. Dougal, *Rev. Sci. Inst.*, **35**, 1501 (1964).
53. I. M. Frank, *Izv. Akad. Nauk SSSR, Ser. Fiz.*, **6**, 3 (1942).
54. H. Motz, *J. Appl. Phys.*, **22**, 527 (1951).
55. W. W. Salisbury, *Opt. Soc. Am.* (Oct. 1962); U.S. Pat. No. 2,634,372, Apr. 7, 1953.
56. J. J. Smith and E. M. Purcell, *Phys. Rev.*, **92**, 1069 (1953).
57. K. Ishiguro and T. Tako, *Opt. Acta*, **8**, 25 (1961).
58. G. Toraldo di Francia, *Nuovo Cim.*, **16**, 61 (1960).
59. H. Rubens, *Pogg. Ann.*, **45**, 238 (1892).
60. A. H. Pfund, *J. Opt. Soc. Am.*, **25**, 351 (1935).
61. M. A. Pittman, *J. Opt. Soc. Am.*, **29**, 358 (1939).
62. J. H. Jaffe, *J. Opt. Soc. Am.*, **41**, 166 (1951).
63. J. H. Jaffe and U. Oppenheim, *J. Opt. Soc. Am.*, **47**, 782 (1957).
64. J. H. Jaffe, H. Goldring, and U. Oppenheim, *J. Opt. Soc. Am.*, **49**, 1199 (1959).
65. R. E. Kagarise and J. W. Mayfield, *J. Opt. Soc. Am.*, **48**, 430 (1958).
66. R. E. Kagarise, *J. Opt. Soc. Am.*, **50**, 36 (1960); **51**, 830 (1961).
67. T. S. Moss, Ref. *18*, p. 13.
68. J. H. Jaffe, "Refraction of gases in the infrared," *Advances in Spectroscopy*, Vol. II, Wiley-Interscience, New York, 1961.
69. F. Legay, *Nuovo Cim.*, **2**, Suppl. 3, 781 (1955).
70. J. H. Jaffe, J. Meiron, and N. Jacobi, *J. Opt. Soc. Am.*, **52**, 8 (1962).
71. J. R. Collins and R. O. Bock, *Rev. Sci. Inst.*, **14**, 135 (1943).
72. D. G. Avery, *Proc. Phys. Soc.* (*London*), **64B**, 1087 (1952); **65B**, 425 (1953).
73. J. Fahrenfort, *Spectrochim. Acta*, **18**, 1103 (1962).
74. J. Fahrenfort and W. M. Visser, *Spectrochim. Acta*, **21**, 1433 (1965).
75. N. J. Harrick, *Internal Reflection Spectroscopy*, Wiley, New York, 1967.
76. J. R. Beattie, *Phil. Mag.*, **46**, 235 (1955); J. R. Beattie and G. K. T. Conn, *ibid.*, 222 (1955).
77. S. Roberts, "Ellipsometry in the measurement of surfaces and thin films," *NBS Misc. Publ. 256*, U.S. Govt. Printing Office, Washington, D.C., 1964, p. 119.
78. T. S. Robinson and W. C. Price, *Proc. Phys. Soc.* (*London*), **66B**, 969 (1953); *Molecular Spectroscopy*, G. Sell, ed., Institute of Petroleum, London, 1955.
79. H. J. Bowlden and J. K. Wilmshurst, *J. Opt. Soc. Am.*, **53**, 1073 (1963).
80. V. K. Miloslavskii, *Opt. Spectry.*, **21**, 193 (1966).
81. B. Welber, *Appl. Opt.*, **6**, 925 (1967).
82. E. V. Loewenstein, *J. Opt. Soc. Am.*, **51**, 108 (1961).
83. J. E. Chamberlain, J. E. Gibbs, and H. A. Gebbie, *Nature*, **198**, 874 (1963).
84. E. E. Bell, *Infrared Phys.* **6**, 57 (1966); E. E. Russell and E. E. Bell, *ibid*, 75 (1966).
85. B. C. Murray and J. A. Westphal, *Sci. Am.*, **213** No. 2, 20 (1965).
86. W. M. Sinton, *Appl. Opt.*, **1**, 105 (1962).
87. J. Connes and P. Connes, *J. Opt. Soc. Am.*, **56**, 896 (1966).
88. G. N. Plass and H. Yates, "Atmospheric phenomenon," *Handbook of Military Infrared Technology*, W. L. Wolfe, ed., U.S. Govt. Printing Office, Washington, D.C., 1965, Chap. 6.
89. J. Yarnell and R. M. Goody, *J. Sci. Inst.*, **29**, 352 (1952).
90. J. T. Houghton, T. S. Moss, and J. P. Chamberlain, *J. Sci. Inst.*, **35**, 329 (1958).
91. J. Strong, *Appl. Opt.*, **6**, 179 (1967).

92. R. C. Jaklevic and J. Lambe, *Phys. Rev. Letters*, **17**, 1139 (1966); *Bull. Am. Phys. Soc.*, **12**, No. 1, 23 (1967).
93. D. J. Scalapino and S. M. Marcus, *Phys. Rev. Letters*, **18**, 459 (1967).

APPENDIX I: PROPERTIES OF FOURIER AND LAPLACE TRANSFORMS

In this appendix we shall relate, without any pretense of proof, some of the properties of Fourier and Laplace transforms. The reader is referred to such textbooks as the work of Papoulis (*1, 2*) for more complete expositions. The following summary is taken from Papoulis.

The Fourier transform of a suitably defined function $f(t)$ is defined,

$$F(\omega) = \int_{-\infty}^{\infty} f(t) e^{-j\omega t} \, dt$$

The function is related to its transform by the inverse relationship

$$f(t) = \frac{1}{2\pi} \int_{-\infty}^{\infty} F(\omega) e^{j\omega t} \, d\omega$$

The notation $f(t) \leftrightarrow F(\omega)$ is sometimes used. The variable ω is real. In many physical applications t plays the role of time and ω is frequency.

Fourier transforms have the following properties:

Linearity. If $f_1(t) \leftrightarrow F_1(\omega)$ and $f_2(t) \leftrightarrow F_2(\omega)$ and a_1 and a_2 are constants, then

$$a_1 f_1(t) + a_2 f_2(t) \leftrightarrow a_1 F_1(\omega) + a_2 F_2(\omega)$$

Symmetry. If $f(t) \leftrightarrow F(\omega)$, then $F(t) \leftrightarrow 2\pi f(-\omega)$.

Scaling. If a is a real constant, then $f(at) \leftrightarrow (1/|a|)F(\omega/a)$.

Shifting. If t_0 is a constant "time" shift, then

$$f(t - t_0) \leftrightarrow F(\omega) e^{-jt_0\omega}$$

Also, if ω_0 is a constant "frequency" shift, then

$$e^{j\omega_0 t} f(t) \leftrightarrow F(\omega - \omega_0)$$

Differentiation.

$$\frac{d^n f(t)}{dt^n} \leftrightarrow (j\omega)^n F(\omega)$$

Also,

$$(-jt)^n f(t) \leftrightarrow \frac{d^n F(\omega)}{d\omega^n}$$

Conjugates. The transform of the complex conjugate of a function is given by $f^*(t) \leftrightarrow F^*(-\omega)$.

Moments. The nth moments of a function,

$$m_n = \int_{-\infty}^{\infty} t^n f(t)\, dt$$

may be found from

$$(-j)^n m_n = \frac{d^n F(\omega)}{d\omega^n}\bigg|_{\omega=0}$$

Convolution. If $f(t)$ is the convolution of $f_1(t)$ and $f_2(t)$,

$$f(t) = \int_{-\infty}^{\infty} f_1(x) f_2(t-x)\, dx = f_1(t) * f_2(t)$$

then

$$f(t) \leftrightarrow F(\omega) = F_1(\omega) F_2(\omega)$$

Also,

$$f_1(t) f_2(t) \leftrightarrow \frac{1}{2\pi} \int_{-\infty}^{\infty} F_1(y) F_2(\omega - y)\, dy$$

Parseval's theorem.

$$\int_{-\infty}^{\infty} |f(t)|^2\, dt = \frac{1}{2\pi} \int_{-\infty}^{\infty} F^2(\omega)\, d\omega$$

Sampling theorem. If $F(\omega) = 0$ for $|\omega| > \omega_c$, then $f(t)$ is uniquely determined from its values at discrete points π/ω_c apart, that is, from the values $f_n = f(n\pi/\omega_c)$, $n = 0, \pm 1, \ldots$.

Also, if $f(t) = 0$ for $|t| > T$, then $F(\omega)$ is uniquely determined from its values at discrete points π/T apart, that is, from the values

$$F_n = F(\pi n/T), n = 0, \pm 1, \ldots$$

In fact,

$$f(t) = \sum_{n=-\infty}^{\infty} f_n \frac{\sin(\omega_c t - n\pi)}{\omega_c t - n\pi}$$

and

$$F(\omega) = \sum_{n=-\infty}^{\infty} F\left(\frac{\pi n}{T}\right) \frac{\sin(\omega T - n\pi)}{\omega T - n\pi}$$

Delta functions have the property

$$\delta(t - t_0) = \begin{cases} 1 & t = t_0 \\ 0 & t \neq t_0 \end{cases}$$

A useful property of delta functions is the *sifting* property,

$$\int_{-\infty}^{\infty} \delta(t - t_0) f(t) \, dt = f(t_0)$$

Some useful Fourier transforms are tabulated in Table A.1.

TABLE A.1

Fourier Transforms

$f(t)$	$f(\omega)$
$\begin{cases} 1 & \lvert t \rvert > T \\ 0 & \lvert t \rvert < T \end{cases}$	$\dfrac{2 \sin \omega t}{\omega}$
$\dfrac{\sin at}{\pi t}$	$\begin{cases} 0 & \lvert \omega \rvert > a \\ 1 & \lvert \omega \rvert < a \end{cases}$
$\begin{cases} 1 - \lvert t \rvert / T & t < T \\ 0 & \lvert t \rvert > T \end{cases}$	$\dfrac{4 \sin^2 \omega T/2}{T\omega^2}$
$\dfrac{\sin^2 at}{\pi a t^2}$	$\begin{cases} 1 - \lvert \omega \rvert / 2a & \lvert \omega \rvert < 2a \\ 0 & \lvert \omega \rvert > 2a \end{cases}$
$\begin{cases} e^{-\alpha t} & t > 0 \\ 0 & t < 0 \end{cases}$	$\dfrac{1}{\alpha + j\omega} = \dfrac{e^{-j \tan^{-1} \omega/\alpha}}{(\alpha^2 + \omega^2)^{1/2}}$
$e^{-\alpha \lvert t \rvert}$	$2\alpha / (\alpha^2 + \omega^2)$
$\begin{cases} e^{-\alpha t} \sin \beta t & t > 0 \\ 0 & t < 0 \end{cases}$	$\beta / [(\alpha + j\omega)^2 + \beta^2]$
$e^{-\alpha t^2}$	$(\pi/\alpha)^{1/2} \exp(-\omega^2/4\alpha)$
$\delta(t - t_0)$, delta function	$e^{-j\omega t_0}$
$e^{j\omega_0 t}$	$2\pi \delta(\omega - \omega_0)$
$\cos \omega_0 t$	$\pi[\delta(\omega - \omega_0) + \delta(\omega + \omega_0)]$
$\sin \omega_0 t$	$j\pi[\delta(\omega + \omega_0) - \delta(\omega - \omega_0)]$
$\displaystyle\sum_{n=-\infty}^{\infty} \delta(t - nT)$, impulse sequence	$\dfrac{2\pi}{T} \displaystyle\sum_{n=-\infty}^{\infty} \delta\left(\omega - \dfrac{2\pi n}{T}\right)$
$\displaystyle\sum_{n=-\infty}^{\infty} A_n \delta(t - nT)$, impulse modulation	$\displaystyle\sum_{n=-\infty}^{\infty} A_n e^{-jnT\omega}$

The unilateral Laplace transform of a suitable function is defined by

$$L(s) = \int_0^\infty e^{-st} f(t)\, dt$$

where s can be a complex variable. The inverse transform is

$$f(t) = \frac{1}{2\pi j} \int_{\alpha-j\infty}^{\alpha+j\infty} e^{st} L(s)\, ds$$

where the integration is carried out along a path in the complex plane to the right of the region of convergence of $L(s)$.

In engineering we usually deal with causal time functions, $f(t) = 0$ for $t < 0$. If the Laplace transform of a causal function converges for pure imaginary values of s, the Fourier transform of the function may be found from the Laplace transform by merely substituting $j\omega$ for s. This is the case of stable, nonoscillating time functions. If $L(s)$ fails to converge for all imaginary values of s, $F(\omega)$ does not exist. If there are isolated imaginary singularities, $L(j\omega)$ must be amended by the addition of certain delta functions associated with oscillating time behavior.

Similarly, the Laplace transform of a causal function may be found from the Fourier transform by substituting s/j for ω if $L(s)$ converges for imaginary values of s.

REFERENCES

1. A. Papoulis, *The Fourier Integral and its Applications*, McGraw-Hill, New York, 1962.
2. A Papoulis, *Systems and Transforms with Applications in Optics*, McGraw-Hill, New York, 1968.

APPENDIX II: SELECTION OF OPERATING PARAMETERS OF INFRARED SPECTROPHOTOMETERS

This appendix is offered as a guide to operators of spectrophotometers. The efficient use of a complex instrument depends on the skill of the user in selecting the values of adjustable parameters. In an infrared spectrophotometer the important parameters are slit width, time constant, and scanning speed. In most instances the parameters should be adjusted in that order.

Slit width (Chapter 7)

The half bandwidth (HBW) of the narrowest band in the spectrum is determined or guessed. The effective bandwidth (EBW) of the monochromator should be no wider than one-fifth of the HBW to keep the peak height error less than 2 or 3%. The slit width required for a desired EBW is calculated from the reciprocal linear dispersion curve usually supplied by the manufacturer. When very narrow slits are used, the effect of broadening by diffraction must be considered. The observed bandwidths (OBW) of the spectral bands are given roughly by the sum of EBW and HBW.

Time constant (Chapter 13)

The time constant is selected to give a reasonable signal-to-noise ratio. The signal-to-noise ratio increases as the square root of the time constant. In a double-beam optical-null instrument the actual time constant departs from the indicated time constant according to the reference beam energy (or servo loop gain). The manufacturer's manual should be consulted for the method of setting and measuring the loop gain.

Scanning speed (Chapter 13)

The spectrum should be scanned at a rate no faster than one-fifth of the OBW of the narrowest band in one time constant to keep the peak height error less than 2%. Note that some manufacturers use period rather than time constant to denote instrument response. One period is usually four time constants. The recorded spectrum is displaced in time by one time constant, and the wavenumber error can be calculated from the time constant and scanning speed. In a spectrum scanned from low to high values of wavenumber, positive errors in peak position are introduced.

Final choice of parameters

The final choice of parameters is made on a trial and error basis, starting with the limiting values calculated above and scanning over a narrow band. Many modern instruments have adaptive servo control of slit width, time constant, or scanning speed. Proper use of these systems permits the operator to scan at a faster average rate.

Author Index

Numbers in parentheses are reference numbers and indicate that an author's work is referred to, although his name is not cited in the text. Numbers in italics show the page on which the complete reference is listed.

Subject Index

624